INTERFACIAL PHENOMENA
EQUILIBRIUM AND DYNAMIC EFFECTS
SECOND EDITION

SURFACTANT SCIENCE SERIES

INTERFACIAL PHENOMENA

EQUILIBRIUM AND DYNAMIC EFFECTS

SECOND EDITION

Edited by

Clarence A. Miller
Rice University
Houston, Texas

P. Neogi
University of Missouri-Rolla
Rolla, Missouri

CRC Press
Taylor & Francis Group
Boca Raton London New York

CRC Press is an imprint of the
Taylor & Francis Group, an **informa** business

CRC Press
Taylor & Francis Group
6000 Broken Sound Parkway NW, Suite 300
Boca Raton, FL 33487-2742

First issued in paperback 2019

ISBN-13: 978-1-4200-4442-3 (hbk)
ISBN-13: 978-0-367-38852-2 (pbk)

Library of Congress Cataloging-in-Publication Data

Miller, Clarence A., 1938-
    Interfacial phenomena : equilibrium and dynamic effects / Clarence A. Miller, P. Neogi. -- 2nd ed.
    p. cm. -- (Surfacant science series ; v. 137)
    Includes bibliographical references and index.
    ISBN 1-4200-4442-7 (alk. paper)
    1. Surface chemistry. I. Neogi, P. (Partho), 1951- II. Title.

QD506.M55 2007
541'.33--dc22
                                        2006051788

Visit the Taylor & Francis Web site at
http://www.taylorandfrancis.com

and the CRC Press Web site at
http://www.crcpress.com

Contents

Preface to Second Edition

Authors

PREFACE TO
SECOND EDITION

Since the first edition was published in 1985, fundamental understanding of interfacial phenomena has advanced significantly although more extensively in some areas than in others. Interest in interfaces and surfactant behavior has increased, and applications have multiplied. While we discuss some new applications in this expanded second edition, we have retained the chief objective of the first edition: a concise summary of the fundamentals with emphasis on equilibrium phenomena followed by chapters on flow, transport and stability of interfaces understandable to graduate students and others entering the field but also useful to researchers whose major focus is not on dynamic interfacial phenomena. Some background in thermodynamics and transport phenomena is assumed. Although all chapters of the first edition have been modified, Chapters IV on surfactants and VI on transport effects have undergone the greatest expansion and increase in scope. Moreover, an entirely new Chapter VIII has been added on mass transfer measurements and key experimental techniques for determination of microstructure in colloidal dispersions and surfactant systems. In all chapters references have been updated and new problems added.

Major changes in individual chapters are as follows:
- Chapter I on interfacial tension. New section on solid-fluid interfaces.
- Chapter II on contact angles. New sections on acid-base interactions and characterization of solid surfaces.
- Chapter III on colloidal dispersions. Electrostatics follows S.I. units. New section on characterization of colloidal dispersions. Section on effects of adsorbed polymer completely rewritten with new material.
- Chapter IV on surfactants. New sections on surfactant/polymer interactions and on chemical reactions in micellar solutions and microemulsions plus expansion and inclusion of new material in almost all other sections.
- Chapter V on interfacial stability. Section on thin liquid films and foam completely rewritten to incorporate recent developments.
- Chapter VI on transport effects on interfacial stability. New material on solubilization rates and formation of intermediate phases during diffusion in surfactant systems; sections on spontaneous emulsification and dynamic surface tension revised and expanded.

- Chapter VII on interfacial dynamics. Sections on drainage of thin liquid films and on dynamic contact lines updated and expanded and new section on very thin films added.
- Chapter VIII on mass transfer and techniques for determining micro-structure. New chapter on experimental techniques as indicated above including static and dynamic light scattering and NMR self-diffusion. Further discussion of mass transfer effects.

Clarence A. Miller
Partho Neogi
May, 2006

Authors

Clarence A. Miller is Louis Calder Professor of Chemical and Biomolecular Engineering at Rice University in Houston, Texas and a former chairman of the department. Before coming to Rice he served on the faculty at Carnegie-Mellon University from 1969-1981. He received B.A. and B.S. degrees from Rice University (1961) and the Ph.D. degree from the University of Minnesota (1969), all in chemical engineering. He has been a Visiting Scholar at Cambridge University, England (1979-80), University of Bayreuth, Germany (1989, 1995) and Delft University, the Netherlands (1995).

Dr. Miller's research interests center on equilibrium and dynamic phenomena in oil/water/surfactant systems, specifically interfacial stability and behavior of emulsions, microemulsions and foams and their application in areas such as detergency, enhanced oil recovery and environmental remediation. He is a Fellow of the American Institute of Chemical Engineers and a member of the American Chemical Society, American Oil Chemists Society, International Association of Colloid and Interface Scientists, and the Society of Petroleum Engineers. He has published numerous research papers and review articles on interfacial phenomena, served on the editorial boards of leading journals in the field, and given invited lectures at conferences, universities and industrial laboratories in many countries.

Partho Neogi is a Professor of Chemical and Biological Engineering at the University of Missouri-Rolla. He received his B. Tech. (Hons.) at the Indian Institute of Technology Kharagpur (1973), M. Tech. at the Indian Institute of Technology Kanpur (1975) and his Ph. D. at Carnegie-Mellon University (1979), all in Chemical Engineering. He joined the University of Missouri-Rolla in 1980, and has been there since.

Dr. Neogi's research area lies in studies of transport at interfaces. They include dynamics of wetting, surfactant systems and electrochemical systems. He has also contributed in thermodynamics and transport in surfactant systems and polymer membranes. He is member of the American Institute of Chemical Engineers and the American Chemical Society.

1 Fundamentals of Interfacial Tension

1. INTRODUCTION TO INTERFACIAL PHENOMENA

An "interface" is, as the name suggests, a boundary between phases. Because interfaces are very thin—in most cases only a few molecular diameters thick—we sometimes tend to think of them as two-dimensional surfaces and neglect their thicknesses. But the third dimension is of great significance as well. Indeed, the rapid changes in density and composition across interfaces give them their most important property, an excess free energy or lateral stress which is usually called interfacial tension.

When three phases are present, three different interfaces are possible, one for each pair of fluids. Sometimes all three interfaces meet, the junction forming a curve known as a contact line. If one phase is a solid, the contact line lies along its surface. In this case, the angle that the fluid interface makes with the solid surface is called the contact angle. Since it reflects the wetting properties of liquids on solids, the contact angle is a second fundamental property important in interfacial phenomena.

Chapters 1 and 2 are devoted to these two fundamental properties which have roles in many diverse situations involving interfaces. Interfacial tension is a key factor influencing the shape of fluid interfaces, and it controls their deformability. Contact angles and wetting properties strongly affect the arrangement of the various phases in multiphase systems. When two fluid phases are present in a porous medium, for instance, as in soils or underground oil reservoirs, contact angle effects determine the positions of the two fluids, that is, which pores are occupied by which fluids. Interfacial tension, on the other hand, determines whether individual globules of one of the fluids, if present, will deform sufficiently to pass through the tortuous pore structure when a pressure gradient is applied.

Interfacial effects are especially important in systems where interfacial area is large. This condition is met when one phase is dispersed in another as small drops or particles. With spherical particles, for example, the area:volume ratio of the dispersed phase is $(3/R)$, where R is the particle radius. Clearly as R decreases with a given volume of the dispersed material present, interfacial area increases. When at least one dimension of each drop or particle decreases to a value in the range of 1 μm or less, we say that a "colloidal dispersion" exists. Foams, aerosols, and emulsions are colloidal dispersions involving fluid interfaces that are familiar from everyday life and are important in applications ranging from food products

1

to cosmetics to drug delivery to detergency. Colloidal dispersions with particle sizes in the range of 5–100 nm have received considerable attention recently in connection with potential applications in nanotechnology to produce advanced materials with special properties. Chapter 3 deals with the behavior of colloidal dispersions.

In multicomponent systems, composition in the interfacial region can differ dramatically from that of either bulk phase. This difference is especially striking when surface-active agents or "surfactants" are present. As their name implies, these substances find it energetically favorable to be located at the interface rather than in the bulk phases. As discussed in Chapter 4, surfactants produce significant decreases in interfacial tensions and alter wetting properties as well, the latter property being of particular importance in detergency. Moreover, surfactant molecules tend to aggregate in solution, forming phases such as micellar solutions, microemulsions, and lyotropic liquid crystals, which are "complex" fluids with microstructures that have interesting and unusual properties. Surfactants and the phases they form are a major factor influencing the stability and behavior of some colloidal dispersions such as emulsions and foams.

The first four chapters thus provide a general background on interfacial phenomena, colloidal dispersions, and surfactants, with emphasis on their equilibrium properties. The remaining chapters deal with the dynamic behavior of interfaces, emphasizing this subject to a much greater degree than most books on interfacial phenomena.

Chapter 5 considers the stability of fluid interfaces, a subject pertinent both to the formation of emulsions and aerosols and to their destruction by coalescence of drops. The closely related topic of wave motion is also discussed, along with its implications for mass transfer. In both cases, boundary conditions applicable at an interface are derived—a significant matter because it is through boundary conditions that interfacial phenomena influence solutions to the governing equations of flow and transport in fluid systems.

Interfacial phenomena involving heat and mass transfer are described and analyzed in Chapter 6. Much of the chapter again deals with stability, in this case the "Marangoni" instability produced by interfacial tension gradients associated with temperature and concentration gradients along the interface. Time-dependent variation of interfacial tension resulting from diffusion, adsorption, and desorption of various species is also discussed.

Some interesting problems in fluid flow where there are significant interfacial effects are analyzed in Chapter 7. Topics considered are the effect of surface viscosity on flow, motion of drops and bubbles, simple coating flows, and spreading of liquid drops on solids. The method of matched asymptotic expansions is used in some of the analyses, and the likely utility of this method in dealing with other flow problems involving interfaces is emphasized. Finally, in Chapter 8 we consider the dynamics of heat and mass transfer, and particularly the many experimental measurements made to understand these. For instance, diffusivities are measured using dynamic light scattering, nuclear magnetic resonance (NMR) spin echo methods, Taylor dispersion, electrochemical techniques, etc. In complex fluids, it takes a great deal of effort in order to understand what these measurements mean.

2. INTERFACIAL TENSION: QUALITATIVE CONSIDERATIONS

We begin with the interface between a pure liquid and its vapor. As indicated in Figure 1.1, the boundary between phases is not a surface of discontinuity where density changes abruptly, but a region of finite thickness where density changes continuously. To be sure, this thickness is, except near the critical point, only a few molecular diameters. Nevertheless, it is an essential feature of interfacial structure.

Surface or interfacial tension of a fluid interface can be viewed in two quite different ways. From a thermodynamic point of view, it is an additional free energy per unit area caused by the presence of the interface. As Figure 1.1 indicates, the density is lower in the interfacial region than in the bulk liquid phase. As a result, the average distance between molecules is greater. Because the molecules attract one another, energy must be supplied to move them apart. System energy per molecule is thus greater at larger average separation distances, and we conclude that the energy per molecule is greater in the interfacial region than in the bulk liquid.

At a liquid-liquid interface, such as exists between a pure hydrocarbon and water, the situation is slightly more complex because the separation distance between adjacent molecules does not vary much in the interfacial region and thus is not the main source of the excess energy of the interface. For simplicity, let us consider a binary system where the bulk phases are nearly pure component A and nearly pure component B. It is clear that in the interfacial region a molecule of A will have more B molecules and fewer A molecules as nearest neighbors than in bulk liquid A. A similar statement can be made about a molecule of B. Thermodynamics teaches that for phase separation to occur in the first place, the attraction between an A and a B molecule must be less than the average of that between two A molecules and two B molecules. Hence the

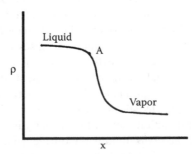

FIGURE 1.1 Density variation in interfacial region.

total attractive interaction per molecule is less in the interfacial region than in the bulk phases, the same result as found for the liquid-vapor interface. We conclude again that the energy per molecule must be greater in the interfacial region than in the bulk fluids.

A brief remark on terminology is in order at this point. In this book the term "interfacial tension" is used as an all-inclusive term applicable to liquid-gas, liquid-liquid, and solid-fluid interfaces. This usage differs from that of some authors who restrict interfacial tension to situations where neither phase is a gas or vapor. On the other hand, the term "surface tension" is used here only when one phase is a gas or vapor, in agreement with the usage of most authors.

Interfacial tension of a fluid interface also has a mechanical interpretation. From this point of view, tension is a force per unit length parallel to the interface (i.e., perpendicular to the local density or concentration gradient). At a given point, this force is the same in all lateral directions along the interface. Accordingly, interfacial tension is the two-dimensional counterpart of pressure in a bulk fluid, which has the same magnitude in all directions in three-dimensional space, although, of course, pressure is compressive instead of tensile. We note that the concepts of interfacial tension obtained from the thermodynamic and mechanical approaches—an energy per unit area and a force per unit length—are dimensionally equivalent. In a similar way, pressure may be thought of as an energy per unit volume or a force per unit area.

Why does a lateral force arise along an interface and why is it a tensile instead of a compressive force? The answer is not a simple one, and we provide a qualitative explanation here only for the basic liquid-vapor interface of Figure 1.1. At such an interface, two factors influence the local stress. One is the familiar kinetic effect due to the thermal motion of molecules, which is responsible for pressure in a dilute gas. Its contribution to the stress is an isotropic pressure that is proportional to the local molecular density and to the absolute temperature. The second factor is attractive interaction among molecules. Each pair of interacting molecules makes a tensile contribution to the stress which is directed along the line joining their centers. The overall tensile stress due to the sum of all such contributions is isotropic in a bulk phase, but anisotropic in an interfacial region (i.e., different in directions parallel to and perpendicular to the density gradient). The reason for this anisotropy is simply that the number and distribution of molecules with which a given molecule can interact is different in the two directions. At point A in Figure 1.1, for example, the molecular density decreases more rapidly toward the vapor phase than it increases toward the liquid phase. As a result, we anticipate that there will be fewer pairs of molecules with their lines of centers nearly parallel to the density gradient than with their lines of centers nearly aligned with some lateral direction (perpendicular to the density gradient). In other words, the tensile stress at A due to molecular interaction will be greater in the lateral direction than in the normal direction.

The net pressure in each bulk fluid is the difference between the kinetic and interaction contributions. Both contributions are much larger in the liquid than in the vapor, owing to the higher molecular density of the former, but their difference is the same in both phases for a plane interface. Similarly the net stress at any point in the interfacial region is the difference between the local kinetic and interaction contributions. It is anisotropic, with its normal component always equal to the bulk fluid pressures for a plane interface, as required for static equilibrium. In view of the argument of the preceding paragraph, we anticipate that the average lateral stress in the interfacial region is less compressive and more tensile than in bulk fluids. The overall result of this effect is what we know as interfacial tension. We note that the anisotropy of stress in the interfacial region implies that, unlike a bulk fluid, an interface can support certain types of shear stress at static equilibrium.

It is noteworthy that, as shown in Section 4, the lateral stress normally becomes tensile in nature in some portion of the interfacial region and is not simply a pressure with a smaller magnitude than in the bulk fluids. Careful experiments have shown that even bulk liquids can sometimes be subjected to tensile stresses (Hayward, 1971). It has been proposed that negative pressures (i.e., tensile stresses) are responsible for the ability of sap to rise to great heights in trees and for the ability of mangrove roots, whose sap contains little salt, to absorb water from the salty waters in which the mangrove tree thrives. This latter situation may be an example of nature's use of the reverse osmosis principle. Of course, negative pressures in bulk liquids are metastable conditions, while interfacial tension is a true equilibrium property.

Whether interfacial tension is developed from thermodynamic (energy) or mechanical (force) considerations, its main effect is that a system acts to minimize its interfacial area. This tendency for interfacial contraction is the reason that a small drop of one fluid in another will, provided gravitational effects are small, be spherical, the shape which minimizes drop area for a given drop volume. But it is essential to recognize that the energy and force arguments lead not simply to qualitatively similar concepts, but to the same quantitative value of interfacial tension, a point which is demonstrated below.

Both approaches are useful. The energy approach relates interfacial tension to thermodynamics and thus allows useful results to be derived (e.g., the Kelvin equation of Example 1.1, which gives the effect of drop size on vapor pressure). The force approach is needed to justify using interfacial tension in boundary conditions involving forces and stresses at interfaces. Such boundary conditions are employed in solving the governing equations of fluid mechanics when fluid interfaces are present.

Some values of interfacial tension in common systems are listed in Table 1.1. The International System (SI) units of millinewtons per meter (mN/m) are used, 1 mN/m being equal to 1 dyne/cm, the units of interfacial tension found in much of the older literature. Note that liquids with higher cohesive energies have larger surface tensions. Thus surface tensions of liquid metals are higher than those of hydrogen-bonded liquids such as water, which in turn are higher than those of nonpolar liquids such as pure hydrocarbons. Phenomenological methods of estimating surface tension are discussed in Section 5 and in Chapter 2 (Section 2).

TABLE 1.1
Representative values of surface and interfacial tensions

Fluid	Temperature (°C)	Surface tension (mN/m)
Silver	1100	878
Mercury	20	484
Sodium nitrate	308	117
Water	20	72.8
Propylene carbonate	20	41.1
Benzene	20	29.0
n-Octanol	20	27.5
Propionic acid	20	26.7
n-Octane	20	21.8
Ethyl ether	20	17.0
Argon	−183	11.9
Perfluoropentane	20	9.9

Fluid pair	Temperature (°C)	Interfacial tension (mN/m)
Mercury-water	20	415
Mercury-benzene	20	357
n-Octane-water	20	50.8
Benzene-water	20	35.0
Ethyl ether-water	20	10.7
n-Octanol-water	20	8.5
n-Butanol-water	20	1.8

Interfacial tensions between immiscible liquids decrease as the liquids become more similar in chemical nature, approaching zero near criticality.

3.   INTERFACIAL TENSION: THERMODYNAMIC APPROACH

As we have seen, the interfacial region is, in fact, three dimensional (i.e., it has a finite thickness). It is very convenient, however, to represent an interface as a mathematical surface of zero thickness because such properties as area and curvature are well defined and because the differential geometry of surfaces is well understood. How can a thermodynamic analysis be developed that reconciles the use of mathematical surfaces with the actual three-dimensional character of the interface?

More than a century ago, Gibbs (1878) introduced surface excess quantities as a first step toward resolving this problem. The basic idea is to choose a reference surface S somewhere in the interfacial region. This surface is everywhere perpendicular to the local density or concentration gradient. Consider a property such as internal energy in the region between surfaces S_A and S_B of Figure 1.2 which are parallel to S but are located in the respective bulk phases. Because the

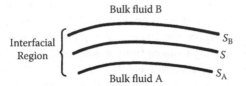

FIGURE 1.2 Interfacial region bounded by parallel surfaces. S is the reference surface.

transition between bulk compositions occurs over a finite thickness, the actual internal energy U of this region differs from the value $(U_A + U_B)$ which would be calculated by assuming that bulk phases A and B extend unchanged all the way to S. The difference is called the surface excess internal energy (U^S) and is assigned to S:

$$U^S = U - U_A - U_B \qquad (1.1)$$

Similarly, surface excess values of other thermodynamic properties can be defined, as can the surface excess number of moles n_i^S of species i:

$$n_i^S = n_i - n_{iA} - n_{iB} \qquad (1.2)$$

Here n_i represents the actual number of moles of i in the region between S_A and S_B, n_{iA} represents the moles of i that would be present in the region between S_A and S if it were occupied by bulk fluid A, and n_{iB} is defined similarly for the region between S_B and S. While independent of the exact positions of S_A and S_B as long as these surfaces are in the bulk fluids, the values of surface excess quantities do depend on the position of S. They may also be either positive or negative.

It is clear from Equations 1.1 and 1.2 that surface excess quantities do take into account the variation of composition and properties across an interfacial region of finite thickness. As we shall see shortly, they can be used to define interfacial tension. Moreover, since all surface excess properties are assigned to the reference surface S, the area and curvature of S can be identified as the corresponding properties of the interface and used, for example, to describe interfacial deformation.

Let us consider further the interfacial region between S_A and S_B under equilibrium conditions. If its shape remains fixed, we suppose that its internal energy U is a function only of its entropy S and the number of moles n_i of species i in the region. Then we can write

$$dU = TdS + \sum_i \mu_i dn_i \qquad (1.3)$$

where $T = (\partial U / \partial S)_{n_i, shape}$ and $\mu_i = (\partial U / \partial n_i)_{s, n_j, shape}$.

Now consider a process where bulk fluid A remains unchanged but some energy or mass is transferred between bulk fluid B and the interfacial region. The entire system consisting of the interfacial region and bulk fluids A and B is presumed to be completely isolated from its surroundings. In this case, an energy balance (the first law of thermodynamics) shows that the total energy of the system remains unchanged. Recalling that there is no change in the energy of bulk fluid A, we have

$$dU_{tot} = 0 = TdS + \sum_i \mu_i dn_i + T_B dS_B + \sum_i \mu_{iB} dn_{iB} \qquad (1.4)$$

Mass balances for the various species require that $dn_i = -dn_{iB}$. In addition, the entropy of such an isolated system must be a maximum if it is a equilibrium, according to the second law of thermodynamics. As a result, we have

$$dS_{tot} = 0 = dS + dS_B. \qquad (1.5)$$

From Equations 1.4 and 1.5, and the mass balances, we find that

$$0 = (T - T_B)\,dS + \sum_i (\mu_i - \mu_{iB})dn_i \ . \qquad (1.6)$$

If this equation is to be satisfied for all possible changes, it is clear that

$$T = T_B, \mu_i = \mu_{iB}, \qquad (1.7)$$

A similar argument can be made for energy or mass transfer between the interfacial region and bulk fluid A. The overall conclusion is that the temperature T and the chemical potential μ_i of each component must be uniform throughout the system.

With the interfacial region still maintained at a constant shape, we can write Equation 1.3 and subtract from it the analogous equations that would apply for its two parts if they were occupied by bulk fluids A and B, respectively. The result is

$$dU^s(\textit{fixed shape}) = TdS^s + \sum_i \mu_i dn_i^s \ . \qquad (1.8)$$

We now consider how dU^s might change for a fluid interface if the reference surface S is deformed. Both the area and curvature of S can change; but if the radii of curvature are much greater than the interfacial thickness, we might expect curvature effects to be small, so that

$$dU^s = TdS^s + \sum_i \mu_i dn_i^{\ s} + \gamma dA \ . \tag{1.9}$$

Here γ is the interfacial tension defined as $(\partial U^s / \partial A)_{S^s, n_i^{\ s}}$.

The Helmholtz free energy F for the interfacial region is defined in the usual way:

$$F = U - TS. \tag{1.10}$$

Subtracting from Equation 1.10 the analogous equations that would apply if the two parts of the interfacial region were occupied by bulk fluids A and B, we obtain

$$F^S = U^S - TS^S. \tag{1.11}$$

Differentiating Equation 1.11 and invoking Equation 1.9, we find

$$dF^S = -S^s dT + \sum_i \mu_i dn_i^{\ s} + \gamma dA \ . \tag{1.12}$$

It is clear from this equation that the interfacial tension can be written in terms of the surface excess free energy as

$$\gamma = \left(\frac{\partial F^s}{\partial A} \right)_{T, n_i^{\ s}} ; \tag{1.13}$$

that is, γ is the change in surface excess free energy produced by a unit increase in area.

Let us consider a given interfacial region and investigate the effect of shifting the reference surface S uniformly toward S_B by some small amount $d\lambda$. Naturally, the free energy F of the overall region is unchanged since no change occurs in its physical state. We have

$$\begin{aligned}
0 = dF &= dF^s + dF_A + dF_B \\
&= -S^s dT + \sum_i \mu_i dn_i^{\ s} + \gamma dA \\
&\quad -S_A dT + \sum_i \mu_i dn_{iA} - p_A dV_A \\
&\quad -S_B dT + \sum_i \mu_i dn_{iB} - p_B dV_B
\end{aligned} \tag{1.14}$$

Now both the total volume V of the interfacial region and the total number of moles n_i of each species are constant. Hence

$$dV = dV_A + dV_B = 0 \qquad (1.15)$$

$$dn_i = dn_i^s + dn_{iA} + dn_{iB} = 0. \qquad (1.16)$$

Invoking Equations 1.15 and 1.16, and recognizing that there is no temperature change, we find that Equation 1.14 simplifies to

$$0 = \gamma dA - (p_A - p_B)dV_A. \qquad (1.17)$$

Let A be the initial area of the reference surface S. Then it is clear that a simple shift in the reference surface gives

$$dV_A = Ad\lambda. \qquad (1.18)$$

Moreover, it can be shown from geometrical considerations that

$$dA = -2H \cdot Ad\lambda, \qquad (1.19)$$

where $(-2H)$ is the sum of the reciprocal radii of curvature of S as measured in any two perpendicular planes containing the local normal to S, that is,

$$2H = -\left(\frac{1}{r_1} + \frac{1}{r_2}\right). \qquad (1.20)$$

The quantity H is often called the mean curvature. The minus sign in Equation 1.20 is a convention; a radius of curvature is deemed positive if its center of curvature is on the A side of S and negative if its center of curvature is on the B side. Evaluation of r_1 and r_2 is considered in Section 6.

While Equation 1.19 will not be proved here, it may be helpful to demonstrate its validity for the special case of a sphere which increases in radius by an amount dr. In this case we have

$$dA = d(4\pi r^2) = 8\pi r \cdot dr. \qquad (1.21)$$

As $2H$ is $(-2/r)$, A is $4\pi r^2$ and $d\lambda$ is dr, it is clear that Equation 1.19 is satisfied.

When Equations 1.18 and 1.19 are substituted into Equation 1.17, the result is

$$p_A - p_B = -2H\gamma. \qquad (1.22)$$

This relationship among the two bulk phase pressures, the mean curvature, and the interfacial tension is fundamental to the study of fluid interfaces. It is often referred to as the Laplace or the Young-Laplace equation and is the basis for several methods of measuring interfacial tension. Note that with the interface curved in the manner shown in Figure 1.2, H is negative and Equation 1.22 implies that $p_A > p_B$. Thus the pressure inside a drop or bubble always exceeds the pressure outside.

A second fundamental equation of interfacial thermodynamics can be derived starting with Equation 1.9. Now U^s, S^s, n_i^s, and A are all extensive variables (i.e., they are proportional to the area of S if the intensive variables of the system, such as temperature and pressure, remain constant). Accordingly, Euler's theorem of homogeneous functions can be applied just as in ordinary bulk phase thermodynamics to obtain

$$U^s = TS^s + \sum_i \mu_i n_i^s + \gamma A \ . \tag{1.23}$$

Differentiating Equation 1.23 and applying Equation 1.9, we find

$$d\gamma = -\Gamma_s dT - \sum_i \Gamma_i d\mu_i \ , \tag{1.24}$$

where Γ_s and Γ_i are defined as (S^s/A) and (n_i^s/A), respectively. This equation is known as the Gibbs adsorption equation and is the analog of the Gibbs-Duhem equation for bulk fluids.

For a two-component system at constant temperature with S chosen in such a way that n_1^s vanishes, Equation 1.24 simplifies to

$$\left(\frac{\partial \gamma}{\partial \mu_2}\right)_T = -\Gamma_2 = -\frac{n_2^s}{A} \ . \tag{1.25}$$

Further simplifying to the case where species 2 is a solute that exhibits ideal behavior in a solvent (species 1), we find

$$\Gamma_2 = -\frac{x_2}{RT}\left(\frac{\partial \gamma}{\partial x_2}\right)_T \ , \tag{1.26}$$

where x_2 is the bulk phase solute mole fraction. We see that the surface excess concentration Γ_2 can be calculated from measurements of interfacial tension as a function of composition. We see, moreover, that if interfacial tension decreases with increasing solute concentration, Γ_2 is positive (i.e., the solute tends to concentrate near the interface). Such solutes are called surface active materials.

The derivation leading to Equation 1.26 involves a particular choice of the reference surface S. We may naturally ask about the sensitivity of quantities such as interfacial tension γ, the surface excess entropy per unit area Γ_S, and the surface excess concentrations Γ_i to the location of the reference surface. For a plane interface, the value of γ is independent of the position of S, as may be shown for instance by differentiating Equation 1.4.v of Problem 1.4 with respect to the radius a of the interface and taking the limit as a becomes very large and the pressure difference $(p_A - p_B)$ very small. Even for an interface that is slightly curved, γ should not vary greatly with small shifts in the position of S.

The surface excess concentrations Γ_i are another matter, however. Let us consider the simple liquid-vapor interface of Figure 1.1 and suppose that the reference surface S is at position A. If the bulk vapor density is small compared to the bulk liquid density, we see that shifting S by even one molecular diameter changes the value of Γ by an amount comparable to the surface concentration along an imaginary plane through the bulk liquid. Clearly this is a very significant change, and it raises questions about results involving Γ_i based on a particular (though reasonable) choice of S.

Fortunately such problems can usually be avoided — at least for plane interfaces. The Gibbs-Duhem equations for the two bulk phases in a binary system at constant temperature are given by

$$dp_A = c_{1A}d\mu_1 + c_{2A}d\mu_2 \qquad (1.27)$$

$$dp_B = c_{1B}d\mu_1 + c_{2B}d\mu_2.$$

For a plane interface, $dp_A = dp_B$, and these equations can be used to eliminate $d\mu_1$ from Equation 1.24. The result is

$$\left(\frac{\partial\gamma}{\partial\mu_2}\right)_T = -\Gamma_2 + \left(\frac{c_{2A} - c_{2B}}{c_{1A} - c_{1B}}\right)\Gamma_1 = -\Gamma_{2,1}. \qquad (1.28)$$

No assumption about the position of the reference surface has been made in deriving this equation. Moreover, since $(\partial\gamma/\partial\mu_2)_T$ is independent of the location of S for a plane interface, the "relative adsorption" $\Gamma_{2,1}$ must be as well. It is, of course, numerically equal to Γ_2 for the choice that makes Γ_1 vanish. Similar definitions of the various relative absorptions $\Gamma_{i,1}$ can be made for multicomponent systems, with i replacing 2 as a subscript in the definition of $\Gamma_{2,1}$ (Defay et al., 1966). It can be shown that the $\Gamma_{i,1}$ are all independent of the reference surface position in this case.

Another useful relationship is obtained by dividing Equation 1.23 by A and rearranging:

$$\gamma = \frac{F^s}{A} - \sum_i \mu_i \Gamma_i .$$

(1.29)

As discussed in Section 10, this equation is used to *define* the interfacial tension of a solid-fluid interface. We note that it avoids complications involving strain energy effects associated with extending the previous definition (given following Equation 1.9) to the case of solids. We note from Equation 1.29 that the interfacial tension is not, in general, equal to the Helmholtz free energy per unit area. The two are the same, however, for the special case of a single-component system with the reference surface chosen to make Γ_1 vanish.

EXAMPLE 1.1 VAPOR PRESSURE OF A DROP

Calculate the vapor pressure of a liquid drop as a function of its radius. Find the increase in vapor pressure for water at 100°C for drop radii of 100 μm, 1 μm, and 0.01 μm. Take the surface tension to be 60 mN/m and the density to be 994 kg/m³ for water at this temperature.

SOLUTION

The Young-Laplace equation (Equation 1.22) for the drop takes the form

$$p_L - p_v = \frac{2\gamma}{r} ,$$

(1.E1.1)

where r is the drop radius. When r varies at constant temperature, this equation can be differentiated to obtain

$$dp_L - dp_v = d\left(\frac{2\gamma}{r}\right) .$$

(1.E1.2)

Since equilibrium is maintained during this process, the changes in the liquid and vapor chemical potentials must be equal:

$$d\mu_L = v_L dp_L = d\mu_v = v_v dp_v.$$

(1.E1.3)

From this equation it is clear that

$$dp_L = \frac{v_v}{v_L} dp_v . \qquad (1.E1.4)$$

Substituting this expression into Equation 1.E1.2, invoking the ideal gas law since vapor pressure is low, and further noting that $v_L \ll v_v$ under these conditions, we find

$$\frac{RT}{v_L} \frac{dp_v}{p_v} = d\left(\frac{2\gamma}{r}\right) . \qquad (1.E1.5)$$

When this equation is integrated using the condition that p_v is the ordinary vapor pressure $p^o{}_v$ for a large pool of liquid in the limit $r \to \infty$, the result is the Kelvin equation

$$p_v = p_v{}^o \exp\left(\frac{2\gamma v_L}{rRT}\right) . \qquad (1.E1.6)$$

The vapor pressure is thus enhanced with decreasing drop radii, although the effect is appreciable only for extremely small drops (see calculation below). This effect is of importance in homogeneous nucleation of a new phase.

For water at 100°C, we find the following values using the properties given above

r (μm)	$p_v/p^o{}_v$
100	1.000007
1	1.0007
0.01	1.073

In the same manner it can be shown that the melting points of very small particles called nanoparticles (diameters less than 30 nm) are significantly decreased. Nanoparticles sinter more easily.

4. INTERFACIAL TENSION: MECHANICAL APPROACH

Our next step is to analyze the interfacial region at static equilibrium from a mechanical point of view. It is necessary to specify the lateral boundary of the region, which will be taken to be a surface S_o everywhere normal to S, S_A, and S_B of Figure 1.2, and to all other surfaces parallel to these three which could be constructed within the region. The result is a closed interfacial region bounded by S_A, S_B, and S_o (Figure 1.3).

Within the interfacial region, tangential and normal pressures p_T and p_N may differ, as discussed previously. In the bulk fluids, of course, the pressure is

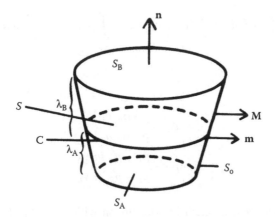

FIGURE 1.3 Pillbox control volume. S_o is perpendicular to S, S_A, and S_B.

isotropic, so that p_N and p_T are equal. Using this information, we apply Newton's second law to the interfacial region to obtain

$$\int_V \rho \hat{F} \, dV + \int_{S_A} p_A \boldsymbol{n} \, dS - \int_{S_B} p_B \boldsymbol{n} \, dS - \int_{S_o} p_T \boldsymbol{M} \, dS = \int_V \rho \boldsymbol{a} \, dV. \quad (1.30)$$

Note that since no shear stresses act on the outer boundary of the interfacial region, the local force on this boundary is everywhere parallel to the local normal, viz., $\boldsymbol{n}$ for S_A and S_B, $\boldsymbol{M}$ for S_o (see Figure 1.3). Accordingly, the net force on the outer boundary is given by the three surface integrals of Equation 1.30. Also $\hat{F}$ is the local body force per unit mass, most commonly gravity, and $\boldsymbol{a}$ is the local acceleration, which must be either a uniform translational acceleration or a uniform rotation consistent with static equilibrium.

Using the basic surface excess concept described previously, we wish to subtract from Equation 1.30 the corresponding equations which would apply if the region between S and S_A were occupied by bulk fluid A and that between S and S_B by bulk fluid B. These equations are

$$\int_{V_A} \rho_A \hat{F} \, dV + \int_{S_A} p_A \boldsymbol{n} \, dS - \int_S p_A \boldsymbol{n} \, dS - \int_{S_{oA}} p_A \boldsymbol{M} \, dS = \int_{V_A} \rho_A \boldsymbol{a} \, dV \quad (1.31)$$

$$\int_{V_B} \rho_B \hat{F} \, dV + \int_S p_B \boldsymbol{n} \, dS - \int_{S_B} p_B \boldsymbol{n} \, dS - \int_{S_{oB}} p_B \boldsymbol{M} \, dS = \int_{V_B} \rho_B \boldsymbol{a} \, dV. \quad (1.32)$$

When they are subtracted from Equation 1.30, the result is

$$\int_V \Delta\rho \hat{F} \, dV + \int_S (p_A - p_B) n \, dS - \int_{S_o} p_T M \, dS = \int_V \Delta\rho a \, dV \,, \qquad (1.33)$$

where

$$\Delta\rho = \begin{matrix} \rho - \rho_A \ \ in \ V_A \\ \rho - \rho_B \ \ in \ V_B \end{matrix}$$

$$\Delta p_T = \begin{matrix} p_T - p_A \ \ in \ V_A \\ p_T - p_B \ \ in \ V_B \end{matrix} \ .$$

If interfacial thickness is small in comparison with the radii of curvature of S, and if the body force at any lateral position is approximately uniform between S_A and S_B, we can write

$$\int_V \Delta\rho \, \hat{F} \, dV = \int_S \left[\int_{\lambda_A}^{\lambda_B} \Delta\rho \, d\lambda \right] \hat{F} \, dS \,. \qquad (1.34)$$

Similarly, if C is the closed curve which forms the outer boundary of S and if m is its outward pointing normal, we have

$$\int_{S_o} \Delta p_T M \, dS = \int_C \left[\int_{\lambda_A}^{\lambda_B} \Delta p_T \, d\lambda \right] m \, ds \,. \qquad (1.35]$$

Finally, a theorem from the differential geometry of surfaces (Weatherburn, 1955) can be used to transform the integral along C to an integral over S:

$$\int_C f \, m \, dS = \int_S \left[\nabla_s f + 2H f \, n \right] dS \,, \qquad (1.36)$$

where $(\nabla_s f)$ is the gradient of f within the surface, for example, $[e_x(\partial f/\partial x) + e_y(\partial f/\partial y)]$ for a plane surface. This "surface divergence" theorem will not be proved here, although demonstration of its validity in one simple situation is the objective of Problem 1.6.

Substitution of Equations 1.34 through 1.36 into Equation 1.33 yields

$$\int_S \left[\Gamma \hat{F} - \Gamma \, a + (p_A - p_B) n + 2H\gamma n + \nabla_s \gamma \right] dS = 0 \qquad (1.37)$$

$$\Gamma = surface\ excess\ mass = \int_{\lambda_A}^{\lambda_B} \Delta\rho\ d\lambda \qquad (1.38)$$

$$\gamma = interfacial\ tension = -\int_{\lambda_A}^{\lambda_B} \Delta p_T\ d\lambda\ . \qquad (1.39)$$

As the extent of S is arbitrary, the integral of Equation 1.37 must itself vanish, which leads to the differential equation of fluid statics:

$$\Gamma\ \hat{F} - \Gamma\ a + (p_A - p_B)n + 2H\gamma n + \nabla_s\gamma = 0\ . \qquad (1.40)$$

In most cases of interest, the surface excess mass Γ is small, so that the acceleration and body force terms may be neglected. Then Equation 1.40 simplifies to two conditions. One of them, $\nabla_s\gamma = 0$, requires that interfacial tension be uniform. The other is the Young-Laplace equation (Equation 1.22), which was obtained previously from thermodynamics for situations where body force and acceleration terms were unimportant. That the same equation (Equation 1.22) results from independent thermodynamic and mechanical derivations implies that interfacial tension must have the same value whether it is defined as in Equation 1.9 from energy considerations or as in Equation 1.39 from force considerations. Simply put, the force and energy definitions of interfacial tension are equivalent, a conclusion emphasized in the work of Buff (1956).

Equation 1.39 also allows us to demonstrate that tangential pressure p_T ordinarily must have negative values at some locations within the interfacial region. Consider a case where interfacial tension γ is about 10 mN/m and interfacial thickness is about 1 nm. It is clear from Equations 1.39 and 1.33 that the average value of $(p_N - p_T)$ in the interfacial region must be on the order of 10^7 N/m^2 or about 100 atm. As p_N must always lie between the bulk fluid pressures, and hence is typically not more than a few atmospheres, it is evident that p_T must have appreciable negative values at some locations within the interfacial region (i.e., the lateral stress must be tensile there).

EXAMPLE 1.2 LOCATING THE REFERENCE SURFACE

The density profiles of two components in a mixture across a flat vapor-liquid interface are given by

$$\rho_i = \frac{1}{2}\ (\rho_{Li} + \rho_{vi}) - \frac{1}{2}\ (\rho_{Li} - \rho_{vi})\ \tanh\left[\frac{2\ (z - Z_i)}{L_i}\right], \qquad (1.E2.1)$$

for $i = 1,2$. Here Z_i and L_i are constants and ρ_{Li} and ρ_{Vi} are the bulk densities at $z = -\infty$ and $+\infty$, respectively. Define suitable reference surfaces and obtain the surface excess masses and surface tension. The excess pressure across the interface is given by

$$\Delta p_T = -\frac{\gamma_o}{L_3\sqrt{2\pi}}\,\exp\left[-\frac{(z-Z_3)^2}{2L_3^2}\right]. \tag{1.E2.2}$$

SOLUTION

The surface tension in all cases is $\gamma = -\int_{-\infty}^{+\infty}\Delta p_T\,dz = \gamma_o$, on integration. From Equation 1.38 we have

$$\Gamma_1 = \int_{Z_o}^{\infty}(\rho_1 - \rho_{V1})\,dz + \int_{-\infty}^{Z_o}(\rho_1 - \rho_{L1})\,dz \tag{1.E2.3}$$

$$= (\rho_{L1} - \rho_{V1})\,(Z_o - Z_1)$$

where Z_0 is the position of the reference surface which has not yet been specified. Γ_2 can be obtained from Equation 1.E2.3 by substituting subscript 2 for 1. One choice of the reference is $\Gamma_1 = 0$, whence Equation 1.E2.3 yields

$$Z_0 = Z_1. \tag{1.E2.4}$$

With this definition of Z_0,

$$\Gamma_{2,1} = (\rho_{L2} - \rho_{v2})(Z_1 - Z_2). \tag{1.E2.5}$$

This approach can be generalized with

$$\Gamma_i = \int_{Z_o}^{\infty}u_i\,(\rho_1,\rho_2)\,(\rho_i - \rho_{Vi})\,dz + \int_{-\infty}^{Z_o}u_i\,(\rho_1,\rho_2)\,(\rho_i - \rho_{Li})\,dz\ , \tag{1.E2.6}$$

where u_i is a suitable weighting function. Instead of using 1.0 as the weighting factor, as in Equation 1.E2.3, we use something else, such as local specific volume or internal energy, made properly dimensionless to bias the surface excess with an end use in mind. Next, Γ_1 can be set to zero to obtain Z_0. Γ_2 becomes $\Gamma_{2,1}$ with this definition of Z_0. Conversely, one could set Γ_2 to zero to get Z_0 (another value) and $\Gamma_{1,2}$.

A mechanical counterpart gives another kind (Gibbs) of dividing surface, which is

$$0 = \int_{-\infty}^{+\infty} z \, (\Delta p_T) \, dz \, , \qquad\qquad (1.E2.7)$$

and leads to the location of the reference surface at $z = Z_3 = 0$. Thus $Z_0 = Z_3 = 0$ and Equation 1.E2.3 yields for this choice of reference surface $\Gamma_1 = - (\rho_{L1} - \rho_{V1})Z_1$ and $\Gamma_2 = - (\rho_{L2} - \rho_{V2})Z_2$. Known as the surface of tension, its location for a curved interface is discussed in Problem 1.4.

Equation 1.E2.1 can be arrived at using van der Waals' approach, where for a single-component fluid, the first term on the right-hand side is not the arithmetic mean density but the value at the critical point, and the thickness,

$$L = (\rho_L - \rho_V)/ \, | \frac{\partial \rho}{\partial z} |_o \, ,$$

is a measure of the thickness of the interface and the derivative is evaluated at the reference surface.

5. DENSITY AND CONCENTRATION PROFILES

Let us return to the density profile shown in Figure 1.1 for a simple liquid-vapor interface. We may ask what determines the shape of the profile (i.e., whether it is broad or sharp). Clearly the answer is that the system chooses the shape that minimizes its free energy. But what factors influence the free energy?

One factor, which favors a sharp profile, is the intrinsically high free energy of material having densities intermediate between those of bulk liquid and bulk vapor. As Figure 1.4 shows, these intermediate free energies (e.g., point Q on the curve between liquid and vapor densities L and V) are higher than those of corresponding mixtures, such as R of liquid and vapor having the same overall specific volume v or density ρ. Indeed, it is precisely this situation that is responsible for the phase separation. As a result of the high intermediate free energies, the system has a tendency to minimize the amount of material with intermediate densities and hence to form a thin interfacial region.

Each point on the curve of Figure 1.4 represents the free energy of a large volume of material with a uniform specific volume v. Each element of material interacts only with other material having the same specific volume v. In the interfacial region, however, density varies rapidly so that each element of material has elements with other values of v within the effective range of intermolecular forces. A correction to the free energy curve of Figure 1.4 is required to account for this effect.

Near a point such as A of Figure I-1 the average density decrease over the effective range of intermolecular forces is greater in the direction of the vapor phase than is the average density increase in the opposite direction. The result is an increase in free energy of the material at A over that predicted by Figure 1.4.

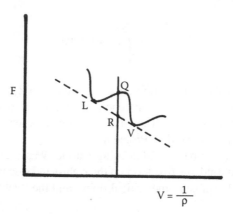

$$V = \frac{1}{\rho}$$

FIGURE 1.4 Dependence of free energy on molar volume v for a fluid below its critical temperature. L and V are the states of the liquid and vapor phases in equilibrium.

The magnitude of the increase is greater for a thinner interface with a sharper density profile. Hence this effect opposes sharp profiles and favors thick interfaces with broad profiles.

These two factors, the intrinsically high free energy of intermediate densities and the correction due to rapid density changes within the interfacial region, thus have opposite effects on the interfacial profile. The actual profile represents a balance between them and corresponds to an interface having a thickness that is finite, but typically quite small. The exception is near a critical point where the interface is thick, gradient energy effects are small, and interfacial tension approaches zero. Van der Waals, and later Cahn and Hilliard (1958), developed a phenomenological approach for determining the interfacial density profile and interfacial tension, taking into account these two factors. A key equation in their development is the following for the local Helmholtz free energy per unit volume f in the interfacial region for a flat interface:

$$
\begin{aligned}
f &= f_o(n) + k_1 \frac{d^2 n}{dz^2} + k_2 \left(\frac{dn}{dz}\right)^2 + \ldots \\
&= f_o(n) + \frac{d}{dz}\left[k_1 \frac{dn}{dz}\right] + \left[k_2 - \frac{dk_1}{dn}\right]\left(\frac{dn}{dz}\right)^2 + \ldots
\end{aligned}
\tag{1.41}
$$

Here, $f_0(n)$ is applicable to a large region having uniform molecular density n, while the remaining terms are necessary to account for the density variations. The parameter $[k_2 - (dk_1/dn)]$, which we shall designate k, is presumed to have a positive value.

As indicated in the discussion following Equation 1.29, the interfacial tension γ is equal to the surface excess Helmholtz free energy per unit area (F^s/A) when the reference surface is chosen to make the surface excess mass Γ vanish. But for this reference surface, the total amounts of liquid and vapor present in the extrapolated bulk phases must equal the amounts of the two phases that would form upon complete separation of the material in the interfacial region at constant total volume. Hence (F^s/A) can be evaluated by integrating the energy change required for complete separation at each local position:

$$\gamma = \frac{F^s}{A} = \int_{-\infty}^{\infty} (f - \varphi_L f_L - \varphi_V f_V)\, dz \ , \tag{1.42}$$

where φ_L and φ_V are the volume fractions of liquid and vapor that would be formed by material at position z and f_L and f_V are the Helmholtz free energy densities of the bulk phases. Invoking Equation 1.41 and noting that $[k_1(dn/dz)]$ vanishes in both bulk fluids, we can write

$$\gamma = \int_{-\infty}^{\infty} \left[\Delta f + k \left(\frac{dn}{dz} \right)^2 \right] dz \tag{1.43}$$

$$\Delta f(n) = f_0(n) - \varphi_L f_L - \varphi_V f_V \ . \tag{1.44}$$

It is readily seen that Δf is simply the distance between the free energy curve and the line LV in Figure 1.4 (e.g., the vertical distance QR). It can also be shown that $\Delta f = f_0 - n\mu + p$, where μ and p are the uniform values of chemical potential and pressure.

The density profile realized in actuality is that which minimizes the integral in Equation 1.43. Since Δf and n are functions of z, the calculus of variations must be invoked (Courant and Hilbert, 1953). The resulting condition which must be satisfied throughout the interfacial region is given by

$$2k \frac{d^2 n}{dz^2} - \frac{d\Delta f}{dn} = 0 \ . \tag{1.45}$$

Multiplying this expression by (dn/dz), integrating the resulting expression, and recalling that Δf and (dn/dz) vanish in the bulk fluids, we obtain

$$\Delta f(n) = k \left(\frac{dn}{dz} \right)^2 \ . \tag{1.46}$$

This equation provides information about the steepness of the density gradient at a position where the density is n. Substituting Equation 1.46 into Equation

1.43 and changing the variable of integration from z to n, we obtain the following expression for the interfacial tension:

$$\gamma = 2 \int_{n_V}^{n_L} [k \ \Delta f(n)]^{1/2} \ dn \ . \tag{1.47}$$

Evaluation of from γ Equation 1.47 requires information on both $\Delta f(n)$ and k. For the former, it is necessary to have an equation of state. For the latter, it is necessary to have some knowledge of how forces between molecules depend on the distance of separation. Attempts to include intermolecular forces and thus predict density profiles and interfacial tensions go back at least as far as van der Waals, who used his equation of state in a calculation of density profiles. Davis (1996) and Davis and Scriven (1981) have summarized work in this area very clearly, including some of their own significant contributions. Among the topics they discuss is the use of density and composition profiles to calculate the stress distribution within the interfacial region. Rowlinson and Widom (1982) also deal with various aspects of the molecular theory of interfacial tension in their book. In particular they discuss key ideas involving critical phenomena, that is, on how the interfacial tension behaves as it vanishes at a critical point. Detailed molecular theories provide a full interpretation not only of the coefficients that appear in Equation 1.41, but also of all the coefficients and terms that appear in the infinite series. In fact, the series can be summed so that one may avoid working with a truncated series (Barker and Henderson, 1982). In some instances the errors from truncation can be significant.

We note that the approach described above can be used for thin films and other "nonuniform" systems where density or composition varies rapidly with position at equilibrium (see Chapter 2). It contrasts sharply with the classical treatment of interfacial thermodynamics given in Section 3, which deals with overall properties of the interfacial region without considering density or concentration profiles or even the existence of molecules and intermolecular forces. An advantage of the earlier approach is that its results are independent of molecular properties. The principal disadvantage is that values of interfacial tension cannot be predicted. Hence the role of molecular theory is, as in other areas of thermodynamics, to provide information beyond that obtainable with the classical approach.

6. EQUILIBRIUM SHAPES OF FLUID INTERFACES

As indicated above, the Young-Laplace equation (Equation 1.22) is one fundamental result obtained from the theory of interfaces. Because this equation relates interfacial tension to the pressure difference between fluids at each point along an interface, it can be used with the equations of hydrostatics to calculate the shape of a static interface. Or, if interfacial shape can be determined experimentally, Equation 1.22 can be used to obtain the interfacial tension. This latter procedure is the basis of several methods for measuring interfacial tension.

Because the mean curvature H appears in Equation 1.22, its application to an interface of arbitrary shape requires considerable knowledge of differential geometry. A few general formulas from differential geometry are given in Chapter 7. Here we restrict attention to relatively simple shapes with considerable symmetry. In particular, we consider the sessile drop method for measuring surface or interfacial tensions. A drop of fluid A is placed on a solid S which is otherwise in contact with fluid B (see Figure 1.5). The materials must be chosen so that A is denser than B and does not spread spontaneously on S, indeed preferably so that the contact angle λ shown in Figure 1.5 is relatively large at the final equilibrium position. For a given λ (and λ is basically a property of the materials present as discussed in Chapter 2), two factors determine the shape of the fluid interface. One is gravity, which favors a large diameter drop of small height, so that the denser fluid A is at its lowest possible position. The other is interfacial tension, which favors a smaller drop of greater height where the surface area is minimized. Indeed, in the absence of gravity, the interface would be a portion of a spherical surface. The actual drop shape reflects a balance between these two effects.

According to the usual equations of hydrostatics, we have in bulk fluids A and B

$$p_A = p_{Ao} + \rho_A g z \tag{1.48}$$

$$p_B = p_{Bo} + \rho_B g z. \tag{1.49}$$

Here the reference pressures p_{AO} and p_{BO} have been chosen for convenience as those at the drop apex O of Figure 1.5. Also, a cylindrical coordinate system with its origin at O has been chosen as shown. If the drop is axisymmetric, the radius of curvature b at O is the same for all orientations, and Equation 1.22 requires that

$$p_{Ao} - p_{Bo} = \frac{2\gamma}{b} . \tag{1.50}$$

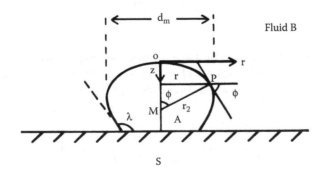

FIGURE 1.5 Sessile drop.

We now want to apply Equation 1.22 to a general point P on the drop where the radii of curvature are r_1 and r_2. Using Equations 1.48 through 1.50 with Equation 1.22, we have

$$\gamma \left(\frac{1}{r_1} + \frac{1}{r_2} \right) = \frac{2\gamma}{b} + (\rho_A - \rho_B)\, gz \;. \tag{1.51}$$

The principal radii of curvature for an axisymmetric drop are given by (Adamson and Gast, 1997)

$$\frac{1}{r_1} = \frac{d^2 z / dr^2}{[1 + (dz / dr)^2]^{3/2}} \tag{1.52}$$

$$\frac{1}{r_2} = \frac{dz / dr}{r[1 + (dz / dr)^2]^{1/2}} \;. \tag{1.53}$$

Here, r_1, the radius of curvature in the plane of Figure 1.5, is given by a well-known expression from analytic geometry. Also, r_2 is the radius of curvature around the drop as measured in a plane perpendicular to that of Figure 1.5 but containing the local normal at P. The center of curvature for r_2 is at point M of Figure 1.5, where the normal meets the drop axis. It is clear from the figure that

$$\sin \phi = \frac{r}{r_2} \;. \tag{1.54}$$

Invoking an identity from trigonometry, we can rewrite this equation in the following form:

$$\frac{1}{r_2} = \frac{1}{r} \frac{\tan \phi}{(1 + \tan^2 \phi)^{1/2}} \;. \tag{1.55}$$

Since the tangent to the drop profile at P makes an angle ϕ with the horizontal, we can replace $\tan \phi$ with (dz/dr) in Equation 1.55, thus obtaining Equation 1.53.

Substituting Equations 1.52 and 1.53 into Equation 1.51 and using the dimensionless coordinates $Z = (z/b)$ and $R = (r/b)$, we find

$$\frac{Z''}{[1 + (Z')^2]^{3/2}} + \frac{Z'}{R[1 + (Z')^2]^{1/2}} = 2 + \beta z \;, \tag{1.56}$$

where

$$\beta = (\rho_A - \rho_B)\ \frac{gb^2}{\gamma}\ . \tag{1.57}$$

For this second-order ordinary differential equation, two boundary conditions are required. They are that both Z and its derivative Z' with respect to R must vanish at the origin $R = 0$.

Equation 1.56 cannot be solved analytically except for certain limiting cases. The numerical solution is straightforward, however, and leads to knowledge of $Z(R)$ for any specified value of β. Bashforth and Adams (1893) first carried out a numerical solution more than a century ago, although complete and accurate tables are now available from more recent calculations using a high-speed computer (see Padday, 1969). In the past, interfacial tension was usually found by measuring on a photographic negative the maximum diameter d_m of the drop (see Figure 1.5) and the vertical distance between the plane of the maximum diameter and the apex and then referring to the tables. Nearly all workers now use a video camera and image analysis techniques. They digitize the drop's image and then use the stored information to locate numerous points on the drop profile. A suitable optimization procedure is then utilized to find the best match between the selected points and the computed solutions to the governing equations (Li et al., 1992).

EXAMPLE 1.3 DIMENSIONS OF A SESSILE DROP

(a) A sessile drop of water is to be formed on a solid surface where the contact λ angle is 90°. Find the maximum diameter, height, and volume of the drop if the parameter β of Equation 1.57 is to have a value of 20. The numerical results of Bashforth and Adams for $\beta = 20$ are summarized in Table 1.2. Take $\gamma = 72$ mN/m and $\rho_A = 1$ g/cm³ for water. Compare the drop dimensions you obtain with those which would exist if the drop volume and contact angle were the same but gravitational effects were negligible.

(b) Repeat the calculation of drop dimensions for $\beta = 20$ and $\lambda = 90°$ for a drop of an aqueous surfactant solution in oil. Take $\rho_A = 1.0$ g/cm³, $\rho_B = 0.8$ g/cm³, and $\gamma = 0.01$ mN/m.

SOLUTION

(a) From Equation 1.57 with $\rho_B = 0$ and the other values as given, we find b = 1.212 cm for the radius of curvature at the drop apex. From the tables with ϕ = 90°, we obtain for the coordinates at the base of the drop $R = 0.51153$, $Z = 0.29623$. Hence the maximum drop radius is $(0.51153 \times 1.212) = 0.620$ cm and the height is $(0.29623 \times 1.212) = 0.359$ cm. The ratio of height to radius is about 0.58, which gives some feeling for the degree of flattening caused by gravity.

TABLE 1.2
Dimensionless coordinates of sessile drop for $\beta = 20$ from Bashforth and Adams tables (1893)

ϕ	$R = r/b$	$Z = z/b$	V/b^3
5	0.08558	0.00370	0.00004
10	0.16261	0.01377	0.00058
15	0.22768	0.02811	0.00235
20	0.28143	0.04498	0.00584
25	0.32571	0.06327	0.01117
30	0.36236	0.08230	0.01828
35	0.39283	0.10167	0.02699
40	0.41824	0.12113	0.03706
45	0.43941	0.14050	0.04827
50	0.45698	0.15965	0.06037
55	0.47145	0.17849	0.07314
60	0.48322	0.19694	0.08635
65	0.49259	0.21492	0.09981
70	0.49984	0.23240	0.11334
75	0.50518	0.24931	0.12676
80	0.50880	0.26561	0.13993
85	0.51087	0.28126	0.15272
90	0.51153	0.29623	0.16502
95	0.51091	0.31049	0.7672
100	0.50914	0.32400	0.19809
105	0.50632	0.33674	0.19809
110	0.50255	0.34869	0.20764
115	0.49795	0.35982	0.21640
120	0.49258	0.37012	0.22435
125	0.48656	0.37959	0.23147
130	0.47995	0.38820	0.23780
135	0.47285	0.39596	0.24333
140	0.46533	0.40285	0.24810
145	0.45746	0.40889	0.25214
150	0.44933	0.41407	0.25549
155	0.44100	0.41841	0.25819
160	0.43254	0.42191	0.26029
165	0.42402	0.42460	0.26184
170	0.41550	0.42649	0.26289
175	0.40704	0.42761	0.26348
180	0.39870	0.42797	0.26367

Dimensionless drop volume (V/b^3) is 0.16502, so that the actual volume is $(0.16502 \times 1.212^3) = 0.294$ cm^3.

In the absence of gravity the drop would be a hemisphere of radius r. Since the volume is $(2\pi\, r^3/3)$ in this case, we can easily find $r = 0.520$ cm. Of course, the ratio of height to radius is now unity.

(b) Repeating the calculations of (a), one finds $b = 3.19 \times 10^{-2}$ cm, $V = 5.38 \times 10^{-6}$ cm^3, a maximum radius of 1.63×10^{-2} cm, and a drop height of 9.46×10^{-3} cm. Clearly, much smaller drops must be used to measure very low tensions. Systems having such low tensions are of interest in certain processes which have been studied for increasing oil recovery from underground reservoirs or for removing organic liquids from ground water aquifers (see Chapter 4).

EXAMPLE 1.4 SHAPE OF A SOAP FILM BETWEEN PARALLEL RINGS

A soap film is formed between parallel rings of radius R as shown in Figure 1.6. The line connecting the centers of the rings is of length $2L$ and is perpendicular to the planes of the rings. The rings are open so that the space within the film is in communication with the atmosphere.

(a) Will the profile of the soap film most nearly resemble curve 1, 2, or 3 in the diagram? Explain.

(b) Derive an equation for the film profile if effects of gravity can be neglected.

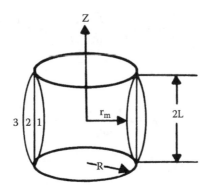

FIGURE 1.6 Hypothetical profiles for a soap film between parallel rings.

SOLUTION

(a) Since the rings are open, the pressure drop across the film is zero. From the Young-Laplace equation (Equation 1.22) we conclude that

$$\frac{1}{r_1} + \frac{1}{r_2} = 0 .$$ (1.E4.1)

That is, the centers of curvature for the two radii r_1 and r_2 at a point on the film must lie on opposite sides of the film. Since the film is axisymmetric, we define r_1 and r_2 as in Equations 1.52 and 1.53. Now the center of curvature for r_2 must, by symmetry, lie on the axis joining the centers of the rings. Hence the center of curvature for r_1 must lie outside the film and profile 1 is the appropriate choice.

(b) This problem is most easily solved by using expressions other than those given in Equations 1.52 and 1.53 for r_1 and r_2. Owing to symmetry, we need consider only the upper half of the film, that is, the region $z > 0$ in Figure 1.6. From Equations 1.53 and 1.54 we have

$$\sin \phi = \frac{dz/dr}{[1 + (dz/dr)^2]^{1/2}} .$$ (1.E4.2)

When this expression is differentiated, the result is

$$\frac{d}{dr} \sin \phi = \frac{d^2z/dr^2}{[1 + (dz/dr)^2]^{3/2}} = \frac{1}{r_1} .$$ (1.E4.3)

Hence we can write

$$\frac{1}{r_1} + \frac{1}{r_2} = \frac{1}{r} \frac{d}{dr} (r \sin \phi) = 0 .$$ (1.E4.4)

Integration of this equation yields

$$r \sin \phi = r_\mathrm{m},$$ (1.E4.5)

where r_m is the minimum radial coordinate of the film. It is reached at the position $z = 0$ exactly halfway between the two rings.

Now, as shown previously in deriving Equation 1.53,

$$\frac{dz}{dr} = \tan\phi = \frac{\sin\phi}{[1 - \sin^2\phi]^{1/2}} \, . \tag{1.E4.6}$$

Replacing $\sin\phi$ by (r_m/r) in accordance with Equation 1.E4.5, we find

$$\frac{dz}{dr} = \frac{r_m/r}{[1 - (r_m/r)^2]^{1/2}} \, . \tag{1.E4.7}$$

If we let $u = (r_m/r)$, this equation becomes

$$\frac{dz}{du} = -\frac{r_m}{u(1 - u^2)^{1/2}} \, . \tag{1.E4.8}$$

Integrating with the boundary condition $r = r_m$ ($u = 1$) for $z = 0$, we find

$$z = r_m \, \ln\left\{ \frac{r}{r_m} + \left[\left(\frac{r}{r_m}\right)^2 - 1 \right]^{1/2} \right\} . \tag{1.E4.9}$$

The unknown quantity r_m can be found by imposing a second boundary condition, that is, $r = R$ for $z = L$. The resulting equation is

$$L = r_m \, \ln\left\{ \frac{R}{r_m} + \left[\left(\frac{R}{r_m}\right)^2 - 1 \right]^{1/2} \right\} . \tag{1.E4.10}$$

The profile shape in this case is called a catenary, the shape of the soap film itself is a catenoid.

7.   METHODS OF MEASURING INTERFACIAL TENSION

Besides the sessile drop method just discussed, various techniques have been used for measuring interfacial tensions. They are discussed extensively by Padday (1969) and Rusanov (1996). Closely related to the sessile drop method is the pendant drop technique (see Figure 1.7). Here a drop of the denser fluid is suspended from a capillary tube. Gravity acts to elongate the drop while interfacial tension opposes elongation because of the associated increase in interfacial area. Equation 1.56 again governs drop shape, except that β now assumes negative

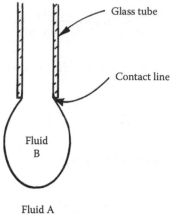

FIGURE 1.7 Pendant drop.

values (since $\rho_A < \rho_B$). Measurements of parameters indicative of drop shape, such as the maximum diameter and the height of the position of maximum diameter above the base of the drop can be used with tables based on the numerical solution of Equation 1.56 to obtain the interfacial tension (Ambwani and Fort, 1979). As mentioned above for the sessile drop method, image analysis techniques are now almost universally used to calculate the interfacial tension using the entire drop profile.

Both sessile and pendant drops are suitable for use in liquid-gas and liquid-liquid systems, provided, of course, that the liquid surrounding the drop is transparent. Changes in drop shape can be followed to determine time-dependent effects on interfacial tension. Both methods have been used to measure low interfacial tensions in liquid-liquid systems, but only the sessile drop works well for tensions below about 0.01 mN/m. Finally, it should be noted that sessile and pendant "bubbles" of the less dense phase can be employed where the whole apparatus is, in effect, turned upside down (Figure 1.8).

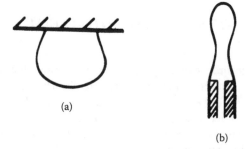

FIGURE 1.8 (a) Sessile bubble. (b) Pendant bubble.

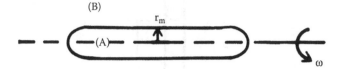

FIGURE 1.9 Schematic diagram of a spinning drop.

A related technique employs a "spinning drop" (Figure 1.9). In this case a drop of the less dense fluid is injected into a horizontal tube containing the denser fluid and the whole system is rotated as shown. In the resulting centrifugal field, the drop elongates along the axis of rotation. Interfacial tension opposes the elongation because of the increase in area and a configuration that minimizes system free energy is reached. The analysis is, in principle, similar to that for the sessile drop, with the gravitational acceleration g replaced by the appropriate acceleration term for a centrifugal field (see Problem 1.10). If the fluid densities ρ_A and ρ_B and the angular velocity ω of rotation are known, interfacial tension can be calculated from the measured drop profile. When drop length is much greater than the radius r_m, the following approximate expression holds:

$$\gamma = \frac{(\rho_B - \rho_A)\,\omega^2 r_m^3}{4}. \tag{1.58}$$

The spinning drop device has been widely used to measure very low interfacial tensions in systems where one phase is a microemulsion (see Chapter 4). Unlike the sessile and pendant drop schemes, no contact between the fluid interface and a solid surface is required. Both the drop and the surrounding fluid layer can also be made rather thin, so that results can be obtained even when the surrounding fluid is somewhat turbid, a frequent occurrence in practical systems. Finally, interfacial tension can, as with the sessile and pendant drops, be followed as a function of time.

A classical technique for measuring interfacial tension is the capillary rise method. As shown in Figure 1.10, a capillary tube of small diameter is partially inserted into a liquid A. Fluid B may be either a gas or an immiscible liquid. If A is the wetting fluid (i.e., if its attractive interaction with the solid is stronger than that of B) (see Chapter 2), liquid A begins to rise in the tube. Gravity, of course, opposes this rise, and an equilibrium height h is ultimately reached where the overall free energy of the system is minimized.

Provided that the contact angle λ measured through A is zero and that the tube radius R is sufficiently small, the meniscus has the shape of a hemisphere of radius R. Hence from the Young-Laplace equation (Equation 1.22) we have

$$p_{Bh} - p_{Ah} = \frac{2\gamma}{R}. \tag{1.59}$$

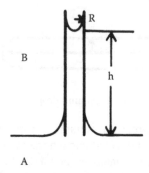

FIGURE 1.10 Capillary rise in a tube of radius R.

If the interface between A and B in the large region outside the tube is flat, we can again invoke the Young-Laplace equation at the height of this interface:

$$p_{B0} - p_{A0} = 0. \tag{1.60}$$

The result of subtracting these equations is

$$(p_{Bh} - p_{Bo}) - (p_{Ah} - p_{Ao}) = \frac{2\gamma}{R} . \tag{1.61}$$

As in the sessile drop analysis discussed previously, the equations of hydrostatics also apply, so that

$$p_{Bh} - p_{B0} = \rho_B g h \tag{1.62}$$

$$p_{Ah} - p_{A0} = \rho_A g h. \tag{1.63}$$

Substituting these equations into Equation 1.61 and solving for γ, we find

$$\gamma = \frac{Rgh}{2} (\rho_A - \rho_B) . \tag{1.64}$$

The capillary rise method provides very accurate values of interfacial tension if used carefully. The capillary tubes must not deviate significantly from a circular shape, must be of known and uniform radius, and must be carefully aligned in the vertical position. They must be cleaned or otherwise prepared to ensure that the contact angle is zero. Finally, Equation 1.64 must be corrected in experiments of high precision to account for meniscus deviation from the hemispherical shape due to gravity effects. Equation 1.56 describes the meniscus shape in this case, since the situation is basically the same as that for a sessile bubble. Hence the numerical solution as obtained for the sessile drop and bubble can be used to make the correction (see Problem 1.13).

Some variants of the capillary rise method have been employed. In the differential capillary rise technique, two tubes of different diameters are used, and the difference in height between the two menisci is measured. In the inclined capillary rise scheme, the tube is inclined at a known angle so that the length Q of the tube occupied by the wetting fluid is greater than the height h of the rise in a vertical tube. This method is useful when the interfacial tension is low because under these conditions the vertical rise h is small and difficult to measure accurately.

A relatively simple way to measure interfacial tensions is the drop weight or drop volume technique. The size of a pendant drop (Figure 1.7) is slowly increased until the drop can no longer be prevented from falling and breaks off. The total weight or volume of a known number of drops is measured.

There is no exact theoretical method for calculating the interfacial tension from the known values of drop volume V and fluid densities ρ_A and ρ_B. Dimensional analysis can be used, however, to find suitable dimensionless variables for an empirical correlation. If viscous effects on drop size are assumed negligible, we may suppose that interfacial tension γ is a function F of the following variables:

$$\gamma = F(V, (\rho_B - \rho_A), g, r), \qquad (1.65)$$

where r is the radius of the capillary tube from which the drop is formed. The outside radius should be used when the drop liquid wets the tube material, the inner radius when it does not. Application of dimensional analysis (see Problem 1.11) yields two dimensionless variables, $(\gamma/\rho_B - \rho_A)gr^2)$ and $(r/V^{1/3})$. If the relationship between these two variables is determined by experiment for one pair of fluids A and B (e.g., air and water), the resulting correlation may be used for other pairs.

Use of a micrometer syringe to feed the drop liquid to the tube makes determination of drop volume straightforward. The tip of the tube should be carefully ground so that it is free from nicks and other irregularities. As with all other methods, evaporation must be prevented in liquid-gas systems to prevent cooling of the interface and a resulting change in interfacial tension. With these provisions, the use of the drop volume method is convenient, and it is widely used. It should be noted that, by its very nature, this method is unsuitable when diffusion or adsorption effects dictate that considerable time is needed for the equilibrium value of interfacial tension to be attained.

Somewhat different are methods of measuring interfacial tension in which the vertical force acting on a solid body is measured during its withdrawal from a liquid. Objects of various shapes have been used, but the most common techniques are those employing a vertical plate (Wilhelmy plate method) and a horizontal circular ring (DuNouy ring method). In the former case, illustrated in Figure 1.11, the force is measured when the lower edge of the plate is at the same

FIGURE 1.11 Schematic diagram of the Wilhelmy plate method for measuring interfacial tension.

elevation H as the large flat interface far from the plate. Under these conditions there is no buoyant force in a liquid-gas system; that is, the liquid pressure acting on the bottom of the plate is equal to the (uniform) pressure in the gas phase. Provided that the contact angle is zero, the total force f on the plate is given by

$$f = mg + \gamma P, \qquad (1.66)$$

where m is the mass of the plate and P its perimeter. Since all quantities except γ in Equation 1.66 are known or can be measured, the tension is readily calculated.

The situation is more complex for the ring (Figure 1.12) since the force on the ring due to the meniscus is not vertical at most positions of the ring above the surface. What is done is to measure the maximum force f_m during withdrawal and to employ dimensional analysis, neglecting viscous effects in the same manner as was done above for the drop volume method. In this case we have

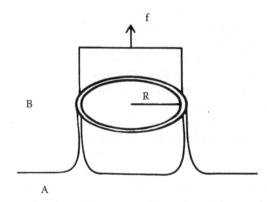

FIGURE 1.12 Schematic diagram of the DuNouy ring method for measuring interfacial tension.

$$\gamma = F(f_{m'} \, (\rho_A - \rho_B), \, g, \, r, \, R). \tag{1.67}$$

Here R is the radius of the ring and r is the radius of the wire used to make the ring. Application of dimensional analysis yields three dimensionless groups: $(\gamma/(\rho_A - \rho_B)g \, R^2)$, $(f_m/(\rho_A - \rho_B) \, gR^3)$, and (r/R). As before, experiments in one (or a few) systems can be undertaken to determine how the first of these groups, a dimensionless interfacial tension, depends on the other two. The results can then be applied in any system of interest.

In using the Wilhelmy plate and DuNouy ring methods, it is important to ensure that the plate is indeed vertical and the ring horizontal during the measurement. One advantage of the Wilhelmy plate method is that, unlike the other techniques discussed in this section, it does not require knowledge of fluid densities. Since the plate is never removed from the interface, and indeed a measurement requires very little interfacial distortion, it can also be used to follow time-dependent changes in interfacial tension. The DuNouy ring method cannot be used for this purpose, however, since relatively large distortions of the interface are involved in detaching the ring.

Also worth mentioning is the maximum bubble pressure technique. Consider the process leading to the formation of the pendant drop shown in Figure 1.7. For simplicity, neglect gravitational effects. When the fluid B just starts to leave the bottom of the tube, its interface with fluid A is flat, and its pressure p_B is, according to the Young-Laplace equation, equal to that of A at this depth. As the volume of B below the end of the tube increases, the radius of curvature of the interface, which is a portion of a sphere, decreases and p_B increases. Eventually the drop of B becomes a hemisphere with $p_B = [p_A + (2\gamma/R)]$, where R is the radius of the tube. Further increases in drop volume lead to configurations similar to that shown in Figure 1.7, with the radius of curvature increasing and p_B decreasing. In other words, a plot of p_B as a function of drop volume exhibits a maximum value equal to the expression given above for the hemispherical shape. By varying the time of drop formation, information can be gained on dynamic effects on the interfacial tension, such as diffusion and adsorption of surface active materials. It is noteworthy that, unlike most of the techniques described previously, the maximum bubble pressure method can be used when gravitational effects are small, owing either to a small density difference between fluids or to small gravitational acceleration (e.g., for experiments carried out in space). Indeed it has been used to measure liquid-liquid interfacial tensions under microgravity conditions (Passarone et al., 1991) and also between dilute aqueous surfactant solutions and a heavy crude oil (bitumen) having nearly the same density as water at temperatures between 60°C and 90°C (Pandit et al., 1995).

If the interfacial tension is known beforehand, the maximum bubble pressure method can be used to determine the diameter of the pores in a porous membrane. Microporous polymer membranes are used to filter large molecules such as proteins, viruses, and bacteria, and the process is called ultrafiltration. The quality of the membrane for medical applications is completely compromised if there is even one large hole that permits a virus to go through. Maximum bubble pressure is used to detect such holes in membranes since the maximum in p_B for the large hole is seen before that for the regular pores. The method actually determines the flow rate of air versus p_B (Kulkarni et al., 1992).

Interfacial tensions that are very low can also be determined by measuring the light scattered from the interface by spontaneously generated capillary waves. Further comment on this method is made in Chapter 4 (Section 8) and again briefly in Chapter 8 (Section 6).

The actual calculation of interfacial tension from experimental data using some of these methods is considered in several of the problems at the end of this chapter. For further discussion of the various techniques, including some not described here, and for extensive tables of the numerical solution of Equation 1.56 and of the empirical correlations for the drop volume and ring methods using dimensionless variables, the reader is referred to the excellent review by Padday (1969) and the book by Rusanov (1996).

EXAMPLE 1.5 CAPILLARY RISE IN AIR-WATER AND OIL-WATER SYSTEMS

Calculate the height h of capillary rise in the air-water system ($\gamma = 72$ mN/m) for tube radii of 0.1 mm, 0.01 mm, and 0.001 mm. Repeat for an oil-water system with $\gamma = 30$ mN/m and $\Delta\rho = 0.15$ g/cm³. Assume that water completely wets the solid surface in both cases.

SOLUTION

Equation 1.64 applies. Results are tabulated below. Note that the smaller tube radii are comparable to pore sizes in many soils and underground oil reservoirs. Since pores of various sizes may be anticipated in a given soil or reservoir, the results suggest that an underground boundary between air and water (water table) or between oil and water is probably not very sharp but has a substantial thickness.

R (mm)	h (cm) (air-water)	h (cm) (oil-water)
0.1	14.7	40.8
0.01	147	408
0.001	1470	4080

8. SURFACE TENSION OF BINARY MIXTURES

In the previous discussion of interfacial thermodynamics (Section 3), the number of moles n_i^s of each species at the interface was distinguished from the number of moles n_{iA} and n_{iB} in the adjacent bulk phases. We anticipate, therefore, that the corresponding interfacial mole fractions x_i^s differ from the bulk phase values x_{iA} and x_{iB}. To confirm this expectation and develop a quantitative expression for the difference, we consider an ideal liquid mixture whose surface is in contact with air or some other dilute gas.

When the temperature and pressure are constant, the chemical potential of each species in an ideal bulk solution is given by

$$\mu_i = \mu_i^o(T, p) + RT \ln x_i . \tag{1.68}$$

If pressure is allowed to vary, we have instead

$$\mu_i = \mu_i^*(T) + \bar{v}_i p + RT \ln x_i , \tag{1.69}$$

where $\bar{v}_i$ is the partial molar volume of species i. In an analogous manner we write for interfacial chemical potential

$$\mu_i^s = \mu_i^{os}(T) + RT \ln x_i^s - \bar{a}_i \gamma , \tag{1.70}$$

where $\bar{a}_i$ is the partial molar area of species i.

Let us consider the special case where temperature and pressure are fixed and the liquid is pure component i. In this case $x_i = x_i^s = 1$ and the surface tension has the pure component value γ_i. Equating the right-hand sides of Equations 1.68 and 1.70 for this case, we find

$$\mu_i^{os} = \mu_i^o + \gamma_i a_i . \tag{1.71}$$

If this expression is used to replace μ_i^{os} in Equation 1.70 and if μ_i^s is set equal to μ_i for a mixture with some arbitrary value of x_i, the result is

$$\ln \frac{x_i^s}{x_i} = \frac{(\gamma - \gamma_i) a_i}{RT} . \tag{1.72}$$

Here $\bar{a}_i$ for the mixture has been taken equal to a_i for the pure component, usually a reasonable assumption. As the surface tension of a mixture normally differs from that of any of its pure components, we conclude that the surface composition does differ from the bulk liquid composition.

For simplicity, let us now restrict our attention to a binary mixture where $a_1 = a_2 = a$. Writing Equation 1.72 for both components, we have

$$\frac{\gamma a}{RT} = \frac{\gamma_1 a}{RT} + \ln \frac{x_1^s}{x_1} = \frac{\gamma_2 a}{RT} + \ln \frac{x_2^s}{x_2} . \tag{1.73}$$

This equation may be rearranged to obtain

$$\frac{x_1^s}{1 - x_1^s} = \frac{x_1}{1 - x_1} \exp\left[\frac{(\gamma_2 - \gamma_1)a}{RT}\right]. \tag{1.74}$$

Thus the surface composition x_1^s can be calculated for any bulk phase composition x_1, provided that the molecular area a and the pure component surface tensions γ_1 and γ_2 are known. We note that if $\gamma_1 < \gamma_2$, Equation 1.74 indicates that $x_1^s > x_1$. That is, the surface is enriched in the species having the lowest surface tension or surface free energy, the expected result.

Once x_1^s has been found, the surface tension γ of the mixture is readily obtained from Equation 1.73. Indeed, use of Equation 1.74 to eliminate x_1^s yields the following explicit expression:

$$\exp\left(\frac{-\gamma a}{RT}\right) = x_1 \exp\left(\frac{-\gamma_1 a}{RT}\right) + x_2 \exp\left(\frac{-\gamma_2 a}{RT}\right). \tag{1.75}$$

When $[|\gamma_1 - \gamma_2| a/RT] \ll 1$, this equation may be divided by $\exp(-\gamma_1 a/RT)$ and the resulting expression expanded in a series. The result in this special case is

$$\gamma \sim x_1\gamma_1 + x_2\gamma_2. \tag{1.76}$$

This formula is often useful because two species that form an ideal mixture are normally very similar chemically and hence have nearly equal surface tensions.

The basic approach described here for ideal solutions can be extended to regular solutions, that is, those where mixing is random but where interaction energies differ significantly for the various species present. For a binary mixture of two species with equal molar volumes it is well known that the regular solution approximation leads to the following expression for the bulk phase chemical potentials:

$$\mu_i = \mu_i^o(T, p) + RT \ln x_i + \alpha (1 - x_i)^2, \tag{1.77}$$

where α is a constant proportional to the change in interaction energy accompanying exchange of a molecule in pure liquid 1 with a molecule in pure liquid 2. It seems plausible and can be shown using a lattice model (Defay et al., 1966) that the corresponding expression for the surface chemical potential is given by

$$\mu_i^s = \mu_i^{os}(T) + RT \ln x_i^s + \alpha\ell (1 - x_i^s)^2 + \alpha m (1 - x_i)^2 - \gamma a. \tag{1.78}$$

Here ℓ is the fraction of the nearest neighbor sites of a surface molecule which are also at the surface and m is the fraction which are in the underlying liquid.

A similar fraction m is in the overlying gas layer and presumed to be unoccupied. The quantities ℓ and m must therefore satisfy the relationship

$$\ell + 2m = 1. \tag{1.79}$$

Note that it has been assumed that $a_1 = a_2 = a$ in writing Equation 1.78, a reasonable step in view of the previous assumption of equal molar volumes.

As before, the quantities μ_i^{os} can be eliminated using the condition that $\gamma = \gamma_i$ for each pure liquid. Then equating the right-hand sides of Equations 1.77 and 1.78 for a general mixture, we find after some manipulation that

$$
\begin{aligned}
\gamma &= \gamma_1 + \frac{RT}{a}\ln\frac{x_1^s}{x_1} + \frac{\alpha\ell}{a}\left[(1-x_1^s)^2 - (1-x_1)^2\right] - \frac{\alpha m}{a}(1-x_1)^2 \\
&= \gamma_2 + \frac{RT}{a}\ln\frac{1-x_1^s}{1-x_1} + \frac{\alpha\ell}{a}\left[(x_1^s)^2 - x_1^2\right] - \frac{\alpha m}{a}x_1^2
\end{aligned} \tag{1.80}
$$

The surface composition x_1^s can be found by equating the two expressions in Equation 1.80 and the resulting value can then be used to calculate the surface tension γ. We note that the basic method can be extended to predict the interfacial tension between liquid phases in a binary system described by regular solution theory (Defay et al., 1966).

Surface solution theories should be consistent with the Gibbs adsorption equation (Equation 1.24), just as the bulk solution theories should be consistent with the Gibbs-Duhem equation. It is readily shown that ideal surface solution theory, as discussed above, satisfies this criterion (see Problem 1.15). However, regular surface solution theory does not, owing to the limitations of modeling the interfacial region by a single composition. This matter is discussed in some detail by Defay et al. (1966).

While the above discussion has dealt with liquid systems, it is noteworthy that surface solution theory is also of interest for solids. For example, the surface composition of metal alloy catalyst particles can differ significantly from the bulk composition. The proper design of the catalyst thus requires knowledge of how bulk and surface compositions are related.

EXAMPLE 1.6 SURFACE TENSION OF IDEAL BINARY SOLUTIONS

Chlorobenzene (1) and bromobenzene (2) form ideal solutions. At 293°K the surface tensions are $\gamma_1 = 33.11$ and $\gamma_2 = 36.60$ mN/m. The area per molecule may be taken as 0.37 nm². Calculate the surface composition x_1^s and the surface tension γ when x_1 has values of 0.25 and 0.50. In the latter case, compare the result with the experimental surface tension of 34.65 mN/m.

SOLUTION

Equations 1.74 and 1.75 apply. The partial molar area a, calculated by multiplying the given area per molecule by Avogadro's number, is 2.228×10^{15} m^2/g-mol. Since $[(\gamma_2 - \gamma_1)a/RT]$ is about 0.32, Equation 1.76 should give fairly good values for γ, but they should differ slightly from those of the general equation (Equation 1.75). Calculated results are shown in the following table. Agreement with the above experimental value is good.

x_1	x_1^s	γ (Equation 1.75) (mN/m)	γ (Equation 1.76) (mN/m)
0.250	0.314	35.62	35.73
0.500	0.579	34.72	34.86

EXAMPLE 1.7 SURFACE TENSION OF REGULAR SOLUTIONS
(DEFAY ET AL., 1966)

Assume that dimethyl ether (1) and acetone (2) form regular solutions at 15°C, where $\gamma_1 = 17.6$ and $\gamma_2 = 23.7$ mN/m. The interchange energy parameter α may be taken as 1.88 k^J/g-mol and the area per molecule as 0.30 nm^2. Calculate the surface tension and surface composition of a liquid mixture having $x_2 = 0.7$. Make two calculations, one for a simple cubic lattice ($\ell = 2/3$, $m = 1/6$) and one for a close packed lattice ($\ell = 1/2$, $m = 1/4$). Compare your results with the predictions of ideal solution theory and with the experimental value of 20.7 mN/m.

SOLUTION

From Equation 1.80 we find by trial that

$$x_1^s = 0.46, \gamma = 21.0 \text{ mN/m for simple cubic lattice, and}$$
$$x_1^s = 0.45, \gamma = 20.8 \text{ mN/m for close packed lattice,}$$

From Equations 1.72 and 1.74 for an ideal solution the corresponding values are

$$x_1^s = 0.40, \gamma = 21.6 \text{ mN/m.}$$

Clearly the calculations based on regular solution theory are in better agreement with the experimental measurement.

9. SURFACTANTS

In the preceding section we saw that the surface of a binary liquid mixture is enriched in the species with the lower surface tension. This result is qualitatively consistent with the Gibbs adsorption equation (Equation 1.25), although, as indicated above, agreement is not quantitative for nonideal mixtures (Defay et al., 1966).

Some materials are so highly enriched at the surface that they are called surface-active agents or "surfactants." These materials are discussed extensively in Chapter 4. Some aspects of Chapters 2 and 3 will be clearer, however, if a brief account of their two key properties is given here.

Water, a highly polar material, resists incorporation of nonpolar entities such as hydrocarbon chains into its structure. One familiar consequence is that pure hydrocarbons are rather insoluble in water. Polar materials, in contrast, have considerable solubility in water. We might ask what happens to a material such as sodium laurate ($C_{12}H_{25}COONa$), a typical soap which has both a hydrocarbon chain and a polar group, when it is added to water. It seems clear that the most favorable configuration would be one where the hydrocarbon chain of each molecule is removed from the water but the polar group remains in the water. As shown in Figure 1.13, this ideal condition can be achieved if the molecules adsorb at the interface between the water and an adjacent fluid as a monomolecular layer or "monolayer." We conclude that sodium laurate and similar "amphiphilic" compounds which have separate nonpolar and polar regions are surfactants.

For most situations where two bulk phases are in equilibrium, the amount of a surface-active contaminant required to form an interfacial monolayer is quite small. For instance, if each contaminant molecule occupies 0.4 nm², a simple calculation shows that 2.5×10^{14} molecules are required to cover an area of 1 cm². But this number of molecules corresponds to a bulk concentration of only about 4×10^{-7} molar when dissolved in 1 cm³ of a solvent. Because of its relatively high surface tension, water is particularly susceptible to surface contamination. Indeed, the surface tension of water is an important indicator of its purity. The possibility of surface contamination and the associated requirement for high purity must be kept in mind when planning, performing, and evaluating experiments involving interfacial phenomena.

Surfactants have another important property. It too is a consequence of the reluctance of water to incorporate hydrocarbon chains. At very low concentrations, dissolved surfactant exists as individual molecules, as with most other solutes. But at a particular concentration it becomes more favorable for the surfactant molecules to form aggregates called micelles. As illustrated in Figure 1.13, the polar groups are all in contact with water while the hydrocarbon chains

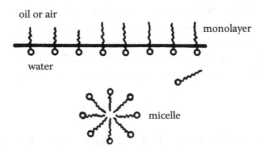

FIGURE 1.13 Idealized diagram showing surfactant behavior.

form the interior of the micelle. For typical water-soluble surfactants such as sodium laurate, the micelles contain perhaps a hundred molecules and are approximately spherical in shape.

Because many types of polar groups exist and because the hydrocarbon portion of a surfactant may take different forms, many types of surfactants exist. Some simple polar groups are carboxylic acid (COOH), carboxylate ion (COO⁻), alcohol (OH), sulfate (SO_4^-), sulfonate (SO_3^-), and amine (NH_3^+). Hydrocarbon chains may be straight or branched, saturated or unsaturated. Aromatic and naphthenic rings may also be present in the hydrocarbon region. Some surfactants have other nonpolar groups such as fluorocarbons. It is evident that, given suitable methods of synthesis, practically an infinite variety of surfactants could be made.

10. SOLID-FLUID INTERFACES

The preceding sections of this chapter dealt with the behavior of fluid interfaces at equilibrium. At least two factors which are not relevant to fluid interfaces must be considered for interfaces between fluids and crystalline solids. One is that interfacial properties such as free energy depend on the orientation of the interface relative to the crystal structure (i.e., which crystal plane is exposed to the fluid). The second is that elastic solids at rest can support anisotropic stresses. As a result, elastic energy effects have to be taken into account in determining the energy change accompanying deformation of an interface, and interfacial tension cannot, in general, be defined as in Equation 1.9. Since a general treatment of the thermodynamics of solids and their interfaces would be lengthy and complex, we limit consideration here to certain special cases of interest.

It is convenient to define interfacial tension by Equation 1.29. Then if interfacial area increases with all intensive properties maintained constant, in particular the temperature, chemical potentials, and interfacial tension, it is clear from Equation 1.29 that

$$dF^s = \gamma dA + \sum_i \mu_i \, dn_i^s . \qquad (1.81)$$

Now suppose that equilibrium exists between a fluid and a single-component solid that are in contact at a plane interface. If the stress in the solid is anisotropic but uniform and if the normal to the interface is along one of the principal directions of stress, it can be shown (see Problem 1.17) that the chemical potential μ in the fluid of the component making up the solid is given by

$$\mu = u_s - Ts_s + p_n v_s, \qquad (1.82)$$

where u_s, s_s, and v_s are the molar internal energy, entropy, and volume of the solid phase and p_n is (the negative of) the normal stress in the solid at the surface. Now, if another interface normal to another principal direction of stress is also exposed to the fluid, a similar argument leads again to Equation 1.82, except that p_n now has the value that exists at the second interface. Thus the values of μ in the fluid required for equilibrium at different interfaces of the same solid are different, and mass transfer will occur until the stress becomes isotropic and equilibrium is reached at all interfaces. If the time scale for this process is relatively short, as is likely for temperatures not far below the melting point, it suffices to limit consideration to the case of isotropic stress. This approach is taken in the following paragraphs. However, one can imagine that some species in a multicomponent solid may diffuse so slowly that the time required to attain equilibrium would be much greater than times of interest for other phenomena. Rusanov (1978) discussed the thermodynamics of a solid where some species are "immobile" in this sense.

Even when the stress is isotropic, there remains the matter of anisotropic interfacial properties. For simplicity we discuss here the special case of a crystal that is a rectangular parallelepiped with dimensions $2h_1$, $2h_2$, and $2h_3$ in equilibrium with a fluid in which it is immersed (see Figure 1.14).

We first seek the values of h_i that minimize the free energy of the system for constant intensive properties and constant volumes of the solid and fluid phases. We have

$$dF = dF_s + dF_f + \sum dF^s , \tag{1.83}$$

where the first two terms represent the free energy changes for the bulk solid and fluid phases and the summation is over all the crystal faces. Since the conditions for use of Equation 1.81 are satisfied, Equation 1.83 can be simplified to

$$dF = \sum_i \mu_i \ (dn_{is} + dn_{if} + \Sigma dn_{ij}^s) + \sum_j \gamma_j dA_j . \tag{1.84}$$

Moreover, the number of moles of each species in the overall system is fixed. Hence the expression in parentheses in this equation vanishes and only the terms involving the interfacial tensions need be considered.

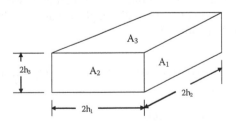

FIGURE 1.14 Diagram of rectangular parallelepiped crystal.

The area A_1 of the crystal face whose normal is in the same direction as the dimension having length $2h_1$ is given by

$$A_1 = 4h_2h_3. \tag{1.85}$$

Similar expressions may be written for areas of the other faces. Crystal volume V_s is given by

$$V_s = 8h_1h_2h_3. \tag{1.86}$$

The method of Lagrange multipliers can be used to minimize F subject to the condition of constant solid volume:

$$dF - \lambda \, dV_s = 0, \tag{1.87}$$

where λ is the (unknown) multiplier. Differentiation of Equations 1.85 and 1.86 and substitution into Equation 1.87 yields

$$0 = \left[\frac{\gamma_1}{h_1} + \frac{\gamma_2}{h_2} - \lambda \right] h_1 h_2 dh_3 + \left[\frac{\gamma_1}{h_1} + \frac{\gamma_3}{h_3} - \lambda \right] h_1 h_3 dh_2$$

$$+ \left[\frac{\gamma_2}{h_2} + \frac{\gamma_3}{h_3} - \lambda \right] h_2 h_3 \, dh_1 \tag{1.88}$$

As the three terms in this equation are independent, each of the three expressions in brackets must vanish, which leads after a little manipulation to

$$\frac{\gamma_1}{h_1} = \frac{\gamma_2}{h_2} = \frac{\gamma_3}{h_3} = \frac{\lambda}{2}. \tag{1.89}$$

That is, the distance of the crystal face from the center of the crystal is proportional to its interfacial tension. Equations 1.89 are known as the Wulff relations.

Now suppose that the solid phase increases slightly at the expense of the fluid phase by a small increase in h_1, all intensive properties as well as h_2 and h_3 being maintained constant. Equation 1.83 still applies, but the following equation is obtained instead of Equation 1.87:

$$dF = -(p_s - p_f) \, dV_s + \Sigma \, \mu_i \, (dn_{is} + dn_{if} + \sum_j dn_{ij}^s)$$

$$+ \sum_j \gamma_j \, dA_j \tag{1.90}$$

where p_s and p_f are the pressures in the solid and fluid phases. As before, the component mass balances require that the second term be zero. Upon substitution of the appropriate expressions for changes in area and volume, one finds

$$(p_s - p_f) = \frac{\gamma_2}{h_2} + \frac{\gamma_3}{h_3} = 2\frac{\gamma_1}{h_1} = 2\frac{\gamma_2}{h_2} = 2\frac{\gamma_3}{h_3} , \qquad (1.91)$$

where Equation 1.89 has been invoked. This equation is the analog of the Young-Laplace equation for an isotropically stressed crystalline rectangular parallelepiped in equilibrium with a surrounding fluid. The generalization of Equations 1.89 and 1.91 to a crystal of arbitrary shape, derived in Chapter 17 of Defay et al. (1966), is given by

$$(p_s - p_f) = 2\frac{\gamma_1}{h_1} = 2\frac{\gamma_2}{h_2} = ...2\frac{\gamma_i}{h_i} . \qquad (1.92)$$

When the crystal is small, the internal pressure can be rather high, according to this equation. Some x-ray diffraction evidence for a reduction in lattice parameters for small crystals, which would be expected at high pressures, has been presented by Nicholson (1955).

We may ask whether Equation 1.91, which has been derived from thermodynamic considerations, also has a mechanical interpretation. Let us consider a small pillbox control volume similar to that in Figure 1.3 which covers only a portion of the crystal face. The unit normal M to the outer pillbox surface S_o has no component in the normal direction for this flat interface. As we have seen, the lack of curvature leads ultimately to equality of the bulk phase pressures for a fluid interface. Since the pressures are not equal for a solid-fluid interface, according to Equation 1.91, shear stresses must act on S_o if the normal component of the momentum balance on the pillbox is to be satisfied. No such shear stresses were present for the fluid interface.

If we extend the control volume to cover the entire crystal face, however, we can see that a mechanical interpretation is possible. Consider the face with area A_1 in Figure 1.14. Forces acting perpendicular to this face are the bulk fluid pressures over the area A_1 and the interfacial tensions γ_2 and γ_3 around the periphery. If gravitational effects are negligible, we obtain

$$(p_s - p_f)A_1 = 4\gamma_2 h_3 + 4\gamma_3 h_2. \qquad (1.93)$$

Substituting Equation 1.85 into this expression, we find

$$(p_s - p_f) = \frac{\gamma_2}{h_2} + \frac{\gamma_3}{h_2} . \qquad (1.94)$$

A similar argument for the face with area A_2 yields

$$(p_s - p_f) = \frac{\gamma_1}{h_1} + \frac{\gamma_3}{h_3} . \qquad (1.95)$$

From these two equations and the corresponding one for the face with area A_3, it is evident that Equation 1.91 must hold (i.e., it can be derived from mechanical considerations). The general result given by Equation 1.92 can also be derived in this way (see Problem 1.18). Indeed, the surface tension of solid wires near their melting temperatures has been determined by finding the load required to overcome surface tension effects and produce elongation (Udin et al., 1949) (see Problem 1.19).

REFERENCES

GENERAL TEXTS ON INTERFACIAL PHENOMENA

Adam, N.K., *The Physics and Chemistry of Surfaces*, 3rd ed., Oxford University Press, New York, 1941.

Adamson, A.W. and Gast, A.P., *Physical Chemistry of Surfaces*, 6th ed., Wiley, New York, 1997.

Boys, C.V., *Soap Bubbles*, Dover Publications, New York, 1959.

Davis, H.T., *Statistical Mechanics of Phases, Interfaces, and Thin Films,* Wiley, New York, 1996.

Davies, J.T. and Rideal, E.K., *Interfacial Phenomena*, 2nd ed., Academic Press, New York, 1963.

Defay, R., Prigogine, I., Bellemans, A., and Everett, D.H., *Surface Tension and Adsorption*, Longmans, London, 1966.

Harkins, W.D., *The Physical Chemistry of Surface Films*, Reinhold, New York, 1952.

Rowlinson, J.S. and Widom, B., *Molecular Theory of Capillarity*, Clarendon Press, Oxford, 1982.

TEXT REFERENCES

Adamson, A.W. and Gast, A.P., *Physical Chemistry of Surfaces*, 6th ed., Wiley, New York, 1997.

Ambwani, D.S. and Fort, T., Jr., Pendant drop technique for measuring liquid boundary tensions, in *Surface and Colloid Science*, vol. 11, Good, R. J. and Stromberg, R. R. (eds.), Plenum, New York, 1979. p. 93.

Barker, J.A. and Henderson, J.R., Generalized van der Waals theories and the asymptotic form of the density profile of a liquid-vapor interface, *J. Chem. Phys.*, 76, 6303, 1982.

Bamberger, S., Seaman, G.V.F., Sharp, K.A., and Brooks, D.E., The effects of salts on the interfacial tension of aqueous dextran poly(ethylene glycol) phase systems, *J. Colloid Interface Sci.*, 99, 194, 1984.

Bashforth, F. and Adams, J.C., *An Attempt to Test the Theory of Capillary Action*, Cambridge University Press, Cambridge, 1893.

Buff, F.P., Curved fluid interfaces. I. The generalized Gibbs-Kelvin equation, *J. Chem. Phys.*, 25, 146, 1956.

Cahn, J.W. and Hilliard, J.E., Free energy of a nonuniform system. I. Interfacial free energy, *J. Chem. Phys.*, 28, 258, 1958.

Courant, R. and Hilbert, D., *Methods of Mathematical Physics*, vol. 1, Wiley, New York, 1953, chap. 4.

Davis, H.T., *Statistical Mechanics of Phases, Interfaces, and Thin Films*, Wiley, New York, 1996.

Davis, H.T. and Scriven, L.E., Stress and structure in fluid interfaces, *Adv. Chem. Phys.*, 49, 357, 1981.

Defay, R., Prigogine, I., Bellemans, A., and Everett, D.H., *Surface Tension and Adsorption*, Longmans, London, 1966.

Gibbs, J.W., On the equilibrium of heterogeneous substances, *Trans. Conn. Acad.*, 3, 108, 1878. Reprinted in *The Scientific Papers of J. Willard Gibbs*, vol. 1, Ox Bow Press, Woodbridge, Conn., 1993, p. 55; originally published by Longmans, Green, London, New York, 1906.

Good, R.J., Thermodynamics of adsorption and Gibbsian distance perameters in two- and three phase systems, *Pure Appl. Chem.*, 48, 427, 1976.

Hayward, A.T.J., Negative pressure in liquids: Can it be. harnessed to serve man?, *Am. Sci.*, 59, 434, 1971.

Hemingway, S.J., Henderson, J.R., and Rowlinson, J.S., The density profile and surface tension of a drop, *Faraday Symp. Chem. Soc.*, 16, 33, 1981.

Kulkarni, S.S., Funk, E.W., and Li, N.N., Ultrafiltration: applications and economics, in *Membrane Handbook*, Ho, W. S.W. and Sirkar, K.K. (eds.), van Nostrand Reinhold, New York, 1992.

Lando, J.L. and Oakley, H.T., Tabulated correction factors for the drop-weight-volume determination of surface and interfacial tensions, *J. Colloid Interface Sci.*, 25, 526, 1967.

Li, D., Cheng, P., and Newmann, A.W., Contact angle measurement by axisymmetric drop shape analysis (ADSA) , *Adv. Colloid Interface Sci.*, 39, 347, 1992.

Murphy, C.L., Thermodynamics of Low Tension and Highly Curved Surfaces," Ph.D. thesis, University of Minnesota, Minneapolis, 1966.

Nicholson, M.M., Surface tension in ionic crystals, *Proc. Roy. Soc. A*, 228, 490, 1955.

Nishioka, G. and Ross, S., A new method and apparatus for measuring foam stability, *J. Colloid Interface Sci.*, 81, 1, 1981.

Ono, S. and Kondo, S., Molecular theory of surface tension in liquids, in *Encyclopedia of Physics*, vol. 10, S. Flugge (ed.), Springer, Berlin. 1960, p. 134.

Padday, J.F., Theory of surface tension, in *Surface and Colloid Science*, vol. 1, E. Matijevic (ed.), Wiley, New York, pp. 39, 101, 151, 1969.

Pandit, A., Miller, C.A., and Quintero, L., Interfacial tensions between bitumen and aqueous surfactant solutions by maximum bubble pressure technique, *Colloids Surfaces A*, 98, 35, 1995.

Passarone, A., Liggieri, L., Rando, N., Ravera, F., and Ricci, E., A new experimental method for the measurement of the interfacial tension between immiscible fluids at zero Bond number, *J. Colloid Interface Sci.*, 146, 152, 1991.

Rowlinson, J.S. and Widom, B., *Molecular Theory of Capillarity*, Clarendon Press, Oxford, 1982.

Rusanov, A.I., On the thermodynamics of deformable solids, *J. Colloid Interface Sci.*, 63, 330, 1978.

Rusanov, A.I., *Interfacial Tensiometry*, Elsevier, Amsterdam, 1996.

Schofield, P. and Henderson, Statistical mechanics of inhomogenous fluids, J.R., *Proc. Roy. Soc. Lond. A*, 379, 231, 1982.

Sugden, S., The determination of surface tension from the rise in capillary tubes, *J. Chem. Soc.*, 119, 1483, 1921.

Udin, H., Shaler, A.J., and Wulff, J., The surface tension of copper, *J. Metals AIMMPE*, 1, 186, 1949.

Vrij, A., Equation for the interfacial tension between demixed polymer solutions, *J. Polymer Sci. A-2: Polymer Physics*, 6, 1919, 1968.

Weatherburn, C.E., *Differential Geometry of Three Dimensions*, vol. 1, Cambridge University Press, Cambridge, 1955.

PROBLEMS

1.1   (a)   Starting with Equation 1.12, show that

$$(\partial S^s / \partial A)_{T,n_i^s} = -(\partial \gamma / \partial T)_{A,n_i^s}.$$

(b)   Use this result to show that

$$\left(\frac{\partial U^s}{\partial A}\right)_{T,n_i^s} = \gamma - T\left(\frac{\partial \gamma}{\partial T}\right)_{A,n_i^s}.$$

(c)   Consider a single-component system with the reference surface chosen to make $n_1^s = 0$. In this case $\gamma = (F^s/A)$, according to Equation 1.29. Also

$$\left(\frac{\partial U^s}{\partial A}\right)_T = \frac{U^s}{A}, \left(\frac{\partial S^s}{\partial A}\right)_T = \frac{S^s}{A}.$$

Except near the critical point, γ is often found to decrease linearly with T in this case. Assuming such behavior applies, sketch the expected behavior of U^s, F^s, and S^s as a function of temperature.

1.2   Is the vapor pressure p_v of a liquid inside a small capillary tube which it completely wets greater or less than the vapor pressure p_v° of a large pool of the same liquid at the same temperature? Calculate (p_v/p_v°) for water at 100°C inside a capillary tube with a radius of 1 μm. Use the values of physical properties given in Example 1.1. The change in vapor pressure for a liquid in a capillary tube or, more generally, a porous medium is usually called the capillary condensation effect.

1.3   For a single-component liquid-vapor interface, it is not possible to vary temperature and pressure independently while maintaining the phases in equilibrium. However, measurement of interfacial tension variation with pressure at constant temperature is possible in a binary system. Good (1976) has shown that such data can be used to obtain useful information about interfacial characteristics in this case.

(a) Use the Gibbs adsorption equation to derive formulas for Γ_2 when Γ_1 = 0 and Γ_1 when Γ_2 = 0 in terms of the pressure coefficient of interfacial tension at constant temperature. For simplicity, assume a liquid-liquid system where $x_2 \ll 1$ in phase 1 and $x_1 \ll 1$ in phase 2.

(b) Let Λ be the distance between the reference surfaces corresponding to $\Gamma_1 = 0$ and $\Gamma_2 = 0$. Show that $\Lambda = - (\partial\gamma/\partial p)_T$ and hence that the pressure variation of interfacial tension provides some information on interfacial thickness.

(c) In the benzene-water system at room temperature, $(\partial\gamma/\partial p)_T$ has been found to be -7×10^{-4} mN/(m·atm). Calculate Λ and Γ for each component at the reference surface, where Γ for the other component vanishes.

1.4   For interfaces with very low interfacial tensions or with small radii of curvature, the assumption made in writing Equation 1.9 is not valid. Consider the special case of a spherical interface and rewrite Equation 1.9, adding a term $(Cd(2/a))$, where a is the radius of the sphere and C is defined by

$$C = \left(\frac{\partial U^s}{\partial(2/a)}\right)_{S^s, A, n_i^s}.$$

(a) Repeat the derivation of Section 3 to obtain the following generalization of the Young-Laplace equation for a spherical interface:

$$p_A - p_B = \frac{2\gamma}{a} - \frac{2(C/A)}{a^2}. \tag{1.4.i}$$

This equation can be generalized (Buff, 1956; Murphy, 1966) to include nonspherical interfaces, the result being

$$p_A - p_B = - 2H\gamma - (4H^2 - 2K)(C/A), \tag{1.4.ii}$$

where K is the Gaussian curvature of the reference surface defined as the product of $(1/r_1)$ and $(1/r_2)$. It can also be shown that the quantity (C/A) has a mechanical interpretation:

$$\frac{C}{A} = \int_{\lambda_A}^{\lambda_B} \Delta p_T \ (\lambda - \lambda_s) d\lambda \ , \tag{1.4.iii}$$

where Δp_T is as defined in Equation 1.33 and λ_s is the coordinate of the reference surface. Murphy (1966) termed $(-C/A)$ the "bending stress" because of this relation to the first moment of the tangential stress distribution. His general derivation yields yet another term in (1.4.i) and (1.4.ii). In the former equation it is proportional to $(1/a^3)$. Finally, we note that under conditions where (1.4.i) or (1.4.ii) holds, Equation 1.39 must be corrected for curvature effects as well:

$$\gamma = -\int_{\lambda_A}^{\lambda_B} \Delta p_T \ (1 - 2H \ (\lambda - \lambda_s)) d\lambda \ . \tag{1.4.iv}$$

(b) Consider the situation of the adjacent diagram where the three spherical surfaces are parallel and occupy a solid angle ω. With the reference surface maintained at $r = a$, but with a_A and a_B allowed to change, calculate the change in Helmholtz free energy for the closed system at a constant temperature and volume when it is deformed so that it occupies a solid angle $(\omega + d\omega)$ and show that the interfacial tension is given by

$$\gamma = (p_A - p_B) \frac{a}{3} + \frac{Q}{a^2} \ , \tag{1.4.v}$$

where Q is a constant. By differentiating this expression, show that (1.4.i) may be written in the following form given by Ono and Kondo (1960):

$$p_A - p_B = \frac{2\gamma}{a} + \frac{d\gamma}{da} \ . \tag{1.4.vi}$$

(c) The reference surface where $(d\gamma/da)$ vanishes (i.e., where $C = 0$) is usually called the "surface of tension," an expression coined by Gibbs. Rewrite (1.4.v) in terms of the radius a_s of the surface of tension and the corresponding interfacial tension γ_s and hence show that

$$\gamma = (\gamma_s/3)[2(a/a_s) + (a_s/a)^2].$$

Note that this expression implies that γ attains a minimum value γ_s at the surface of tension.

These methods for describing surface tension do not appear to work at very large curvatures (Hemingway et al., 1981; Schofield and Henderson, 1982).

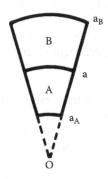

PROBLEM 1.4 Figure for Problem 1.4(b).

1.5   A drop of a perfectly wetting liquid is placed between two parallel plates, as shown in the accompanying figure. As the plates are moved together, the diameter of the wetted region increases. For the case $D \gg h$:

(a) Derive an equation for net force acting on each plate.

(b) Calculate the value of the force for water ($\gamma = 72$ mN/m) when drop volume is 2 cm³ and $h = 1$ mm. This force is a major factor contributing to the adhesion of surfaces.

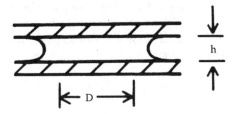

PROBLEM 1.5 Figure for Problem 1.5.

1.6   The surface divergence theorem was stated in Equation 1.36 without proof. Except for the last term of the integrand, it is basically what one would expect in view of the more familiar three-dimensional relationship:

$$\int_s \mathbf{n} \, f \, dS = \int_v \nabla f \, dV \, .$$

Some insight into the origin of the last term in Equation 1.36 may be gained by considering a special case where (1) f is uniform so ∇f vanishes and (2) the surface is cylindrical, having a radius R and a length L and enclosing an angle of $2\theta_1$. Show that Equation 1.36 is satisfied for this case.

1.7   Nishioka and Ross (1981) have proposed the following method for measuring the decrease with time in the surface area of a foam, which provides a good indication of foam stability. Foam is generated in a chamber of fixed volume which is maintained at constant temperature. The pressure in the container, but outside the foam itself is monitored as a function of time as the foam decays. The pressure p_∞ corresponding to complete decay of the foam is obtained by injecting a defoaming agent at the end of the experiment.

(a) Use the Young-Laplace equation (Equation 1.22) and the ideal gas equation to derive the following equation of state for a foam:

$$n_f RT = p_e V_f + \frac{2\gamma A}{3} \, ,$$

where n_f is the number of moles of gas in the foam, p_e is the pressure external to the foam, A and V_f are the foam area and volume, and γ is the surface tension of the foam lamellae.

(b) Assuming constant γ and T and imposing requirements that the total number of moles of gas in the vessel (both internal and external to the foam) and the total volume V_g of this gas remain constant, show that for times t_1 and t_2

$$3V_g(p_{e2} - p_{e1}) + 2\gamma(A_2 - A_1) = 0.$$

(c) Use the information that $A = 0$ for $p_e = p_\infty$ to show that

$$A = \frac{3V_g}{2\gamma} (p_\infty - p_e) \, .$$

1.8 At small saturations, the wetting fluid in a porous medium forms "pendular rings" near the points of contact of individual grains. To simplify the calculations we will consider the junction between two planes as sketched instead of between two spheres. The fluid interface between A and B may be taken to be a portion of a circle which, assuming zero contact angle, is tangent to the planes at A and B. The angle between the planes is 2α. Calculate the area of region OAB occupied by the wetting fluid and the pressure difference between the fluids for the following four cases:

(a) $\qquad \alpha = 10°, \overline{OA} = 10 \ \mu\text{m}$

(b) $\qquad \alpha = 10°, \overline{OA} = 1 \ \mu\text{m}$

(c) $\qquad \alpha = 30°, \overline{OA} = 10 \ \mu\text{m}$

(d) $\qquad \alpha = 30°, \overline{OA} = 1 \ \mu\text{m}$

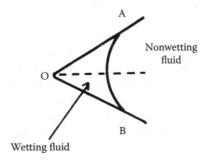

PROBLEM 1.8 Figure for Problem 1.8.

1.9 A very long solid body having a square cross section floats at an interface with the configuration sketched. Determine h and α at equilibrium in terms of a, densities ρ_A and ρ_B, and interfacial tension γ_{AB}. (It is not necessary to solve explicitly for h and α from the final equations you obtain.) Hint: The differential equation of interfacial statics need not be solved if one is interested only in finding h and α (and not details of the interfacial profile). One can instead write a horizontal force balance for the entire fluid interface on one side of the solid.

That surface tension can keep a small solid body from sinking is important to certain insects and, from a more practical point of view, to those interested in using flotation to separate valuable ores from material containing no minerals. Note that the basic method of solution can be used even if the fluid surface does not contact the solid along an edge; for example, if one had a long cylinder floating at the surface. Note also that the surface tension can not help keep the solid at the surface unless the equilibrium contact angle (on a plane surface) differs from zero; in other words, the mechanism is ineffective if the liquid spreads spontaneously on the solid.

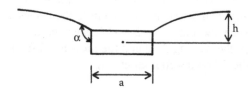

PROBLEM 1.9 Figure for Problem 1.9.

1.10   As indicated in Section 7, one method of measuring interfacial tension between two fluids is the spinning drop device. A drop of the less dense fluid A is placed in the denser fluid B, and both fluids are rotated at a constant angular velocity ω (Figure 1.9).

(a) Neglecting gravity and assuming an axisymmetric drop, derive a differential equation for the drop profile z as a function of radial coordinate r, interfacial tension , fluid densities ρ_A and ρ_B, and the angular velocity ω. Recall that the principal radii of curvature are given by Equations 1.52 and 1.53. Your final equation should be in terms of dimensionless variables and should contain only a single dimensionless group involving the various parameters.

(b) Write the boundary conditions needed to solve your differential equation.

(c) For high angular velocities, the drop becomes long and thin (i.e., $z_m \gg r_m$, where z_m and r_m are coordinates at the point of maximum drop diameter). Under these conditions it can be shown that $(r_m/r_0) = 1.5$, where r_0 is the radius of curvature at the origin 0. Use this information to derive Equation 1.58. (Note that you do not need to solve the differential equation to do this.)

1.11   Equation 1.65 lists variables upon which the interfacial tension γ is thought to depend when the drop volume method is used. Assume that the dependence takes the form

$$\gamma = BV^a(\rho_A - \rho_B)^b g^c r^d,$$

where B is a dimensionless constant. By requiring dimensional consistency of this equation, show that

$$\frac{\gamma}{(\rho_A - \rho_B)\, g\, r^2} = B \left(\frac{r}{V^{1/3}}\right)^{-3a}.$$

Thus two dimensionless groups are obtained as stated in the text.

1.12   The drop volume method is being used to measure the interfacial tension between water ($\rho_A = 1.00$ g/cm³) and benzene ($\rho_B = 0.88$ g/cm³). For a capillary tube with an outside diameter of 4 mm, the volume of each drop is found to be 0.273 cm³. Calculate the interfacial tension. Note that the usual formulation of the equation takes the form

$$\frac{\gamma r}{(\rho_A - \rho_B) g V} = \left(\frac{\gamma}{(\rho_A - \rho_B) g r^2}\right)\left(\frac{r}{V^{1/3}}\right)^3 = F\left(\frac{r}{V^{1/3}}\right)$$

where the function F is obtained from experiments and tabulated (see Table 1.3 for selected values in the range of interest here). The reason for this formulation is that the first expression is related to the ratio of the upward force on the drop produced by interfacial tension to the effective weight of the suspended drop. As is clear from dimensional analysis, it is not necessary to consider this ratio explicitly in solving

TABLE 1.3
Partial list of correction factors for the drop volume method

$r/V^{1/3}$	Correction factor (F)
0.300	0.2166
0.301	0.2168
0.302	0.2169
0.303	0.2171
0.304	0.2173
0.305	0.2175
0.306	0.2176
0.307	0.2178
0.308	0.2180
0.309	0.2182
0.310	0.2183

Values from Lando and Oakley (1967).

for γ. One needs only to determine the relationship between the two pertinent dimensionless variables. But since the function F as defined above is the quantity tabulated, it is easiest to follow the usual procedure.

1.13   As indicated in Section 7, it is necessary to correct for the nonspherical shape of the meniscus in accurate work using the capillary rise technique.

(a) Show that if h is the height to the bottom of the meniscus and b is the radius of curvature there,

$$\frac{2\gamma}{b} = (\rho_A - \rho_B)\, gh .\qquad(1.13.i)$$

Hence show that

$$\frac{R}{(bh)^{1/2}} = \frac{R}{b}\left(\frac{\beta}{2}\right)^{1/2},\qquad(1.13.ii)$$

where R is the tube radius and β is as defined by Equation 1.57.

(b) The procedure commonly used to make the correction is from Sugden (1921). For any β, the numerical solution of Equation 1.56 yields a value of (r/b) for $\phi = 90°$. Sugden tabulated (r/b) for $\phi = 90°$ as a function of $(r/b)(\beta/2)^{1/2}$. A portion of his table is shown in Table 1.4. Show that the entry for $(r/b) = 0.9631$ in Table 1.4 corresponds to a value β of 0.249.

(c) Sugden's table is used to find r in the following way. Because the contact angle is zero, $r = R$ and $\phi = 90°$ at the point where the meniscus meets the tube. A successive approximation scheme is used. First, γ is calculated from (1.13.i) with $b = R$ (spherical interface approximation). This value of γ is then used to calculate $(R/b)(\beta/2)^{1/2}$. The table then gives a value of (R/b) from which an improved value of b can be obtained. This improved b is substituted into Equation 1.13.i to get an improved γ, and the process is repeated until the same value of is obtained on successive iterations. Calculate the surface tension of toluene ($\rho = 0.866$ g/cm³) if $h = 0.838$ cm in a capillary tube of radius 0.0777 cm.

1.14   Show that for an ideal dilute solution the surface tension is given approximately by

---

TABLE 1.4
Selected values from Sugden's table of
corrections for the capillary rise method (1921)

$(r/b)(\beta/2)^{1/2}$	(r/b)
0.30	0.9710
0.31	0.9691
0.32	0.9672
0.33	0.9652
0.34	0.9631
0.35	0.9610
0.36	0.9589
0.37	0.9567
0.38	0.9545
0.39	0.9522
0.40	0.9498

$$\gamma_1 - \gamma = \frac{RT}{a} x_2 \left[H \exp\left(\frac{\gamma a}{RT}\right) - 1 \right],$$

where γ_1 is the surface tension of the pure solvent, a is the area of solute and solvent molecules, and H is a constant that must be determined experimentally. Recall that the following expression applies in a dilute ideal solution in a bulk phase:

$$\mu_2 = \mu_2^\infty (T, p) + RT \ln x_2 \quad (x_2 << 1).$$

Here μ_2^∞ is not the chemical potential of pure 2 at T and p. The chemical potential μ_1 of the solvent in such a solution is given by Equation 1.68. Assume that the surface is an ideal dilute solution in your derivation.

1.15   Show that the Gibbs adsorption equation for constant temperature is satisfied by ideal surface solution theory as discussed in Section 8. Note that $\Gamma_i = (n_i^s/a)$, for $i = 1, 2$.

1.16   Consider an aqueous solution of dextran (dex) and polyethylene glycol (PEG). One important rule about polymer solutions is that two polymers are rarely compatible with one another, and phase separation into dex-rich and PEG-rich phases take place. (This is the reason why it is difficult to develop solid polymer blends.) Using the gradient theory of Section 5 and properties of polymers, Vrij (1968) showed that the interfacial tension in such phase separated systems is very low (≤ 1 mN/m), irrespective of the common solvent content. In the water-dex-

PEG system, commonly called an "aqueous two-phase system," the water content can exceed 85% by weight. The low interfacial tension and high water content make the pair of liquids suitable for liquid-liquid extraction of proteins. What experimental method can be used to measure the interfacial tension (Bamberger et al., 1984)? Will the thickness of the interface be large or small? Provide a simple estimate.

1.17   We wish to derive Equation 1.82 for the particular situation described there. Consider a process in which a solid having a thickness dx is deposited at the solid-fluid interface, which has an area A. Let the change in thickness of the solid initially present be dl_s and that of the fluid phase be dl_f. The area and total volume of the system are fixed, so that the sum of these three differential lengths is zero. Also the system is isolated from its surroundings.

(a) Show that the total change dU in internal energy for the process is given by $dU = [TdS_f - pA\,dl_f + \mu\,dn_f] + [TdS_s - p_n A dl_s] + u_s(Adx/v_s)$, where dn_f is the change in the number of moles in the fluid of the component making up the (pure) solid and S_s is the entropy of the solid initially present.

(b) Set $dU = 0$, as required by the first law of thermodynamics for an isolated system, and impose the requirements of constant volume (see above), conservation of mass, and $dS = 0$, which follows from applying the second law to an isolated system. Note that p_n in the solid must equal the fluid pressure p since we have taken the interface as perpendicular to one of the principal directions of stress. That is, there can be no shear stresses in this case of the type discussed in the next to last paragraph of Section 10. These steps, combined with the result of part (a) will yield Equation 1.82.

1.18   (a) In Chapter 17 of Defay et al. (1966), it is shown from geometrical considerations that the following equation holds for each face of an arbitrary crystal:

$$A_i = (1/2)\sum_j h_j \frac{\partial A_j}{\partial h_j}.$$

Show that this expression holds for the rectangular parallelepiped crystal of Figure 1.14.

(b) Show that the normal force balance on face i of an arbitrary crystal has the form

$$(p_s - p_f)A_i = \sum_k \gamma_k \sin\theta_{ik} l_{ik} \; ,$$

where θ_{ik} is the angle between faces i and k, and l_{ik} is the length of the line of intersection (if any) between the faces i and k. If Equation 1.89 is valid, one can replace γ_k in this equation by $(\gamma_i h_k/h_i)$. Show that $(l_{ik} \sin\theta_{ik})$ is equal to $(\partial A_k/\partial h_i)$, substitute this result into the above expression, and invoke the equation given in part (a) to obtain

$$(p_s - p_f) = (2\gamma_i/h_i).$$

1.19   A load having weight W is attached to the bottom of a thin, vertical metal wire at a temperature not far below the melting point. Both internal pressure p_s and interfacial tension γ are presumed to be isotropic under these conditions. If the weight of the wire itself is negligible, p_s must also be uniform. Make a vertical force balance for a length 1 of the wire for the case where W has the precise value required to keep the wire from elongating or contracting. Then apply the Young-Laplace equation (which is applicable when interfacial tension is isotropic) to show that $W = \pi r \gamma$, where r is the radius of the wire. What value would γ have if $r = 0.1$ mm and $W = 4 \times 10^{-4}$ N.

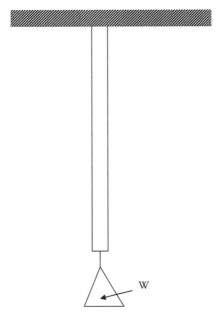

PROBLEM 1.19 Figure for Problem 1.19.

2 Fundamentals of Wetting, Contact Angle, and Adsorption

1. INTRODUCTION

One is familiar with the ways in which small amounts of liquid on a solid surface behave, not only from laboratory studies but also from everyday observations made of our immediate surroundings. One knows that drops of water on an inclined solid surface can remain stationary or can move under gravity. If a waxed paper is withdrawn from water, the thin sheet of water formed initially will break up into beads. On the other hand, if the waxed paper is withdrawn from oil, the sheet of oil formed will drain and grow thinner, but will not form beads. Leaving aside the effects of gravity, two types of behavior are observed on a given solid surface, one where the liquid forms beads and the other where it forms a film. It is easy to assume that liquids which have strong affinities for a solid will form films such that the liquid-solid contact is maximized, while those that have weaker affinities will collect themselves into beads.

The above affinity is referred to as the wettability. It is important for phenomena ranging from flow in packed columns, underground oil reservoirs, and other porous media to condensation of liquids on solid surfaces to flotation schemes for concentrating mineral ores. In recent years, precise control of wettability has been used to confine fluids to desired flow paths in some microfluidic devices.

It is apparent from the above discussion that the success of adhesives and coatings depends strongly on the wettabilities of these materials on the adherent or solid substrate. If the coating material does not wet the substrate, not only would it be difficult to apply, it would also eventually peel away as some print on the surfaces of plastics does. Adhesives are not allowed to behave in this fashion even under severe conditions. The analysis of adhesive behavior and determination of adhesive bond strength require knowledge of the factors that influence the energy of an interface between two materials.

2. YOUNG'S EQUATION

Since the configurations of a wetting liquid and a nonwetting liquid on a solid surface differ substantially in their interfacial areas, any quantitative description of wettability must involve interfacial energies in a major way. Consider the

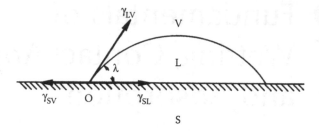

FIGURE 2.1 Configuration analyzed in Section 2. The phases are vapor (V), liquid (L), and solid (S). The contact line is perpendicular to the diagram at 0.

axisymmetric bead shown in Figure 2.1. There are three interfaces: the solid (S)-liquid (L), the liquid-vapor (V) and the solid-vapor, and consequently there are three interfacial tensions γ_{SL}, γ_{SV}, and γ_{LV}. The line common to the three phases (the basal circle of the drop) is called the common line or the contact line. The angle that the vapor-liquid interface makes with the solid surface is called the contact angle. If it is assumed that the interfacial tensions can be taken as forces even for the solid-fluid interfaces, the force balance at point O on the contact line in the horizontal direction is

$$\gamma_{SV} = \gamma_{SL} + \gamma_{LV} \cos \lambda,$$

which is rewritten in the form

$$\gamma_{LV} \cos \lambda = \gamma_{SV} - \gamma_{SL}. \tag{2.1}$$

This is the Young-Dupre equation or simply Young's equation. It should be noted that if no solid is present and a contact line is formed by the junction of the three fluid phases, the interfaces will assume an orientation such that the vector sum of the three interfacial tensions acting along them is zero.

A more rigorous derivation of Equation 2.1 for a solid surface can be found using interfacial thermodynamics. Small changes in Helmholtz's free energy can be written by generalizing Equation 1.14 for multiple interfaces. If temperature and the individual phase volumes are constant, the result is

$$dF = \sum_i \mu_i (dn_i^{LV} + dn_i^{SV} + dn_i^{SL})$$

$$+ \sum_i \mu_i (dn_i^{S} + dn_i^{L} + dn_i^{V}) \tag{2.2}$$

$$+ \gamma_{SV} dA_{SV} + \gamma_{LV} dA_{LV} + \gamma_{SL} dA_{SL}$$

As the number of moles remains unchanged,

$$(dn_i^{LV} + dn_i^{SV} + dn_i^{SL}) + (dn_i^S + dn_i^L + dn_i^V) = dn_{iT} = 0,$$

and one has

$$dF = \gamma_{SV}dA_{SV} + \gamma_{LV}dA_{LV} + \gamma_{SL}dA_{SL}. \qquad (2.3)$$

Further, geometry dictates that

$$dA_{SV} = - dA_{SL}. \qquad (2.4)$$

If the drop is a spherical cap, it can be shown that

$$dA_{LV} = \cos \lambda \, dA_{SL} \qquad (2.5)$$

Combining Equations 2.3 through 2.5, one has at constant volume of the drop:

$$\left(\frac{\partial F}{\partial A_{SL}} \right) = \gamma_{SL} - \gamma_{SV} + \gamma_{LV} \cos \lambda. \qquad (2.6)$$

At equilibrium, F has to be a minimum with respect to A_{SL} (i.e., the left-hand side of Equation 2.6 is zero and Equation 2.1 is obtained). Note that this derivation uses the principle of minimizing the free energy. It is known that on minimization of total energy, independent sets of force balances are obtained and consequently the fact that Young's equation can be derived in this way is appropriate.

More detailed derivations of Equation 2.1 are available (Buff and Saltzburg, 1957; Johnson, 1959). They uncover some basic limitations of Young's equation which are well worth noting. These are that (a) Equation 2.1 is valid when there is no adsorption at the interfaces. Adsorption and some of its effects on contact angles are discussed later. (b) The interfacial tensions that appear in Equation 2.1 are those that are evaluated far from the contact line. The interfacial tensions at the contact line may have different values, however, and indeed may not be well defined owing to overlap of portions of the three interfacial regions. All such deviations near the contact line can be lumped together in the form of the additional effect of line tension. Gibbs (1961) anticipated this effect when he argued that just as the interface separating two bulk fluids has a surface excess free energy and is in a state of tension, as discussed in Chapter 1, so may a line at the junction of three fluids have a linear excess (or deficiency) of free energy and be in a state of tension (or compression) (Kerins and Widom, 1982; Toshev et al., 1988; Vignes-Alder and Brenner, 1985; see also Gaydos and Neumann, 1987). A line tension may also exist when one of the three phases is a solid, as in Figure 2.1, although caution must be used in making mechanical interpretations for reasons similar to those given in Chapter 1 for solid interfaces. In any case, the estimated magnitudes and reliable experimental values of the line tensions are very small, and they are frequently negative (e.g., for surfactant solutions joining thin liquid films and for small drops of water condensing on n-hexadecane) (Platikanov et al., 1980a,b; Toshev et al., 1988; see also Gaydos and Neumann, 1987). Owing to the small values of the line tension, the measured contact angles are those predicted by Young's equation except when the contact line radii are small (see Problem 2.5).

Some interesting conclusions are reached from Young's equation. Since the interfacial tensions in Equation 2.1 are equilibrium properties, λ is also an equilibrium property and is consequently referred to as the equilibrium contact angle. For Equation 2.1 to be meaningful, $|\cos \lambda| = |(\gamma_{SV} - \gamma_{SL})/\gamma_{LV}| < 1$, which makes it necessary to examine the physical implication of this constraint. Assume that the drop sizes are small such that gravity forces can be neglected. In that case the shape of the vapor-liquid interface is that of a spherical cap. For drops of constant volume, as λ decreases to zero (or $\cos \lambda$ increases to +1), the drops approach a thin film configuration. As λ increases to $180°$ (or $\cos \lambda$ decreases to -1), the drops become spherical, having only one point of contact with the solid at $\lambda = 180°$. Thus an ordering is established in the spectrum of available λ. For $\lambda = 0°$, the liquid wets the solid surface completely (i.e., it is a wetting liquid). As λ increases beyond $90°$, the partially wetting liquid becomes nonwetting, and it ultimately becomes completely nonwetting for $\lambda = 180°$.

While this concludes the discussion of the nature of nonwetting liquids, more remains to be said about wetting liquids. These completely spread out on the solid surface and consequently the process must be accompanied by a decrease in the free energy. This means that for wetting liquids, the left-hand side of Equation 2.6 is negative. It is also seen in experiments that when a wetting liquid spreads on a solid surface, a zero contact angle is maintained (Bascom et al., 1964; Fox and Zisman, 1950; Johnson and Dettre, 1966). A spreading coefficient is defined as

$$S_{L/S} = \gamma_{SV} - \gamma_{LV} - \gamma_{SL}. \tag{2.7}$$

Substituting Equation 2.7 into Equation 2.6 with $\cos \lambda = 1$, one has $(\partial F/\partial A_{SL}) = -S_{L/S}$, that is, the spreading coefficient for a wetting liquid is positive. For a nonwetting liquid, one finds on substituting Equation 2.1 into Equation 2.7 that $S_{L/S} = \gamma_{LV}(\cos\lambda - 1)$. Thus the spreading coefficient is negative here. Consequently the knowledge of interfacial tensions alone is sufficient to predict the wettability of a liquid on a solid surface using Equation 2.7. Unfortunately the solid-fluid interfacial tensions γ_{SV} and γ_{SL} cannot ordinarily be measured, although methods of obtaining γ_{SV} for nonpolar, polar, and crystalline solids are discussed in Sections 4 and 5, and Problem 1.19.

The spreading coefficient $S_{L/L'}$ for a liquid L on another liquid L' with which it is not miscible is given by Equation 2.7 with the subscript S replaced by L'. As the liquids equilibrate, their surface tensions may change. At equilibrium one must have $S_{L/L'} \leq 0$, a relationship known as Antonov's rule. If it is not satisfied, one has $\gamma_{L'V} \geq \gamma_{LV} + \gamma_{L'L}$. That is, the system could reduce its free energy if some L formed at a (plane) interface between L' and vapor, so that the system would not have been initially in a state where its free energy was minimized. It is not uncommon for the initial spreading coefficient to be positive, while the equilibrium spreading coefficient is negative, so that the film of the spreading liquid ultimately breaks up into lenses (see Problem 2.6). Such behavior is called "autophobic" and can also be seen on solid surfaces, as discussed below.

3. WORK OF ADHESION AND WORK OF COHESION

Consider a cylindrical column of two materials A and B as shown in Figure 2.2a. If the column is torn apart at the A-B interface (Figure 2.2b), then the work done per unit interfacial area is

$$W_{AB} = \gamma_A + \gamma_B - \gamma_{AB}. \tag{2.8}$$

This is the work of adhesion. As the name implies, it is an important concept in the theory of adhesive materials, where it is necessary to have a measure of the strength of the adhesive to the adherent. Equation 2.8 follows from the definition of interfacial tension as the work done in creating a unit area.

If the column is made entirely of the same material A, then the work done per unit area in tearing the column is

$$W_{AA} = 2\gamma_A, \tag{2.9}$$

which is the work of cohesion. It represents the condition of failure in the bulk of a material.

If A is a liquid (L) and B is a solid (S), then if $W_{AA} > W_{AB}$, one has from Equations 2.8 and 2.9, $\gamma_L > (\gamma_S - \gamma_{SL})$. This condition is satisfied only if the contact

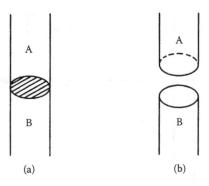

(a)　　　　　　　(b)

FIGURE 2.2 Configuration analyzed in Section 3: (a) before separation of A and B; (b) after separation.

angle is nonzero (Equation 2.1 is used). Consequently one may conclude that if an adhesive does not spread spontaneously, it can come unstuck ("adhesive failure") before the failure of the material itself ("cohesive failure").

To return to the definitions of the work of adhesion and the work of cohesion, one notes an interesting feature, that is, that the work done in moving apart molecules can be related to interfacial tensions. The important energy that is dependent on the intermolecular distances is the intermolecular potential. On assuming that the work done is solely the change in intermolecular potential on tearing apart a column of material, one finds a means for relating interfacial tensions to molecular effects. A simple form of intermolecular potential between molecules i and j is

$$\phi_{ij} = -\varepsilon_{ij} \left\{ \frac{\sigma_{ij}}{|r_{ij}|} \right\}^{6} \quad for \quad |r_{ij}| > \sigma_{ij}$$

$$= +\infty \qquad for \quad |r_{ij}| \le \sigma_{ij} \tag{2.10}$$

where ε_{ij} is a characteristic energy, σ_{ij} is a characteristic length, and $|r_{ij}|$ is the distance between the centers of mass of the molecules i and j. σ_{ij} is the center-to-center distance at contact between molecules i and j.

The first part of the potential, which is negative and varies inversely with the sixth power of the intermolecular distance, is called the van der Waals or the London dispersion potential. It represents the attraction between the two molecules. The second part is positive and represents repulsion. The fact that it is infinite implies that the molecules are rigid spheres and cannot penetrate each other. A sketch of the dimensions involved is shown in Figure 2.3, as well as the form of Equation 2.10. For i = j, the distance of closest approach (when the spheres touch one another) is σ_{ii}, which is also the diameter of the molecule.

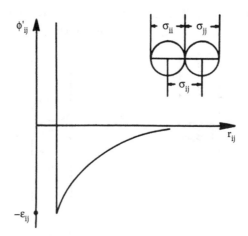

FIGURE 2.3 A model for the intermolecular potential.

When $i \neq j$, one often takes $\sigma_{ij} = (\sigma_{ii} + \sigma_{jj})/2$. However, sometimes it is more convenient to define $\sigma_{ij}^3 = (\sigma_{ii}^3 + \sigma_{jj}^3)/2$. Both forms, it should be noted, are approximate. For the dispersion energy ε_{ij}, a simple rule for mixing is $\varepsilon_{ij} = (\varepsilon_{ii}\varepsilon_{jj})^{1/2}$ (Prausnitz et al., 1986).

Equation 2.10 is confined to nonpolar molecules. More detail on intermolecular potentials is given elsewhere (Israelachvili, 1991; Prausnitz et al., 1986). Equation 2.10 can be written in a more convenient form:

$$\Phi_{ij} = \frac{-n_i n_j \beta_{ij}}{|r_{ij}|^6} \quad for \ |r_{ij}| > \sigma_{ij}$$
$$= +\infty \quad for \ |r_{ij}| < \sigma_{ij} \tag{2.11}$$

where n_i and n_j are moles per unit volume of the i and j species. β_{ij} is a new constant, and Φ_{ij} has the dimensions of energy/[(volume of i)(volume of j)].

With this form of interaction energy, Hamaker (1937) determined that the attraction between the A and B phases per unit area of a planar interface is given by $(-\pi n_A n_B \beta_{AB})/(12\,\sigma_{AB}^2)$. The derivation of a more general form where A and B are separated by a distance h is given in Chapter 3. The potential energy in that case is given by Equation 3.15, and can be written here as $(-\pi n_A n_B \beta_{AB})/(12h^2)$. The smallest value of h is σ_{AB}, the smallest distance of approach of the first lines of the A and B molecules, which are considered to be hard spheres.

If A and B are the same, then the work done in tearing apart the system at a plane $A'B'$ is

$$W_{AA}^d = 2\gamma_A^d = \text{(interaction energy when the two pieces are separated by an infinite distance)} - \text{(interaction energy when the two pieces are continuous)}.$$

Since there is no interaction at infinite distance ($h = \infty$),

$$W_{AA}^d = 2\gamma_A^d = \frac{\pi n_A n_A \beta_{AA}}{12\sigma_{AA}^2}. \tag{2.12}$$

The superscript d indicates that only the contribution of the dispersion potential to the interfacial tension and work of cohesion has been included. If the molecules are nonpolar, $W_{AA} = W_{AA}^d$ and $\gamma_A = \gamma_A^d$. Padday (1969) examined the above relation after adjustments were made for the improvement of the intermolecular potentials, improved liquid theories, etc. He found the agreement with measured values of liquid-vapor interfacial tensions to be excellent.

Along the same lines one obtains the work of adhesion as

$$W_{AB}^d = \frac{\pi n_A n_B \beta_{AB}}{12\sigma_{AB}^2}. \tag{2.13}$$

One of the consequences of the above equation is that if $\sigma_{AA} \simeq \sigma_{BB} = \sigma$, that is, if the molecules are of comparable size, then $W_{AB}^d \simeq 2(\gamma_A^d \gamma_B^d)^{1/2}$, as $\sigma_{AB} \simeq \sigma$. If A and B interact through dispersion alone, then $W_{AB} = W_{AB}^d \simeq 2(\gamma_A^d \gamma_B^d)^{1/2}$. Substituting this result into Equation 2.8, one has

$$\gamma_{AB} = \gamma_A + \gamma_B - 2(\gamma_A^d \gamma_B^d)^{1/2} . \tag{2.14}$$

For a given pair of liquids, the first three terms can be measured experimentally, and if one is nonpolar then $\gamma_A^d = \gamma_A$, and hence γ_B^d of a polar liquid can be experimentally determined. This equation is due to Fowkes (1963) and constitutes a special case of the more general equation of Girifalco and Good (1957), in which the last term of Equation 2.14 is mulitplied by a correction factor. It is noteworthy that Equation 2.14 is derived when the interaction between A and B occurs through dispersion only. This is true when at least one of A and B is nonpolar. When both are nonpolar, $\gamma_A = \gamma_A^d$ and $\gamma_B = \gamma_B^d$, and Equation 2.14 becomes

$$\gamma_{AB} = \gamma_A + \gamma_B - 2(\gamma_A \gamma_B)^{1/2}. \tag{2.15}$$

Equation 2.15 may be viewed as an equation of state for nonpolar materials giving γ_{AB} as a function of γ_A and γ_B. Li et al. (1989) and Li and Neumann (1992) argue that this equation is but a limiting case of a universal equation of state which holds for more polar materials as well. However, this view has been challenged by others (Johnson and Dettre, 1989; Morrison, 1989).

4. PHENOMENOLOGICAL THEORIES OF EQUILIBRIUM CONTACT ANGLES

If Equation 2.14 is substituted into Equation 2.1, one obtains

$$\cos \lambda = -1 + 2(\gamma_S^d \gamma_L^d)^{1/2} / \gamma_L . \tag{2.16}$$

For the purposes of investigating Equation 2.16, it is assumed here that the same solid is used, that is, γ_S^d is a constant, and the contact angles of a number of nonpolar liquids (i.e., $\gamma_L^d = \gamma_L$) are measured. Equation 2.16 becomes

$$\cos \lambda = -1 + 2(\gamma_S^d / \gamma_L)^{1/2} . \tag{2.17}$$

One notes that as $\gamma_L \to \infty$, $\cos \lambda \to -1$, or that such a liquid is completely nonwetting. As γ_L is decreased, one finds that for $\gamma_L = \gamma_S^d$, $\cos \lambda = +1$. Liquids with surface tensions smaller than this value completely wet the given solid surface. The fact that $\gamma_L < \gamma_S^d$ represents wetting liquids can be verified on combining Equation 2.14 with Equation 2.7, thus

$$S_{L/S} = -2\gamma_L + 2(\gamma_S^d \gamma_L)^{1/2} \tag{2.18}$$

for $\gamma_L = \gamma_L^d$. Obviously $S_{L/S} = -2\gamma_L[1 - (\gamma_S^d/\gamma_L)^{1/2}]$ is negative if $\gamma_L > \gamma_S^d$ (non-wetting) and positive (wetting) if $\gamma_L < \gamma_S^d$. Thus Equation 2.17 needs to be rewritten as

$$\cos \lambda = -1 + 2(\gamma_S^d/\gamma_L)^{1/2}, \quad \gamma_L > \gamma_S^d$$
$$= +1 \qquad\qquad , \quad \gamma_L < \gamma_S^d \tag{2.19}$$

In the second part of Equation 2.19, the fact that wetting liquids always have λ equal to zero (Bascom et al., 1964; Fox and Zisman, 1950; Johnson and Dettre, 1966) has been used.

If it is assumed that $|\gamma_L - \gamma_S^d|$ is small, then Equation 2.19 can be expanded in a Taylor series about $\gamma_L = \gamma_S^d$ and only the first term retained. One has

$$\cos \lambda = 1 - [(\gamma_L - \gamma_S^d)/\gamma_S^d], \quad \gamma_L > \gamma_S^d$$
$$= 1 \qquad\qquad , \quad \gamma_L < \gamma_S^d \tag{2.20}$$

Zisman (1964) measured the values of the equilibrium contact angles for various liquids on a given solid surface. The results of Fox and Zisman (1950) for n-alkanes on Teflon are shown in Figure 2.4. The plot is a straight line with

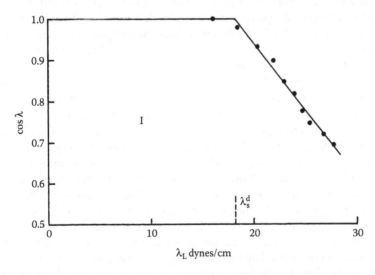

FIGURE 2.4 The Zisman plot for n-alkanes on Teflon (Fox and Zisman, 1950). The liquids with $\cos \lambda = 1$ are wetting liquids. The error bar is shown in the inset.

a constant slope, in agreement with Equation 2.20. The value of γ_L at which the line intersects the horizontal line $\cos \lambda = 1$ is known as the critical surface tension of the solid. From Equation 2.20, this is equal to γ_S^d. For solids that are nonpolar, $\gamma_S = \gamma_S^d$, and consequently the Zisman plot constitutes one of the very few ways of determining γ_S. In all cases the critical surface tension is a very good way of arriving at an estimate for γ_S.

Inspection of Figure 2.4 reveals that the slope of the line is (–0.03) and thus not equal to $(-1/\gamma_S^d) = -0.056$, as predicted by Equation 2.20. This indicates that Equation 2.14 is only a first approximation and that the correction factor of Girafalco and Good (1957) is needed for greater accuracy. From a physical point of view some have suggested that the surface tension of the dry solid should be modified by an adsorption term (Johnson and Dettre, 1966, 1993), a feature which will be taken up later.

Liquid surface tensions γ_L are typically in the range of 20 to 70 mN/m. Solids surfaces may be divided into high energy surfaces (approximately $\sim$ 500 mN/m and more; e.g., glass, metals, etc.) and low energy surfaces (approximately $\sim$ 20 to $\sim$ 40 mN/m; e.g., hydrocarbons, polymers, etc.). From these values and the above results one may conclude that high energy surfaces are almost always wettable. Such a generalization cannot be made for low energy surfaces. The latter are usually not wet by water, since water has a high surface tension of 72 mN/m. It is noteworthy that fluorocarbon surfaces have very low critical surface tensions. For instance, Teflon (polytetrafluoroethylene) has a critical surface tension of about 18 mN/m and consequently is wet by very few liquids. Polyethylene, in contrast, has a critical surface tension of about 31 mN/m and is wet by many organic liquids. Extensive compilations of contact angles and critical surface tensions have been made by Adamson and Gast (1997, pp. 365–368).

To put matters into perspective, we note that Equation 2.16 can be derived only because the work of cohesion was obtained independently, that is, through consideration of intermolecular energies of interaction. This work was expressed in terms of interfacial tensions, and Young's equation (Equation 2.1) was thus simplified to Equation 2.16. An alternate method exists for relating the equilibrium contact angle λ to molecular effects. Consider Figure 2.5, where a gas (G)-liquid (L)-solid (subdivided into S_A and S_B) three-phase region is shown and O is the contact line. The liquid is wedge shaped and has a plane gas-liquid interface. Q is a point on the gas-liquid interface. Φ is the potential energy of interaction at Q, per unit volume of liquid, with the gas phase G, the liquid phase L, and the solid phases S_A and S_B. Now a datum potential Φ_F is defined in the following way. If the contact line region were not present, there would have been only the gas and liquid phases, with the two separated by the original G-L interface, and its extrapolation as shown by the dashed line in Figure 2.5. Thus G + S_A would then be the gas phase, and L + S_B would be the liquid phase. Φ_F for this situation is obviously a constant related to the energy at a plane gas-liquid interface. The energy of interaction with the gas phase can be neglected for there are too few molecules in the gas phase. Hence

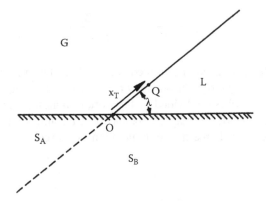

FIGURE 2.5 The construction used for determining the molecular potential at the point Q. The solid has been partitioned into regions S_A and S_B.

$\Phi - \Phi_F$ = (interaction of Q with S_B as a solid – that with S_B as a liquid)
+ (interaction of Q with S_A as a solid).

With the form of interaction energy given by Equation 2.11, the above relation can be expressed in terms of integrals over the respective volumes, that is,

$$\Phi - \Phi_F = \int_{V_{SB}} \frac{n_L(n_L\beta_{LL} - n_S\beta_{SL})}{|r|^6}\, dV - \int_{V_{SA}} \frac{n_L n_S\beta_{SL}}{|r|^6}\, dV\,, \qquad (2.21)$$

where $|r|$ is the radial distance from point Q. The volume integrations are performed for the profile shown in Figure 2.5 in a cylindrical coordinate system with O on the axis of the cylinder. The result is

$$\Phi - \Phi_F = \frac{\pi}{12x_T^3}(n_L^2\beta_{LL} - n_L n_S\beta_{SL})G(\lambda) - \frac{\pi}{12x_T^3}n_L n_S\beta_{SL}G(\pi - \lambda)\,, \qquad (2.22)$$

where the function G is

$$G(\alpha) = \mathrm{cosec}^3\alpha + \cot^3\alpha + \frac{3}{2}\,\cot\,\alpha \qquad (2.23)$$

and α is the angle of wedge. x_T is the distance of the point Q from zero. If the system is at equilibrium, then the hydrodynamic force $[-d(\Phi - \Phi_F)/dx_T]$ along the gas-liquid interface (locus of Q) must be zero, leading to the result

$$\frac{G(\pi - \lambda)}{G(\lambda)} + 1 = \frac{n_L^2 \beta_{LL}}{n_L n_S \beta_{SL}} . \tag{2.24}$$

Equation 2.24 defines the equilibrium contact angle λ. From Equations 2.12 and 2.13 with $\sigma_{SS} \sim \sigma_{LL}$, the right-hand side of Equation 2.24 is equal to W_{LL}^d / W_{SL}^d, that is, the work of cohesion divided by the work of adhesion. Thus for nonpolar systems, the right-hand side may be obtained from Young's equation (Equation 2.1) as $2/(1 + \cos \lambda)$. If the above assumption is correct, one should have an identity in

$$\frac{G(\pi - \lambda)}{G(\lambda)} + 1 = \frac{2}{1 + \cos \lambda} . \tag{2.25}$$

Equation 2.25 is not an identity. The left-hand side and the right-hand side of the equation are shown in Figure 2.6 with bold and dashed lines, respectively; the agreement is good. The plots form mirror images about $\lambda = \pi$. Equation 2.24 was

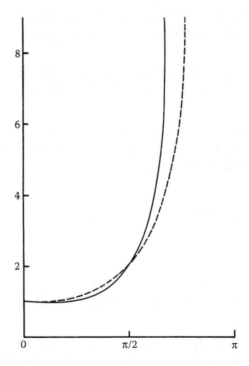

FIGURE 2.6 The left-hand side and the right-hand side in Equation 2.25 are shown with bold and dashed lines, respectively, for different values of λ.

derived by Miller and Ruckenstein (1974) and in a somewhat different way by Jameson and del Cerro (1976).

Thus Equation 2.24 provides an approximation to Young's equation. The important contribution of Equation 2.22 is that it is valid (within its degree of approximation) in nonequilibrium situations as well and provides a means for obtaining an appropriate hydrodynamic driving force in the liquid for those cases where λ in Figure 2.5 differs from the equilibrium contact angle.

The question that one asks now is if Hamaker forces are the only forces that one needs to take into account, and obviously there are others. These can be classified into short-range and long-range forces. An important short-range force is considered in the next section. Among long-range forces, the electrostatic forces are important and are considered in detail in Chapter 3. The more restrictive $\Phi - \Phi_F$ used here can be replaced by the disjoining pressure Π and Hamaker constant A_H as defined later in this book (Chapter 3, Section 2). Derjaguin and Frumkin considered systems with small contact angles, which allow the use of disjoining pressure at the interface, since the interface is only slightly slanted. A balance of forces leads to the Derjaguin-Frumkin equation (Churaev, 1994):

$$\cos \lambda = 1 - \frac{1}{\gamma_L} \int_{h_0}^{\infty} \Pi(h') \, dh' \, , \qquad (2.26)$$

where h or h' is the film thickness. In general,

$$\Pi = -\frac{\pi^2 (n_L n_S \beta_{SL} - n_L^2 \beta_{LL})}{6\pi h^3} \, ,$$

where $\pi^2 (n_L n_S \beta_{SL} - n_L^2 \beta_{LL}) > 0$ for a wetting liquid. Substituting into Equation 2.26, one has

$$\cos \lambda = 1 + \frac{\pi (n_L n_S \beta_{SL} - n_L^2 \beta_{LL})}{12 \gamma_L h_0^2} \, .$$

The right-hand side being greater than one provides no solution for λ, implying a lack of equilibrium, an expected result for a wetting liquid. In Equation 2.26, h_0 is the thickness of the thin film that lies ahead of the contact line. Such a film could very well be an adsorbed film and an additional constraint is often required to quantify h_0. Equation 2.26 releases the earlier restriction on the use of specified forces but adds another in that the contact angles will have to be small. This equation will be encountered again in the section on foams in Chapter 5. At present it suffices to say that for a contact angle to exist (i.e., for the liquid to be nonwetting), the integral has to be positive. This is possible only if there is some

term in the disjoining pressure that disfavors a thinning effect because it increases
the interaction energy.

5. ACID-BASE INTERACTION

Whereas there has been steady progress in using London dispersion–Lifshitz–van
der Waals–Hamaker-type forces to explain interfacial effects, there has always
existed a great deal of confusion regarding the polar effects. It appears now to
be more certain that polar forces are not long range. This led investigators to look
in detail at the very short-range effects that characterize the polarity of molecules
in terms of acidity and basicity, and alternatively by their electron donor and
acceptor roles. That is, these locations for the polar bonds can be determined
from the way the electrons are distributed in the molecules. These effects of
electron transfer between two adjacent sites on two molecules of the same or
dissimilar species occur at about 0.2 nm separation. The result is seen macro-
scopically as a release of heat when acid and base groups interact. In fact, it is
possible to measure acid-base interaction from the heat of mixing if one knows
how to subtract the effects of van der Waals attraction. It is possible to relate this
heat to the work of adhesion. The latter is written as a sum,

$$W_{12} = W_{12}^d + W_{12}^{AB} , \qquad (2.27)$$

where the superscript AB denotes the acid-base interaction. Note that W_{12} is
calculated from measured values of γ_1, γ_2, and γ_{12} when both are liquids and from
the equilibrium contact angle and the surface tension of the liquid when one of
them is a solid. W_{12}^d is $2(\gamma_1^d \gamma_2^d)^{1/2}$, also an experimentally accessible quantity (see
discussions following Equation 2.14). It is now assumed that

$$W_{12}^{AB} = fN(-\Delta H_{12}^{AB}) , \qquad (2.28)$$

where ΔH_{12}^{AB} is the enthalpy difference between the contacted and separated states
and W_{12}^{AB} is the free energy difference between separated and contacted states
(i.e., the reverse order). N is the number of acid-base groups at the interface per
unit area and f is supposed to be close to one. That is, the acid groups of one
phase interact with the basic groups of the other phase only along the interface.
For benzene-water, the heat effects are known, it is reasonable to set f to one,
and the value for N can also be approximately provided if it is assumed that one
molecule of benzene binds with one molecule of water (Fowkes, 1990). Water-
benzene interfacial tension can be predicted with as good an accuracy as the
experimental data. Although this success of the formulation by Fowkes in liquid-
liquid systems is a great achievement, the unsatisfactory feature is the uncertainty
in the value of f. For solids, using Equation 2.28 and van't Hoff's rule,

$$N(-\Delta H_{S1}^{AB}) = -T^2 \left(\frac{d\dfrac{W_{S1}^{AB}}{T}}{dT} \right) = \frac{1}{f} W_{S1}^{AB} . \qquad (2.29)$$

Thus

$$f = \left(1 - \frac{d\ln W_{S1}^{AB}}{d\ln T} \right)^{-1} . \qquad (2.30)$$

Vrbanac and Berg (1990) found f to be significantly less than one, and their data also indicate that polyethylene has acid-base activity!

An important reason why problems arise in characterizing surfaces is that solid surfaces can have more than one type of acid group. Thus a more detailed knowledge of the surface constitution is required (see, for instance, the discussion in Section 9), which has not yet been achieved. Similarly an added method for characterizing groups would be valuable. Steric effects are also involved. It is well known now that polymer chain segments at an interface rotate to bring out their polar groups when they are exposed to a polar liquid or to shield the polar groups when the liquid is nonpolar (Andrade, 1988). This feature has become very important in evaluating the surface properties of artificial organs. There, proteins adsorb on the surface and lead to the formation of blood clots. Deaths from heart attacks and strokes are more common than from organ rejection. Due to the effects just described, a hydrocarbon probe provides information on the characteristics of such surfaces which is of limited use for describing interactions with blood plasma.

Another formulation is by van Oss et al. (1987). They take the acid-base components in the surface tension to be

$$\gamma_i^{AB} = 2(\gamma_i^{\oplus}\gamma_i^{\ominus})^{1/2} , \qquad (2.31)$$

with the new superscripts denoting an acid property and a basic property of the same species. The geometric mixing rule in the above equation is reasonable, but not rigorously justified. This is the conventional polar contribution to the surface tension. Of course, a species could be purely acidic, in which case its basic contribution is zero. In the interfacial tension one has

$$\gamma_{ij}^{AB} = 2\left(\sqrt{\gamma_i^{\oplus}} - \sqrt{\gamma_j^{\oplus}} \right)\left(\sqrt{\gamma_i^{\ominus}} - \sqrt{\gamma_j^{\ominus}} \right) . \qquad (2.32)$$

The resulting generalization for Equation 2.14 is given in Problem 2.4. Finally, for a liquid against a solid surface,

$$\gamma_i(1 + \cos \lambda_i) = 2\left(\sqrt{\gamma_i^d \gamma_S^d} + \sqrt{\gamma_i^{\oplus} \gamma_S^{\ominus}} + \sqrt{\gamma_i^{\ominus} \gamma_S^{\oplus}}\right). \qquad (2.33)$$

Most often one wishes to characterize a solid surface, which as shown in Equation 2.33 is done through three unknowns — γ_S^d, $\gamma_S^{\oplus}$, and $\gamma_S^{\ominus}$ — and thus needs three equations, namely Equation 2.33 for the three liquids denoted by i = 1, 2, and 3. The following data can be used:

water: $\gamma_L = 72.8$, $\gamma_L^d = 21.8$, $\gamma_L^{\oplus} = 25.5$, $\gamma_L^{\ominus} = 25.5$;

glycerol: $\gamma_L = 64$, $\gamma_L^d = 34$, $\gamma_L^{\oplus} = 3.92$, $\gamma_L^{\ominus} = 54.7$;

diiodomethane: $\gamma_L = 50.8$, $\gamma_L^d = 50.8$, $\gamma_L^{\oplus} = 0.00$, $\gamma_L^{\ominus} = 0.00$;

all in millinewtons per meter.

The values of $\gamma_L^{\oplus}$ and $\gamma_L^{\ominus}$ in the above table were obtained as follows. The investigators assumed that the acid and base contributions for water were equal, and hence $\gamma_L^{\oplus} = \gamma_L^{\ominus} = 25.5$ was obtained. Then they made measurements of contact angles of water and a second liquid on different monofunctional (either $\gamma_S^{\oplus}$ or $\gamma_S^{\ominus}$ is zero) solids to calculate the ratio $\gamma_W^{\ominus} / \gamma_L^{\ominus}$ or $\gamma_W^{\oplus} / \gamma_L^{\oplus}$ using Equation 2.33, and showed this ratio to be independent of the solids used. To get the acid and base contributions of the second liquid, the acid and base contributions of water have to be known, hence the need to make the assumption on water mentioned earlier. Berg (1993) reviewed this area in detail.

One application by Good et al. (1998) illustrates the importance of this approach. They analyzed the published results (Horbett et al., 1985) on cell growth at solid-water/culture medium interfaces using this approach. The solid was a copolymer of hydroxymethyl acrylate (HEMA) and ethyl methacrylate (EMA). The relative ratio was changed and it was found that cells grew only when EMA content was more than 50%. Contact angle data on surfaces with different HEMA:EMA ratios for a number of liquids including the above list were reported, but the data in that form were not capable of clarifying the mechanisms of cell attachment and growth. However, Good et al. showed that under their acid-base analysis, the calculated values of $\gamma_S^{\oplus}$ became zero and stayed there as EMA content went above 50% (see Problem 2.4). This leads to an entirely new way of looking at cell adhesion and growth at interfaces.

The acid-base theory has been criticized by some who have provided data to show that the acid-base theory can sometimes give very unreliable results (Kwok et al., 1994).

6. CONTACT ANGLE HYSTERESIS

Experimental measurements of equilibrium contact angles may be accomplished photographically. There are no formulas to apply. In some cases, notably the sessile drop, the equipment used to measure the interfacial tensions also serves to measure the contact angles, provided that in such systems well-defined contact lines exist. If interfacial tensions are known from separate measurements, the Wilhelmy plate and capillary rise methods may be used to obtain contact angles. Although the basic arrangements are simple, a great many precautions are necessary for sufficiently accurate results (Good, 1979; Neumann and Good, 1979). An overview of methods using computers and image analysis is given by Li et al. (1992). A particular method exists where a flat plate is partially immersed in a liquid and tilted until the clinging meniscus disappears. The angle of the tilt describes the contact angle. The method is sketched in Figure 2.7a,b. For very low contact angles, approximately $\sim 1°$, interference techniques are needed. These low angles are encountered in foam films, which are discussed in Chapter 5.

The first experimental data on contact angles were quite confusing. If they were at all reproducible, they could not be explained. Although in general they were supportive of the theory, the data did a great many things other than that anticipated by the theory. In what followed, virtually everything was put under doubt as a matter of principle. It was argued that some of the conditions under which the experiments were performed could have led to spurious results (e.g., vibrations in the system). Alternately it was said that the theory was incomplete and a list of a dozen physical effects was advanced which, it was claimed, should be incorporated into the theory. Obviously, in such a case, both arguments could be true, a point that appeared to be quite unclear given the polemics of those times. Thanks to the meticulous and very provocative work by Zisman, Johnson, Dettre, Good, and others, the reasons behind the anomalous behavior are now well known. The initial inability to explain contact angle hysteresis stemmed from the usual simplifying assumption that solid surfaces are perfectly smooth, impermeable, pure, homogeneous, etc. Obviously these terms apply only to an idealized solid surface, reasonable for some purposes, but quite a naive description of most solids used in the experiments in interfacial phenomena.

A description of an experiment is necessary to explain the contact angle hysteresis. A flat plate is shown immersed in a pool of liquid in Figure 2.7a. A meniscus forms near the contact line. The plate is rotated until the meniscus disappears; the inclination of the plate to the horizontal liquid surface is the equilibrium contact angle λ, as shown in Figure 2.7b. The contact line is at point P on the solid. If the plate is dipped more into the liquid until the point P' is reached on the solid, the new contact angle there is seen to be λ_a. This is called the advancing contact angle because the contact line has advanced to a new position over a dry solid surface (Figure 2.7c). The plate is withdrawn to a point where the contact line is at P''. The contact angle there is found to be λ_r (Figure 2.7d). This is called the receding contact angle because the contact line has

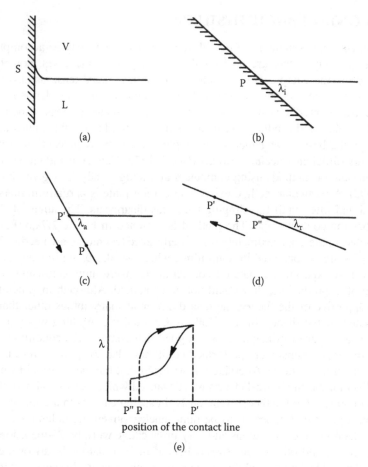

FIGURE 2.7 The solid surface in (a) is turned to obtain a horizontal vapor (V)-liquid (L) interface in (b). (c) The advancing contact angle λ_a is shown, and (d) shows the receding contact angle λ_r. (e) The measured values of λ are shown to illustrate the "hysteresis."

receded into the region previously occupied by the liquid. On repeated dipping and withdrawing, the plate λ_a and λ_r may reach steady values but are seen not to be equal to one another (Figure 2.7e). Johnson and Dettre (1969) discussed why the term hysteresis came to be applied. It is only of historical interest. However, the understanding of why λ_a is different from λ_r is of importance. The main reasons are discussed below.

6.1   IMPURITIES ON THE SURFACE

It has been argued previously that an ordinary glass surface is completely wet by water. A casual experiment will show this is not the case. Since glass has a high surface energy, it is wet by almost all substances, and consequently small amounts

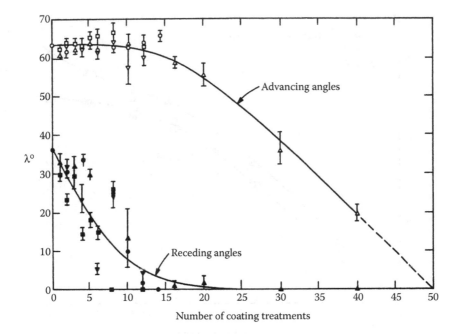

Number of coating treatments

FIGURE 2.8 The advancing and receding contact angles of water on titania-coated glass after treatment with trimethyloctadecylammonium chloride are shown as a function of coating treatments with 1.1% polydibutyl titanate. Reprinted with permission from Dettre and Johnson (1965). Copyright 1965 American Chemical Society.

of grease (and dirt) wet the surface and adhere to it. The waxy material brings down the interfacial tension to sufficiently low values that it is not wet by water. Glass cleaning procedures have been discussed by Lelah and Marmur (1979). It needs to be treated with a strong oxidizing agent such as nitric acid, acidified dichromates, etc. These help to make the impurities water soluble. Potassium hydroxide (KOH) has a similar action but with a different chemistry. The solubility of impurities in organic solvents like propanol is also exploited. Use of very dilute hydrofluoric acid (HF) to cut silicone grease is also practiced. These procedures serve to show the importance of contamination and how difficult it is to avoid.

Surface contamination, which gives rise to a patchy or heterogeneous surface, can also cause contact angle hysteresis. In Figure 2.8, where the advancing and the receding contact angles are shown for water on a glass surface (titania-coated glass treated with waxy trimethyloctadecyl ammonium chloride) as a function of the number of coating treatments with 1% polydibutyl titanate (Dettre and Johnson, 1965). It is seen that about 50 coating treatments are required to remove or recoat the contaminant and make the surface sufficiently homogeneous that there is no hysteresis.

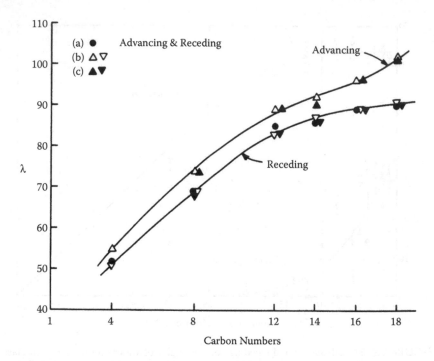

FIGURE 2.9 The advancing and receding contact angles of water are plotted against the carbon number of n-alkyl platinum for amines adsorbed (a) from aqueous solutions, (b) from melts, and (c) from cetane solutions. Data from Baker et al. (1952) and Shafrin and Zisman (1949, 1954).

6.2   EFFECT OF ADSORPTION

It is noteworthy that the impurities discussed in Section 6.1 are in adsorbed form. Here, a special type of adsorption effect is analyzed, that is, it is postulated that due to adsorption γ_s for determining the advancing contact angle is different from that of the receding contact angle. Baker et al. (1952) and Shafrin and Zisman (1949, 1954) coated the surface of platinum with alkyl amines by (a) adsorbing them from aqueous solutions, (b) adsorbing them from melts, and (c) adsorbing them from an n-hexadecane solutions. The advancing and receding contact angles for water on the above three surfaces are shown in Figure 2.9, plotted against the carbon number of the n-alkyl amines. A striking feature is that the advancing and receding contact angles for water in case (a) are the same. It is also seen that the advancing and receding contact angles for cases (b) and (c) fall on only two curves, the "advancing" curve and the "receding" curve labeled in the figure. Case (a) falls on the "receding" curve. Interesting conclusions can be drawn.

It is known that the amine group adsorbs onto the platinum surface with the hydrocarbon tails directed away from the platinum surface, as shown in Figure 2.10. Under the bulk water, the water molecules penetrate the tail region and the nature of the advancing contact angle is as shown in Figure 2.10a. Now, for the

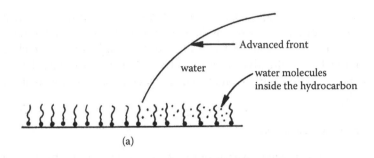

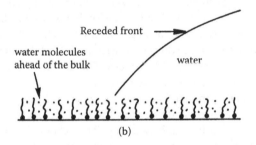

FIGURE 2.10 Schematic diagrams explaining the nature of the results shown in Figure 2.9.

receding contact angle, the situation is as shown in Figure 2.10b. The "solid surface" has alkyl amines adsorbed on it as in Figure 2.10a, but with water molecules among the hydrocarbon tails. Obviously the appropriate values of γ_S in Figures 2.10a and 2.10b are different, and the advancing and receding contact angles differ. However, if the amine is adsorbed from aqueous solution, water molecules will be interdispersed in the hydrocarbon tail region everywhere. Consequently, in this case the advancing and receding contact angles will be the same (same γ_S) and will be equal to the receding contact angles obtained in the other cases (Figure 2.10b).

One attempts to generalize the above effect by rewriting Equation 2.1 as

$$\gamma_{LV} \cos \lambda = (\gamma_{SV} - \pi_e) - \gamma_{SL}, \tag{2.34}$$

where π_e is the surface pressure of the adsorbed material on the solid surface. The adsorbed material changes the interfacial tension of the solid from γ_{SV} to $\gamma_{SV} - \pi_e$. Thin films whose thickness is greater than the length of a single molecule also affect γ_{SV}, with $-\pi_e$ closely related to the "disjoining pressure" (see Chapter 5). The π_e term is usually unimportant for nonwetting liquids, but becomes very important as the wettability increases.

In this connection a special class of systems where γ_L is close to the critical surface tension of the solid needs to be mentioned. It is found that in such cases the liquid wets the dry solid but will not wet the solid surface, which has the vapor of the liquid already adsorbed on it (Hare and Zisman, 1955). This phenomenon is called autophobicity. The contact angles so formed are usually very small.

6.3 SURFACE ROUGHNESS

Vibrations in contact angle measuring systems are known to affect the experimental results. Investigators found that contact angle hysteresis is made more acute as steps are taken to eliminate vibrations (i.e., the data become more irreproducible). In contrast, the problem is minimized by introducing controlled amounts of vibration in the system (del Giudice, 1936; Fowkes and Harkins, 1940; Phillipoff et al., 1952). Eventually Johnson and Dettre (1964, 1969) showed that vibrations or the kinetic energy of the drop very seriously affect the measured contact angles in experiments on rough solid surfaces. The degree of roughness in most cases is small, but sufficient to give rise to hysteresis.

Johnson and Dettre calculated the surface free energy of a drop lying on a rough surface. The roughness was idealized as concentric sinusoidal grooves. The liquid-vapor interface was taken to be a spherical cap. The scheme is shown in Figure 2.11. It is seen that there is an apparent equilibrium contact angle ϕ, which is the angle of the drop with the horizontal direction. Its significance lies in the fact that although the actual equilibrium contact angle λ remains the same, ϕ varies depending on the position of the contact line. It is also this angle, ϕ, that is measured experimentally by extrapolating the observed shape of the liquid-vapor interface to meet the horizontal direction.

Johnson and Dettre were able to show that of all possible positions of the foot of the drop inside a given groove, there was only one where the free energy

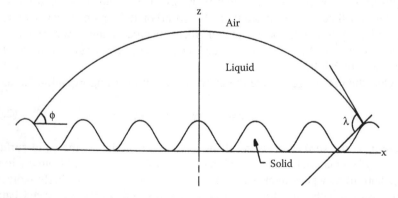

FIGURE 2.11 A drop on an idealized rough surface. Reprinted with permission from Johnson and Dettre (1964). Copyright 1964 American Chemical Society.

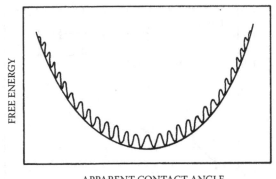

APPARENT CONTACT ANGLE

FIGURE 2.12 In Figure 2.11 it was shown that the drop can have many configurations. Being symmetric, each configuration has a fixed apparent contact angle Φ and a total free energy. It can be seen that for the foot of the drop lying inside any groove there is a minimum. Two neighboring minima (metastable equilibrium positions) are separated by an energy barrier that is highest near the overall minimum. Consequently this overall minimum is the most difficult to attain without sufficient vibrations. Reprinted with permission from Johnson and Dettre (1969).

was a minimum. At that point the local contact angle was the equilibrium value λ (see Problem 2.18). The minima in two neighboring grooves were separated by an energy barrier. For a drop of a given volume, the values of the free energy minima for different grooves were different and the system had one global minimum.

They concluded that although the drop had only one position favored energetically (i.e., where the position of the foot gave rise to the overall free energy minimum), in practice, the foot could get trapped in any groove between two energy barriers. How and where such an entrapment occurs depends on the vibrations in the system (i.e., on the kinetic energy of the drop available to overcome the energy barriers between adjacent positions of metastable equilibrium) (see Figure 2.12). It is noteworthy that a controlled amount of vibration in the system can lead the foot of the drop eventually to the position of the overall minimum energy. The ϕ may still differ from λ, but it is at least reproducible and connected to the thermodynamics of the system. A sketch of the energy barriers near the region of the contact line is shown in Figure 2.12. Note that the energy barriers are largest near the equilibrium position. Hitchcock et al. (1981) prepared surfaces that were rough because of regular geometric protrusions. Two length scales, amplitude and wavelength, can be defined on such surfaces. It was found that all anomalous behaviors decreased with a decreasing amplitude:wavelength ratio. For a simple treatment showing that the roughness will affect the apparent contact angle, see Problem 2.6.

The contact angle hysteresis of water on a rough wax surface is shown in Figure 2.13 as the surface is made smoother by annealing. Beyond the seventh

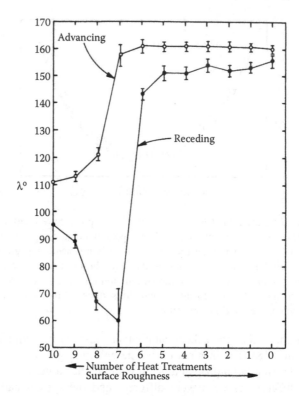

FIGURE 2.13 Water contact angles on a tetrafluoroethylene (TFE)-methanol telomer wax surface as a function of roughness. Reprinted from Johnson and Dettre (1964) with permission. Copyright 1964 American Chemical Society.

annealing treatment, the difference between the advancing and receding contact angles progressively vanishes. Prior to the seventh treatment, the surface is so rough that solid-liquid contact is broken at numerous locations by air gaps, and the above arguments must be modified. Very large contact angles are seen for these conditions (see Problem 2.19), and such "superhydrophobic" surfaces are a topic of considerable interest and current research.

Besides the above three important features, there are many others, some of which are mentioned briefly below. One question that is often asked is if the horizontal force balance at the contact line leads to Young's equation, what happens to the normal component $\gamma_{LV} \sin \lambda$? It appears that such a component does exist and tends to distort the solid surface. Since in most cases the magnitude of this force is small and the modulii of elasticity of the solids are large, no significant distortion can be observed. (However, Bailey (1957) has actually observed the distortion of the solid near the contact line region of a mercury drop on a mica sheet 1 μm thick.)

Liquid imbibition into the solid substrate can also alter the results (Yuk and Jhon, 1986, 1987). As discussed in Chapter 1, crystal structures affect the surface energies and hence the contact angles, even on low energy polymer surfaces (Fort, 1965).

It appears that although one knows why hysteresis can occur, or even why it occurs in a given system, it is a formidable job to get rid of it entirely. Extremely smooth and homogeneous surfaces can rarely be prepared, and certainly the means or methods are not available to investigators outside this select field, say to an investigator looking at a fluid mechanics problem involving contact angles, or to a polymer chemist who needs to test the efficacy of a new coating material. Nevertheless, contact angle measurements remain of importance for the following reasons. First, in spite of all the difficulties discussed earlier, the contact angle turns out to be a robust property. With care it can generally be reproduced to within $\pm 2°$, and certainly within $\pm 5°$ among investigators (Good, 1979; Neumann and Good, 1979). Consequently contact angle measurements continue to provide useful information on surface characteristics. Even in difficult systems such as crude oil and porous reservoir rocks, great attention is paid to measuring wettability in meaningful terms (Anderson, 1986; Yang et al., 1998). Second, experimental observations involving equilibrium contact angles indirectly as a parameter can be correlated without references to advancing or receding, hysteresis, etc., if the data were all obtained using a fixed experimental procedure, although by different investigators. An example is provided by Hoffman (1975) in his correlations for dynamic contact angles. The equilibrium values used are not explained, but can be assumed to be advancing contact angles because the experiments deal with advancing menisci. That is, it is quite reasonable to expect to correlate similar sets of data without being hampered by hysteresis effects.

There has also been some discussion that the contact angle generally observed is the extrapolation of the profile of the bulk drop to the solid surface ("macroscopic" contact angle). The real contact angle ("microscopic") is quite different. Differential ellipsometry can measure molecular dimensions quite well and show that the microscopic contact angles are indeed quite different (Heslot et al., 1992).

7.   ADSORPTION

Previously the influence of adsorbed material on contact angles has been seen to be important. In general, the study of adsorption has been prompted by its importance in catalysis and reactions at interfaces. The Langmuir-Hinshelwood model for the kinetics of such reactions is popular and depends on the adsorption-desorption rates (Boudart and Djéga-Mariadassou, 1984). The nature of adsorption also needs to be studied to determine the viability of chromatographic separation processes for a given system. Its attraction is that it is one of the very few means for separating similar components, such as isomers, azeotropes, etc., that are traditionally "difficult" (Holland and Liapis, 1983).

As noted previously, the adsorbed amount Γ is the surface excess. The concentration of a species changes smoothly across an interface attaining bulk values in the interiors of the two phases. If, for simplicity, the bulk concentrations are extrapolated to the dividing surface, it leaves one with excess material that is assumed to reside at the dividing surface. This is the surface excess as seen in Chapter 1. Consider a gas-solid interface where the gas is ideal and the solid is impermeable. Now, the gas molecules are attracted toward the solid because of the intermolecular forces. However, these forces are not felt in the bulk of the gas far away from the interface. At equilibrium, the chemical potential of the gas in the bulk is $\mu^o + RT\ell n\, n_B$ and that near the interface is $\mu^o + RT\ell n\, n + \Phi$. Here, n is the gas density and n_B is the density in the bulk. μ^o is the standard state chemical potential. Φ is the attraction potential between the gas and the solid and $\Phi(z) \to 0$, $n(z) \to n_B$ as $z \to \infty$, where z is the perpendicular distance from the interface. At equilibrium, the two chemical potentials are equal, leading to $n = n_B \exp(-\Phi/kT)$. Following the definition of Γ, one has

$$\Gamma = \int_0^\infty (n - n_B)\, dz \qquad (2.35)$$

or

$$\Gamma = n_B \int_0^\infty [\exp(-\Phi/RT) - 1]\, dz\,.$$

As the integral is a constant, Equation 2.35 reduces to Henry's law

$$\Gamma = Hp, \qquad (2.36)$$

where H is the Henry's law constant given by

$$H = \frac{1}{RT} \int_0^\infty [\exp(-\Phi/RT) - 1]\, dz$$

and the ideal gas law has been used to replace n_B with the pressure p. It is found that Henry's law is valid for moderately small pressures.

It is seen that n deviates most from n_B at the interface. Thus the gas molecules that make up the surface excess concentration Γ reside very close to the interface. One may now assume a model where the adsorbed gas molecules are supposed to be stuck to the solid surface. A quantity θ, the fractional surface coverage, is thus defined as

$$\theta = \sigma^o \Gamma, \qquad (2.37)$$

where σ^o is the projected area of a molecule adsorbed at the interface.

7.1   Langmuir Adsorption Isotherm

When adsorption is such that $\theta < 1$, the rate of adsorption is, neglecting interaction between adsorbed molecules, proportional to $p(1 - \theta)$, where $(1 - \theta)$ is the fractional available space for adsorption and p is the gas pressure. The model is similar to that describing the kinetics of a bimolecular reaction. The rate of desorption is proportional to θ or the concentration of the adsorbed species. At equilibrium

$$k_a p(1 - \theta) = k_d \theta,$$

where k_a and k_d are the adsorption and desorption rate constants. Rearranging the above, one has

$$\theta = \frac{(k_a / k_d)p}{1 + (k_a / k_d)p} \tag{2.38}$$

or

$$\Gamma = \frac{[k_a / (k_d \sigma^\circ)]p}{1 + (k_a / k_d)p}, \tag{2.39}$$

which is the Langmuir adsorption isotherm. At small pressures, Henry's law is obtained:

$$\Gamma = \left(\frac{k_a}{k_d \sigma^\circ} \right) p. \tag{2.40}$$

Note that from Equation 2.38, $\theta < 1$.

7.2   The Brunauer-Emmett-Teller Isotherm

It is sometimes seen that as $p \to p^\circ$, the saturation pressure, the adsorbed amount becomes infinite. This seems to imply that adsorption turns into condensation at some stage. In the Brunauer-Emmett-Teller (BET) adsorption isotherm, it is assumed that the first layer is an adsorption layer. Multiple layers are built on it due to condensation. The activation energy for the first layer is the heat of adsorption, Q. The activation energy for adsorbing a layer on an n-stack layer ($n > 1$) is the heat of condensation, Q_v.

Thus the equilibrium for the first layer is given by

$$ap\theta_o = b\theta_1 e^{-Q/RT}, \tag{2.41}$$

and for others by

$$ap\theta_{n-1} = b\theta_n e^{-Q_v/RT} \ , \ n > 1 \tag{2.42}$$

In Equations 2.41 and 2.42, the left-hand sides denote the rates of adsorption, which are proportional to the pressure and the unoccupied surface. The right-hand sides denote the rates of desorption, which are proportional to the occupied surface. The rate constants are a for adsorption and $[b \exp(- Q/RT)]$ and $[b \exp(- Q_v/RT)]$ for desorption in the two cases ($n = 1$ and $n > 1$). θ_n is the fractional area covered by the n-stack layer. The volume adsorbed per unit area is proportional

to $\displaystyle\sum_{n=1}^{\infty} n\theta_n$, thus

$$\frac{v}{v_m} = \frac{\displaystyle\sum_{n=0}^{\infty} n\theta_n}{\displaystyle\sum_{n=0}^{\infty} \theta_n} \ , \tag{2.43}$$

where v_m is a reference quantity. The summation can be carried out using Equations 2.41 and 2.42 to yield

$$\frac{v}{v_m} = \frac{Ky}{(1-y)[1+(K-1)y]} \ , \tag{2.44}$$

where $K = \exp[(Q - Q_v)/RT]$ and $y = (a/b)p \exp(Q_v/RT)$. One notes in Equation 2.38 that at $y = 1$, $v = \infty$. This is precisely the condition necessary to have complete condensation at $p = p^\circ$. Substituting $p = p^\circ$ and $y = 1$ into the definition for y, one has

$$\frac{a}{b}p^\circ e^{Q_v/RT} = 1 \ and \ hence \ y = p / p^\circ \ .$$

Equation 2.44 is the BET isotherm. For small y (i.e., $1 - y \sim 1$) and large K (i.e., $K-1 \sim K$), the Langmuir isotherm is obtained. Henry's law (Equation 2.36), the Langmuir isotherm (Equation 2.38), and the BET isotherm (Equation 2.44) are plotted in Figure 2.14, with $\sigma^\circ Hp^\circ = p^\circ k_a/k_d = K$. Among them, they describe wide ranges of adsorption behavior.

Experimentally v is measured in volumes at standard temperature and pressure (STP). If the adsorption is only a saturated monolayer, then $\theta_1 = 1$ and all other θ_i are zero. In that case, one has $v = v_m$, allowing one to interpret v_m as the volume

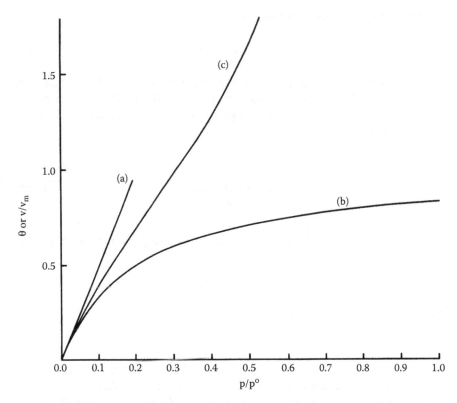

FIGURE 2.14 (a) Henry's law, (b) the Langmuir adsorption isotherm, and (c) the BET isotherm plotted with parameters such that (b) and (c) reduce to (a) at low p/p^o.

of gas at STP in cubic centimeters required to cover the surface of the adsorbent completely and with no excess material. Thus

$$v_m = \frac{S v^o}{N_A \sigma^o}, \tag{2.45}$$

where $v^o = 22400$ cm³/mole and N_A is Avogadro's number. S is the surface area of the adsorbent and σ^o is the projected area of one molecule of the adsorbed gas. Hence from the experimental data, v_m and K can be determined (see Problem 2.17 for a useful method). Then S can be found from Equation 2.45 if a good estimate of σ^o is available. Consequently the BET equation has been widely used to estimate S, knowledge of which is invaluable in catalysis.

Adsorption of a component from a liquid follows similar behavior. In fact, similar adsorption equations are used. An important case is where a component is adsorbed from the bulk liquid onto a solid surface. The previous equations are altered by replacing the pressure p with c, the concentration of the adsorbed species in the bulk phase.

The surface pressure is defined as

$$\pi = \gamma^\circ - \gamma, \qquad (2.46)$$

where γ° is the interfacial tension of the pure substance and γ is that after adsorption. For liquid-fluid interfaces, π is particularly significant, as it can be measured directly (see Chapter 4). Using the Gibbs adsorption equation (Equation 1.24), one has

$$\pi = \int_0^c \Gamma \, d\mu . \qquad (2.47)$$

Assuming that $\mu = \mu^\rho + RT\ell nc$ and that Henry's law applies in the form $\Gamma = Hc$, one has

$$\pi = RTHc = RT\, \Gamma,$$

which on rearrangement yields

$$\pi\sigma = RT, \qquad (2.48)$$

where $\sigma = \Gamma^{-1}$ is the interfacial area per adsorbed molecule. Equation 2.48 is reminiscent of the ideal gas law, $pv = RT$, where π is the two-dimensional analog of the pressure p and σ is the analog of specific volume v. Consequently models are built in more complicated systems along the lines of appropriate equations of state in three dimensions (Gaines, 1966). It should be noted that the analogy is not exact. For instance, negative values of π, unlike negative pressures p, are physically meaningful and are known to occur in strong electrolyte solutions. Adamson and Gast (1997, pp. 599–755) provide a more comprehensive discussion on the subject of adsorption.

8. DENSITY PROFILES IN LIQUID FILMS ON SOLIDS

The density profile at a vapor-liquid interface was considered briefly in Chapter 1 (Section 5). Prominent in that discussion was the positive contribution to the local free energy density produced by sharp gradients in density (cf. Equation 1.41). The buildup of a thin film on a solid surface is also influenced by this gradient energy (Cahn, 1977; Teletzke et al., 1982). As we shall see, gradient effects sometimes prevent the continuous film buildup predicted by conventional adsorption isotherms as the bulk density of the vapor approaches its saturation value. Moreover, when both liquid and vapor are present, the liquid spreads spontaneously on the solid for temperatures sufficiently near the critical temperature, but has a finite equilibrium contact angle at lower temperatures.

Suppose that a solid surface is placed in a bulk vapor phase having a pressure p_B and molecular density n_B below the saturation values at the existing temperature. Attraction between the solid and molecules of the vapor favors buildup of an adsorbed film. However, the additional energy associated with the density gradient between the film and bulk vapor phase opposes film formation, an effect not considered in developing the BET adsorption isotherm of the preceding section.

The local Helmholtz free energy density corresponding to Equation 1.41 in such a film is

$$f = f_o(n) + \frac{d}{dz}\left[k_1 \frac{dn}{dz}\right] + k\left(\frac{dn}{dz}\right)^2 + n\Phi(z) , \qquad (2.49)$$

where $\Phi(z)$ is the interaction potential between solid and gas as in Section 6. Since the temperature T and the external pressure p_B are fixed, the Gibbs free energy G of the system must be minimized at equilibrium. Evaluating the local Gibbs free energy at each position relative to that of the same number of molecules at T and p_B, we obtain

$$G = \int_0^\infty \left[\Delta f + k\left(\frac{dn}{dz}\right)^2 + n\Phi(z)\right] dz . \qquad (2.50)$$

Since the term in dn/dz will lead to a flux, it has been set to zero at the wall as appropriate in a closed system and

$$\Delta f = f_o(n) + p_B - \frac{n}{n_B}(f_o(n_B) + p_B) . \qquad (2.51)$$

Recognizing that $[(f_o(n_B) + p_B)/n_B]$ is the chemical potential μ_B, we see that Equation 2.50 is of the same form as the expression given following Equation 1.43. Invoking the calculus of variations, we find that the density profile that minimizes G must satisfy the following condition:

$$2k\frac{d^2n}{dz^2} - \frac{\partial \Delta f}{\partial n} - \Phi(z) = 0 . \qquad (2.52)$$

Finally, we substitute Equation 2.51 into this expression to obtain

$$2k\frac{d^2n}{dz^2} - \frac{df_o(n)}{dn} - \Phi(z) + \mu_B = 0 . \qquad (2.53)$$

Teletzke et al. (1982) used Equation 2.50 and the Peng-Robinson equation of state (Peng and Robinson, 1976) to calculate density profiles for various values of the temperature T and bulk density n_B. They found that below the saturation density n_B° relatively thick films can form near the solid when the temperature is below but not too far from the critical temperature T_c of the fluid. Under these conditions the gradient energy contribution is relatively small, as would be expected in the neighborhood of the critical point. For densities above n_B°, where the bulk phase is a liquid-vapor mixture, the liquid completely wets the solid (i.e., there is no equilibrium contact angle).

Below a "surface critical temperature," T_{cs}, which is less than T_c, two profiles with the same total free energies are possible at certain bulk densities slightly less than n_B°. One is a relatively thin film with but a small enhancement in density near the solid. It has both a relatively low attractive interaction with the solid and a low gradient energy. The other solution is a thicker film having a greater density near the solid and consequently both a greater attractive interaction and a larger gradient energy. In principle, films with these two profiles could coexist on the same surface. Again the liquid wets the solid for bulk densities exceeding the saturation value n_B°.

Below a still lower temperature, T_{cw}, the "critical wetting temperature," only a thin adsorbed film exists for $n_B < n_B^\circ$ because the gradient energy is so high that development of thicker films is energetically unfavorable. For densities above n_B°, a finite contact angle exists where liquid and vapor phases meet the surface. Thus T_{cw} marks the boundary between the domains where the liquid has zero and nonzero contact angles with the solid. The spreading coefficient, $S_{L/S}$, also vanishes here, and this is sometimes used as an alternative definition of surface criticality.

The salient features of the theory of wetting transition and some experimental observations have been reviewed by de Gennes (1985). The experiments, which are performed solely on liquid-liquid systems, do not always conform with the theoretical predictions because of trace impurities, slow equilibration and fluctuations, and the solubility of one liquid in another. Schmidt (1988) reports the results of studies on water against a homologous series of fluorocarbons which support all of the theoretical findings.

9. CHARACTERIZING SOLID SURFACES

Low energy solid surfaces are conventionally characterized using Zisman plots and critical surface tension. For more details on such surfaces and for high energy surfaces, adsorption isotherms are obtained. In the case of adsorption of symmetric molecules on homogeneous surfaces, the adsorption isotherms even show the multilayer buildup step by step (Putnam and Fort, 1975). Such results are of considerable interest in building up fundamental adsorption laws (Nicholson and Parsonage, 1982; Steele, 1983). Real surfaces are considerably more complex. Heterogeneities can arise because of surface chemical groups (Moller and Fort, 1975) or because of surface defects (Koestner et al., 1983; Somorjai and Zaera,

1982). In general, adsorption, and hence catalysis, is characterized as taking place at sites, each site having a different adsorption energy:

$$\Gamma(E) = \frac{[K_o e^{-E/RT} / \sigma_o]p}{1 + [K_o e^{-E/RT}]p} , \qquad (2.54)$$

where the equilibrium constant from Equation 2.39 is $k_a/k_d = K = K_o \exp(-E/RT)$. The net adsorption is

$$\Gamma = \int_0^\infty \Gamma(E) f(E) \, dE , \qquad (2.55)$$

where $f(E)$ is a distribution of sites with energy E. It is suggested that one common isotherm called the Freundlich adsorption isotherm, $\Gamma = Kp^v$ is a consequence of such heterogeneities, where the basic isotherm follows Langmuir's model. The mathematics of how to obtain the distribution of energies, $f(E)$, for active sites given the overall response is important and has been the subject of considerable investigation (Boudart and Djéga-Mariadassou, 1984; Brown and Travis, 1983; House, 1983). Unique results are not yet available and the energy distributions that are sometimes obtained using these procedures are yet to be related to the physical details discussed earlier. Additional complications arise as catalytic surfaces are available in the form of surfaces within porous media, usually random porous media, but sometimes structured (crystalline), as in zeolites. In such systems, what constitutes the surface area depends on the size and shape of the adsorbent molecules. Obviously, very large molecules cannot enter small pores, and the adsorption isotherms are modified to address this feature (Sircar and Myers, 1986).

The advent of high performance materials containing joints between metals and ceramic materials, a growing variety of electronic materials, shape selective catalysts, etc., requires more details than can be obtained by adsorption. For instance, surface composition is important, which, as mentioned in Chapter 1, may be different from the bulk composition. In addition, knowledge of the surface structure (surface crystallography) and surface electronic states is very important. There are a number of books (Clark and Feast, 1975; Feldman and Mayer, 1986; Somorjai, 1981; Walls, 1989) and a series titled *Progress in Surface Science* (Pergamon Press, New York) that address the new experimental methods used to determine surface properties. Usually the surface under vacuum or as a gas-solid system is exposed to ions, electron beams, x-rays, or other radiation such as infrared or ultraviolet, and the reflected beams or ejected particles are analyzed for inventory, energies, wavelengths, or interference. The relatively recent addition of scanning, tunneling, and atomic force microscopy now allows one to "see" molecules on a site. Details of the methods, interpretation of results, and discussion of where they are best applied can be found in the above references and many more in this area.

REFERENCES

GENERAL REFERENCES

Berg, J.C. (ed.), *Wettability*, Marcel Dekker, New York, 1993.
Gould, R.F. (ed.), *Contact Angle, Wettability and Adhesion*, Advances in Chemistry Series, vol. 43, American Chemical Society, Washington, D.C., 1964.
Mittal, K.L. (ed.), *Contact Angle, Wettability, and Adhesion*, VSP, Utrecht, The Netherlands, 1993.
Mittal, K.L. and Anderson, H.R., Jr. (eds.), *Acid-Base Interactions: Relevance to Adhesion Science and Technology*, VSP, Utrecht, The Netherlands, 1991.

TEXT REFERENCES

Adamson, A.W. and Gast, A.P., *Physical Chemistry of Surfaces*, 6th ed., Wiley Interscience, New York, 1997.
Anderson, W.G., Wettability literature survey. II: Wettability measurements, *J. Petrol. Technol.*, 38, 1246, 1986.
Andrade, J.D. (ed.), *Polymer Surface Dynamics*, Plenum Press, New York, 1988.
Bailey, A.I., in *Proceedings of the Second International Congress on Surface Activity*, J.H. Schulman (ed.), vol. 3, Butterworths, London, 1957, p. 189.
Baker, H.R., Shafrin, E.G., and Zisman, W.A., The adsorption of hydrophobic monolayers of carboxylic acids, *J. Phys. Chem.*, 56, 405, 1952.
Bascom, W.D., Cottington, R.L., and Singleterry, C.R., Dynamic surface phenomena in the spontaneous spreading of oils on solids, in *Contact Angle, Wettability, and Adhesion*, Advances in Chemistry Series, vol. 43, American Chemical Society, Washington, D.C., 1964, p. 355.
Berg, J.C., The role of acid-base interactions in wetting and related phenomena, in *Wettability*, J.C. Berg (ed.), Marcel Dekker, New York, 1993, p. 75.
Boudart, M. and Djéga-Mariadassou, G., *Kinetics of Heterogeneous Catalytic Reactions*, Princeton University Press, Princeton, New Jersey, 1984.
Brown, L.F. and Travis, B.J., Optimal smoothing of site-energy distributions from adsorption isotherms, in *Fundamentals of Adsorption*, A.L. Myers and G. Belfort (eds.), Engineering Foundation, New York, 1983, p. 125.
Buff, F.P. and Saltzburg, H., Curved fluid interfaces. II. The generalized Neumann formula, *J. Chem. Phys.*, 26, 23, 1957.
Cahn, J.W., Critical point wetting, *J. Chem. Phys.*, 66, 3667, 1977.
Clark, D.T. and Feast, W.J., Applications of electron spectroscopy to studies of structure and bonding in polymeric systems, *J. Macromol. Sci.: Rev. Macromol. Chem.*, C 12(2), 191, 1975.
Churaev, N.V., Contact angles and surface forces, *Colloid J.*, 56, 631, 1994.
de Gennes, P.G., Wetting: statics and dynamics, *Rev. Mod. Phys.*, 57, 827, 1985.
del Giudice, G.R.M., The bubble machine for flotation testing, *Eng. Mining J.*, 137, 291, 1936.
Dettre, R.H. and Johnson, R.E., Jr., Contact angle hysteresis. IV. Contact angle measurements on heterogeneous surfaces, *J. Phys. Chem.*, 69, 1507, 1965.

Dyson, D.C., Contact line stability at edges: comments on Gibbs's inequalities, *Phys. Fluids*, 31, 229, 1988.

Emmett, P.H. and Brunauer, S., The use of low temperature van der Waals adsorption isotherms in determining the surface area of iron synthetic ammonia catalysts, *J. Am. Chem. Soc.*, 59, 1553, 1937.

Feldman, L.C. and Mayer, J.W., *Fundamentals of Surfaces and Thin Films*, North Holland, New York, 1986.

Fort, T., Jr., The wettability of a homologous series of nylon polymers, in *Contact Angle, Wettability, and Adhesion*, Advances in Chemistry Series, vol. 43, American Chemical Society, Washington, D.C., 1965, p. 302.

Fowkes, F.M., Additivity of intermolecular forces at interfaces. I. Determination of the contribution to surface and interfacial tensions of dispersion forces in various liquids, *J. Phys. Chem.*, 67, 2538, 1963.

Fowkes, F.M., Quantitative characterization of the acid-base properties of solvents, polymers and inorganic surfaces, *J. Adhesion Sci. Technol.*, 4, 669, 1990.

Fowkes, F.M. and Harkins, W.D., The state of monolayers adsorbed at the interface solid—aqueous solution, *J. Am. Chem. Soc.*, 62, 3377, 1940.

Fox, H.W. and Zisman, W.A., The spreading of liquids on low-energy surfaces. I. Polytetrafluoroethylene, *J. Colloid Sci.*, 5, 514, 1950.

Franses, E.I., Siddiqui, F.A., Ahn, D.J., Chang, C.-H., and Wang, N.-H.L., Thermodynamically consistent equilibrium adsorption isotherms for mixtures of different-sized molecules, *Langmuir*, 11, 3177, 1995.

Gaines, G.L., *Insoluble Monolayers at Liquid-Gas Interfaces*, Wiley-Interscience, New York, 1966.

Gaydos, J. and Neumann, A.W., The dependence of contact angles on drop size and line tension, *J. Colloid Interface Sci.*, 120, 76, 1987.

Gibbs, J.W., *The Scientific Papers*, vol. 1, Dover, New York, 1961, p. 288.

Girifalco, L.A. and Good, R.J., A theory for the estimation of surface and interfacial energies. I. Derivation and application to interfacial tension, *J. Phys. Chem.*, 61, 904, 1957.

Good, R.J., Contact angles and the surface free energy of solids, in *Surface and Colloid Science*, vol. 11, R.J. Good and R.R. Stromberg (eds.), Plenum Press, New York, 1979, p. 1.

Good, R.J., Islam, M., Baier, R.E., and Meyer, A.E., The effect of surface hydrogen bonding (acid-base interaction) on the hydrophobicity or hydrophilicity of copolymers: variation of contact angles and cell adhesion and growth with composition, *J. Dispersion Sci. Technol.*, 19, 1163, 1998.

Hamaker, H.C., The London-van der Waals attraction between spherical particles, *Physica*, 4, 1058, 1937.

Hare, E.F. and Zisman, W.A., Autophobic liquids and the properties of their adsorbed films, *J. Phys. Chem.*, 59, 335, 1955.

Harkins, W.D., *The Physical Chemistry of Surface Films*, Reinhold, New York, 1952.

Heslot, F., Cazabat, A.M., Fraysse, N., and Levinson, P., Experiments on spreading droplets and thin films, *Adv. Colloid Interface Sci.*, 39, 129, 1992.

Hitchcock, S.J., Carroll, N.T., and Nicholas, M.G., Some effects of substrate roughness on wettability, *J. Mater. Sci.*, 16, 714, 1981.

Hoffman, R.L., A study of the advancing interface. I. Interface shape in liquid-gas systems, *J. Colloid Interface Sci.*, 50, 228, 1975.

Holland, C.D. and Liapis, A.I., *Computer Methods for Solving Dynamic Separation Problems*, McGraw-Hill, New York, 1983.

Horbett, T.A., Schway, M.B., and Ratner, B.D., Hydrophilic-hydrophobic copolymers as cell substrates: effect on 3T3 cell growth rates, *J. Colloid Interface Sci.*, 104, 28, 1985.

House, W.A., Adsorption on heterogeneous surfaces, in *Colloid Science*, vol. 4, Special Periodical Report, Royal Society of Chemistry, London, 1983, p. 1.

Israelachvili, J.N., *Intermolecular and Surface Forces*, 2nd ed., Academic Press, New York, 1991.

Jameson, G.J. and del Cerro, M.C.G., Theory for equilibrium contact-angle between a gas, a liquid and a solid, *J. Chem. Faraday Trans. I*, 72, 883, 1976.

Johnson, R.E., Jr., Conflicts between Gibbsian thermodynamics and recent treatments of interfacial energies in solid-liquid-vapor, *J. Phys. Chem.*, 63, 1655, 1959.

Johnson, R.E., Jr. and Dettre, R.H., Contact angle hysteresis, I, II, in *Contact Angle, Wettability, and Adhesion,* Advances in Chemistry Series, vol. 43, American Chemical Society, Washington, D.C., 1964, pp. 112, 136.

Johnson, R.E., Jr. and Dettre, R.H., The wettability of low-energy liquid surfaces, *J. Colloid Interface Sci.*, 21, 610, 1966.

Johnson, R.E., Jr. and Dettre, R.H., Wettability and contact angles, in *Surface and Colloid Science*, vol. 2, E. Matijevic (ed.), Wiley-Interscience, New York, 1969, p. 85.

Johnson, R.E., Jr. and Dettre, R.H., An evaluation of Neumann's "surface equation of state," *Langmuir*, 5, 293, 1989.

Johnson, R.E., Jr. and Dettre, R.H., Wetting of low energy surfaces, in *Wettability*, J.C. Berg (ed.), Marcel Dekker, New York, 1993, p. 1.

Kerins, J. and Widom, B., The line of contact of three fluid phases, *J. Chem. Phys.*, 77, 2061, 1982.

Koestner, R.J., van Hove, M.A., and Somorjai, G., Molecular structure of hydrocarbon monolayers on metal surfaces, *J. Phys. Chem.*, 87, 203, 1983.

Kwok, D.Y., Li, D., and Neumann, A.W., Evaluation of the Lifshitz-van der Waals/acid-base approach to determine interfacial tensions, *Langmuir*, 10, 1323, 1994.

Lelah, M.D. and Marmur, A., Wettability of soda-lime glass: the effect of cleaning procedures, *Am. Chem. Soc. Bull.*, 58, 1121, 1979.

Li, D., Cheng, P., and Neumann, A.W., Contact angle measurement by axisymmetric drop shape analysis (ADSA), *Adv. Colloid Interface Sci.*, 39, 347, 1992.

Li, D., Gaydos, J., and Neumann, A.W., The phase rule for systems containing surfaces and lines. 1. Moderate curvature, *Langmuir*, 5, 1133, 1989.

Li, D. and Neumann, A.W., Equation of state for interfacial tensions of solid-liquid systems, *Adv. Colloid Interface Sci.*, 39, 299, 1992.

Miller, C.A. and Ruckenstein, E., The origin of flow during wetting of solids, *J. Colloid Interface Sci.*, 48, 368, 1974.

Moller, P.J. and Fort, T., Jr., Structural analysis of graphite fiber surfaces. I. Mass spectroscopy and low temperature adsorption of N_2 and Ar, *Colloid Polymer Sci.*, 253, 98, 1975.

Morrison, I.D., On the existence of an equation of state for interfacial free energies, *Langmuir*, 5, 540, 1989.

Myers, A.L. and Prausnitz, J.M., Thermodynamics of mixed-gas adsorption, *AIChE J*, 11, 121, 1965.

Neumann, A.W. and Good, B.J., Techniques of measuring contact angles, in *Surface and Colloid Science*, vol. 11, R.J. Good and R.R. Stromberg (eds.), Plenum Press, New York, 1979, p. 31.

Nicholson, D. and Parsonage, N.G., *Computer Simulation and Statistical Mechanics of Adsorption*, Academic Press, New York, 1982.

Padday, J.F., Theory of surface tension, in *Surface and Colloid Science*, vol. 1, E. Matijevic (ed.), Wiley-Interscience, New York, 1969, pp. 44–59.

Patankar, N.A., On the modeling of hydrophobic contact angles on rough surfaces, *Langmuir*, 19, 1249, 2003.

Peng, D.Y. and Robinson, D.B., A new two-constant equation of state, *Ind. Eng. Chem. Fund.*, 15, 59, 1976.

Phillippoff, W., Cooke, S.R.B., and Caldwell, D.E., Contact angles and surface coverage, *Mining Eng.*, 4, 283, 1952.

Platikanov, D., Nedyalkov, M., and Nasteva, V., Line tension of Newton black films. II. Determination by the diminishing bubble method, *J. Colloid Interface Sci.*, 75, 620, 1980a.

Platikanov, D., Nedyalkov, M., and Scheludko, A., Line tension of Newton black films. I. Determination by the critical bubble method, *J. Colloid Interface Sci.*, 75, 612, 1980b.

Prausnitz, J.M., Lichtenthaler, E.G., and de Azevedo, E.G., *Molecular Thermodynamics of Fluid-Phase Equilibria,* 2nd ed., Prentice-Hall, Englewood Cliffs, New Jersey, 1986.

Putnam, F.A. and Fort, T., Jr., Physical adsorption of patchwise heterogeneous surfaces. I. Heterogeneity, two-dimensional phase transitions, and spreading pressure of the krypton-graphitized carbon black system near 100.deg.K, *J. Phys. Chem.*, 79, 459, 1975.

Schmidt, J.W., Systematics of wetting at the vapor-liquid interface, *J. Colloid Interface Sci.*, 122, 575, 1988.

Shafrin, E.G. and Zisman, W.A., Hydrophobic monolayers adsorbed from aqueous solutions, *J. Colloid Sci.*, 4, 571, 1949.

Shafrin, E.G. and Zisman, W.A., Hydrophobic monolayers and their adsorption from aqueous solution, in *Monomolecular Layers,* H. Sobotka (ed.) American Association for the Advancement of Science, Washington, D.C., 1954

Somorjai, G.A., *Chemistry in Two-Dimensions: Surface*, Cornell University Press, Ithaca, New York, 1981.

Somorjai, G.A. and Zaera, F., Heterogeneous catalysis on the molecular scale, *J. Phys. Chem.*, 86, 3070, 1982.

Sircar, S. and Myers, A.L., Characteristic adsorption isotherm for adsorption of vapors on heterogeneous adsorbents, *AIChE J*, 32, 650, 1986.

Steele, W.A., Computer simulation of physisorption. Summary — Theory and Models, in *Fundamentals of Adsorption*, A.L. Myers and G. Belfort (eds.), Engineering Foundation, New York, 1983, pp. 597, 743.

Teletzke, G.F., Scriven, L.E., and Davis, H.T., Gradient theory of wetting transitions, *J. Colloid Interface Sci.*, 87, 550, 1982.

Toshev, B.V., Platikanov, D., and Scheludko, A., Line tension in three-phase equilibrium systems, *Langmuir*, 4, 489, 1988.

van Oss, C.J., Chaudhury, M.K., and Good, R.J., Monopolar surfaces, *Adv. Colloid Interface Sci.*, 28, 35, 1987.

Vignes-Alder, M. and Brenner, H., A micromechanical derivation of the differential equations of interfacial statics. III. Line tension, *J. Colloid Interface Sci.*, 103, 11, 1985.

Vogler, E.A., Practical use of concentration-dependent contact angles as a measure of solid-liquid adsorption. 1. Theoretical aspects, *Langmuir*, 8, 2005, 1992.

Vrbanac, M.D. and Berg, J.C., The use of wetting measurements in the assessment of acid-base interactions at solid-liquid interfaces, *J. Adhesion Sci. Technol.*, 4, 255, 1990.

Walls, J.M., *Methods of Surface Analysis*, Cambridge University Press, New York, 1989.

Yang, S.-H., Hirasaki, G.J., Basu, S., and Vaidya, R., Statistical analysis on paramenters that affect wetting for the crude oil/brine/mica system, *J. Pet. Sci. Eng.*, 33, 203, 2002.

Yuk, S.H. and Jhon, M.S., Contact angles on deformable solids, *J. Colloid Interface Sci.*, 110, 252, 1986.

Yuk, S.H., and Jhon, M.S., Temperature dependence of the contact angle at the polymer-water interface, *J. Colloid Interface Sci.*, 116, 25, 1987.

Zisman, W.A., Relation of the equilibrium contact angle to liquid and solid constitution, in *Contact Angle, Wettability and Adhesion*, Advances in Chemistry Series, vol. 43, American Chemical Society, Washington, D.C., 1964, p. 1.

PROBLEMS

2.1   The data on the following interfacial tensions have been compiled by Girifalco and Good (1957):

Hydrocarbons	γ_{hc} (hydrocarbon)	$\gamma_{w\text{-}hc}$ (water-hydrocarbon)
n-Hexane	18.4	51.1
n-Heptane	20.4	50.2
n-Octane	21.8	50.8
n-Decane	23.9	51.2
n-Tetradecane	25.6	50.2
Cyclohexane	25.5	50.2
Decalin	29.9	51.4

where γ is in millinewtons per meter (mN/m). Predict the values of the water-hydrocarbon interfacial tensions using Fowkes' equation (Equation 2.14) after appropriate assumptions. Compare the results with the experimental data given in the table. The surface tension of water is 72.8 mN/m.

2.2   In the above case, can a general result be quoted for water-hydrocarbon interfacial tensions? Perfluorodibutyl ether, $(C_4F_9)_2O$, and perfluorodibutyl amine, $(C_4F_9)_3N$, have surface tensions of 12.2 and 16.8 mN/m, respectively. Their interfacial tensions with water are 51.9 and 25.6 mN/m. Make an appropriate assumption about the polar nature of the ether and calculate the interfacial tension at the ether-n-heptane interface. Compare this with the experimental value (Girifalco and Good, 1957) of 3.6 mN/m.

If a similar assumption is made for the amine, what will the interfacial tension be at the amine-n-heptane interface? Compare with the experimental value of 1.6 mN/m. What are the reasons for the discrepancies?

2.3   Obtain an expression for the work of adhesion W_A for nonwetting liquids using the Zisman equation (Equation 2.20). What will be the value of W_A^* when $\gamma_L = \gamma_S^d$? Show that W_A has a maximum. Find its value W_A^{**} and the ratio W_A^{**}/W_A^*. Show that this extremum, which can be computed from experimental values, is seen not to exist according to Fowkes equation. Explain.

2.4   The contact angle data of Good et al. (1998) on HEMA-EMA surfaces are,

---

TABLE 2.1
Contact angles (in degrees) of various liquids on HEMA-EMA copolymers

Percentage EMA	Water	Glycerol	Diiodomethane	α-Bromonaphthalene	α-Methylnaphthalene
100	74	72	46	—	9
80	67	68	41	16	5
60	68	66	40	23	15
50	53	58	40	28	13
40	60	56	39	25	11
20	54	50	35	22	7
0	59	44	36	23	12

The properties of the liquids are given by,

TABLE 2.2
Surface energy parameters (mJ/m²)

	γ_{lv}	γ^d	$\gamma^\oplus$	$\gamma^\ominus$	γ^{AB}
Water	72.8	21.8	25.5	25.5	51.0
Glycerol	64.0	34.0	3.92	57.4	30.0
Diiodomethane	50.8	50.8	(0)	(0)	(0)
α-Bromonaphthalene	45.0	45.0	(0)	(0)	(0)
α-Methylnaphthalene	39.3	39.3	(0)	(0)	(0)

---

where γ_i^{AB} is the acid-base contribution. Calculate the surface energy parameters for the solids at various compositions. The generalization of Equation 2.14 making use of Equation 2.32 is

$$\gamma_{ij} = \gamma_i^d + \gamma_j^d - 2\sqrt{\gamma_i^d \gamma_j^d} + 2(\sqrt{\gamma_i^+} - \sqrt{\gamma_j^+})(\sqrt{\gamma_i^-} - \sqrt{\gamma_j^-})$$

$$= \gamma_i + \gamma_j - 2\sqrt{\gamma_i^d \gamma_j^d} - 2(\sqrt{\gamma_i^+ \gamma_j^-} + \sqrt{\gamma_i^- \gamma_j^+})$$

2.5   Consider a drop on a solid surface where the effects of line tension have to be accounted for. Change Equations 2.2 through 2.6 appropriately to accommodate the energy τL, where τ is the line tension and L is the length of the contact line. If it is assumed that the drop has a profile of a spherical cap, show that the following equation results:

$$\left(\frac{\partial F}{\partial A_{SL}}\right) = \gamma_{SL} - \gamma_{SV} + \gamma_{LV} \, \cos \hat{\lambda} + (\tau/r),$$

where r is the radius of the basal circle and $2\pi r = L$. Show that at equilibrium this contact angle $\hat{\lambda}$ is greater than λ of Equation 2.1 if τ is positive. Use the data given below for pentadecane on Teflon (Gaydos and Neumann, 1987) to calculate τ if the surface tension is 26.9 mN/m.

λ°	r (mm)
50.3	1.30
49.7	1.56
49.4	2.31
48.9	2.015
48.8	1.28
47.9	1.61
47.5	3.76
47.4	2.64
45.2	4.245
42.8	5.54

2.6   It is well-known that the roughness of a solid surface affects the equilibrium contact angle. Modify Equations 2.2 through 2.6 using Wenzel's approach to include the effects of surface roughness. Note the following:

(i)   The surface area of the solid-liquid interface is A_{SL} and equal to rA_{SL}^{*}, where r is the roughness factor and A_{SL}^{*} is the area of the solid-liquid interface projected on a horizontal plane.

(ii)   The quantity r is a constant and $dA_{SL} = r \, dA_{SL}^{*}$.

(iii) One also has the geometric relation that $dA_{LV} = \cos \hat{\lambda} \cdot dA_{SL}^{*}$, where $\hat{\lambda}$ is the contact angle on a rough surface.

Show with a sketch the significance of $\hat{\lambda}$ and compare it to λ of Equation 2.1. For $r = \sec \lambda$, a phenomenon called wicking occurs: describe physically what would be observed?

2.7   When a drop rests on a liquid surface, mutual solubilities, which are very often extremely small, can have significant roles. The following data (Harkins, 1952) are available

Compound	γ (mN/m)
Benzene	28.9
Benzene saturated with water	28.8
Water	72.8
Water-benzene interface	35.0
Water saturated with benzene	62.2

A similar approach can be used to calculate the apparent contact angle $\hat{\lambda}$ for smooth surfaces with small-scale heterogeneities.

In an experiment, a drop of dry benzene is put on the surface of pure water. First, the benzene becomes saturated with water, then the water becomes saturated with benzene. Calculate the spreading coefficients in all three stages and from the results describe the course of the physical process.

2.8   Very little is known about the right-hand side of Equation 2.1, and Cahn (1977) assumed it to be independent of temperature to the first approximation. On the other hand, the surface tension of a liquid decreases linearly with temperature, and can be written as $\gamma_L = a + bT$, where a and b are constants. Assume that this equation is valid at the critical temperature T_c as well. Use this to obtain a relation between a and b. Substitute into Equation 2.1. Use now the definition of the critical wetting transition temperature T_{cw} to show that

$$\cos \lambda = \frac{(T_c - T_{cw})}{(T_c - T)}, \qquad (2.8.\text{i})$$

that is, transition to wetting takes place before the critical point is reached and the progress scales inversely with temperature difference, Cahn's results.

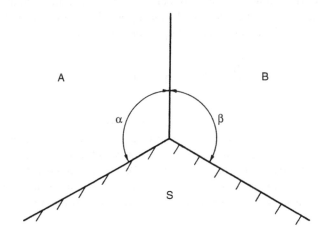

PROBLEM 2.9 Figure for Problem 2.9.

2.9   Determine the condition for (stable) equilibrium when a fluid interface intersects a solid along an edge of the latter. Assume that the angles between the fluid interface and the solid are α in fluid A and β in fluid B (see accompanying figure). Comment briefly on the cases $\alpha + \beta <$

π, $\alpha + \beta = \pi$, and $\alpha + \beta > \pi$. Find the range of stable α and β for an edge where $\alpha + \beta = 250°$ and α (equilibrium) = α_e = $40°$ for a flat surface. This situation was first considered by Gibbs and has been discussed by Dyson (1988) in more detail.

2.10   In deriving Equation 2.47 it was necessary to express the chemical potential in terms of the concentration of the absorbed species in the bulk solution.

(a) The solubility of n-pentane in water is so low that it cannot be measured with simple experiments. On the other hand the surface tension is significantly altered by adsorption of n-pentane. Consequently the surface tension of water is measured as a function of the pressure of pentane in the gas phase. Derive Γ as a function of the pressure in terms of the measured variables. Assume that the gas phase is ideal.

(b) Normal cetyl alcohol is nonvolatile and insoluble in water. How would one determine or control its surface concentration on water? (See Gaines, 1966).

2.11   A surfactant is sometimes used to make a liquid wet a solid surface, such as during dyeing of fabrics. To obtain an equation relating the equilibrium contact to the surfactant concentration, differentiate Equation 2.1 with respect to the chemical potential of the adsorbent. Use Equation 1.24 at a constant temperature to obtain

$$\frac{\gamma_{LV} \sin \lambda}{RT} \frac{d\lambda}{d\ln C} = \Gamma_{SV} - \Gamma_{SL} - \Gamma_{LV} \cos \lambda . \qquad (2.11.\text{i})$$

If λ is less than $90°$, obtain the constraint on the nature of adsorption that will cause λ to decrease. These results are due to Vogler (1992). It is easy to understand why Γ_{SV} is not zero for a receding case, but for some reason it is also not zero in the case of an advancing contact line, even though it is difficult to imagine how surfactant could have crept ahead of the drop on the dry surface (which the results indicate it does).

2.12   The π–α relation at small values of σ is often given by the equation

$$\pi(\sigma - \sigma_o) = f(T), \qquad (2.12.\text{i})$$

where the right-hand side denotes a known function of temperature. What does the constant σ_o signify? From the answer to this question, is it possible to adapt the van der Waals equation of state to a two-dimensional $\pi - \sigma$ system? Argue that the system described by Equation 2.12.i is a liquid, whereas Equation 2.48 describes a gas. Show that the adapted van der Waals equation encompasses both.

2.13   In the Gibbs adsorption equation for a two-component system $d\gamma = -\Gamma_1 d\mu_1 - \Gamma_2 d\mu_2$, the left-hand side is a total differential. Consequently it is equivalent to $d\gamma = (\partial\gamma/\partial\mu_1)d\mu_1 + (\partial\gamma/\partial\mu_2)d\mu_2$, leading to the constraint that $(\partial\Gamma_1/\partial\mu_2) = (\partial\Gamma_2/\partial\mu_1)$. Show that in a binary system and ideal solution, this constraint is satisfied for the Henry's law case, but not in the natural extension of Langmuir model.

Obtain the constraints under which the binary Freundlich model satisfies the above conditions:

$$\Gamma_i = \frac{a_i c_i^{n_l}}{1 + b_i c_i^{n_i} + b_j c_j^{n_j}}, \qquad (2.13.i)$$

where i and j are the two components. Construction of valid isotherms is important (Franses et al., 1995; Myers and Prausnitz, 1965) because of their application in chromatographic separation. In particular, one would like to know if one species is preferentially adsorbed in one region of operation and the second species in another, and if such behaviors can be related to shape, interaction energies, etc.

2.14   If the Langmuir isotherm is written in the form $\Gamma = \Gamma_o c/(c + b)$, show that the equation of state corresponding to Equation 2.42 is given by $\pi = (RT/\sigma_o)\, \ell n\, [1 + (c/b)]$. The result is often called the Szyszkowski equation. Here, $\sigma_o = \Gamma_o^{-1}$.

2.15   A porous medium is sometimes modeled as a bundle of capillary tubes of circular cross section. Suppose that an air-filled porous medium is contacted with a liquid at atmospheric pressure. For what range of contact angles will liquid spontaneously enter the porous medium? When entry is not spontaneous, develop an expression for the liquid pressure required for entry in terms of the surface tension of the liquid, the contact angle, and tube diameter. In view of your answers, why do you think that fluorinated compounds are often used for waterproofing fabrics?

2.16   A porous medium with a distribution of pore sizes contains equal volumes of a wetting and a nonwetting fluid. With which fluid will most, if not all, of the small pores be filled? Why?

2.17   Contact angle hysteresis is one mechanism that hinders motion of fluid drops in a porous medium. Find an expression for the pressure drop required to initiate motion of a drop in a capillary tube of radius a if the advancing and receding contact angles of the drop material are λ_a and λ_r. Evaluate for $a = 10\ \mu m$, $\lambda_a = 120°$, $\lambda_r = 20°$, and $\gamma = 30$ mN/m.

2.18 Consider a rough surface with a profile given by

$$\hat{z} = \hat{z}_o\left(1 + \cos\frac{2\pi\hat{x}}{\hat{x}_o}\right); \ \hat{x} = \frac{x}{V_o^{1/3}}; \ \hat{z} = \frac{z}{V_o^{1/3}} ,$$

where V_o is the volume of the drop.

It can be shown that when $(\hat{z}_o / \hat{x}_o) = 0.1$, Wenzel's factor r, defined in Problem 2.5, is 1.092. Suppose that the drop has an equilibrium contact angle λ_o of $45°$ on a perfectly smooth surface.

(a) Find the apparent equilibrium contact angle $\hat{\lambda}$ on the rough surface. Also, calculate the drop dimensions $\hat{h} = (h/ V_o^{1/3})$ and $\hat{a} = (a/ V_o^{1/3})$ at equilibrium from the following formulas for a spherical segment:

$$\hat{a}^2 = \frac{\frac{6}{\pi} - \hat{h}^3}{3\hat{h}}; \ \hat{r} = \frac{\hat{a}^2 + \hat{h}^2}{2\hat{h}}; \ \sin\theta = \frac{\hat{a}}{\hat{r}} ,$$

where r is the radius of curvature of the segment, h is the maximum height of the segment, and a is the radius on a solid surface. Finally, calculate the free energy, $\hat{F} = (F / \gamma_{LG} V_o^{2/3})$, at equilibrium and find $(\hat{a}/\hat{x}_o)$, the ratio of the drop radius to the wavelength of the roughness for the case $\hat{x}_o = 3.6 \times 10^{-3}$.

(b) Show that the change in free energy when the contact line moves slightly along the wavy surface at a constant drop volume is given by

$$d\hat{F} = d\hat{A}_{LG} - \cos\lambda_o\ d\hat{A}_{SL} ,$$

where $d\hat{A}_{LG} = (dA_{LG} / V_o^{2/3})$ and $d\hat{A}_{SL} = (dA_{SL} / V_o^{2/3})$.

Show further that

$$d\hat{A}_{SL} = 2\pi\ \hat{x}\ d\hat{x}\ (1 + \tan^2\alpha)^{1/2} ,$$

where $\tan \alpha = (d\hat{z}/d\hat{x})$ along the rough surface

$$d\hat{A}_{LG} = 2\pi\hat{x}\, d\hat{x}\, (1 + \tan^2 \alpha)^{1/2}\, \cos \lambda',$$

where λ' is the local angle between the solid-fluid and the liquid-gas interfaces.

(c) Combine the results of (b) to show that

$$\frac{d\hat{F}}{d\hat{x}} = 2\pi\hat{x}\left(1 + 4\pi^2\left(\frac{z_o}{x_o}\right)^2 \sin^2 \frac{2\pi\hat{x}}{x_o}\right)^{1/2} (\cos(\theta + \alpha) - \cos \lambda_o),$$

where θ is the local angle of the fluid interface with the horizontal. Show that $\hat{F}$ has both a local maximum and a local minimum for $(\hat{a}/\hat{x}_o)$ between 328.5 and 329.0. Note that these extrema occur when $\lambda' = \lambda_o$ (i.e., when the local contact angle has the equilibrium value λ_o).

PROBLEM 2.19 Reprinted with permission from Patankar (2003). Copyright 2003 American Chemical Society.

2.19   In Figure 2.13, it is seen initially that with decreasing roughness the agreement between the advancing and receding contact angles worsens. This suggests that in this region the roughness is so large that air is trapped in the grooves under the drop. The air is eventually driven off after the roughness is sufficiently decreased. The approach of Problem 2.6 is applicable when the entire irregular solid surface beneath a drop of liquid is in contact with the solid. Obviously for very rough surfaces, where the equilibrium contact angle λ is large, this may not happen. Roughness need not be random, but can deliberately be made periodic, as in grooved surfaces. Suppose that the drop cannot penetrate the grooves and the contact lines remain pinned along both edges of each groove, as shown in the accompanying figure (see Problem 2.9 with regard to pinning at the edges). For such a case, show, using an approach similar to that of Equations 2.2 through 2.6, that the apparent contact angle λ_a is given by

$$\cos \lambda_a = \phi_s \cos \lambda + \phi_s - 1,$$

where ϕ_s is the fraction of the area of the base of the drop where liquid and solid are in contact. Suppose that $r = 1.3$ (see Problem 2.6) and $\phi_s = 0.05$. Plot the apparent contact angle for both types of rough surfaces as a function of $\cos \lambda$ over the entire range $[-1,1]$. For what value of λ_a do the curves intersect and what is the value of λ_a there? It can be shown that the configuration with the lower value of λ_a has the lower free energy (Patankar, 2003). Hence, for fixed values of r and ϕ_s, the grooves are filled for surfaces with some values of λ and empty for others.

2.20   Show that the BET isotherm (Equation 2.44) can be rearranged to give

$$\frac{y}{v(1-y)} = \frac{1}{Kv_m} + \frac{(K-1)y}{Kv_m} . \qquad (2.20.i)$$

Thus a straight line can be expected if $y/[v(1-y)]$ is plotted as a function of y. The slope and intercept of this line can be used to calculate K and v_m.

Obtain K and v_m for the data of Emmett and Brunauer (1937) for argon on alumina-iron catalyst at $-183°C$ (1096 mmHg).

p (torr)	v (cm^3 at STP/g solid)
21	70
45	93
90	120
175	135
245	155
330	175
405	200
485	220
540	245

3 Colloidal Dispersions

1. INTRODUCTION

Colloidal dispersions are those having particles or drops with at least one dimension greater than about 1 nm but less than about 1 μm. These systems are classified as emulsions when a liquid phase is dispersed in a second liquid, suspensions when a solid phase is dispersed in a liquid medium, foams when a gas is dispersed in a liquid, or aerosols when liquid droplets or solid particles are dispersed in gas. Other combinations are less common.

Viscous, sticky, or waxy materials are easier to dispense in the form of emulsions, as are solids in suspended form. Consequently numerous consumer products are greatly influenced by the knowledge of how to make stable colloidal dispersions. Breaking such dispersions also has many interesting applications. In secondary oil recovery, for instance, petroleum is flushed from underground oil fields with water. The material that is extracted is frequently in the form of an emulsion: oil-in-water or water-in-oil, depending on the relative amounts of the two liquids. As refinery feed streams should be free of water, it is necessary to know how to break the emulsion into the two bulk phases.

Pollutants are also found in the form of colloidal dispersions: the haze in the atmosphere is the result of pollutants dispersed as aerosols. Oils or oily materials are emulsified in the wastewater from refineries and chemical plants and have to be removed before discharging the effluent into rivers or seas.

These examples illustrate the diversity and widespread occurrence of colloidal dispersions in systems of practical interest. Thus it is pertinent to consider their behavior and properties.

Most colloidal dispersions are thermodynamically unstable. That is, system free energy is reduced if the dispersed material is collected into a single bulk phase. If the surface free energy is on the order of 20 ergs/cm^2 (2×10^{-10} J/m^2), which corresponds to $\gamma = 20$ mN/m, one has a surface free energy of [2×10^{-10} $(6/d)$ Φ] J/m^3. Here d is the particle diameter, on the order of 1 μm, and Φ is the volume fraction, on the order of 0.1. With these values, the interfacial free energy is around 1.2×10^{-7} J/m^3, or in terms of the thermal energy per particle, 1.52×10^7 kT at 27°C. Such a value constitutes an overwhelming and positive contribution to the free energy of formation of the dispersion. Consequently the system attempts to go back into separate bulk phases. It is a matter of experience, however, that some colloidal dispersions are stable from a practical point of view, in that they remain unchanged upon standing for time periods of weeks, months, or even years. Technically they are metastable. It is possible, in principle, for them to attain a lower energy state, but a large energy barrier must be overcome before this state can be reached. If the energy barrier is sufficiently high, the dispersed state can be maintained almost indefinitely. Of course, not all colloidal dispersions are stable in this sense. Many are patently unstable and separate quickly upon standing.

For many reasons, it is important to understand the causes of stability in colloidal systems. Manufacturers of cosmetics and paints, for example, want long-term stability so that their products do not separate into two or more bulk phases during the time period between manufacture and use. The operator of an oil field, on the other hand, needs to know how to rapidly destabilize the emulsion of oil and water produced by his wells.

Whether a colloidal dispersion is stable in the above sense depends ultimately on the interactions among the particles or drops. In the following sections we review the main types of interactions, present some equations for predicting their magnitude, and show how to use these equations to determine conditions for stability of colloidal dispersions. Some colloidal dispersions are stable in the true sense in that a solution is formed (and not two phases). As examples we cite proteins, discussed later in brief, and surfactant aggregates such as micelles, discussed extensively in Chapter 4.

First, a brief comment on terminology. When a dispersion is unstable, "flocculation" occurs. That is, the particles or drops aggregate to form clusters that move as a unit. If the dispersed phase is fluid, the drops in a cluster may coalesce to form a single drop. It is possible to have flocculation without coalescence in such systems.

2. ATTRACTIVE FORCES

Attractive forces between particles favor flocculation and oppose stability. Of prime interest in colloidal systems are attractive forces between particles due to attraction between the individual molecules comprising each particle. We consider here only London–van der Waals forces, which are the chief source of attraction between molecules for nonpolar or slightly polar materials. These forces stem from induced dipole–induced dipole interactions. That is, a dipole that develops transiently in one molecule as a result of continual motion of its electrons induces a dipole in the second molecule. Although the time-averaged dipole moments are zero, the time-averaged interaction energy is not. The expression for the potential energy of interaction between a unit volume of species 1 and another of species 2 has the form discussed in Chapter 2:

$$\varphi_{12} = -\frac{n_1 n_2 \beta_{12}}{r^6} , \qquad (3.1)$$

where n_1 and n_2 are the moles per unit volume of species 1 and 2, β_{12} is a constant, and r is the distance between the two molecules. For large separation distances (a few tens of nanometers), "retardation effects" must be considered

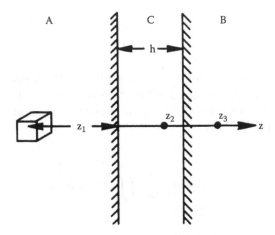

FIGURE 3.1 Interacting semi-infinite slabs A and B with a separating medium C.

(i.e., the time required for electromagnetic radiation to travel between interacting molecules). Under these conditions it can be shown that φ_{12} varies inversely with r^7. More extensive but nevertheless simplified discussion of the origin of London–van der Waals forces is given by Hiemenz and Rajagopalan (1997) and Israelachvili (1991).

The attractive interaction between colloidal particles is, to a first-order approximation, the sum of the attractive interaction between all pairs of molecules chosen such that one molecule of the pair is in each particle. The simplest situation is that shown in Figure 3.1 in which two semi-infinite bodies of material are separated by a gap of thickness h. We suppose in our initial derivation that the two bodies are made up of the same pure substance and that the gap is empty.

Let us first calculate the total energy of interaction ϕ_B between a unit volume in region A on the z-axis at coordinate $(-z_1)$, as shown in Figure 3.1, and the entire region B. Using a cylindrical coordinate system. We find

$$\varphi_B = -\int_h^\infty \int_0^\infty \frac{2\pi n^2 \beta \, r \, dr \, dz}{[r^2 + (z + z_1)^2]^3} . \tag{3.2}$$

It is convenient to integrate first with respect to r. The result is

$$\varphi_B = -\frac{\pi n^2 \beta}{2} \int_h^\infty \frac{dz}{(z + z_1)^4} .$$

Integrating with respect to z we obtain

$$\varphi_B = -\frac{\pi n^2 \beta}{6(h + z_1)^3} \; . \tag{3.3}$$

Equation 3.3 applies for a unit volume in region A. If we integrate this result over all of region A (i.e., between $z_1 = 0$ and $z_1 \to \infty$) we obtain the total interaction energy between the two semi-infinite regions. Actually it is convenient to calculate the energy φ_{AB} per unit area:

$$\varphi_{AB} = \lim_{R \to \infty} \frac{1}{\pi R^2} \int_0^R \int_0^\infty 2\pi r \varphi_B \, dr \, dz_1 \; . \tag{3.4}$$

Carrying out the integration we find that

$$\varphi_{AB} = -\frac{\pi n^2 \beta}{12 h^2} = -\frac{A_H}{12 \pi h^2} \; . \tag{3.5}$$

Here the Hamaker constant, A_H, is given by $\beta \pi^2 n^2$. It is a key parameter characterizing the strength of the attractive forces.

It should be noted that the attractive interaction between two large bodies decreases much more slowly as they are separated than does the interaction between two molecules. According to Equations 3.1 and 3.5, the interaction is proportional to the inverse square of the separation distance in the former case and the inverse sixth power in the latter case. For colloidal particles, London–van der Waals interactions are significant for separation distances h up to about 100 nm (1000 Å).

Hamaker (1937) used a similar procedure to derive an expression for the interaction energy between two spherical particles. He found that the attractive energy φ_{AB}^S for two identical spheres of radius R with centers a distance $2R + h$ apart is given by

$$\varphi_{AB}^S = -\frac{A_H}{12} \left[\frac{1}{\xi^2 + \xi} + \frac{1}{\xi^2 + 2\xi + 1} + 2 \ln \frac{\xi^2 + 2\xi}{\xi^2 + 2\xi + 1} \right], \tag{3.6}$$

where $\xi = (h/R)$. When the spheres are close together so that $\xi \ll 1$, Equation 3.6 simplifies to

$$\varphi_{AB}^S = -\frac{A_H}{12 \xi} \; . \tag{3.7}$$

In this last case, note that the decrease in interaction energy with separation distance h is even slower than that given by Equation 3.5 for semi-infinite bodies. In the opposite extreme, where $(h/R) \gg 1$, Equation 3.6 predicts that φ_{AB}^{S} is proportional to h^{-6}, as would be expected from the fundamental attraction law (Equation 3.1).

Another useful result is that for two plates of finite thickness d separated by a distance h. The derivation leading to Equation 3.5 is readily modified for this case and yields the following expression for the interaction energy φ_{AB}^{P} :

$$\varphi_{AB}^{P} = -\frac{A_H}{12\pi}\left[\frac{1}{h^2} + \frac{1}{(h+2d)^2} - \frac{2}{(h+d)^2}\right] \tag{3.8}$$

Clearly Equation 3.8 simplifies to Equation 3.5 in the limit of very thick plates, where $d \to \infty$. Finally, Equations 3.6 and 3.7 can be generalized to the case of spheres of unequal radii R_1 and R_2. The results are

$$\varphi_{AB}^{S} = -\frac{A_H}{12}\left[\frac{b}{\zeta^2 +(b+1)\zeta} + \frac{b}{\zeta^2 +(b+1)\zeta+b} +2\,\ell n\frac{\zeta^2 +(b+1)\zeta}{\zeta^2 +(b+1)\zeta+b}\right], \tag{3.9}$$

where $\zeta = (h/R_1)$ and $b = (R_2/R_1)$, and

$$\varphi_{AB}^{S} = -\frac{A_H R_1 R_2}{6h(R_1 + R_2)} \quad \textit{for } h \ll R_1,\, R_2 \,. \tag{3.10}$$

Numerical evaluation of attractive forces using the above formulas requires that a value be assigned to the Hamaker constant A_H. Hiemenz and Rajagopalan (1997) use an order of magnitude argument to estimate β and conclude that A_H should be in the range of 10^{-13} to 10^{-12} erg (10^{-20} to 10^{-19} J) for most materials.

A complicating factor ignored thusfar is the effect of the medium separating the two particles. We return to the situation of Figure 3.1. Let us find the excess energy of a unit volume in region A at a position $(-z_1)$ compared to a unit volume in an infinite bulk phase of A. We imagine starting with such an infinite phase. Now we replace molecules of A with those of B in the region $z > h$ and molecules of A with those of C in the region $0 < z < h$. Using the derivation leading to Equations 3.2 and 3.3, we find that the excess energy of our unit volume caused by these changes is

$$\varphi_A' = -\frac{\pi n_A (n_C \beta_{AC} - n_A \beta_{AA})}{2} \int_0^h \frac{dz}{(z+z_1)^4} - \frac{\pi n_A (n_B \beta_{AB} - n_A \beta_{AA})}{2} \int_h^\infty \frac{dz}{(z+z_1)^4}$$

$$= -\frac{\pi n_A (n_C \beta_{AC} - n_A \beta_{AA})}{6 z_1^3} - \frac{\pi n_A (n_B \beta_{AB} - n_A \beta_{AA})}{6(h+z_1)^3}$$

$$(3.11)$$

Similar expressions can be derived for a unit volume of C at position z_2 and a unit volume of B at position z_3 of Figure 3.1. The results are

$$\varphi_C' = -\frac{\pi n_C (n_A \beta_{AC} - n_C \beta_{CC})}{6 z_2^3} - \frac{\pi}{6} \frac{n_C (n_B \beta_{BC} - n_C \beta_{CC})}{(h-z_2)^3} \qquad (3.12)$$

and

$$\varphi_B' = -\frac{\pi}{6} \frac{n_B (n_A \beta_{AB} - n_C \beta_{BC})}{z_3^3} - \frac{\pi}{6} \frac{n_B (n_C \beta_{BC} - n_B \beta_{BB})}{(z_3 - h)^3}. \qquad (3.13)$$

We now wish to integrate Equations 3.11 through 3.13 for z_1, z_2, and z_3 having ranges corresponding to regions A, C, and B, respectively, in Figure 3.1. We also want to calculate φ_{ACB}, the excess energy of the actual situation compared to that where the thickness h of region C becomes very large. Now φ_{ABC} is the sum of the three expressions obtained by integrating Equations 3.11 through 3.13 less the corresponding sum of these integrals in the limit $h \to \infty$, with the final result halved to correct for double counting of the interaction energy between each pair of molecules in separate regions. Using this procedure we find that

$$\varphi_{ACB} = -\frac{1}{12\pi h^2} \left[\pi^2 n_A n_B \beta_{AB} + \pi^2 n_C^2 \beta_{CC} - \pi^2 n_A n_C \beta_{AC} - \pi^2 n_B n_C \beta_{BC} \right]. \qquad (3.14)$$

The expression in brackets in this equation is the effective Hamaker constant for two particles of A and B separated by a medium C.

If both particles are of the same material A, Equation 3.14 simplifies to

$$\varphi_{ACA} = -\frac{1}{12\pi h^2}[A_{HAA} + A_{HCC} - 2A_{HAC}] = -\frac{(A_H)_{eff}}{12\pi h^2}. \qquad (3.15)$$

For London–van der Waals forces, a frequently used approximation is

$$\beta_{AC} \cong (\beta_{AA}\beta_{CC})^{1/2}. \tag{3.16}$$

With this approximation, Equation 3.15 becomes

$$\varphi_{ACA} \cong -\frac{1}{12\pi h^2}[A_{HAA}^{1/2} - A_{HCC}^{1/2}]^2 . \tag{3.17}$$

In this case, the effective Hamaker constant is always positive (i.e., the interaction between like particles is an attractive one). But the strength of the attraction is usually less than if the dispersing medium C were absent, and effective Hamaker constants not much greater than 10^{-14} erg (10^{-21} J) have been reported. We note that effective Hamaker constants of the forms indicated in Equations 3.14 and 3.15 may be used in any of the expressions of Equations 3.5 through 3.10 to account for the continuous phase C of a colloidal dispersion.

Since about 1970, considerable attention has been focused on a different and more rigorous method for calculating attractive forces. Sometimes called Lifshitz theory (Dzyaloshinskii et al., 1960), this method assumes that particle separation is sufficiently large that each appears as a continuous medium to the other. Thus concepts from the continuum theory of electrodynamics can be employed. Unlike the Hamaker theory, it does not assume that the interaction energies are large compared to the electronic energies. The results show that the interaction energy does drop as h^{-2} at small separations, but as h^{-3} at large separations. A readable review of the subject is provided by Parsegian (1975) and a more detailed discussion is provided by Mahanty and Ninham (1976). The theory requires energy absorption spectral data, and the latter authors show how it can be approximated from very accessible measurements. Detailed information on a hydrocarbon and water are given. Thus the greatest problem with using the simpler Hamaker procedure described above, that is, the uncertainty in what value to use for the Hamaker constant, is avoided. It can be shown that under many conditions the effective Hamaker constant A_{Heff} of Equation 3.14 can be replaced by (Israelachvili, 1991, p. 184)

$$A_{Heff} = \frac{3}{4}kT\left(\frac{\varepsilon_A - \varepsilon_C}{\varepsilon_A + \varepsilon_C}\right)\left(\frac{\varepsilon_B - \varepsilon_C}{\varepsilon_B + \varepsilon_C}\right)$$
$$+ \frac{3h\nu_e}{8\sqrt{2}} \frac{(n_A^2 - n_C^2)(n_B^2 - n_C^2)}{(n_A^2 + n_C^2)(n_B^2 + n_C^2)\{(n_A^2 + n_C^2)^{1/2} + (n_B^2 + n_C^2)^{1/2}\}} ,$$

where ε_i and n_i are the dielectric constant and the refractive index of the medium i, h is Planck's constant, and ν_e is the mean electronic absorption frequency in the ultraviolet spectrum (assumed to be approximately the same for the media

A, B, and C). Clearly A_{Heff} is positive when media A and B are identical, the same result as found from Equation 3.17.

Parsegian (1975) presents calculations for the polystyrene-water and polystyrene-vacuum systems for which the required spectral data are available. For simplicity he considered the situation of Figure 3.1 with polystyrene in the two semi-infinite regions A and B. Results of his calculations are shown in Figures 3.2 through 3.4. As Figure 3.2 shows, the attractive energy is about an order of magnitude less when the space between the polystyrene particles is occupied by water instead of a vacuum, at least for separation distances less than 1000 Å (100nm), which is the range of greatest interest.

Figure 3.3 shows that attraction is even less in salt water than in pure water. Since the Hamaker constant depends on separation distance, the form of Equation

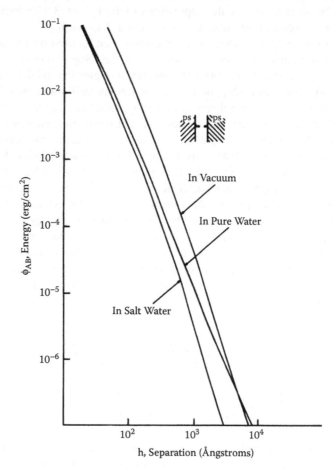

FIGURE 3.2 Interaction potential of two semi-infinite polystyrene media separated by a vacuum, pure water, and 0.1 M salt solution. Reprinted from Parsegian (1975) with permission.

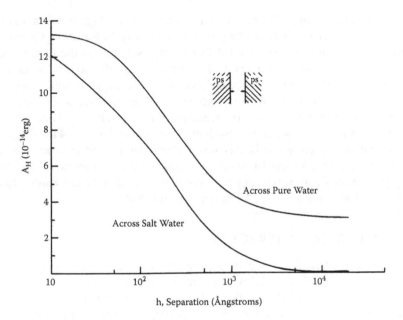

FIGURE 3.3 Hamaker constant calculated from Equation 3.5 and the results shown in Figure 3.2. Reprinted from Parsegian (1975) with permission.

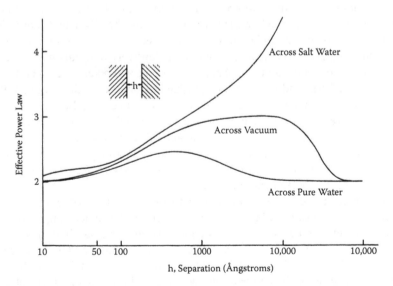

FIGURE 3.4 Effective power (n) law calculated from a modified version of Equation 3.5 (i.e., $\phi_{AB} = -A_H / 12\pi h^n$) and using the values from Figure 3.2. The value for n as a function of separation distance h is chosen to keep A_H constant. Reprinted from Parsegian (1975) with permission.

3.5 obtained by Hamaker's method is not quite correct. Figure 3.4 makes this point more clearly by showing the power of h required in Equation 3.5 to force it to agree with the results of the Lifshitz theory. Note that in the vacuum case, the exponent increases from about 2 to nearly 3 as the separation distance increases from 50 to 1000 Å. Indeed, Equation 3.5 is often used, with h^2 replaced by h^3 for large separation distances, since the latter dependence is obtained when the integration is performed with the retardation effect mentioned above included. When pure water is the separating medium, we see from Figure 3.4 that Equation 3.5 should not be corrected in this way for retardation, and the curve for salt water indicates that, except for small separation distances, Equation 3.5 is not a very good approximation with any power of h. We note that the salt water solution in these calculations was 0.1 M sodium chloride (NaCl).

3. ELECTRICAL INTERACTION

An electrical interaction force exists between colloid particles that plays an important role in colloid stability. It is found that solid surfaces electrify in the presence of an electrolyte solution. This behavior is mainly due to unequal adsorption of the electrolyte ions on the surface, giving it a net positive charge or a net negative charge. The charge varies with the electrolyte content of the solution. For instance, silver iodide particles are positively charged in the presence of a large excess of silver ions and negatively charged in the presence of a large excess of iodide ions. In more precise terms, it is the electrical potential at the particle surfaces that is determined in this case by the silver and iodide ions, which are called "potential-determining ions." Because of its use in film photography, silver iodide (AgI) has been studied extensively and thus has played an important role in understanding the nature of charge and potential at the interface. Since adsorbed Ag^+ and I^- are indistinguishable from the original Ag or I, there is no need to distinguish the potentials below the layer of adsorbed ions from above the layer of adsorbed ions, which is a complicating feature when the adsorbed ions are foreign in nature. For metal oxide particles, the surface potential varies with solution pH, and H^+ is the potential determining ion. For oil-in-water emulsions, the sign of the surface charge is that of adsorbed surface-active ions (e.g., negative for ionized carboxylic acids).

A charged surface attracts counterions from the solution while repelling ions of like sign. The result is a layer near the surface that has a net charge. Although this layer is diffuse (i.e., starts at the wall and extends outward into the fluid), its main portion is confined to the region close to the surface. The local charge separation gives rise to a capacitance, and the system can be modeled as a capacitor made up of the charged interface and an infinitely thin layer of counterions (Verwey and Overbeek, 1948), hence the name of this region—the "electrical double layer." Since the charge separation cannot be measured, indirect means must be used to characterize the phenomenon, as discussed below.

We begin with a simple situation when ion concentrations and the electric potential vary only in the z direction (e.g., near a single plate or between two parallel plates having uniform surface potentials). The key variable is the electrostatic potential ψ, which is defined as the work done on a unit charge in bringing it from infinity to its present position against a force. Hence in one dimension (z direction), this force is $-d\psi/dz$. The force on a unit charge is defined as the charge times the electrostatic field, making the electrostatic field also equal to $-d\psi/dz$ here. Application of a fundamental law from electrostatics (Jackson, 1975) on the small element in Figure 3.6 takes the form

$$A\left(-\varepsilon\varepsilon_o \frac{d\psi}{dz}\right)\Big|_z - A\left(-\varepsilon\varepsilon_o \frac{d\psi}{dz}\right)\Big|_{z+\Delta z} + \rho_e(A\Delta z) = 0 , \qquad (3.18)$$

where the fields act in the z direction on the two faces of area A, which are separated by a distance Δz. ρ_e is the charge density, ε_o is the permittivity of a perfect vacuum, and ε is the dielectric constant (i.e., the ratio of the permittivity of the material to ε_o). Dividing by $A\Delta z$ and taking the limit as Δz goes to zero, one has the Poisson equation for constant ε:

$$\frac{d^2\psi}{dz^2} = -\frac{1}{\varepsilon\varepsilon_o}\rho_e . \qquad (3.19)$$

The above balance can also be made on a surface. In Figure 1.2 it leads to

$$A\left[-\varepsilon_A\varepsilon_o \frac{d\psi_A}{dz}\right]_{\lambda_A} - A\left[-\varepsilon_B\varepsilon_o \frac{d\psi_B}{dz}\right]_{\lambda_B} + \rho_e A (\lambda_B - \lambda_A) = 0 . \quad (3.20)$$

In the limit where $(\lambda_B - \lambda_A)$ goes to zero, if we make $\rho_e(\lambda_B - \lambda_A)$ go to a nonzero limit of σ, the surface charge density, the Gauss equation results:

$$-\varepsilon_A\varepsilon_o \frac{d\psi_A}{dz} + \varepsilon_B\varepsilon_o \frac{d\psi_B}{dz} = -\sigma . \qquad (3.21)$$

It is possible to show that Equation 3.21 is an expression of global electroneutrality (see Problem 3.2); that is, the net charge imbalance due to the diffuse double layer is balanced at the surface.

At equilibrium, the net force on each small element of fluid such as that shown in Figure 3.6 must be zero. That is, pressure forces acting on the two surfaces of area A perpendicular the z axis must balance the electrical body force, which is the product of the net electrical charge of the element and the local electrical field. This balance implies

$$pA\big|_z - pA\big|_{z+\Delta z} + (\rho_e A \Delta z)\left(-\frac{d\psi}{dz}\right) = 0 \ . \tag{3.22}$$

Dividing by $A\Delta z$ and taking the limit as $z \to 0$, we find

$$\frac{dp}{dz} + \rho_e \frac{d\psi}{dz} = 0 \ . \tag{3.23}$$

Hence, pressure varies with position in the electrical double-layer region, where an electrical body force exists, just as it varies with position in a static pool of liquid, where gravitational body force exists. In the latter case, pressure increases with depth in accordance with the familiar rules of hydrostatics. Substitution of Equation 3.19 into Equation 3.23 yields

$$\frac{dp}{dz} = \varepsilon\varepsilon_o \frac{d\psi}{dz}\frac{d^2\psi}{dz^2} = \frac{\varepsilon\varepsilon_o}{2}\frac{d}{dz}\left[\left(\frac{d\psi}{dz}\right)^2\right]. \tag{3.24}$$

In order to evaluate ψ, we find from Equation 3.19 that knowledge of ρ_e is necessary. It can be obtained from the Boltzmann distribution for a symmetric electrolyte with bulk concentration c_o:

$$n_+ = N_A c_o \exp(e_o v\psi/kT), \ n_- = N_A c_o \exp(e_o v\psi/kT). \tag{3.25}$$

Here n_+ and n_- are the number of cations and anions per unit volume, respectively, and ψ has been taken as zero in the bulk solution. Problem 3.9 indicates how these expressions can be derived. Since $\rho_e = ve_o n_+ - ve_o n_-$, one has

$$\rho_e = 2vN_A c_o e_o \sinh (e_o v\psi/kT), \tag{3.26}$$

and the resulting Poisson-Boltzmann equation obtained on substituting Equation 3.26 into Equation 3.19 is

$$\frac{d^2 u}{dz^2} = \left(\frac{2v^2 e_o^2 N_A c_o}{\varepsilon\varepsilon_o kT}\right) \sinh u = \kappa^2 \sinh u \tag{3.27}$$

$$u = \frac{ve_o\psi}{kT} \ , \tag{3.28}$$

where v is the magnitude of the ion valence, e_o is the electronic charge, κ^{-1} is the Debye length, a measure of double-layer thickness, N_A is Avogadro's number, k

is Boltzmann's constant, and T is the absolute temperature. Now, if we multiply Equation 3.27 by du/dz and rearrange, we find

$$\frac{d}{dz}\left[\left(\frac{du}{dz}\right)^2\right] = 2\,\kappa^2 \sinh\,u\,\frac{du}{dz}\,, \qquad (3.29)$$

Integration of this equation to obtain the potential distribution in the electrical double-layer region near a single plate is considered in Problem 3.2. Some useful formulas applicable to double layers near single-plane and spherical surfaces are given in Table 3.1.

Of particular interest here is the electrical interaction between two parallel plates separated by a distance h (Figure 3.5). When h is small enough, the double layers of the two plates overlap and interaction occurs. Combining Equations 3.24 and 3.29 and making use of Equation 3.28, we obtain a relationship between the pressure distribution and the potential distribution in the region between interacting plates:

$$\frac{dp}{dz} = \varepsilon\varepsilon_o\kappa^2\left(\frac{kT}{\nu e_o}\right)^2 \sinh\,u\frac{du}{dz}\,. \qquad (3.30)$$

Integrating between the centerline position $z = 0$ in Figure 3.5, where $u = u_c$ and $p = p_c$, and the bulk phase outside the double layer, where $u = 0$ and $p = p_b$, we obtain the following result:

$$p_c - p_b = \varepsilon\varepsilon_o\kappa^2\left(\frac{kT}{\nu e_o}\right)^2 (\cosh\,u_c - 1)\,. \qquad (3.31)$$

If the first limit of integration is the particle surfaces where $u = u_o$ and $p = p_o$ instead of the centerline position, the result is

$$p_o - p_b = \varepsilon\varepsilon_o\kappa^2\left(\frac{kT}{\nu e_o}\right)^2 (\cosh\,u_o - 1)\,. \qquad (3.32)$$

Thus fluid pressure at both the centerline and the particle surfaces exceeds the bulk phase pressure, an indication that the interaction due to overlap of double layers having the same charge is a repulsive one, the expected result.

In the next section we will look at the combined effect of attractive and electrical interaction. For this purpose it is desirable to know the potential energy E_e of electrical interaction. Let F_e be the force per unit area in excess of bulk

TABLE 3.1

Formulas for electrical double layers near single particles

Plane interface

1. Gouy-Chapman solution for a symmetric electrolyte of valence υ:

$$u = 2 \; \ell n \left(\frac{1 + \eta e^{-\kappa z}}{1 - \eta e^{-\kappa z}} \right)$$

$$\sigma = \frac{2 \varepsilon \varepsilon_o \kappa kT}{\upsilon e_o} \sinh \left(\frac{\upsilon e_o \psi_o}{2 \, kT} \right)$$

$$\eta = \tanh \left(\frac{\upsilon e_o \psi_o}{4 \, kT} \right)$$

2. Small potential approximation for general electrolyte mixture with κ given by Equation 3.37:

$$\psi = \psi_o \, e^{-\kappa z}$$

$$\sigma = \varepsilon \varepsilon_o \kappa \psi_o$$

Spherical interface (double layer outside spherical particle)

1. Small potential (Debye-Huckel) approximation for general electrolyte as above:

$$\psi = \frac{\psi_o R}{r} e^{-\kappa(r-R)}$$

$$\sigma = \varepsilon \varepsilon_o \psi_o \, (1 + \kappa R)$$

2. Result for $\kappa R > 1$ including first correction to Gouy-Chapman solution to account for curvature (Evans and Ninham, 1983; Loeb et al., 1961). Note that no exact solution is available for the general Poisson-Boltzmann equation ($\nabla^2 u = \kappa^2 \sinh u$) in spherical coordinates:

$$\sigma = \frac{2 \varepsilon \varepsilon_o \kappa kT}{\upsilon e_o} \left[\sinh \left(\frac{\upsilon e_o \psi_o}{2 \kappa T} \right) + \frac{2\eta}{\kappa R} \right]$$

fluid pressure p_b that the fluid exerts on the particle surfaces of Figure 3.5, with a positive F_e denoting a repulsive force. It is known from electrostatics that

$$F_e = (p_o - p_b) - \frac{\varepsilon \varepsilon_o}{2} \left(\frac{d\psi}{dz} \right)^2 \Big|_{z=\frac{h}{2}} . \tag{3.33}$$

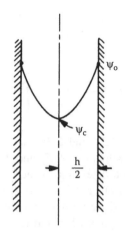

FIGURE 3.5 Electrical potential distribution in an interacting double layer between two identical slabs.

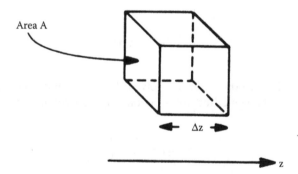

FIGURE 3.6 A small rectangular element in a fluid is shown. The pressure forces in the z-direction act on the surface of area A. The electrical body forces are proportional to the volume ($A\Delta z$).

The last term in this equation is an electrical stress (negative pressure) that can be viewed as resulting from the attraction between dipoles induced in the solvent by the electric field and lying adjacent to one another in the direction of the field, here the z direction. The electrical stress term can be evaluated by integrating Equation 3.24 between the centerline $z = 0$ and the particle surface $z = (h/2)$, noting that $d\psi/dz$ vanishes at the centerline. When the result is substituted into Equation 3.33, we find

$$F_e = p_c - p_b.$$ (3.34)

The potential energy E_e per unit area is related to F_e by

$$\frac{dE_e}{dh} = -F_e.$$ (3.35)

If the energy is taken as zero when the separation distance between the particles is large, integration of Equation 3.35 yields

$$E_e = -\int_\infty^h F_e \, dh = \int_h^\infty F_e \, dh \ . \tag{3.36}$$

We see from Equations 3.31 and 3.35 that knowledge of the centerline potential u_c is needed for evaluation of F_e and E_e. The potential distribution between parallel plates can be determined exactly in terms of elliptic integrals (Overbeek, 1952; Verwey and Overbeek, 1948). It is more convenient, however, to use simpler expressions, and we present two approximate solutions.

The first approximation applies when surface potential ψ is small in magnitude (i.e., $u \ll 1$). In this case, the Poisson-Boltzmann equation (Equation 3.27) may be linearized by expanding the hyperbolic function and retaining only the first term of the series:

$$\frac{d^2\psi}{dz^2} = \kappa^2\psi \ . \tag{3.37}$$

This differential equation is readily solved. Since the potential distribution is symmetric about the centerline, the solution may be written as

$$\psi = \psi_o \frac{\cosh \kappa z}{\cosh (\kappa h / 2)} \ . \tag{3.38}$$

In this case, Equation 3.31 simplifies to

$$p_c - p_b \cong \frac{\varepsilon\varepsilon_o\kappa^2\psi_c^2}{2} \ . \tag{3.39}$$

From Equations 3.34, 3.38, and 3.39 we have

$$F_e = p_c - p_b = \frac{\varepsilon\varepsilon_o\kappa^2\psi_o^2}{2 \cosh^2(\kappa h / 2)} \ . \tag{3.40}$$

If ψ_o is independent of separation distance h, we invoke Equation 3.36 to find

$$E_e = \varepsilon\varepsilon_o\kappa\psi_o^2 \left[1 - \tanh (\kappa h / 2)\right] \ . \tag{3.41}$$

It is clear from this equation that electrical interaction is significant only when the separation distance h does not greatly exceed the Debye length, κ^{-1}. The latter is about 10 nm (100 Å) in a 0.001 M aqueous solution of a univalent electrolyte at room temperature and about 1 nm (10 Å) in a 0.1 M solution. Thus double-layer thickness decreases with increasing electrolyte concentration.

The surface charge density σ can be obtained from Equation 3.21 for a solid with $\varepsilon_B = 0$ and $\varepsilon_A = \varepsilon$, and one has

$$\sigma = \varepsilon \varepsilon_o \kappa \psi_o \tanh (\kappa h/2). \tag{3.42}$$

Thus we may rewrite Equation 3.40 as

$$F_e = p_c - p_b = \frac{\sigma^2}{2 \varepsilon \varepsilon_o \sinh^2 (\kappa h / 2)}. \tag{3.43}$$

If σ is independent of h, integration in accordance with Equation 3.36 yields

$$E_e = \frac{\sigma^2}{\varepsilon \varepsilon_o \kappa} [\coth (\kappa h / 2) - 1]. \tag{3.44}$$

The constant potential and constant charge density results given by Equations 3.41 and 3.44 are limiting cases of behavior that may occur as colloidal particles approach one another. In the constant potential case, the approach is slow enough that equilibrium of the potential determining ion is maintained between the surface and bulk solution. Adsorption or desorption occurs as necessary to maintain the equilibrium potential ψ_o. The opposite extreme is the constant charge density case where the particles approach so rapidly that no adsorption or desorption has time to occur. Clearly, intermediate situations are possible as well when the time constant for adsorption or desorption and double-layer relaxation are comparable to the approach time of the particles.

It is readily shown that Equations 3.38 through 3.44 apply to two identical particles of small surface potential even when the bulk solution contains anions and cations of different valence. In this case, the inverse Debye length κ is given by

$$\kappa^2 = \frac{e_o^2 N_A}{\varepsilon \varepsilon_o kT} \sum_i v_i^2 c_{io}, \tag{3.45}$$

where the sum is over all ionic species in the bulk solution whose valences are v_i and bulk concentrations c_{io}.

The other widely used approximation for calculation of u_c in Equation 3.31 is not restricted to small surface potentials, but it is restricted to small degrees of overlap of the double layers between the two particles. In this case, the centerline potential is assumed to be small ($u_c \ll 1$) and approximately equal to the sum of the potentials due to the two double layers individually, neglecting interaction. The solution of the Poisson-Boltzmann equation for a single double layer can be used to develop the following expression applicable for small u_c:

$$u_c = 8\eta \, e^{-\kappa h/2}, \tag{3.46}$$

where $\eta = \tanh(\upsilon e_o \psi_o / 4kT)$. Combining this equation with Equations 3.31 and 3.34, we find

$$F_e \cong 32 \, \varepsilon\varepsilon_o \kappa^2 \left(\frac{kT}{\upsilon e_o}\right)^2 \eta^2 \, e^{-\kappa h}. \tag{3.47}$$

Using Equation 3.36, we can calculate the interaction energy E_e when the surface potential ψ_o is independent of separation distance:

$$E_e = 32 \, \varepsilon\,\varepsilon_o \kappa \left(\frac{kT}{\upsilon e_o}\right)^2 \eta^2 \, e^{-\kappa h} = \frac{64 N_A c_o kT}{\kappa} \eta^2 \, e^{-\kappa h}. \tag{3.48}$$

No exact general solution exists for the interaction between electrical double layers of spherical particles. Derjaguin has developed an approximation useful when particle radius R is much greater than the Debye length κ^{-1} (i.e., $\kappa R \gg 1$). In this case, the radius of curvature of each surface is much greater than the separation distance between surfaces when the particles are close enough to interact appreciably, i.e., h/R << 1. Derjaguin's idea was to model each spherical surface as a central flat disk and a series of surrounding flat rings, as indicated in Figure 3.7. The equations developed above for flat surfaces could then be used to calculate forces between the central disks of the two surfaces and between the various pairs of corresponding rings. When the small overlap approximation given by Equation 3.47 is used with Derjaguin's basic scheme to calculate these forces, the total repulsive force F_e^S between the spheres is found to be

$$F_e^S = \frac{64\pi R c_o N_A kT}{\kappa} \eta^2 \, e^{-\kappa h}. \tag{3.49}$$

If the surface potential of the two spheres remains constant during their approach, the corresponding energy of interaction is

$$E_e^S = \frac{64\pi R c_o N_A kT}{\kappa^2} \eta^2 \, e^{-\kappa h} = 32\pi R \varepsilon \, \varepsilon_o \left(\frac{kT}{\upsilon e_o}\right)^2 \eta^2 \, e^{-\kappa h}. \tag{3.50}$$

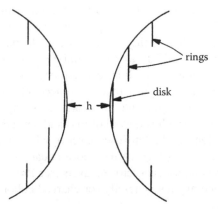

FIGURE 3.7 Approximation used to calculate the electrical energy of interaction between surrounding rings for two spheres by a central disk and surrounding rings for $h/R \ll 1$.

We note that F_e^S and E_e^S have units of force and energy since they represent the entire interaction between spheres. In contrast, F_e and E_e for the parallel plate case are the force and energy of interaction per unit area.

Derjaguin's approach and the small overlap approximation have also been used for the case of two spheres having different radii and surface potentials. The resulting interaction energy is found to be

$$E_e^S = 64\pi\varepsilon \left(\frac{kT}{ve_o} \right)^2 \left(\frac{R_1 R_2}{R_1 + R_2} \right) \eta_1 \eta_2 \, e^{-\kappa h} . \tag{3.51}$$

As before, the surface potentials have been taken as independent of separation distance h. Note that E_e^S is positive and the interaction is repulsive when η_1 and η_2 have the same sign. In this case, both surfaces have charges of the same sign. But as might be expected, the interaction energy is negative, indicating an attractive interaction when η_1 and η_2 have opposite signs, so that one surface has a positive charge and the other a negative charge.

Finally, for completeness, we give the corresponding electrical interaction energy E_e per unit area between flat plates with different surface potentials. The small overlap approximation is employed to obtain

$$E_e^P = 64 \frac{c_o N_A kT}{K} \eta_1 \eta_2 \, e^{-\kappa h} . \tag{3.52}$$

We emphasize that Equations 3.46 through 3.52 apply only for the case of a symmetric electrolyte. But they are useful for practical purposes when the valence v is that of the counterion (i.e., the ion whose charge is opposite that of the surfaces when both have charges of the same sign).

So far it has been implicitly assumed that the Poisson-Boltzmann equation (Equation 3.27) applies throughout the liquid phase near the solid-liquid interface and that ψ_o is the potential at the interface itself. A major limitation of this approach is that Equation 3.27 neglects the finite size of ions and thus ignores the fact that centers of ions in the fluid can approach no closer to the interface than the ionic radius. A first approximation to account for this effect involves dividing the double layer region into two parts: a "compact" or "Stern" layer adjacent to the interface which is devoid of ions and has a thickness on the order of molecular dimensions, and a "diffuse" layer in which Equation 3.27 applies. With this model, the potential ψ_o of the above equations is at the (imaginary) surface separating the compact and diffuse layers (see Problem 3.3). The problem of locating the surface applies generally for charged colloids including the AgI sols discussed earlier.

4.   COLLOIDS OF ALL SHAPES AND SIZES

Everett (1988) briefly described the procedures for making inorganic colloids, including gold sols and silver halide sols. Essentially the sols are precipitated through reactions induced by boiling or by added reagents. For laboratory purposes α-alumina and γ-alumina particles, glass spheres, etc., are commercially available. The methods for making nanoparticles vary and have been put into a generalized procedure by Wang et al. (2005). The thiol-capped gold nanoparticles have formed the workhorse of research in nanoparticles. Akyl thiols are alkyl chains with a sulfur end that reacts with a gold surface atom. Once they cover the entire surface, they become hydrophobic, and according to some (Collier et al., 1998), form true solutions in toluene or hexane. The akyl thiols are discussed again in Chapter 4 (Section 6) as self-assembled monolayers (SAMs). Monodispersed colloid particles of various materials have been produced (Matijevic, 1976).

Latex particles are more difficult to make and are made by emulsion polymerization for smaller particles and suspension polymerization for the larger ones. These procedures need surfactants (typically ionic sodium dodecyl sulfate), which eventually give the resulting particles a charge. Sulfuric acid groups are seen as well. Synthesis of organic nanoparticles is discussed in Chapter 4 (Section 13).

When the colloidal suspension is titrated with sodium hydroxide (NaOH), its conductivity decreases and reaches a minimum when the surface acid groups have all been neutralized. From material balances it is possible to calculate the total number of acid head groups initially present, and eventually the area per head group, which is seen to vary from 3 to 4 nm^2/group. When there are two acid groups present (sulfuric and carboxylate), the conductivity plots show a feature similar to having two minima. These steps are often required to prepare well-characterized colloids to perform controlled experiments.

Sometimes it is necessary to strip the colloidal suspension of not only the dissolved ions but also the adsorbed ones. The ions in the dispersion medium/water are removed by washing followed by centrifugation and decantation, and more slowly, but down to very low concentrations, by dialysis. Mixed ion exchange removes practically all the adsorbed ions, which makes the suspension unstable, and some surfactant, typically nonionic, has to be added to give it stability.

Latex particles are generally made of polystyrene, but the composition can vary widely, including one that has the same density as water. Just as acid groups give the particles negative charge, amines can be included to give them positive charge. The sizes of latex particles range from below 0.1 to 5 μm. Those that are commonly available for laboratory use are extremely monodispersed, with standard deviations of about 3%, which corresponds to approximately 0.005 μm for particles of order 0.1 μm. There are many features of both fundamental and practical interest discussed in the book by Lovell and El-Aasser (1997); the book includes very significant contributions by Vanderhoff, El-Aasser, and others in this area.

The sizes and shapes are determined using transmission electron microscopy (TEM) over a diffraction grating replica. Dynamic light scattering can also be used to obtain the dimensions (see Chapter 8).

The approximate experimental determination of ψ_o is based on measurement of the velocity of a charged particle in a solvent subjected to an applied voltage. Such a particle experiences an electrical force that initiates motion. Since a hydrodynamic frictional force acts on the particle as it moves, a steady state is reached, with the particle moving with a constant velocity U. To calculate this electrophoretic velocity U theoretically, it is, in general, necessary to solve Poisson's equation (Equation 3.19) and the governing equations for ion transport subject to the condition that the electric field is constant far away from the particle. The appropriate viscous drag on the particle can be calculated from the velocity field and the electrical force on the particle from the electrical potential distribution. The fact that the sum of the two is zero provides the electrophoretic velocity U. Actual solutions are complex, and the electrical properties of the particle (e.g., polarizability, conductivity, surface conductivity, etc.) come into play. Details are given by Levich (1962) (see also Problem 7.8).

The electrophoretic velocity is given by

$$U = \frac{4f \zeta \varepsilon \varepsilon_o E}{\mu}, \qquad (3.53)$$

where f is a function of κR and R is the particle radius. For small values of κR, $f \to 1/6$, which is the Huckel limit, and for large κR, $f \to 1/4$, which is von Smoluchowski's result (see Problem 7.8). For conducting spheres, f decreases from 1/6 to 0 with increasing κR. Values of f for various κR have been calculated (O'Brien and White, 1978; Overbeek, 1952).

An important feature of the detailed analysis is that the potential ζ in the above equation is the value that exists at the surface where the no-slip boundary condition applies. Since the first layer of solvent molecules and any adsorbed counterions present are normally rather strongly bound to the surface, it is plausible to assume that they move with the particle. Hence the surface separating the particle and these bound molecules from the remaining liquid is likely close to that separating the compact and diffuse portions of the electrical double layer. If these surfaces coincide, $\zeta = \psi_o$, and ψ_o can be obtained from the above equation for an experiment where f is known and where U, E, and μ are measured. If they do not coincide, the potential ζ determined in this way is not equal to ψ_o.

It is clear from this discussion that the assumption involved in equating the measured potential ζ with the potential ψ_o at the inner boundary of the diffuse layer is open to question. Nevertheless, it is widely made because it is the commonly available method of estimating ψ_o from a relatively simple experiment. The electrophoretic mobility of AgI is shown as a function of Ag^+ concentration in water in Figure 3.8.

Use of the electroacoustic method (O'Brien et al., 1995) to determine both potential and particle size is both new and holds considerable promise.

Some colloids where the details of the charged groups are known, as in proteins, not only allow an accurate estimation of charge but also provide an inventory of which groups are charged by what amount from the absorption of light in the ultraviolet range. This can be done as a function of pH. The thermodynamics of dissociation in the presence of electrostatic double layers allows one

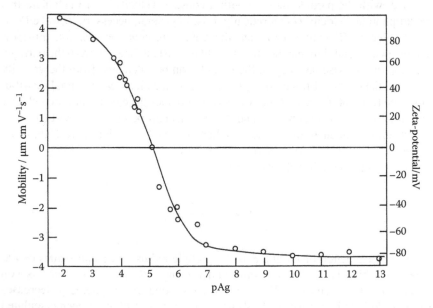

FIGURE 3.8 Electrophoretic mobility, U/E, of AgI as a function of $pAg^+ = -\log_{10}[Ag^+]$. Note that the concentration of Ag^+ can be use to calculate the concentration of I, and vice versa, in the solution using the solubility product. The potentials are calculated values. Reproduced with permission from Ottewill et al. (1978).

to calculate the pK values of individual groups (Tanford, 1961). It is expected that some of these concepts can be used in colloids as well.

Formation of emulsions has been reviewed by Walstra (1983). In contrast to solid colloids, there appears to be much less control of the sizes of emulsion drops. Mechanical devices called homogenizers can reduce fat globules in milk to sizes where they take much longer to flocculate and cream to the top. Homogenizers capable of making nearly uniform size drops are beginning to appear.

The size distributions of small drops and particles generally follow the lognormal distribution

$$p(\ln a) = \frac{1}{(\ln \sigma)\sqrt{2\pi}} e^{-\frac{(\ln a - \ln \mu)^2}{2(\ln \sigma)^2}} , \tag{3.54}$$

where a is the droplet diameter, μ is the mean, σ^2 is the variance, and the phase space is given by d ln a. The mean of the distribution,

$$\ln \mu = \int_{-\infty}^{+\infty} \ln a .p \, d\ln a , \tag{3.55}$$

is actually its geometric mean,

$$\mu = \left(\prod_{i=1}^{N} a_i \right)^{1/N} , \tag{3.56}$$

where N is the total number of particles. Lognormal distribution compresses the range of radii and the data on emulsions are generally well represented with this distribution (Hazlett and Schechter, 1988; Kerker, 1969).

Another distribution encountered is the exponential distribution

$$p(a) = \frac{1}{\mu} e^{-a/\mu} . \tag{3.57}$$

It describes processes with no memory.

The aerosol that is left in the upper atmosphere by aircraft has been a source of concern because of the role it plays in the destruction of ozone. The size distribution of the aerosol droplets in exhaust at the point of discharge from the engine into the atmosphere during the flight is lognormal. Because of the speed of the exhaust, measurements can be taken only far downstream, and such measurements cannot be taken on supersonic jet engines. Some vapor condenses on the droplets and the measured distribution is exponential (Whitefield et al., 1996).

Monodispersed drops are prepared by taking advantage of the fact that jets break at a natural frequency. If a disturbance of such a frequency is imposed then monodispersed drops result (Aniskin et al., 1998). Bibette (1991) showed how monodispersed emulsions can be prepared. The technique consists of adding a surfactant to an oil-in-water emulsion. Because of the changes in interparticle forces produced by the formation of additional micelles, a phase separation into droplet-rich and droplet-lean phases takes place along the lines discussed in Section 6 for colloidal suspensions containing nonadsorbing polymers. The oil droplet-rich phase is separated and fractionated again by adding more surfactant, and eventually the oil droplet-rich phase that is produced is iridescent because it is monodispersed and packed with a crystalline order. It appears that droplets that are too large to show Brownian motion (greater than 2 μm) do not phase separate easily and hence cannot be fractionated.

5.   COMBINED ATTRACTIVE AND ELECTRICAL INTERACTION: DLVO THEORY

In the absence of adsorbed polymeric molecules, which are discussed below, colloid stability is governed in many cases by the combination of London–van der Waals attractive forces and the repulsive forces produced by double-layer overlap. This concept is the basis of the famous DLVO theory, developed independently in the late 1930s and early 1940s by Derjaguin and Landau in the Soviet Union and by Verwey and Overbeek in the Netherlands.

To find the total interaction energy between two particles we simply add the contributions φ and E_e from attractive and electrical effects. For example, the total interaction energy per unit area E_T between identical semi-infinite blocks can be found from Equations 3.5 and 3.48:

$$E_T = -\frac{A_H}{12\pi h^2} + 64 N_A c_o kT \; \kappa^{-1} \, \eta^2 \, e^{-\kappa h} . \tag{3.58}$$

Figure 3.9 shows variation of E_T with separation distance for several values of surface potential ψ_o with all other parameters fixed. Several general features of these curves are of interest. In the first place, E_T becomes large and negative for small values of h. Attractive forces dominate under these conditions, as a brief inspection of Equation 3.58 indicates. The curve must eventually reach a minimum, since short-range repulsive forces that come into play when the particles are virtually in contact have been ignored. The existence of this "primary" minimum shows that the particles will adhere to one another if they can ever be brought sufficiently close together.

Most of the curves in Figure 3.9 exhibit a maximum at somewhat larger values of h. The increase in the height of this maximum with increasing surface potential demonstrates that it is produced by electrical repulsion. It amounts to an energy barrier that must be surmounted as particles approach if they are to

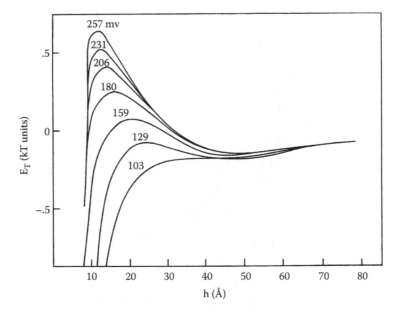

FIGURE 3.9 Total interaction energy for two semi-infinite flat plates for various surface potentials and $\kappa = 10^7$/cm and $A_H = 2 \times 10^{-12}$ ergs. E_T is expressed in terms of the thermal energy, kT, for a 4.0 nm^2 area. Equation 3.58 has been used. Reprinted from Hiemenz and Rajagopalan (1997, Figure 13.7), courtesy of Marcel Dekker, Inc.

adhere. In the limit $\psi_o = 0$, Equation 3.58 predicts that E_T is inversely proportional to $(-h^2)$.

Another feature of most of the curves in Figure 3.9 is a shallow "secondary" minimum at a separation distance somewhat larger than that of the maximum. As Equation 3.58 suggests, the electrical forces diminish more rapidly than the attractive forces with increasing h. Thus the interaction energy is small, but negative, at large separation distances where attractive forces dominate. As h decreases and electrical forces become important, E_T reaches a minimum and begins to increase.

Figure 3.10 illustrates the effect of varying the Hamaker constant A_H. As would be expected, increasing A_H lowers the height of the maximum in the curve. Figure 3.11 shows the dramatic changes caused by varying electrolyte concentration, and hence the inverse Debye length κ. These curves are for spherical particles, so that E_T^s is the sum of the expressions given by Equations 3.6 and 3.50. Adding electrolyte decreases the Debye length, and hence double-layer thickness, with the result that electrical forces do not become significant until the particles are closer together. When they finally do become important, attractive interaction is greater and the height of the maximum in the E_T^s curve is less. Adding sufficient electrolyte removes the maximum altogether, as Figure 3.11 shows.

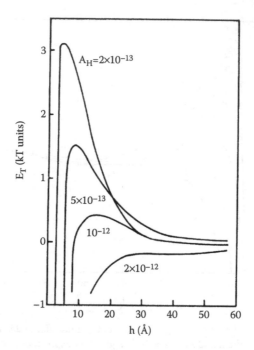

FIGURE 3.10 E_T is shown here with varying A_H. The values of $\psi_o = 103$ mV and $\kappa = 10^7$/cm have been used. Reprinted from Hiemenz and Rajagopalan (1997, Figure 13.6), courtesy of Marcel Dekker, Inc.

Curves of this type provide a basic understanding of colloid stability. Attractive forces favor flocculation and oppose stability, while electrical repulsion has the opposite effect. Flocculation occurs if particles can reach the separation corresponding to the primary minimum. The single most important factor influencing flocculation is the height of the maximum in the E_T curve. If it is nonexistent or small, particles have little trouble reaching the primary minimum as a result of ordinary thermal motion and the dispersion is unstable. But if the maximum is sufficiently high—several times the effective energy of thermal motion, kT—particles almost never have sufficient energy to surmount this energy barrier and the dispersion is usually stable. The qualifying word "usually" is needed because flocculation can, in principle, occur at separations corresponding to the secondary minimum. Such flocs cannot persist, however, unless the depth of the secondary minimum is several times kT. This condition is usually not satisfied, with the result that no flocculation takes place even though the E_T curve has a secondary minimum.

Several general conclusions can be drawn concerning colloid stability. Larger particles are more likely to flocculate since attractive forces are greater (see Figure 3.12). This assertion is most easily justified from Equation 3.7. At a separation distance h comparable to the Debye length κ^{-1}, the dimensionless parameter ξ decreases with increasing particle radius R. According to Equation 3.7, the attrac-

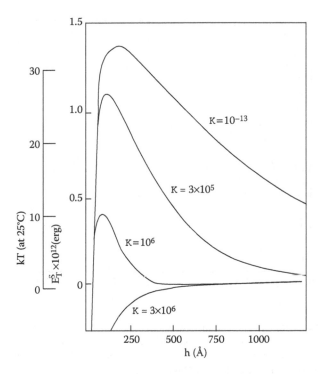

FIGURE 3.11 E_T is shown here with varying electrolyte strengths (κ per centimeter). The sphere radii are 1000 Å, $A_H = 10^{-12}$ ergs, and $\psi_o = 25$ mV. Reprinted from Verwey and Overbeek (1948) with permission.

tive energy is greater for small values of ξ. Reducing the magnitude of the surface potential by adjusting the concentration of the potential-determining ion in solution also promotes flocculation. For instance, the pH can be adjusted for metal oxide particles where H^+ is the potential-determining ion. Finally, adding electrolyte strongly promotes flocculation, as indicated previously. Changes in the opposite direction (e.g., a decrease in electrolyte content) can be made if a stable dispersion is desired.

A major success of the DLVO theory has been its ability to predict the very large and striking effect of counterion valence on colloid stability. To see this, we focus on the maximum in the E_T curve. We first use Equation 3.58 to find the separation distance h_m at which (dE_T/dh) vanishes. The result is

$$\frac{A_H}{6\pi h_m^3} = 64\, N_A\, c_o kT\, \eta^2\, e^{-\kappa h_m} . \tag{3.59}$$

As a reasonable initial estimate of the condition separating stable and unstable dispersions we set $E_T = 0$ at $h = h_m$. Accordingly we have

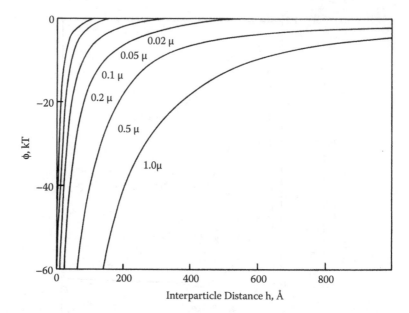

FIGURE 3.12 Hamaker interaction potential between two particles as a function of the particle size. For nanoparticles, the reach of attractive forces becomes extremely small and nanoparticles are stable even without the electrostatic repulsion. Reprinted from Sato and Ruch (1980), courtesy of Marcel Dekker, Inc.

$$\frac{A_H}{12\pi h_m^2} = 64 N_A c_o kT \kappa^{-1} \eta^2 e^{-\kappa h_m} . \tag{3.60}$$

If we take the ratio of these equations, we find

$$\kappa h_m = 2. \tag{3.61}$$

Substitution of Equation 3.61 into Equation 3.59 yields

$$c_o = \frac{A_H}{3072\pi N_A kT \eta^2 e^{-2}} \kappa^3 = B' \kappa^3 . \tag{3.62}$$

Using the definition of κ given in Equation 3.27, this equation can be written as

$$c_o = \frac{B}{\nu^6} , \tag{3.63}$$

where B is a constant. The electrolyte concentration c_o for which E_T vanishes at h_m is referred to as the critical flocculation concentration (CFC).

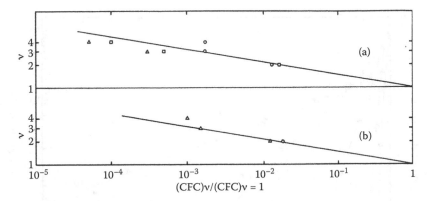

FIGURE 3.13 The Schulze-Hardy rule for colloids is shown with bold lines with slopes of 1/6: (a) negatively charged particles and (b) positively charged particles. From data compiled in Overbeek (1952).

According to Equation 3.63, the concentration required to produce flocculation in a colloidal dispersion with a (nearly) constant surface potential is inversely proportional to the sixth power of the valence. That is, the concentration of a symmetric electrolyte containing divalent ions required for flocculation should be only 1/64 of that required with monovalent ions. We note that since the double layer of a negatively charged colloidal particle contains mostly cations, it suffices, for practical purposes, to consider only the valence of the counterion in applying Equation 3.63. Hence the requirement of a symmetric electrolyte may be relaxed. In Figure 3.13, Overbeek's (1952) compilation of the CFC data for a wide range of flocculation electrolytes has been plotted for negatively charged colloid particles as well as for positively charged colloid particles. The CFC data used are the averages over a number of flocculating electrolytes with the same value of ν. Equation 3.63 is known as the Schulze-Hardy rule.

It should be noted that an assumption of the derivation leading to Equation 3.63 is that η does not change greatly when electrolyte valence ν is varied at constant surface potential ψ_o. Inspection of the definition of η given in Equation 3.46 reveals that this condition is satisfied only for sufficiently large ψ_o. For small values of ψ_o, the dependence of the CFC on valence is weaker than predicted by the Schulze-Hardy rule.

In the late 1970s, the surface forces apparatus was developed for actually measuring forces between molecularly smooth mica sheets immersed in a liquid and separated by distances ranging from a few angstroms to a few thousand angstroms (Israelachvili, 1991; Ninham, 1980). These measurements show that, in the absence of adsorbed polymeric molecules, DLVO theory is valid for separation distances h exceeding about 30 Å (3 nm).

Figure 3.14 shows measured forces in potassium nitrate (KNO_3) solutions of various concentrations. At large separation distances, attractive forces are small and the force decreases exponentially with increasing separation, as predicted by electrical double-layer theory (Equation 3.47). Moreover, at the two lowest KNO_3

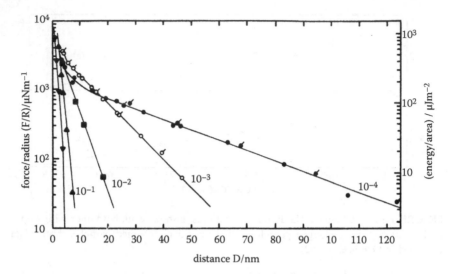

FIGURE 3.14 Experimental results of direct measurements of repulsive forces F as a function of separation D between two crossed mica cylinders of radius R in aqueous KNO_3 solution (concentrations marked in mol/dm^{-3}). The right-hand ordinate gives the interaction energy per unit area for two parallel plates, with A calculated according to the Derjaguin approximation. The results in 10^{-4} to 10^{-1} mol/dm^{-3} solutions are for the same pair of mica sheets. The points with tails in 10^{-4} and 10^{-3} mol/dm^{-3} solutions are for a different pair of sheets cut from the same sheet as the first pair. In 1 mol/dm^{-3} KNO_3, the force was attractive above 4 nm. Reproduced from Israelachvilli and Adams (1978) with permission.

concentrations, the measured slopes are within 10% of the predicted values. At the two higher KNO_3 concentrations, measured and predicted slopes agree within 25%. In all cases the surface potential ψ_o is found from the data to be about 75 mV.

At 0.1 M KNO_3, the attractive force can be calculated as the difference between the measured force and the electrical force at small separation distances, in the range of 40 to 70 Å, where attractive forces are appreciable. The results are consistent with the formulas given above, with a value of 2.2×10^{-20} J for the Hamaker constant A_H.

Other techniques for measuring forces have also been developed. For example, with total internal reflection microscopy, one can measure directly the mean potential of interaction between a single colloidal particle and a flat plate. This technique can detect smaller forces than can the surface forces apparatus, although at 1 nm, its limit on separation distance is not as small (Prieve, 1999). Recently atomic force microscopy has also been used to measure forces on a colloidal scale.

At separation distances of less than about 30 Å, "solvation forces" and other effects of molecular structure sometimes cause large deviations from the predictions of DLVO theory. At very small separation distances, for instance, oscillatory forces are sometimes observed that are evidently associated with the ease of packing molecules between the two surfaces (Christenson et al., 1982; Israelachvili, 1991; Ninham, 1980). At certain definite separation distances, molecular dimensions are such that an integral number of molecular layers fits precisely into the available space, and the energy has a local minimum. At intermediate separations, the solvent structure must be distorted to occupy the available space, and energy increases. When the continuous phase is water, the solvation forces are called hydration forces, which possess additional properties.

Flocculation as discussed in this section is a process by which suspended drops or particles aggregate in a continuous medium. Although free energy is reduced, the system remains thermodynamically unstable. In some suspensions, however, separation into two thermodynamically stable phases can occur. For example, both theory and experiment show that transformation from a disordered or "fluid" phase to an ordered or "solid" phase occurs in uniform suspensions of noninteracting hard spheres when the volume fraction reaches a value of approximately 0.50.

A similar transformation occurs in dilute suspensions of uniformly charged spheres in a medium of low ionic strength. The solid phase, which can form at volume fractions as low as 0.001, is iridescent and reverts to liquid phase on dilution or addition of electrolyte (Hachisu et al., 1973; Hiltner and Krieger, 1969). Since each sphere is surrounded by a thick electrical double layer and since repulsion prevents significant overlap of double layers of adjacent particles, the spheres can be considered to have a much larger effective radius and hence a much larger volume fraction than the nominal values. The iridescent phase forms when the effective volume fraction reaches approximately 0.5. The fractionation of emulsion droplets into a phase containing monodispersed droplets, discussed at the end of Section 4 (Bibette, 1991), is also based on these principles of phase separation.

6. EFFECT OF POLYMER MOLECULES ON THE STABILITY OF COLLOIDAL DISPERSIONS

Polymers have been used to both stabilize and destabilize colloidal dispersions. Their use in stabilizing colloidal dispersions in nonaqueous liquids is particularly important because, owing to the low dielectric constants of such liquids, the concentration of ions is very low and the electrostatic stabilizing forces minimal. Since London–van der Waals forces are attractive, stabilization provided by polymers may be the only means to prevent flocculation. In contrast to electrostatic forces in DLVO theory, polymeric stabilization forces often are significant at particle separations of a few tens of nanometers, well beyond the range of attractive forces.

For purposes of discussion, polymers are separated into two classes: adsorbing and nonadsorbing polymers. If the polymer chain segments prefer the solvent to the solid surface, then the polymer will not adsorb to a significant extent. Note that zero adsorption is entropically disallowed because the surface will try to adsorb some of everything to maximize the randomness; that is, all polymers adsorb to some extent. If the chain segments favor the solid surface to the solvent,

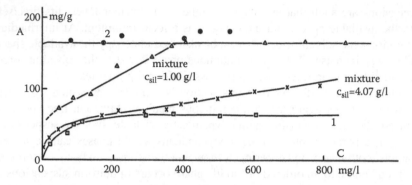

FIGURE 3.15 Adsorption data from Cohen Stuart et al. (1980) for polyvinyl pyrrolidone in water on silica. The effect of polydispersivity is also shown. The weight fraction of component 2 in the mixture is 0.25, $M = 1440$ for component 1 and 10^6 for component 2. Reprinted with permission from John Wiley & Sons.

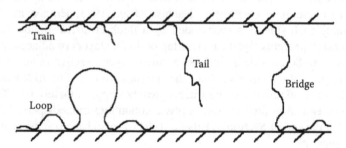

FIGURE 3.16 Schematic view of polymer adsorption.

then the polymer will adsorb. The important feature here is if one chain segment adsorbs and the rest of the molecule floats, then at a later time the floating part will make another contact with the surface and at that location another chain segment will get adsorbed and eventually all of them will get adsorbed (except for those segments that are sterically prevented, which would be more prevalent at high polymer concentrations). One consequence of this is that it takes very little polymer to saturate the surface and the adsorption isotherm leaps to the saturated value from a value of zero over a very small concentration of polymer in the bulk. As Figure 3.15 shows, this effect is enhanced in polymers of large molecular weights (curves 1 and 2). This means that washing away the adsorbed amount is difficult, requiring a lot of clean solvent, and that the kinetics of the process are very slow since many chain segments have to desorb at the same time. For polymers of high molecular weight adsorption is often irreversible for most practical purposes.

The manner in which the polymer molecules adsorb can be classified into *trains*, where not all segments adsorb, but chain segments at a reasonable frequency do; *loops*, where the successive adsorbed segments are somewhat far apart and the intervening segments form a loop which lies in the solvent phase; and *tails*, which are anchored at one point with a long segment floating in the solvent (see Figure 3.16). More details of polymer adsorption are given by Fleer et al. (1993). Lin and Blum (1997) determined using nuclear magnetic resonance (NMR) that the actual distribution of the trains, loops, tails is different from the predicted equilibrium distribution because the system finds it kinetically impossible to equilibrate.

When two surfaces approach each other, the polymer configurations undergo some changes. For adsorbing polymers, the protruding chain segments (loops and tails) of one surface encroach into the domain of those on the other surface, resulting in a decrease in the number of conformations that these structures can take. That decreases the entropy of the polymer and hence increases the free energy:

$$\Delta G = \Delta H - T \Delta S \qquad (3.64)$$

where, for simplicity, the enthalpy change ΔH can be taken to be zero. The excess free energy per unit volume (J/m^3) or the force per unit area on the plates (N/m^2) is increased. This increase in free energy is equivalent to a repulsion between the plates. The contribution to the total energy per unit area E_T of the plate is obtained by integrating this force from infinity to the separation distance h; that is, we calculate the work done against this force in bringing the plates to a separation distance h. Of course, the net effect is that, under the right conditions, the increase in free energy is sufficient to provide stabilization. Other structural features also come in. At low polymer concentrations, a tail from one plate may also adsorb on the other, forming a bridge, as shown in Figure 3.16 (Patel and Tirrell, 1989). Bridges lead to flocculation, since they tend to anchor one colloidal particle to another, and even colloids entropically stabilized by polymer adsorption will eventually flocculate.

Nonadsorbing polymers between two plates also show a decrease in entropy when the two plates move toward one another. As the available space is reduced, the random excursions that the polymer chain was taking begin to see restrictions and the entropy decreases. However, another factor comes into play. In the nonadsorbing system, the polymer does not "like" the solid surface, and when, in addition, its entropy drops, it has no reason to stay in the gap, and the reservoir becomes a more favored place. There is another source of entropy that has to be considered. Under ideal solution theory in dilute systems, the chemical potential of the polymer is written approximately as $\mu_p = \mu_p^o + kT \ln \varphi$, where the first term on the right is the standard state chemical potential and the term in the logarithm of the volume fraction of the polymer φ is due to the entropy of mixing. The entropy contribution is due to the randomness with which the polymer and the solvent molecules are dispersed, and increases the chemical potential of the polymer in the gap. This is called the osmotic effect. Conversely, if some polymer is squeezed out of the gap into the reservoir, its chemical potential in the gap goes down, a feature that opposes further polymer migration to the reservoir. Which effect wins? It appears that if the polymer has a very large molecular weight (about 10^6), then it is very difficult for it to squeeze into the gap; that is, the overall effect is that the concentration of the polymer chain segments there is very low. Thus the force opposing the Hamaker attraction between particles is very small and "depletion flocculation" results. It is well known that solutions of such high molecular weight polymers and colloidal particles often phase separate into colloid rich–polymer lean and colloid lean–polymer rich phases simply because the polymer cannot get in between the interstices. This is seen not only in regular colloidal systems (Vrij, 1976), but also in microemulsions, discussed in Chapter 4, where the droplets are at thermodynamic equilibrium, unlike in regular emulsions (Qutubuddin et al., 1985). However, the osmotic pressure wins at moderate molecular weights and "depletion stabilization" results.

Adsorbing polymers also show an osmotic effect that is a little different because the adsorbed polymers cannot bodily move with respect to the solid surfaces, but the solvent still can. When two plates containing adsorbed polymers approach one another and the overlap of the polymer chain segments begins, the local polymer chain segment density and its chemical potential go up, but the chemical potential of the solvent goes down. The solvent from the reservoir rushes into the gap to dilute the system and the plates move apart. These effects also carry enthalpic contributions, which are discussed separately later.

Quantification of polymer entropies is done through the statistics of random walk (Flory, 1953). The model is based on a drunk trying to walk in one dimension! Because of his state, the next step the drunk takes could be to the right or to the left with equal probability, but his stride remains of identical length and at every step he waits for the same length of time. The key probability density function is that of end-to-end displacement x, that is, the distance between the beginning and the end, which for a linear polymer in one dimension is Gaussian:

$$p_x = \frac{1}{\sigma\sqrt{\pi}} e^{-x^2/(2\sigma^2)} . \tag{3.65}$$

In three dimensions, the above equation is only slightly modified. Here the term σ is the square root of the variance and is on the order of $\ell N^{1/2}$, where ℓ is the length of a step and N is the number of repeat units. The length ℓ in a flexible chain can be taken to be the mean distance between adjacent repeat units. When the chain is more rigid, ℓ becomes the persistence length, which is the length over which the chain segment can be approximated by a straight line. Of course, N has to be corrected accordingly. It appears that due to cancellation of errors, the Gaussian distribution holds in melts (see de Gennes, 1979, p. 54).

In solutions, instead of calculating the variance, one calculates the radius of gyration R_g, which is found to be smaller than the measured values. The reason is that the drunk can retrace his footsteps, but a polymer chain cannot penetrate itself. Hence the random walk is insufficient and one has to analyze what are called self-avoiding random walks. In some sense the self-avoiding feature is like a repulsion and the polymer coil stretches out, but R_g still correlates with $\ell N^{1/2}$. The differences, however, are sometimes monumental: a self-avoiding random

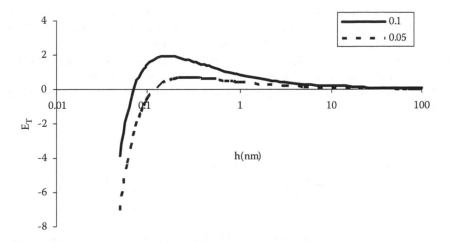

FIGURE 3.17 Equation 3.66 is plotted with $RT = 2478$ J/g·mol, $\rho_p = 0.8$ g/cm^3, $N = M/104$, $\ell = 5 \times 10^{-8}$ cm, $M = 45,000$, $A_H = 10^{-21}$ J, and the polymer volume fractions $\varphi = 0.05$ and 0.1. E_T has been reexpressed as per kT and per 4 nm^2.

walk in one dimension is not statistical because there is only one configuration. Typical experimental values of R_g are on the order of 40 nm or more (Tanford, 1961, p. 308), at which distance the colloidal particles cannot feel the attraction due to Hamaker forces. Thus an important contribution of nonadsorbing polymers to stabilization of colloids is that the repulsion is felt at large distances where the attractive forces are minimal. There is no secondary minimum as in DLVO theory.

The mathematics of self-avoiding random walks is quite complicated and needs computer simulations. De Gennes (1979) has obtained a number of analytical approximations using his scaling principles, including the case of polymers confined between plates. If the result of Daoud and de Gennes (1977) for a nonadsorbing (single) polymer is used to calculate E_T, the total interaction energy between plates, one has

$$E_T = -\frac{A_H}{12\pi h^2} + \frac{3RT\,\varphi\rho_p N\ell}{2M}\left(\frac{\ell}{h}\right)^{5/3}, \qquad (3.66)$$

where the first term on the right-hand side is due to Hamaker interaction and M is the molecular weight of the polymer. Equation 3.66 is plotted in Figure 3.17, where the values of the constants closely approximate those of polystyrene in dibutyl phthalate. The peaks are very shallow because the decay of the repulsion due to polymer with increasing separation distance is very slow. The polymer volume fraction φ between the plates is not the same as in the reservoir. The difference can be calculated by equating the chemical potentials in the gap to those in the bulk for the solvent and the polymer.

Detailed calculations also exist for adsorbing polymers. They also show that for different types of configurations the chain segments extend from a single plate to distances that scale as $\ell N^{1/2}$ (Hesselink, 1971; Hesselink et al., 1971). Evans and Napper (1977) and Vincent et al. (1980, 1986) provide expressions for an elastic contribution to E_T. When the surfaces approach one another, the polymer chain segments see elastic deformations at very small interplate distances. For a sphere

$$V_{el} = \left(\frac{2\pi R}{M} \varphi \delta^2 \rho_p \right) \left\{ \frac{h}{\delta} \ln \left[\frac{h}{\delta} \left(\frac{3 - h/\delta}{2} \right)^2 \right] - 6 \ln \left[\frac{3 - \frac{h}{\delta}}{2} \right] + 3(1 - h/\delta) \right\}, \quad (3.67)$$

where R is the radius of the sphere and δ is the thickness of the adsorbed layer.

In more modern calculations, the distinction between adsorbing and non-adsorbing polymers is not made. The enthalpic contribution not included in Equation 3.67 is accounted for by using χ, the Flory-Huggins coefficient. It represents the interaction energy between a polymer chain segment and a solvent and is almost always adverse (χ is positive). Similar interaction energy can also be defined between a polymer segment and the surface, χ_s. Scheutjens and Fleer (1982) give complete results of random walk simulations for a lattice between two plates. The walk is not self-avoiding. They use the Flory-Huggins equation in the bulk under which the chemical potentials (compared to their standard states) of the polymer and the solvent are

$$\Delta \mu_p / kT = 1 - \varphi - r(1 - \varphi) + \ln \varphi + r\chi(1 - \varphi)^2 \quad (3.68)$$

$$\Delta \mu / kT = \varphi - \varphi/r + \ln(1 - \varphi) + r\chi\varphi^2 \quad (3.69)$$

where r is the number of monomers in the linear polymer chain — and equal to the total number of steps taken by the walker. They show that the concentration of the polymer is not uniform across the slit width, whether it is adsorbing or nonadsorbing. Their results for adsorbing polymers are shown in Figure 3.18 and 3.19. The term Δf_p (in the figures) is the increase in the potential per lattice site on the surface. Thus it contributes $n\Delta f_p$ to E_T, where n is the number of sites per plate and on the order of 100 or more in colloids. H is the number of lattice sites separating the two plates, making the separation distance $h = H\ell$. It is very useful to add an axis in $H/2R_g$, since it tells us of the first-order effect of increased molecular weight. Only repulsion is seen in Figure 3.18, but some attraction is seen in Figure 3.19 where the polymer concentration is very low. A tentative conclusion is that at high polymer coverages, the tails are effective in providing repulsion, but not at low polymer coverages, where the bridging brings about an attraction. One may also say that at high coverages the tails cannot penetrate the

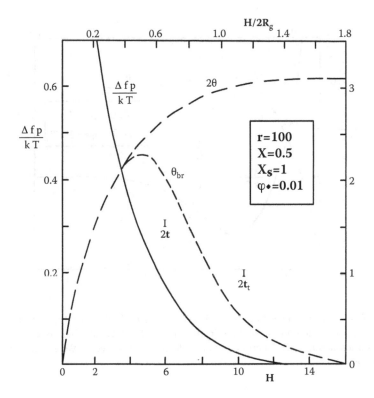

FIGURE 3.18 The amount of polymer between two plates (dashed curves, right-hand scale) and the free energy of interaction per lattice site (full curve, left-hand scale) as a function of plate separation H for adsorbing polymer. The upper abscissa scale gives the interplate separation in terms of the free coil diameter R_g. The adsorbed amount between two plates 2θ and the part due to bridging chains θ_{br} are given in equivalent monolayers. Twice the overall root mean square thickness 2t of the adsorbed layer on a single plate and the same due to the tails only ($2t_t$) are indicated in the figure. $r = 100$, $\chi = 0.5$, $\chi_s = 1$, volume fraction of the polymer in the reservoir $\varphi_* = 0.01$, hexagonal lattice. Reprinted from Scheutjens and Fleer (1982) with permission from Elsevier.

adsorbed layer to form bridges. A dimensionless adsorption energy χ_s has been used which favors adsorption for $\chi_s > 0$.

Figure 3.20 shows nonadsorbing systems with $\chi_s = 0$. At low separations, the polymer is squeezed out and the result is a net attraction. Similarly, at large molecular weights (r and R_g), attraction will dominate. The repulsion, and hence the colloidal stability, is best at intermediate polymer concentrations and intermediate molecular weights.

The Flory-Huggins coefficient χ plays an important role. If this coefficient is above a critical value, then phase separation into a polymer-rich phase and polymer-lean phase takes place. Since the polymer-solvent interaction is adverse, decreased temperature increases the importance of this interaction energy as the random thermal motion is decreased, and the temperature-concentration phase

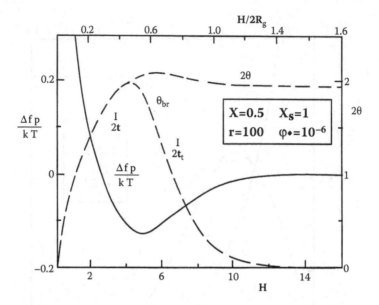

FIGURE 3.19 The same as in Figure 3.18, for $\varphi_* = 10^{-6}$. In this case there is an attractive region at plate separations around the R_g, the free coil radius, and there is a maximum in the adsorbed amount 2θ. Reprinted from Scheutjens and Fleer (1982) with permission from Elsevier.

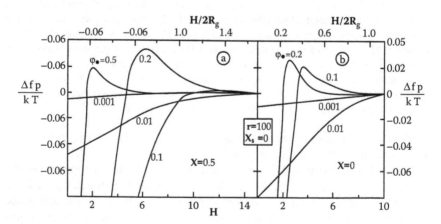

FIGURE 3.20 The free energy of interaction per surface site between two plates in the presence of nonadsorbing polymer at various bulk solution volume fractions φ_*. The left-hand figure shows the results for the Θ solvent ($\chi = 0.5$); on the right-hand side are the data for an athermal solvent ($\chi = 0$). $r = 100$, $\chi_s = 0$, hexagonal lattice. Reprinted from Scheutjens and Fleer (1982) with permission from Elsevier.

diagram shows that an upper consolute temperature exists (i.e., below this temperature a polymer-rich phase separates from a solvent-rich phase). The Flory-Huggins coefficient χ can also be increased by introducing some amount of a nonsolvent. For a polymer of infinite molecular weight, the Θ-temperature designates the point of incipient instability below which two phases are formed. For this case (Flory, 1953, p. 523)

$$\chi - 1/2 = -\psi \, (1 - \Theta/T), \qquad (3.70)$$

where ψ is a constant and taken to be positive, and Θ is the Θ – temperature. Thus $\chi = 1/2$ is the point of incipient instability where $T = \Theta$.

Flory also showed that the left-hand side in Equation 3.70 is proportional to the polymer segment–solvent interaction in dilute solutions. When this term is negative, the polymer "likes" the solvent. In colloids, the enthalpic contribution to the "osmotic" effects in its simplest form for a sphere becomes

$$V_{os} = \frac{4\pi R}{v_s} \varphi^2 \, (1/2 - \chi) \, (\delta - h/2)^2 \, , \qquad (3.71)$$

where v_s is the molar volume of the solvent. Note that when $(\chi - 1/2)$ is negative, V_{os} is positive, indicating a repulsion between spheres and implying colloid stability. When χ approaches $1/2$ from below, both colloid and polymer instability are seen for a polymer with an infinite molecular weight.

There is yet another way of describing the stability of a colloidal suspension in the presence of nonadsorbing polymers. Here global free energy is computed based on interparticle forces that incorporates the presence of polymers (Napper, 1983). One consequence of such a model is that it is possible to predict phase separation, discussed earlier. As the polymer concentration is increased, a polymer rich–colloid lean phase separates from a polymer lean–colloid rich phase at some critical polymer concentration. As the polymer molecular weight is increased, the size of the random coil increases and the critical polymer concentration decreases. For nonadsorbing polymers, Gast et al. (1983) assume a potential where, at small interparticle separations, the excluded polymer creates a suction due to the osmotic pressure difference. Some of the predictions made compare well with the experiments, as shown in Figure 3.21. Rao and Ruckenstein (1985) extended this treatment to adsorbing polymers. They assumed that in addition to adsorbed polymer, the system also had unadsorbed polymer. The special contribution of the adsorbed layer on pair potentials were the elastic effects (Equation 3.67) and osmotic effects (Equation 3.71). Two cases were considered, one in which the free polymer could penetrate the adsorbed layer and one in which it could not. The actual experimental data on polymer concentrations at incipient stability fall between the two limits.

In effect, polymers can stabilize or destabilize a colloidal suspension. The main feature responsible for this is an entropic one, and this mode of stabili-

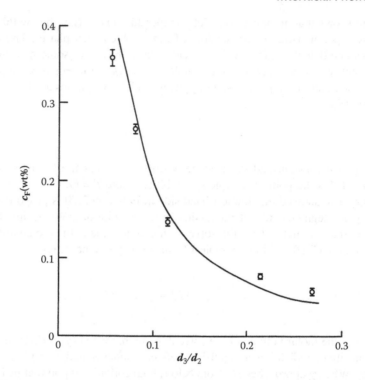

FIGURE 3.21 Comparison of the predictions for the onset of destabilization, c_F (wt %) as a function of polymer size to colloid diameter ratio, d_3/d_2 with the data of Sperry et al. (1981) for polystyrene particles (4.3×10^{-5} cm and potential of –80 mV) in an aqueous solution of hydroxymethylcellulose with a molecular weight of 63,800 to 438,800 at an ionic strength of 0.01 M. Reprinted from Gast et al. (1983). Reproduced with permission from the Royal Society of Chemistry.

zation is often called entropic stabilization. Enthalpic effects can also play a significant role. The forces between two particles due to the polymers are felt at much larger separation distances than the attraction due to London–van der Waals forces, which is the key to their stability. The process of stabilization-destabilization can be effected even in systems where electrolytes cannot be added or are available at very high concentrations and in nonaqueous systems. We note that entropically stabilized dispersions tend to flocculate on cooling since entropy effects decrease with decreasing temperature. But dispersions stabilized by enthalpic effects tend to flocculate on heating because stabilizing interaction effects are eventually outweighed by the destabilizing entropic effects. Polymers of large molecular weight (greater than 10^6) are effective as flocculants and require only small amounts. These polymers are fragile, however, and the simple act of stirring to dissolve them can tear them (Mace, 1990). Many experimental results on stability in the presence of polymers have been described; see Napper (1982), Napper and Hunter (1972), Parfitt and Peacock (1978), Sato and Ruch (1980), and Vincent (1974).

Some special polymers are also of interest. Often, charged polymers are used for flocculating a colloidal suspension (e.g., in removing small solid particles or liquid drops from wastewater before it is discharged to the environment). An important part of the mechanism is that the polymers adsorb on a colloid and effectively neutralize the charge. If the molecular weight of the polymer is sufficiently high, then bridging can also occur (Pefferkorn, 1995). Such polyelectrolytes are often used instead of inorganic salts of trivalent iron and aluminum, which cause flocculation by decreasing double-layer thickness, because much smaller quantities of additive are required. Even in such cases, however, indications exist that surface charge neutralization is not the whole story. Gregory (1973) found that flocculation occurred over a wider range of conditions than would be expected based on charge neutralization. He suggested that particle surface charge was nonuniform, being positive in the vicinity of adsorbed polymer molecules, but remaining negative elsewhere. Attraction between positive and negative regions of adjacent particles would promote flocculation in this case.

Yet another kind of polymer that is used to alter colloid stability is the block copolymer. From the previous discussion, a polymer for which the dispersion medium is a "good" solvent is desired for purposes of stabilization in both nonpolar and aqueous systems. On the other hand, adsorption on particle surfaces is favored if the dispersion medium is a "poor" solvent. One method of resolving these conflicting requirements is to use a block copolymer with a chain having some blocks of a monomer that adsorbs strongly on the colloidal particles and "anchors" the polymer and blocks of another monomer that interacts strongly with the solvent and makes up the loops and tails. It has been seen in emulsions that a block copolymer, where one block is soluble in the oil phase and the other in the polar phase, stabilizes the system when a random copolymer of similar composition cannot (Riess et al., 1971). It suggests that loops and tails are effective in stabilizing an emulsion but trains are not. Thus block copolymers are important stabilizers in numerous practical applications.

7. KINETICS OF COAGULATION

Colloidal particles exhibit random Brownian motion as a result of which a net diffusive flux can be generated. If the thermodynamic force on each particle which causes such a flux is the gradient of chemical potential $-\nabla\mu_c$, it must, under steady conditions, be balanced by frictional forces. These forces are taken to be hydrodynamic drag forces and are obtained from Stokes' law.

Making the balance, one obtains

$$6\pi\mu R U = -\nabla\mu_c.$$

Substituting ($\mu_c^\circ + kT \ell n\ c$) for μ_c, where μ_c° is the standard state chemical potential, one has

$$c\mathbf{U} = -\frac{kT}{6\pi\mu R} - \nabla c \; .$$

This equation can be rewritten as

$$J = -D\nabla c, \tag{3.71}$$

where $J = c\mathbf{U}$ represents the diffusive flux and the diffusion coefficient D is given by

$$D = \frac{kT}{6\pi\mu R} \; . \tag{3.72}$$

Equation 3.72 is known as the Stokes-Einstein equation, which is reconsidered in detail in Chapter 8.

If external forces act on the particles, as electrical fields will for charged colloidal particles, Equation 3.71 is modified to

$$J = -D\nabla c + cMF, \tag{3.73}$$

where F is the force on a single particle and M is its mobility. M is usually taken as (D/kT), and the force F is usually written as the negative of the gradient of a scalar potential ϕ. Equation 3.73 thus becomes

$$J = -D\left[\nabla c + \frac{c}{kT}\nabla\varphi\right] . \tag{3.74}$$

For equilibrium situations, the flux is zero, and it follows from Equation 3.73 that the concentration distribution is of the Boltzmann type:

$$c = c_\infty \exp\left[-\frac{\varphi}{kT}\right] . \tag{3.75}$$

A simple case where flocculation takes place among single colloidal spheres of initial concentration c_∞ is considered below. The concentration decreases with time because the collisions between two spheres lead to flocculation. The analysis is restricted to collisions between two individual particles and their consequences. Hence the solution is valid only at short times.

We consider diffusion in the vicinity of a single fixed colloidal particle. If the quasi-steady-state approximation is satisfactory, the conservation equation simplifies to

$$0 = - \nabla \bullet J = - \frac{1}{r^2} \frac{\partial}{\partial r} (r^2 J_r) , \qquad (3.76)$$

so that $r^2 J_r$ is a constant. Here J_r is the flux in the radial direction using a spherical coordinate system with the center of the fixed particle as the origin. We write, on integrating the above equation,

$$z = -4\pi r^2 J_r, \qquad (3.77)$$

where z is a constant and represents the total number of particles moving toward the fixed single colloidal particle. Substituting Equation 3.74 into Equation 3.77 and rearranging $\exp(\Phi/kT) [dc/dr + (c/kT) (d\Phi/dr)]$ as $[d (ce^{\Phi/kT})/dr]$, we have

$$\int_0^{c_\infty} d (ce^{\frac{\varphi}{kT}}) = c_\infty = \frac{3\mu Rz}{2kT} \int_{2R}^{\infty} \frac{1}{r^2} \exp \left(\frac{\varphi}{kT} \right) dr . \qquad (3.78)$$

In Equation 3.78 we have assumed that as $r \to \infty$, φ tends to some constant value assumed to be its datum, and that c goes to some uniform value c_∞. These assumptions imply that as $r \to \infty$, $J_r \to 0$, as may be verified from Equation 3.77 for a constant z.

It has also been assumed that the distance of closest approach between two single colloidal spheres is $2R$, where $2R$ is the hard sphere diameter of the particles. At the distance of closest approach, the particles are assumed to have flocculated and the concentration of the colloid particles that have approached the fixed colloid is zero.

The quantity z can be evaluated from Equation 3.78. It represents the collision frequency of any one colloid particle. In the given system, the total number of collisions per second is $(1/2 \ zc_\infty)$, the factor $1/2$ being included to avoid double counting. If every collision gives rise to flocculation, then the initial depletion rate in a monodispersed suspension is twice the collision rate, since each collision removes two particles:

$$-\frac{dc_\infty}{dt} = zc_\infty = k_2 c_\infty^2 . \qquad (3.79)$$

Here k_2 is a second-order reaction rate constant given by

$$k_2 = \frac{4 \ kT}{3\mu} \left[\int_1^{\infty} \frac{1}{r^{*2}} \exp \left(\frac{\varphi}{kT} \right) dr * \right]^{-1} , \qquad (3.80)$$

where $r^* = r/2R$ and r is the center-to-center distance between spheres. As discussed below, the integral in this equation is, under many conditions, proportional to exp $[\varphi_{max}/kT]$, where φ_{max} is the potential energy at the maximum of an interaction curve such as those in Figures 3.9 through 3.11. Since k_2 is a rate constant, φ_{max} is analogous to the activation energy in chemical kinetics.

Equation 3.80 is often modified to account for the fact that we have assumed in the derivation that the central particle remains fixed. In fact, it moves due to diffusion, and each particle of a pair may be considered to diffuse toward the other. It turns out that to account for the relative motion, $2D$ should be used instead of D alone, and the value of k_2 becomes twice that given by Equation 3.79 (see Equation 3.80).

This method of formulation by von Smoluchowski and Fuchs is limited to small concentrations of particles. Then the fixed particle can at most feel the presence of one other particle, and φ is equal to the sum of the van der Waals attraction and the electrical double-layer repulsion potential, or E_T^s, as discussed in previous sections. In this limit it is also legitimate to model the reaction as a second-order reaction (i.e., only two-particle collisions can occur and the higher body collisions are virtually nonexistent). In aerosols, which are colloidal dispersions in air, there is no significant electrical repulsion between particles. Hence the effect of interparticle forces on the initial coagulation rate is negligible, and we find

$$k_2 = \frac{8\,kT}{3\mu}.$$

(3.81)

Equation 3.81 is the "fast coagulation" limit of the general expression for "slow coagulation" represented by Equation 3.80, in which it has been implicitly assumed that φ provides a potential energy barrier hindering coagulation. For this case, the appropriate value of k_2 is twice that given by Equation 3.80, as discussed earlier.

The obvious restriction on the theory that flocs are made out of only two colloidal spheres can be relaxed (Overbeek, 1952; Sheludko, 1966), but the mechanism of flocculation is still based on two-particle (floc) collision. Consequently the improved solutions (see Problems 3.4 and 3.5) are still restricted to small dilutions, but valid over longer times. Some of the issues in concentrated colloid suspensions have been considered in a *Faraday Discussions of the Chemical Society* (1983).

The use of binary collisions also fails when the distance over which two flocs (or single particles) first feel the presence of one another due to their interaction potential is much greater than the average center-to-center distance among flocs. For then the term φ, previously taken as the interaction potential between two particles, has to be modified to account for the presence of other particles, even for a two-particle collision. Such effects are observed in nonaqueous media (Albers and Overbeek, 1959) and can occur in concentrated systems as well.

Comparing Equations 3.80 and 3.81, we find that the term

$$W = \int_1^\infty \frac{1}{r^{*2}} \exp\left(\frac{\varphi}{kT}\right) dr^* \qquad (3.82)$$

in Equation 3.80 decreases the flocculation rate. W is called the stability ratio. As mentioned previously, φ is approximated as the interaction potential between two spheres. For systems that are difficult to flocculate, the function φ has a maximum, as discussed in Section 4. Under these conditions, the integrand of Equation 3.82 has a sharp maximum and the integral W is mainly composed of the area under this peak. This allows an asymptotic expansion for W in terms of φ_{max}. Either the approximate expression for W given in Problem 3.6 or Overbeek's (1952) approximate form $W \sim (2\kappa R)^{-1} \exp [\varphi_{max}/kT]$ can also be used together with φ as E_T^S to obtain

$$\log W = K_1 + K_2 \log c. \qquad (3.83)$$

This linear form is verified by experiment (see Figure 3.22). Further, $\log W = 0$ at $c = $ CFC as required. It is also seen from Figure 3.22 that the existence of a highly stable colloidal dispersion requires that W be at least as great as 10^4. In Figure 3.22, W is plotted against the concentrations c of the counterions, where these have valences of 1, 2, and 3. Although Equation 3.82 cannot be used to predict the Schulze-Hardy rule, CFC versus v values obtained from the figure are in reasonable agreement with the rule.

Einarson and Berg (1993) have attempted to explain the data on flocculation kinetics of latex particles with a block copolymer adsorbed on them. The polymer was polyethylene oxide (PEO)/polypropylene oxide (PPO). PPO is water insoluble and forms the part that adsorbs on the latex; PEO forms streaming tails into water. Some charge effects remain after the polymer adsorption. The total potential is DLVO plus elastic plus osmotic effects. After fitting the model to the experimental data, they were able to calculate the value of δ, which they called the adlayer thickness. Their data on the stability ratio of latex with and without the polymer and as a function of NaCl concentration are shown in Figure 3.23. Note that the polymer stabilizes the colloid by almost one order of magnitude in NaCl concentration. That is, polymers may be necessary to maintain stability in aqueous media containing substantial electrolyte.

It was anticipated by von Smoluchowski that Brownian diffusion is not the only mechanism by which one particle can come into contact with another. There can be convection in the system as well. Convection aids flocculation when it enhances the relative velocity between nearby particles, making it easier for them to surmount the maximum in the potential energy curve and reach the primary minimum corresponding to flocculation. Application of a velocity gradient (shear) is especially effective in the flocculation of relatively large particles (Hiemenz and Rajagopalan, 1977; Zeichner and Schowalter, 1977, 1979). It can also break up flocs associated with the secondary minimum of the particle interaction curve (Zeichner and Schowalter, 1977, 1979).

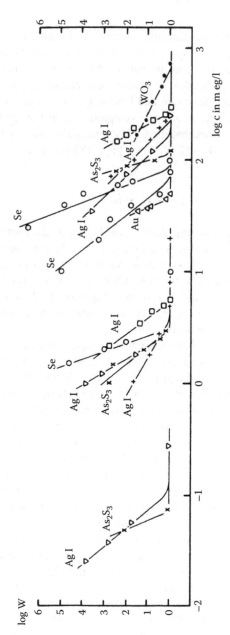

FIGURE 3.22 Relation between rate of flocculation and concentration of electrolyte as determined experimentally. The three groups of curves relate to monovalent, divalent, and trivalent electrolytes (from the right). Reproduced from Overbeek (1952) with permission.

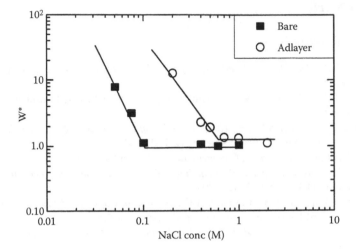

FIGURE 3.23 Stability ratio for bare (black squares) and polymer-coated (white circles) latex particles. Reproduced from Einarson and Berg (1993) with permission from Elsevier.

Another important problem that has attracted the attention of a host of investigators (Ruckenstein and Prieve, 1973; Saville, 1977; Spielman, 1977; Tien and Payatakes, 1979) is particle collection. In the deep bed filtration of colloidal particles, one seeks to describe the interaction and the collision between one colloidal particle and one grain of the packing material that forms the bed. The latter, called the collector, is immobile. The liquid containing the suspended colloidal particles flows past the collectors and flocculation of the colloid particles with the grains of the packing material is called particle capture. The particles are brought to the collector surface both by convection and diffusion.

From fluid mechanical calculations, if $v^{(o)}$, the velocity field around the collector in the absence of the colloidal particles is known, then v_p, the velocity field of the particles, can also be found, at least to a good approximation. Moreover, the particle diffusion coefficient becomes position dependent since the Stokes drag force used to derive Equation 3.72 is modified by the presence of the nearby collector surface. The conservation equations (Equations 3.75 and 3.77) must therefore be replaced with the appropriate one:

$$\nabla \bullet (c v_p) = \nabla \bullet \left[D(r) \left\{ \nabla c + \frac{c}{kT} \nabla \varphi \right\} \right]. \qquad (3.84)$$

Equation 3.84 is solved for a single collector with the same boundary conditions as before. The flux to a single collector can thus be calculated. In some cases it is possible to solve Equation 3.84 using the bulk value of D given by Equation 3.72 and formulate a suitable boundary condition that includes particle-collector interactions and hydrodynamic effects in the immediate vicinity of the wall (Ruckenstein and Prieve, 1973).

REFERENCES

GENERAL REFERENCES

Derjaguin, B.V., *The Theory of Stability of Colloids and Thin Films*, Consultants Bureau, New York, 1989.
Derjaguin, B.V., Churaev, N.V., and Muller, V.M., *Surface Forces*, Consultants Bureau, New York, 1987.
Evans, D.F. and Wennerström, H., *The Colloidal Domain*, 2nd ed., Wiley, New York, 1999.
Fleer, G.J., Cohen Stuart, M.A., Scheutjens, J.M.H.M., Cosgrove, T., and Vincent, B., *Polymers at Interfaces*, Chapman and Hall, London, 1993.
Hiemenz, P.C. and Rajagopalan, R., *Principles of Colloid and Surface Chemistry*, 3rd ed., Marcel Dekker, New York, 1997.
Hunter, R.J., White, L.R., and Chan, D.Y.C., *Foundations of Colloid Science*, vols. 1 and 2, Oxford University Press, New York, 1987, 1992.
Israelachvili, J.N., *Intermolecular and Surface Forces*, 2nd ed., Academic Press, New York, 1991.
Kruyt, H.R. (ed.), *Colloid Science*, vols. 1 and 2, Elsevier, Amsterdam, 1952.
Levich, V.G., *Physico-Chemical Hydrodynamics*, Prentice-Hall, Englewood Cliffs, New Jersey, 1962.
Lyklema, J. (ed.), *Fundamentals of Interface and Colloid Science*, vols. 1 and 2, Academic Press, New York, 1991.
Napper, D.H., *Polymeric Stabilization of Colloidal Dispersions*, Academic Press, New York, 1983.
Russel, W.B., Saville, D.A., and Schowalter, W.R., *Colloidal Dispersions*, Cambridge University Press, Cambridge, 1989.
Shchukin, E.D., *Colloid Chemistry*, Elsevier, Amsterdam, 2001.
Verwey, E.J.W. and Overbeek, J.T.G., *Theory of Stability of Lyophobic Colloids*, Dover, Mineola, New York, 1999 [originally published by Elsevier, Amsterdam, 1948].

TEXT REFERENCES

Albers, W. and Overbeek, J.T.G., Stability of emulsions of water in oil: I. The correlation between electrokinetic potential and stability, *J. Colloid Sci.*, 14, 501, 510, 1959.
Aniskin, S.V., Protod'yakonov, I.O., and Ionov, V.A., Experimental study of hydrogen sulfide desorption using a monodisperse – droplet generator, *Russian J. Appl. Chem.*, 71, 1158, 1998.
Bibette, J., Depletion interactions and fractionated crystallization for polydisperse emulsion purification , *J. Colloid Interface Sci.*, 147, 474, 1991.
Christenson, H.K., Horn, R.G., and Israelachvili, J.N., Measurement of forces due to structure in hydrocarbon liquids, *J. Colloid Interface Sci.*, 88, 79, 1982.
Cohen Stuart, M.A., Scheutjens, J.M.H.M., and Fleer, G.J., Polydispersity effects and the interpretation of polymer adsorption isotherms, *J. Polym. Sci. Polym. Phys. Ed.*, 18, 559, 1980.
Collier, C.P., Vossmeyer, T., and Heath, J.R., Quantum dot superlattices, *Annu. Rev. Phys. Chem.*, 49, 371, 1998.
Daoud, M. and de Gennes, P.G., Statistics of macromolecular solutions trapped in a small pore, *J. Phys. (Paris)*, 38, 85, 1977.

de Gennes, P.G., *Scaling Concepts in Polymer Physics*, Cornell University Press, Ithaca, New York, 1979.

Dzyaloshinskii, I.E., Lifshitz, E.M., and Pitaevski, L.P., Van der Waals forces in liquid films, *Soviet Phys. JETP*, 37, 161, 1960.

Einarson, M.B. and Berg, J.C., Electrosteric stabilization of colloidal latex dispersions, *J. Colloid Interface Sci.*, 155, 165, 1993.

Evans, D.F. and Ninham, B.W., Ion binding and hydrophobic effect, *J. Phys. Chem.*, 87, 5025, 1983.

Evans, R. and Napper, D.H., Theoretical prediction of the elastic contribution to steric stabilization, *J. Chem. Soc. Faraday Trans. I*, 73, 390, 1977.

Everett, D.H., *Basic Principles of Colloid Science*, Royal Society of Chemistry, London, 1988, p. 210.

Faraday Discussions of the Chemical Society, Concentrated Colloidal Dispersions, Royal Society of Chemistry, London, 1983.

Fleer, G.J., Cohen Stuart, M.A., Scheutjens, J.M.H.M., Cosgrove, T., and Vincent, B., *Polymers at Interfaces*, Chapman and Hall, London, 1993.

Flory, P.J., *Principles of Polymer Chemistry*, Cornell University Press, Ithaca, New York, 1953, p. 402.

Gast, A.P., Hall, C.K., and Russel, W.B., Phase separations induced in aqueous colloidal suspensions by dissolved polymer, *Faraday Discuss. Chem. Soc.*, 76, 189, 1983.

Gregory, J., Rates of flocculation of latex particles by cationic polymers, *J. Colloid Interface Sci.*, 42, 448, 1973.

Hamaker, H.C., The London-van der Waals attraction between spherical particles, *Physica*, 4, 1058, 1937.

Hachisu, S., Kobayashi, Y., and Kose, A., Phase separation in monodisperse latexes, *J. Colloid Interface Sci.*, 42, 342, 1973.

Hazlett, R.D. and Schechter, R.S., Stability of macroemulsions, *Colloids Surfaces*, 29, 53, 71, 1988.

Hesselink, F.T., Theory of the stabilization of dispersions by adsorbed macromolecules. I. Statistics of the change of some configurational properties of adsorbed macromolecules on the approach of an impenetrable interface, *J. Phys. Chem.*, 75, 65, 1971.

Hesselink, F.T., Vrij, A., and Overbeek, J.T.G., Theory of the stabilization of dispersions by adsorbed macromolecules. II. Interaction between two flat particles, *J. Phys. Chem.*, 75, 2094, 1971.

Hiemenz, P.C. and Rajagopalan, R., *Principles of Colloid and Surface Chemistry*, 3rd ed., Marcel Dekker, New York, 1997.

Hiltner, A. and Krieger, I.M., Diffraction of light by ordered spheres, *J. Phys. Chem.*, 73, 2386, 1969.

Israelachvili, J.N., *Intermolecular and Surface Forces*, 2nd ed., Academic Press, New York, 1991.

Israelachvilli, J.N. and Adams, G.E., Preparation and characterization of monodispersed metal hydrous oxide sols, *J. Chem. Soc. Faraday Trans. I*, 74, 975, 1978.

Jackson, J.D., *Classical Electrodynamics*, 2nd ed., Wiley, New York, 1975.

Kerker, M., *The Scattering of Light*, Academic Press, New York, 1969, p. 351 ff.

Kruyt, H.R. and van Arkel, A.E., The velocity of flocculation of selenium sol. I. Flocculation by potassium chloride. & II. Flocculation by means of barium chloride, *Rec. Trav. Chim.*, 39, 656, 1920; 40, 169, 1921.

Kruyt, H.R. and van Arkel, A.E., The velocity of coagulation of selenium sols, *Kolloid.-Z.*, 32, 29, 1923.

Levich, V.G., *Physico-Chemical Hydrodynamics*, Prentice-Hall, Englewood Cliffs, New Jersey, 1962, p. 472 ff.

Lin, W.-Y. and Blum, F.D., Segmental dynamics of bulk and adsorbed poly(methyl acrylate)-d_3 by deuterium NMR: effect of adsorbed amount, *Macromolecules*, 30, 5331, 1997.

Loeb, A.L., Overbeek, J.T.G., and Wiersema, P.H., *The Electrical Double Layer Around a Spherical Colloid Particle*, MIT Press, Cambridge, MA, 1961, p. 37.

Lovell, P.A. and El-Aasser, M.S. (eds.), *Emulsion Polymerization and Emulsion Polymers*, John Wiley & Sons, Chichester, 1997.

Mace, G.R., Specifier's guide to polymer feed system, *Pollution Eng.*, 22, 75, 1990.

Mahanty, J. and Ninham, B.W., *Dispersion Forces*, Academic Press, New York, 1976.

Matijevic, E., Preparation and characterization of monodispersed metal hydrous oxide sols, *Prog. Colloid Polym. Sci.* 61, 24, 1976.

Napper, D.H., Polymeric stabilization, in *Colloidal Dispersions*, J.W. Goodwin (ed.), Royal Society of Chemistry, London, 1982, p. 99 ff.

Napper, D.H., *Polymeric Stabilization of Colloidal Dispersions*, Academic Press, New York, 1983.

Napper, D.H. and Hunter, R.J., *MTI International Reviews of Science, Physical Chemistry, Surface Chemistry and Colloids*, series 1, vol. 7, M. Kerker (ed.), Butterworths, London, 1972.

Ninham, B.W., Long-range vs. short-range forces. The present state of play, *J. Phys. Chem.*, 84, 1423, 1980.

O'Brien, R.W., Cannon, D.W., and Rowlands, W.N., Electroacoustic determination of particle size and zeta potential, *J. Colloid Interface Sci.*, 173, 406, 1995.

O'Brien, R.W. and White, L.R., Electrophoretic mobility of a spherical colloidal particle, *J. Chem. Soc. Faraday II*, 74, 1607, 1978.

Ottewill, R.H., Billett, D.F., Gonzalez, G., Hough, D.B., and Lovell, V.M., The variation of contact angle at the silver iodide-liquid-vapor interface with the charge on the solid surface, in *Wetting, Spreading and Adhesion*, J.F. Padday (ed.), Academic Press, London, 1978, p.183.

Overbeek, J.T.G., Kinetics of flocculation, in *Colloid Science*, vol. 1, H.R. Kruyt (ed.), Elsevier, Amsterdam, 1952, p. 245 ff.

Parfitt, G.D. and Peacock, J., Stability of colloidal dispersions in nonaqueous media, *Surface and Colloid Science*, vol. 10, E. Matijevic (ed.), Plenum Press, New York, 1978, p. 163 ff.

Parsegian, V.A., Long range van der Waals forces, in *Physical Chemistry: Enriching Topics from Colloid and Surface Science*, H. van Olphen and K. Mysels (eds.), Theorex, La Jolla, California, 1975, p. 27 ff.

Patel, S.S. and Tirrell, M., Measurement of forces between surfaces in polymer fluids, *Annu. Rev. Phys. Chem.*, 40, 597, 1989.

Pefferkorn, E., The role of polyelectrolytes in the stabilisation and destabilisation of colloids, *Adv. Colloid Interface Sci.*, 56, 33, 1995.

Prieve, D.C., Measurement of colloidal forces with TIRM, *Adv. Colloid Interface Sci.*, 82, 93, 1999.

Qutubuddin, S., Miller, C.A., Benton, W.J., and Fort, T., Jr., Effects of polymers, electrolytes and pH on microemulsion phase behavior, in *Macro- and Microemulsions*, ACS Symposium Series, Shah, D.O. (ed.), American Chemical Society, Washington, D.C., 1985, p. 223.

Rao, I.V. and Ruckenstein, E., Phase behavior of mixtures of sterically stabilized colloidal dispersions and free polymer, *J. Colloid Interface Sci.*, 108, 389, 1985.

Riess, G., Periard, J., and Banderet, A., Emulsifying effects of block and graft copolymers-oil in oil emulsions, in *Colloidal and Morphological Behavior of Block and Graft Copolymer*, G.E. Molau (ed.), Plenum Press, New York, 1971, p. 173.

Ruckenstein, E. and Prieve, D.C., Rate of deposition of Brownian particles under the action of London and double-layer forces, *J. Chem. Soc. Faraday II*, 69, 1522, 1973.

Sato, T. and Ruch, R., *Stabilization of Colloidal Dispersions by Polymer Adsorption*, Marcel Dekker, New York, 1980.

Saville, D.A., Electrokinetic effects with small particles, *Annu. Rev. Fluid Mech.*, 9, 321, 1977.

Scheutjens, J.M.H.M. and Fleer, G.J., Effect of polymer adsorption and depletion on the interaction between two parallel surfaces, *Adv. Colloid Interface Sci.*, 16, 361, 1982.

Sheludko, A., *Colloid Chemistry*, Elsevier, Amsterdam, 1966, p. 208 ff.

Sperry, P.R., Hopfenberg, H.B., and Thomas, N.L., Flocculation of latex by water-soluble polymers: Experimental confirmation of a nonbridging, nonadsorptive, volume-restriction mechanism, *J. Colloid Interface Sci.* 82, 62, 1981.

Spielman, L.A., Particle capture from low-speed laminar flows, *Annu. Rev. Fluid Mech.*, 9, 297, 1977.

Tanford, C., *Physical Chemistry of Macromolecules*, Wiley, New York, 1961, p. 526.

Tien, C. and Payatakes, A.C., Advances in deep bed filtration, *AIChE J.*, 25, 737, 1979.

Tuorila, P., The rapid and slow coagulation of polydispersed systems: gold and alumina dispersions, *Kolloidchem. Beihefte*, 22, 191, 1926.

Verwey, E.J.W. and Overbeek, J.T.G., *Theory of Stability of Lyophobic Colloids*, Dover, Mineola, New York, 1999 [originally published by Elsevier, Amsterdam, 1948].

Vincent, B., The effect of adsorbed polymers on dispersion stability, *Adv. Colloid Interface Sci.*, 4, 193, 1974.

Vincent, B., Edwards, J., Emmett, S., and Jones, A., Depletion flocculation in dispersions of sterically-stabilised particles ("soft spheres"), *Colloids Surfaces* 18, 261, 1986.

Vincent, B., Luckham, P.F., and Waite, F.A., The effect of free polymer on the stability of sterically stabilized dispersions, *J. Colloid Interface Sci.*, 73, 508, 1980.

Vrij, A., Polymers at interfaces and the interactions in colloidal dispersions, *Pure Appl. Chem.*, 48, 471, 1976.

Walstra, P., Formation of emulsions, in *Encyclopedia of Emulsion Technology*, vol. 1, P. Becher (ed.), Marcel Dekker, New York, 1983, p. 57.

Wang, X., Zhuang, J., Peng, Q., and Li, Y., A general strategy for nanocrystal synthesis, *Nature (London)*, 437, 121, 2005.

Whitefield, P.D., Hagan, D.E., and Lilenfeld, H.V., *Impact of Aircraft Emissions upon the Atmosphere — International Colloquium*, Paris, October 1996.

Zeichner, G.R. and Schowalter, W.R., Effects of hydrodynamic and colloidal forces on the coagulation of dispersions, *J. Colloid Interface Sci.*, 71, 237, 1979.

Zeichner, G.R. and Schowalter, W.R., Use of trajectory analysis to study stability of colloidal dispersions in flow fields, *AIChE J.*, 23, 243, 1977.

PROBLEMS

3.1   Use Equations 3.6 and 3.50 to calculate the number of extrema in the interaction potential E_T^S between two identical spheres in an aqueous solution. A_H is 10^{-14} ergs or greater and ψ_o ranges from 5 to about 200 mV. Further,

$N_A = 6.023 \times 10^{23}$ molecules/mole
$e_o = 1.602 \times 10^{-19}$ C
$k = 1.380 \times 10^{-23}$ J/K
e_oV = charge on an electron times 1 volt = 1.602×10^{-19} J.

Note that 1 J/1 C = 1 V. Permittivity of vacuum ε_o is 8.854×10^{-12} $C^2/(N \cdot m^2) = C^2/(J \cdot m)$ = farad/m. The relative permeability or dielectric constant ε (often with a subscript r) is about 78.5 for water and 42.5 for glycerol. These are two commonly available materials with high dielectric constants.

For this problem, use the values of $T = 300$ K, $R = 80$ nm, and $c = 0.001$ mole/ℓ and 0.1 mole/ℓ. Show first that for 1:1 electrolyte in water, the Debye length works out to be $\kappa^{-1} = 0.304/c^{1/2}$ nm, where c is the concentration in moles per liter.

3.2   Integrate Equation 3.29 to calculate u as a function of z for a double layer emanating from a single plate. For $\upsilon = 1$, plot u as a function of κz. If $\psi_o = 300$ mV and $c_o = 0.0001$ mole/ℓ, what is the value of c_- at $z = 0$? Can an explanation be offered for this result?

Derive Equation 3.46 from the results.

Assume that the interface in Figure 1.2 is flat, of unit area, and the z-coordinate is normal to the interface. Integrate the Poisson equation (Equation 3.19) in the two phases in the absence of external fields and use the Gauss relation of Equation 3.21 to show that

$$\int_{-\infty}^{0} \rho_{eA}\, dz + \int_{0}^{\infty} \rho_{eB}\, dz + \sigma = 0 .$$

What is the significance of this result?

3.3   Consider an electrical double layer made up of a compact layer with a dielectric constant ε' and a thickness δ and a diffuse layer with dielectric constant ε. As explained in the text, δ is approximately the radius of a hydrated ion. Calculate ψ_o, the potential at the junction of the two layers, in a 0.001 M aqueous solution of a univalent electrolyte at 25°C where the potential ψ_s on the solid surface is 137 mV. Further, $\varepsilon' = 10$ and $\delta = 0.25$ nm.

3.4 If c_k is the number density of flocs with k particles in them, it can be shown that for rapid coagulation,

$$c_k = \frac{c_\infty (t/T^*)^{k-1}}{\left(1+\dfrac{t}{T^*}\right)^{k-1}},$$

where $T^* = (3\mu/4kTc_\infty)$ and c_∞ is the number density in the original monodispersed system. Furthermore, the total number density is given by

$$\sum_{k=1}^{\infty} c_k = \frac{c_\infty}{\left(1+\dfrac{t}{T^*}\right)}.$$

Calculate T^* for each of the two sets of the data reported by Tuorila (1926).

	t (sec)	Number of flocs $\times 10^{-8}$
Gold sol	0	20.20
	30	14.70
	60	10.80
	120	8.25
	240	4.89
	480	3.03
Kaolin	0	5.0
	105	3.90
	180	3.18
	255	2.92
	335	2.52
	420	2.00
	510	1.92
	600	1.75
	1020	1.54
	2340	1.15

How would you plot the data to get a straight line?

3.5 Overbeek (1952) argues that

$$W = \frac{1}{2\kappa R} \exp(\varphi_{max}/kT).$$

Calculate φ_{max} from Equations 3.7 and 3.47 and obtain Equation 3.68.

3.6   (a) The stability ratio W is given by Equation 3.67. Let us assume that the main contribution to W is for values of r^*, where φ/kT is near its maximum value (φ_m/kT). In particular, use a Taylor series to write

$$\frac{\varphi}{kT} \cong \frac{\varphi_m}{kT} - p^2(r - r_m)^2 \,,$$

where

$$p^2 = -\frac{1}{2kT}\frac{d^2\varphi}{dr^2}\Big|_{r_m} \,.$$

Recalling that for $\kappa R \gg 1$, the separation distance h_m for $\varphi = \varphi_m$ is much less than the particle radius R and assuming that ph_m is relatively large, show that

$$W \cong \frac{2R\pi^{1/2}}{pr_m^2}\exp\left(\frac{\varphi_m}{kT}\right). \qquad (3.6.\text{i})$$

Hint: Manipulate the integral to get the usual form used in defining the error function.

   (b) Taking $\kappa = 10^6/\text{cm}^{-1}$ and other values as in Figure 3.11, calculate W from Equation 3.6.i and from Overbeek's approximate expression given in Problem 3.5.

3.7   Equation 3.79 can be written as

$$-\frac{dc_1}{dt} = \frac{1}{WT^* c_\infty}c_1^2$$

using the notation of Problem 3.4. For slow coagulation, one may also assume that the total number of particles is the same as in Problem 3.4, with WT^* replacing T^*. Analyze the data of Kruyt and van Arkel

(1920, 1921, 1923) on the coagulation of selenium sol of 52 μm diameter particles with 50 mmols/ℓ potassium chloride (KCl).

t (hours)	Number of particles per cm³ × 10⁻⁸
0	33.5
0.25	32.3
22.5	28.6
42.5	19.1
67.5	14.6
187.5	7.5
239	7.5
335	4.7
1008	1.46

Obtain the value of W if T^* is calculated to be 20 sec.

3.8   Why are river waters so muddy while sea water is virtually devoid of particulate material? Suggest a reason for delta formation at the confluence.

3.9   In an electrochemical system, the chemical potential μ_i of the ith ion is replaced by the electrochemical potential η_i, where

$$\eta_i = \mu_i + z_i e_o \psi \qquad (3.9.i)$$

and z_i is the algebraic charge of the ith ion.

(a) Using the method described in Section 7, show that the flux of the ith ion is

$$J_i = -D\left[\nabla c_i + \frac{c_i e_o z_i}{kT}\nabla\psi\right]. \qquad (3.9.ii)$$

This is known as the Nernst-Planck equation. Compare this with Equation 3.73 and interpret the factor $e_o z_i \nabla\psi$. Show also that the Boltzmann distributions given following Equation 3.21 are obtained at equilibrium when the fluxes are zero.

(b) Using the definition of γ from Equation 1.13, the Gibbs adsorption equation (Equation 1.24) at constant temperature, and Equation 3.9.i, show that

$$\Delta F^S = -\int_0^{\psi_0} \sigma(\psi_o')\,d\psi_o' \,, \qquad\qquad (3.9.iii)$$

where ΔF^s is the change in the surface excess free energy per unit area if the interface acquires charge, all other things remaining the same. Using the results of Problem 3.2, show that the change is negative (i.e., that the electrical double layer forms spontaneously).

4 Surfactants

1. INTRODUCTION

Surfactants have long been known to us in the form of soaps, which are salts of naturally occurring fatty acids. Some of their properties have long been recognized, for example, the ability of alkali to convert fatty acids to soaps, the usefulness of sodium soaps for cleaning, and the precipitation of insoluble compounds in hard water (bathtub ring). The advent of petrochemicals has brought synthetic detergents, which are less readily precipitated. The most widely used of these today include (sodium) linear alkylbenzene sulfonates (LASs), alkyl sulfates, and ethoxylated long-chain alcohols. Such surfactants with straight hydrocarbon chains are preferred for common washing applications because they are more readily biodegradable than their branched-chain counterparts and hence are less likely to accumulate in rivers and lakes where they could cause undesirable foaming.

While cleaning is the largest and most widely known application of surfactants, they are also specialty chemicals used in a variety of products including agricultural chemicals, personal care products, pharmaceuticals, and oilfield chemicals. For about 15 years beginning around 1970, much attention was directed toward the possible use of surfactants for increasing the recovery of petroleum from underground formations (Shah and Schechter, 1977). Although such processes have not as yet proved to be economically attractive, their prospects have improved with increasing oil prices. Moreover, they simulated research that revealed much of the information on microemulsions presented later in this chapter. In recent years the microstructure provided by surfactant aggregates in micellar solutions, microemulsions, and lyotropic liquid crystals has attracted attention because of possible applications in separation processes (Scamehorn and Harwell, 1989) such as removal of organic and metal contaminants from waste water streams, in controlled release of drugs, in modifying rheologic properties of fluids, in enhancing the rates and selectivities of certain chemical reactions, and in nanoparticle synthesis. The ability of surfactants to form ordered films at solid-liquid interfaces is also of interest for some high-technology applications. Self-assembly of surfactants and polymers in solution and at interfaces is expected to become more important in processing as applications of nanotechnology increase (Texter and Tirrell, 2001).

It should be emphasized that the importance of surfactants in biological systems has for many years been a strong motivation for developing an improved understanding of the equilibrium phase behavior and other aspects of surfactant systems. Phospholipids, the primary component of cell membranes, are a type of surfactant. So are bile salts, which play a major role in digestive processes.

165

Surfactants also occur as lining material in the lungs. Diseases associated with malfunctioning of these tissues or processes often involve surfactant behavior.

A brief introduction to surfactants is given in Chapter 1. Because water has a high surface tension, many materials, both soluble and insoluble, display some degree of surface activity, i.e., they reduce the surface tension of water. (The strong electrolytes at high concentrations form notable exceptions.) Surfactants of practical interest reduce surface tension dramatically at low concentrations and, in addition, possess aggregation or structure-forming capabilities. A surfactant molecule has a nonpolar tail and a polar, often ionizable head group. The tail, usually a long hydrocarbon chain, but sometimes a fluorocarbon chain or a permethylated siloxane group, acts to reduce solubility in water while the head group has the opposite effect. Many surfactants have tails with only one hydrocarbon chain, but molecules with two chains joined to the same polar group are also common and indeed are the rule in biological membranes. Block copolymers such as those of ethylene oxide (EO) and propylene oxide (PO) behave as surfactants since an EO chain is quite polar and acts as a head group, while a PO chain acts as a tail. These compounds and other amphiphilic block copolymers have received considerable attention in recent years (Alexandridis and Lindman, 2000).

Surfactants are often classified according to their head groups as anionic, cationic, nonionic, and zwitterionic or amphoteric. Anionic surfactants, which make up about 60% of surfactant production in the United States, are mainly sulfonates such as LASs, alkyl or alcohol ether sulfates, and carboxylates (e.g., soaps). About 30% of surfactant production consists of nonionic surfactants, chiefly alcohol and alkylphenol ethoxylates. Cationic surfactants, e.g., alkyl amines and quaternary ammonium compounds such as $R\text{-}N\,(CH_3)_3^+Br^-$, which are less sensitive to pH changes than primary amines, make up most of the remaining surfactant production, with zwitterionics being less than 1%. Many zwitterionic surfactants become ionic with a suitable change in pH. For instance, the oxygen atom of amine oxides is protonated at low pH, thereby making the surfactant cationic. Information on surfactant chemistry is provided by Jönsson et al. (1998) and Rosen (2004), among others. Individual volumes of the *Surfactant Science Series* (see general references at the end of the chapter) provide more details on various types of surfactants.

Most surfactants with hydrocarbon chains do not significantly reduce the surface tension of hydrocarbon liquids. The situation is quite different, however, for surfactants with fluorocarbon chains and siloxane surfactants, which do exhibit significant surface activity when added to such liquids.

2.   MICELLE FORMATION

Figure 4.1 shows the surface tension of potassium laurate, a soap whose tail contains 12 carbon atoms, as a function of its concentration (Roe and Brass, 1954). A notable feature in the plot is the abrupt change in slope at a particular concentration. In the vicinity of this concentration, a host of properties of the

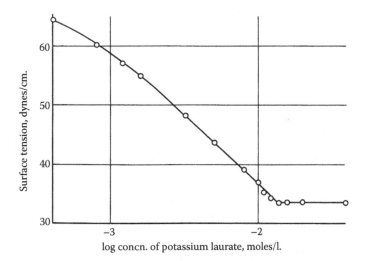

FIGURE 4.1 Surface tension as a function of surfactant concentration. The break in the slope is seen distinctly. Reprinted with permission from Roe and Brass (1954). Copyright 1954 American Chemical Society.

bulk solution also change their rate of variation with concentration, e.g., density, osmotic pressure, electrical conductivity, light scattering properties, capability of dissolving nonpolar compounds, etc. The light scattering experiments show that aggregates, which are known as micelles, form beyond this "critical micellar concentration" (CMC). Addition of surfactant to the solution beyond the CMC produces, for the most part, only more micelles, i.e., the concentration of the dissolved monomer does not increase significantly. Consequently the surface tension remains nearly constant, as shown in Figure 4.1. Values of surface tension of 25 to 40 mN/m at concentrations greater than the CMC are typical.

Block copolymers are more complex. Only a few remarks are made below, and immediately after we revert to the more common surfactants, a plan that is followed in the rest of the book. A linear hydrophilic polymer such as polyethylene oxide (PEO) is attached at one of its terminals to a more hydrophobic polymer such as polypropylene oxide (PPO). The result is an amphiphile PEO-PPO, called a diblock copolymer. Similarly, PEO-PPO-PEO is a triblock copolymer, another common block copolymer. The blocks range up to hundreds of repeat units. The insoluble block can crystallize or form glass. Ionic blocks are also available, although the nonionic block copolymers are more common.

In water, PEO-PPO-PEO shows two breaks in the slopes of surface tension versus surfactant concentration plots compared to one in more common pure surfactants (Figure 4.1). The break at lower concentration has been accepted by Alexandridis et al. (1994) as CMC, and has been described by Yu et al. (1992) as the point of "monomolecular micelle formation." The break point at higher concentration beyond which the surface tension does not change has been accepted as the CMC by Yu et al. (1992), who describe it as the point of

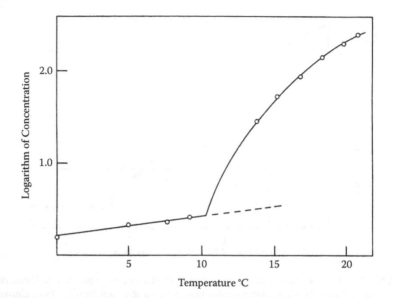

FIGURE 4.2 Solubility of SDS in water. The break in the slope represents the Krafft point. Reprinted from McBain and Hutchinson (1955) with permission.

"polymolecular micelle formation." It is not clear why they are different, but the polymer configurational changes at the water-air interface are assumed to play a role. Because of the surfactants' polymeric nature, the CMCs are very low, ranging from about 10^{-7} to 10^{-2} moles/l, and decrease rapidly with increases in temperature.

In Figure 4.2, the concentration-temperature diagram is shown for sodium dodecyl sulfate (SDS) in water. The solid line denotes the solubility limit; the dashed line is the solubility that is expected based on extrapolation of the behavior at low temperatures. The rate of solubility increases with temperature changes abruptly at about 10°C, which is known as the Krafft point. Below this temperature, no micelles are present and the solid surfactant, in which the hydrocarbon chains are rather rigid, is formed at the solubility limit. However, above the Krafft point, micelles whose hydrocarbon chains are much more flexible form at concentrations above the dashed line, which represents the CMC. At much higher concentrations (not shown), the solubility limit is ultimately reached and a liquid crystalline phase also having flexible chains separates. Such phases are discussed in Section 4.

The Krafft point can be lowered by making it more difficult for the hydrocarbon chain to crystallize and form a solid phase. Decreasing chain length, using branched instead of straight chains, incorporating an EO chain, and adding a second surfactant with a significantly different chain length are common methods of accomplishing this objective and thereby promoting micelle formation at lower temperatures. For example, the Krafft point of sodium alkyl sulfates decreases by about 7°C for each carbon atom removed from the tail. Adding three EO

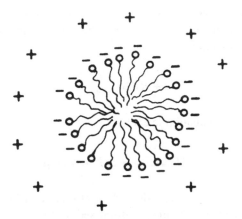

FIGURE 4.3 A schematic view of micellar aggregation of anionic surfactants.

groups to sodium hexadecyl sulfate reduces the Krafft point by about 25°C. The effect of ion type is less tractable. SDS, a common detergent, has a Krafft point of 10°C. On the other hand, potassium dodecyl sulfate has a Krafft point of 45°C, which makes it an insoluble salt at working temperatures, and practically useless (Missel et al., 1982). A table of Krafft points can be found in Rosen's book (2004).

Surfactant molecules form micelles because this aggregation of perhaps 50 to 100 molecules largely removes the hydrocarbon chains from contact with water and the overall structure as seen by water is hydrophilic. A section of a spherical micelle is shown schematically in Figure 4.3. The conventional explanation is that in this form the hydrocarbon chains are shielded from water. Whereas basically correct, this picture is an oversimplification. Detailed consideration of micelle geometry indicates that the chains are more randomly arranged throughout the micelle interior than shown in Figure 4.3 (Ben-Shaul and Gelbart, 1994) and they are not totally shielded by the head groups in typical systems.

It is evident that for micelles to exist the Gibbs free energy of their formation has to be negative. That micelles form only above the CMC indicates that there are both positive and negative contributions to the Gibbs energy, and only above the CMC do the latter prevail. One may write $\Delta G = \Delta H - T\Delta S$ at constant temperature to separate an energy contribution ΔH from an entropic contribution ΔS. One expects that if the hydrophobic tails of surfactants are shielded, ΔH will decrease, as the hydrocarbon and water are incompatible, but hydrocarbon tails among themselves are not. This decrease is partly offset by the fact that head groups repel one another when brought close together (e.g., owing to like electrical charges). Noting that ΔS is a measure of randomness, one might expect ΔS to be negative on micellization, at least as far as the surfactant molecules are concerned. Thus one supposes that only when ΔH is sufficiently negative can micellization be realized. These effects are entered as "primary" and "electrostatic" in Table 4.1. As will be discussed shortly, this simple view is incomplete.

The thermodynamics of micelle formation can be analyzed with varying degrees of complexity. The simplest is to consider that a phase separation into micellar and aqueous phases occurs at a single concentration, the CMC. Requiring that the chemical potential of the surfactant be the same in the two phases, one obtains for a pure nonionic surfactant

$$\mu^o + RT \ln x_{CMC} = \mu_m^o . \tag{4.1}$$

Rearrangement of this equation leads to

$$RT \square n x_{CMC} = \Delta \bar{G}^o = \mu_m^o - \mu^o . \tag{4.2}$$

The phase separation model is particularly useful in describing micelles formed in solutions of surfactant mixtures, as discussed below.

Experimental data indicate that the changes in rates of variation of physical properties near the CMC actually occur over a narrow range of concentrations and not discontinuously at a single concentration. Hence a chemical reaction or mass action model should be more realistic than the phase separation model for describing the thermodynamics of micellization. That is, we consider that micelle formation occurs as follows:

$$N \square M_1^- + N \alpha \square^+ \rightleftarrows M_N^{-N(1-\alpha)} .$$

Here N monomers M_1^- aggregate to form a micelle M_N, the charge of which is determined by the number $N\alpha$ of the counterions C^+ it binds with. From thermodynamics

$$K' = \frac{a_N}{a_C^{N\alpha} a_1^N} , \tag{4.3}$$

where K' is the equilibrium constant, a is the activity, and the subscripts N, C, and 1 denote the micelles, counterions, and single amphiphiles. If the activity coefficients are taken to be one, then after some rearrangement one has

$$K = \frac{C_N}{C_C^{N\alpha} C_1^N} , \tag{4.4}$$

where C represents concentration, and finally

$$\frac{C_N^*}{C_C^{*N\alpha} C_1^{*N}} = 1 , \tag{4.5}$$

where $C^* = C / K^{\frac{1}{1-N(1+\alpha)}} = C / C_{CMC}$. The reason for this definition of CMC is readily seen. When the surfactant concentration is less than this value, both C_1^*

and C_c^* are less than one. Since N is large, the denominator of Equation 4.5 is very small, and so is C_N^*. Thus few micelles are present and the system is below the CMC. In contrast, when $C_1^* > 1$, the denominator of Equation 4.5 and C_N^* are very large, most of the surfactant is present in micelles, and the system is above the CMC. Figure 4.4 shows dimensionless monomer C_1^* and micelle C_N^* concentrations as a function of total surfactant concentration C_T^* $(=C_1^* + N\,C_N^*)$ for N=2 and N=40. In the former case micelles (dimers) begin to appear when C_T^* is only about 0.1, and both monomer and micelle concentrations rise gradually. For N=40 C_1^* and C_T^* are alomost equal until they reach a value of about 0.8, at which point there is a sharp change in slope of the C_1^* curve, and micelles begin to appear. As Figure 4.4 indicates, micelles begin to form at lower values of C_T^* when salt is added.

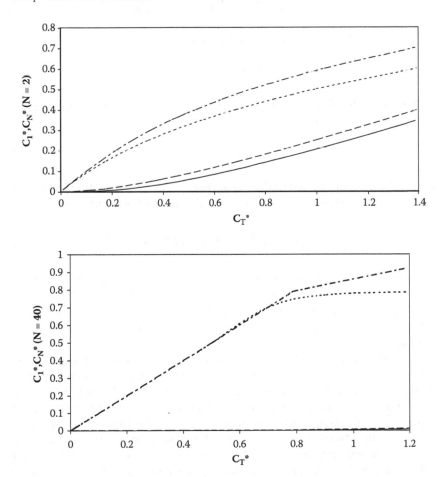

FIGURE 4.4 Monomer C_1^* and micelle C_N^* concentrations depicted as a function of total surfactant concentration for (a) N = 2 and (b) N = 40. In both cases dash-dot and dotted lines are C_1^* with no added electrolyte ($C_c^* = C_1^*$) and with added salt ($C_c^* = C_1^* + 5.0$) respectively. Bold and dashed lines are C_N^* in the absence of and with added salt.

If it is assumed that the activity coefficients are one, that the concentration of micelles is constant at the CMC, and that the standard state for the micelles is chosen to make $a_N = 1$ under these conditions,

$$K' = \frac{1}{x_C^{N\alpha} x_1^N} , \qquad (4.6)$$

where x denotes a mole fraction. Immediately below the CMC, $x_1 \sim x_{CMC}$, the mole fraction of surfactant ions at the CMC. When there are no added electrolytes, $x_C = x_1 = x_{CMC}$, hence

$$- \ln K' = N(1 + \alpha) \ln (x_{CMC}). \qquad (4.7)$$

If there is complete charge neutralization, $\alpha = 1$ and

$$- \ln K' = 2N \ln x_{CMC}. \qquad (4.8)$$

If there is no binding with the counterions, or for the case of the nonionic surfactants, $\alpha = 0$ and

$$- \ln K' = N \ln x_{CMC}. \qquad (4.9)$$

If $\Delta \bar{G}°$ is the standard state Gibbs free energy change per mole of monomer, one obtains the following expression when Equation 4.9 is applicable:

$$\Delta \bar{G}° = -(RT / N) \ln K' = RT \ln x_{CMC} . \qquad (4.10)$$

This equation, which holds for nonionic surfactants, is consistent with Equation 4.2 for the phase separation model. It can be generalized by using Equation 4.7 instead of Equation 4.9 and provides a method for obtaining $\Delta \bar{G}°$ as a function of temperature. Then $\Delta \bar{H}°$ can be obtained by well-known thermodynamic relationships (see Problem 4.1).

As indicated above, $\Delta \bar{H}°$ is expected to be negative. From the experimental results, this is found not to be the case at low temperatures in many systems. Indeed, analysis of the data shows that negative values of $\Delta \bar{G}°$ for these conditions stem from large and positive values of the standard state entropy change $\Delta \bar{S}°$ (Shinoda, 1963). The source of the positive entropy changes was clarified by Ben-Naim (1971). Using a combination of thermodynamic reasoning and experimental data, he obtained the contribution δA^{HI} of "hydrophobic interactions" to the free energy of dimerization of simple hydrocarbons in various solvents (see Figure 4.5). The arrangement of the solvent molecules is considered to be ordered over a short range, but random over long ranges. If one molecule of a different species is inserted in the solvent, then the order of arrangement of

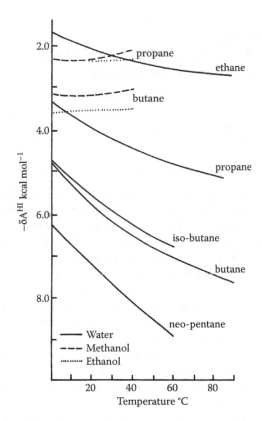

FIGURE 4.5 The free energy of dimerization of simple hydrocarbons in various solvents. The behavior in water is remarkably different. Reprinted from Ben-Naim (1971) with permission.

the solvent molecules around the probe changes, in a manner that decreases the entropy. This decrease in entropy is less if the mismatch between the sizes of the probe and the solvent molecules decreases. The change in entropy also becomes more favorable if two probe molecules dimerize to produce a larger body. Figure 4.5 shows that hydrophobic interactions favor dimerization and that the effect is greater in water than in other solvents. It should be noted that the term "hydrophobic interactions" survives in the literature even though the physical effect it describes is now known to be chiefly an entropic one.

O'Connell and Brugman (1977) provide a simple molecular model where instead of emphasizing the very polar nature of water, as did most previous investigators, they emphasize the size mismatch between the probe and the water molecules. Water, it is seen, is a solvent that is commonly available and which has the smallest size and hence the greatest mismatch with any probe. The entropy that is lost in gathering singly dispersed surfactant molecules into a micelle is more than offset by the decrease in the mismatch effects caused by removing hydrocarbon tails from contact with water (i.e., the hydrophobic effect). Accord-

TABLE 4.1
A short list of entropic and enthalpic
contributions to micellization

	$\Delta \bar{H}^\circ$	$\Delta \bar{S}^\circ$	$\Delta \bar{G}^\circ$	
Primary	–	–	–	Large
Electrostatic	+		+	Small
Hydrophobic		+	–	Large
"Ice"	+	+	0	None

ing to these authors, the hydrophobic effect is roughly a function of the volume and the surface area of the probe, where both are roughly proportional to the carbon number n_c of the tail in the case of surfactant molecules. This makes $\Delta \bar{G}^\circ$ proportional to n_c, to the first approximation. See Example 4.1 and Problem 4.2.

There is one other effect that gives rise to an increase in entropy on formation of micelles. It is known that water molecules form clusters around hydrophobic molecules (Franks, 1975). Since these form a structure "ice", that is, have an order, their entropies are lower than in the bulk. On the formation of micelles, the hydrocarbon tails are no longer in contact with water, that is, the molecules from the cluster are released into the water phase with an increase in entropy. Of course, there can be no structure formation at high temperatures, and this effect disappears there (Anacker, 1970). This effect is entered as "ice" in Table 4.1, where it is assumed that as it is zero in the melting of ice, in micellization the corresponding term must be very small.

Two special cases are considered. At high temperatures, ice cannot form. Table 4.1 shows that once ice stops forming when the temperature is increased, $\Delta \bar{H}^\circ$ will become more negative (in the absence of the last entry in the table), but $\Delta \bar{G}^\circ$ will not be affected. Evans and Wightman (1982) and Shinoda et al. (1987) investigated micellization at high temperatures and observed that $\Delta \bar{H}^\circ$ did become very large and negative as the temperature was increased. Much less variation is observed in the $\Delta \bar{G}^\circ$ values. It follows that $T\Delta \bar{S}^\circ$ also becomes large and negative, but the changes in $\Delta \bar{S}^\circ$ are seen to be less. For SDS at 25°C, $\Delta \bar{G}^\circ$, $\Delta \bar{H}^\circ$, and $T\Delta \bar{S}^\circ$ are –5.78, –4.99, and +0.77 kcal/mol and at 95.5°C they are –5.35, –11.34, and –6.0 kcal/mol (Evans and Wightman, 1982). In the other case, water is substituted with another polar solvent such as hydrazine, which has molecules that are small enough to give rise to hydrophobic effects comparable to water, but ice does not form. The fact that there is no ice gives rise to about the same values of $\Delta \bar{G}^\circ$ (T) and CMC during micellization in hydrazine as in water, but both $\Delta \bar{H}^\circ$ and $T\Delta \bar{S}^\circ$ are more negative for hydrazine than for water (Ramadan et al., 1983).

While the mass action model of Equations 4.3 through 4.10 is an improvement over the phase separation model, it clearly has significant shortcomings. The aggregation number N, for instance, is a parameter that must be determined experimentally or otherwise specified, that is, it does not arise from the analysis itself. Moreover, there is no firm basis for assuming that only aggregates with a single value of N are formed. Indeed, one would expect a distribution of aggregates containing various numbers of monomers.

These issues can be addressed in a more general model. For simplicity, we limit consideration here to nonionic surfactants. Monomer and N-mer are related by the following expression:

$$NM_1 = M_N.$$

Since, at equilibrium, the chemical potential of the surfactant must be the same in all aggregates, we have for the case of an ideal mixture in the aqueous solution:

$$N \, \mu_1^o + N \, RT \, \ln x_1 = [N\mu_N^o] + RT \, \ln(x_N / N) , \qquad (4.11)$$

where, for convenience, the standard chemical potential and the mole fraction of the N-mer have been expressed in terms of individual molecules, that is, x_N is the mole fraction of the total surfactant molecules present as N-mers with x_N/N, the actual mole fraction of N-mers in solution. Rearrangement yields

$$(x_N / N) = \left\{ x_1 / \exp[(\mu_N^o - \mu_1^o) / RT] \right\}^N . \qquad (4.12)$$

Hence x_N can be calculated in terms of the monomer concentration, provided that the dimensionless standard free energy difference on the right-hand side of Equation 4.12 is known. However, this free energy difference cannot be obtained from classical thermodynamics alone, but must be derived from some molecular model of micelle formation.

One such model was originally developed by Nagarajan and Ruckenstein (1977). Various improvements and extensions have been summarized in a review article by the same authors (Nagarajan and Ruckenstein, 1991). Their expression for the standard free energy for the formation of an N-mer includes contributions from (a) transfer of hydrocarbon chains from water to the micelle interior; (b) constraints not present in the bulk liquid hydrocarbon phase on hydrocarbon chain configuration in a micelle; (c) free energy of the interface between water and the hydrophobic micelle interior; and (d) repulsion between adjacent head groups at the micelle surface. Figure 4.6 shows aggregate size distributions calculated for one surfactant using this method. At low total surfactant concentrations (lower curve), most of the surfactant is present as monomer. At higher concentrations (upper curve), a maximum in size distribution appears for an aggregation number near 50 (i.e., micelles of this size are favored). The middle curve represents the situation near CMC, where the maximum is just starting to appear. Clearly this method provides a more complete description of micellization than the phase separation and mass action models. However, it requires more information (e.g., on the nature of repulsion between head groups) and more computational effort.

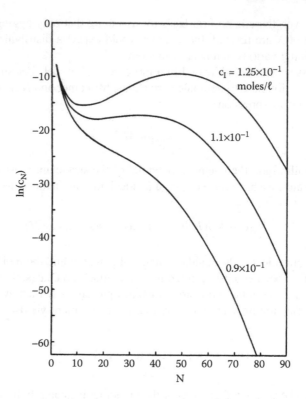

FIGURE 4.6 The calculated distribution of micelles of different aggregation numbers N of sodium octyl sulfate in water. All concentrations are in moles per liter. Reprinted from Nagarajan and Ruckenstein (1977) with permission.

Another model of micelle formation that also includes interaction between micelles is that of Puvvada and Blankschtein (1992).

EXAMPLE 4.1 ESTIMATION OF MICELLE AGGREGATION NUMBER AND CMC

In one simple model of micellization (Israelachvili et al., 1976), the standard free energy difference per surfactant molecule between N-mer and monomer is given by

$$\mu_N^o - \mu_1^o = c_1 + \gamma_m a + (c_2 / a) . \qquad (4.E1.1)$$

In this equation, which assumes that N is large enough for the micelle to have an "interface" with the aqueous solution and a nonpolar "interior," c_1 is a negative constant denoting the free energy change of transfer of the hydrocarbon chain from water to the micelle interior, γ_m is the interfacial tension, and c_2 is a positive constant reflecting the repulsion between head groups. Find the value of the area

a per surfactant molecule that minimizes this free energy difference and estimate the mean aggregation number N_o of the spherical micelles formed. Evaluate N_o when v_H, the volume of the hydrocarbon portion of the surfactant molecule, is 0.323 nm³, γ_m = 50 mN/m, c_2 = 1.625×10^{-38} J·m²/molecule², and c_1 = $-7.0 \times 10^{-21}n_c$ J/molecule, where n_c is the number of carbon atoms in the hydrocarbon chain, which we take to be 12 in this case. Also estimate the value of x_1 at the CMC at 300°K.

SOLUTION

As the area a per surfactant molecule increases, interfacial energy increases, but repulsive energy decreases. It is easily shown that the right-hand side of Equation 4.E1.1 is minimized for

$$a = a_o = (c_2/\gamma_m)^{1/2}. \tag{4.E1.2}$$

This is the preferred head group area based on energetic considerations. With the above values for γ_m and c_2, the resulting value of a_o is 0.57 nm².

For a spherical micelle of aggregation number N we have

$$Nv_H = (4/3)\pi R^3 \tag{4.E1.3}$$

$$Na = 4\pi R^2. \tag{4.E1.4}$$

Eliminating R between these equations, we find

$$a = [36\pi\, v_H^2\, /\, N]^{1/3}. \tag{4.E1.5}$$

For small values of N, this equation yields $a > a_o$, which is energetically unfavorable. When N reaches a particular value N_o, a becomes equal to a_o. For larger values of N, $a < a_o$, which is also unfavorable energetically. (A permissible solution in this case is a nonspherical shape, such as ellipsoidal, for which $a = a_o$.) Since entropic considerations favor small values of N, we anticipate that micelles will be distributed around a mean aggregation number slightly less than N_o. With the values of various parameters given above, we find N_o = 63.

In this case we have $\exp[-(\mu_N^o - \mu_1^o)/RT]$ = 679. For x_1 = 1×10^{-3}, Equation 4.12 predicts that the mole fraction of N-mers (x_N/N) is 2.56×10^{-11} for N = 63, and even smaller for larger values of N (i.e., the size distribution curve is qualitatively similar to the lower curve of Figure 4.6), and the surfactant concentration is below the CMC. In contrast, for x_1 = 1.43×10^{-3}, x_N/N is 0.156 for N = 63 — more than two orders of magnitude greater than x_1. Hence the surfactant concentration is above the CMC, and the size distribution has a maximum, as in the upper curve in Figure 4.6. The CMC in this case occurs when x_1 is about 1.33×10^{-3}, with x_N/N for N = 75 having a similar value.

According to this approach, the CMC is comparable to $\exp[(\mu_N^o - \mu_1^o)/RT]$ for $N = N_o$, which is about 1.47×10^{-3} for the above example. It is clear from Equation 4.E1.1 that this analysis predicts a decrease in CMC as the surfactant chain length n increases (i.e., as c_1 becomes more negative). On the other hand, the CMC is predicted to increase with increasing interfacial energy γ_m and with increasing repulsion c_2 between head groups. These results are expected based on the physical description of micelle formation given above and are in agreement with experimental data.

Tanford (1980) has a different expression for repulsion in Equation 4.E.1 given by c_2/a^2 due to their electrostatic charges. Note the different dependence on a. His surface energy term is also proportional to $a - a_o$ and not just a, where a_o is the area covered by the polar group of the surfactant molecule.

EXAMPLE 4.2 CMC OF NONIONICS

If $\Delta \bar{G}^\circ$ is primarily determined by the carbon number n_c of the tail, show why at 25°C the CMC of SDS can be as high as 8×10^{-3} M and that of the pure nonionic $C_{12}E_5$ is only 6.4×10^{-5} M.

SOLUTION

Use Equation 4.7 with $\alpha = 0.5$ for SDS and Equation 4.9 for $C_{12}E_4$, and for the same values of n_c the two surfactants have the same $\Delta \bar{G}^\circ$. Hence

$$(x_{SDS})^{2-0.5} = x_{C_{12}E_5} .$$

In dilute solutions, it leads to

$$\left(\frac{8 \times 10^{-3}}{55.6} \right)^{1.5} = \frac{9.06 \times 10^{-5}}{55.6},$$

where 55.6 is the molar density of water. That is, the predicted CMC value for the nonionic surfactant is a little higher than the experimental value of $C_{12}E_5$ and a little lower than for $C_{12}E_8$ (this simple approach does not include the effect of the EO chain length).

As mentioned previously, micelle formation in block copolymers is considerably more complex, and not all features are fully understood. Nevertheless, some key aspects are understood and have been used to construct models with reasonable success. For an A-B diblock, where B is hydrophobic, the core is made of B, and A meanders into water, the solvent, forming a region called the corona. The total energy is expressed much like that for detergent micelles: a surface energy at the core (B)-solvent interface, a mixing term for the corona, and an elastic energy of deformation of the polymer chains (Munch and Gast, 1988). Of course, the energy of singly dispersed copolymers is also included.

The model includes two key parameters in the form of ratios between lengths/sizes of A and B chains, and an energy parameter $\chi_{BS}N_B$, where χ_{BS} is the Flory-Huggins interaction coefficient between the B-mer and the solvent molecule and N_B, the number of repeat units in B. Distinct CMCs are observed as the concentration of block copolymer is increased, and CMCs decrease with increasing values of $\chi_{BS}N_B$, which denotes the incompatibility between A and B.

So far we have dealt with the thermodynamics of micelle formation. However, micelles continually form and dissolve in a micellar solution and exchange surfactant molecules with the aqueous phase. Such behavior has been investigated by various experiments where one variable (e.g., temperature) is suddenly changed and the system response is followed (e.g., by turbidity or light scattering) [see the review by Lang and Zana (1987) for further discussion of measurement techniques]. For micellar solutions of a single surfactant, two relaxation times corresponding to two apparent first-order steps are typically observed. This is surprising in view of the fact that the reaction scheme shown after Equation 4.2 for describing the thermodynamics predicts a very complicated kinetics. One relaxation time τ_1 is short, ranging, for instance, from about 10^{-8} seconds for sodium hexyl sulfate to about 10^{-3} seconds for sodium hexadecyl sulfate. It is sufficiently short that it cannot be measured in most experiments. On a logarithmic scale, τ_1^{-1} shows a linear increase with increasing surfactant concentration. The second relaxation time, τ_2, is typically in the milliseconds to seconds range. On a logarithmic scale, τ_2^{-1} is linearly dependent on surfactant concentration, at first decreasing with increasing concentration and then increasing (Kahlweit, 1982).

The kinetics of micellization-demicellization have been analyzed by Aniansson and Wall (1974, 1975) for nonionic surfactants. Nevertheless, they appear to apply to data for ionic systems as well. The short relaxation time is thought to be due to exchange of single surfactant molecules from a micelle. The long relaxation time τ_2 depends on the surfactant structure and on additives, and deals with the process that leads to a change in the number of micelles. Here, stepwise micelle dissociation seems to be the chief relaxation mechanism for ionic surfactants not far above CMC with no added electrolyte. However, coalescence and breakup of surfactant aggregates is also important for nonionic surfactants and for ionic surfactants at sufficiently high counterion concentrations, where electrical repulsion between aggregates is reduced. Data on τ_2 from Kahlweit (1982) are shown in Figure 4.7.

For ionic surfactants, the addition of short-chain alcohols decreases τ_2 (i.e., increases τ_2^{-1}). For medium-chain alcohols, τ_2^{-1} initially increases with increasing alcohol concentration, then reaches a maximum and decreases as additional alcohol is incorporated into the micelle interior instead of the palisade layer. The addition of hydrocarbons, which are solubilized in the micelle interior, also produces a decrease in τ_2^{-1}. In Chapter 6 we call attention to the fact that even though τ_2 is quite small, its effect can be felt even in slow processes such as solubilization, wave motion, etc. It is noteworthy that micelles of block copolymer, PEO-PPO-PEO, also show these two time constants, which have similar interpretations (Goldmints et al., 1997).

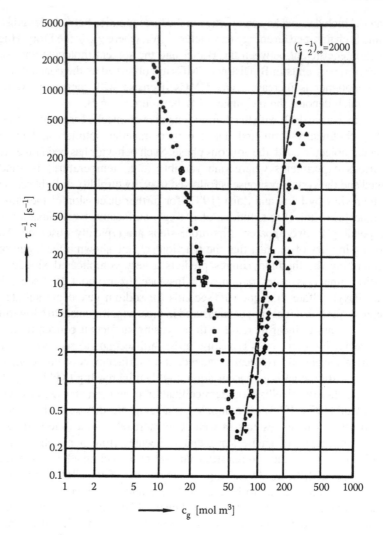

FIGURE 4.7 τ_2^{-1} versus surfactant (SDS) concentration at different values of counterion (Na^+) concentrations. Different symbols denote different concentrations of sodium perchlorate, including one which is zero. Reprinted with permission from Lessner et al. (1981). Copyright 1981 American Chemical Society.

3. VARIATION OF CMC FOR PURE SURFACTANTS AND SURFACTANT MIXTURES

As the preceding example indicates, micelle formation can be promoted by increasing the length of the surfactant's (straight) hydrocarbon chain or by reducing the repulsion between the head groups. As an illustration of the chain length effect, the measured CMC of sodium alkyl sulfates at 40°C decreases from 140 mM for the C_8 compound to 8.6 mM for the C_{12} compound to 0.58 mM for the

C_{16} compound. Increasing salinity decreases the effective repulsion between ionic head groups. Accordingly, the CMC of SDS decreases from about 8 mM to about 1 mM with the addition of 0.1 M sodium chloride (NaCl). As would be expected, the repulsion between head groups is considerably less for nonionic than for ionic surfactants and the CMCs are much lower—about 0.064 mM for $C_{12}E_5$, ethoxylated dodecanol with five EO groups, or about two orders of magnitude lower than for SDS, which has the same chain length (see Example 4.2). For the nonionics, an increase in EO number increases the repulsion between head groups having a given separation distance and thereby slightly increases CMC. In general, nonionics are available commercially as mixtures (with a distribution of EO numbers) and the repulsion between neighboring head groups is much less than for ionics. Both reasons are cited to explain why nonionic CMCs are less easily identified, as shown in Figure 4.8.

For binary surfactant mixtures, one might expect that micelles formed at the CMC of the mixture would be enriched in the less hydrophilic surfactant (i.e., the one with the lower CMC). Analysis based on the phase separation model confirms this expectation. The simplest approach is to assume an ideal mixture in the micellar phase (i.e., activity of each species equal to its mole fraction). With this assumption one obtains the following expression for the ratio (x_{1m}/x_{2m}) of the two surfactants in the micellar phase at the CMC (see Problem 4.3):

$$(x_{1m}/x_{2m}) = x_{2c}/(nx_{1c}), \qquad (4.13)$$

where n is the molar ratio of species 2 to species 1 in the overall mixture and x_{1c} and x_{2c} are the CMCs of the pure surfactants. Clearly, if species 1 is less hydrophilic, so that $x_{1c} < x_{2c}$, it is enriched in the micellar phase relative to the aqueous solution.

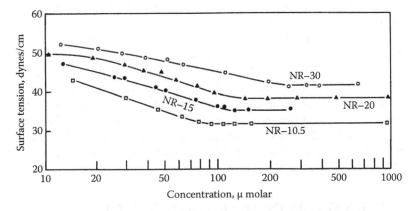

FIGURE 4.8 Surface tension shown as a function of nonylphenol concentration and the average number of moles of EO per mole of alkyphenol (NR). The CMCs are less well marked (compare with Figure 4.1) because of the dispersed surfactants. Reprinted from Hsiano et al. (1956) with permission. Copyright 1956 American Chemical Society.

The CMC x_c of the mixture is given by (see Problem 4.3)

$$x_c = x_{2c}(1 + n)/[n + (x_{2c}/x_{1c})].$$ (4.14)

Beyond the CMC, the ratio of the two surfactants in both the aqueous phase (i.e., water with dissolved monomer) and the micelles changes as more surfactant is added. Eventually most of the surfactant is in the micelles, which then must have the overall composition of the mixture. The implication of the composition changes beyond the CMC is that the surface tension does not remain constant, as in Figure 4.1 for a pure surfactant, although a constant value is approached when the micelles contain nearly all the surfactant. Indeed, a common way to check the purity of a surfactant is to check that the surface tension changes abruptly to a constant value at the CMC without exhibiting a minimum.

For many binary surfactant mixtures, the ideal solution model discussed above and in Problem 4.3 is inadequate. The most common way of handling these deviations is to apply the regular solution model given by Equation 1.80 to the micellar phase:

$$\mu_{im}^o = \mu_{im}^o + RT \ \ln x_{im} + \beta(1 - x_{im})^2 .$$ (4.15)

For mixtures of nonionic and ionic surfactants, the interaction energy parameter β is negative (i.e., micellization is favored to a greater extent than expected from ideal solution theory). Nonionic surfactant molecules are located between ionic surfactant molecules, thereby increasing the average distance between adjacent ionic head groups on the micelle surface and reducing electrical repulsion. As a result, the CMC at a given composition of the mixture is less than predicted by ideal solution theory. Indeed, in some cases the plot of CMC as a function of mixture composition exhibits a minimum (Figure 4.9), which would not occur for an ideal solution. Mixtures of anionic and cationic surfactants exhibit even greater deviations from ideal behavior. Frequently $\beta/RT < -10$.

In contrast, mixtures of hydrocarbon and fluorocarbon surfactants have positive values of β and the CMC-composition curve is above that of an ideal solution. That is, it is energetically unfavorable to mix hydrocarbon and fluorocarbon chains. In some cases, two different types of micelles coexist, one type containing predominantly the hydrocarbon surfactant, the other mainly the fluorocarbon surfactant. Further discussion of surfactant mixtures can be found in the books edited by Holland and Rubingh (1992) and Scamehorn (1986), in particular in the chapter by Holland in the former reference.

4. OTHER PHASES INVOLVING SURFACTANTS

So far we have dealt mainly with the formation of small micelles at relatively low surfactant concentrations in aqueous solutions. For a typical ionic surfactant, the micelles remain more or less spherical over a substantial concentration range

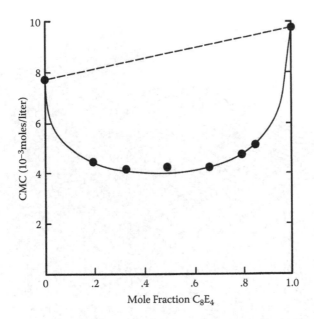

FIGURE 4.9 CMCs in binary anionic/nonionic mixtures of SDS and C_8E_4. The data are fitted to the regular solution theory model with $\beta = -3.1$, and the dashed line is the prediction from ideal solution theory. Reprinted with permission from Holland [Holland and Rubingh (1982)]. Copyright 1982 American Chemical Society.

above the CMC, but ultimately become rodlike as the concentration continues to increase. At surfactant concentrations of perhaps 20% to 30% by weight, a new phase appears that is birefringent and highly viscous. X-ray diffraction experiments demonstrate that this phase consists of many long, parallel, rodlike micelles arranged in a hexagonal array (Figure 4.10). The micelle interiors are apparently rather fluid, resembling liquid hydrocarbons in many respects, but somewhat more ordered. This phase is a "liquid crystal" (i.e., it possesses substantial order but is not truly crystalline). It is usually called the normal hexagonal or simply the hexagonal phase. Occasionally it is referred to as the middle phase, an old term that comes from the soap-making industry.

At even higher surfactant concentrations, the arrangement of surfactant molecules into bilayers becomes favorable and another birefringent, but somewhat less viscous liquid crystalline phase, known as the lamellar phase, forms. Figure 4.11 shows the microstructure of this phase, which was called the neat phase by soap makers. Because the basic structure of biological membranes is a bilayer of phospholipid molecules, which are surfactants, the lamellar phase has sometimes been used as a "model system" to gain insights regarding membrane behavior.

Figure 4.12 shows the effects on phase behavior of both surfactant concentration and temperature for an ionic surfactant. The liquid crystalline phases melt at sufficiently high temperatures. A similar diagram for a particular nonionic

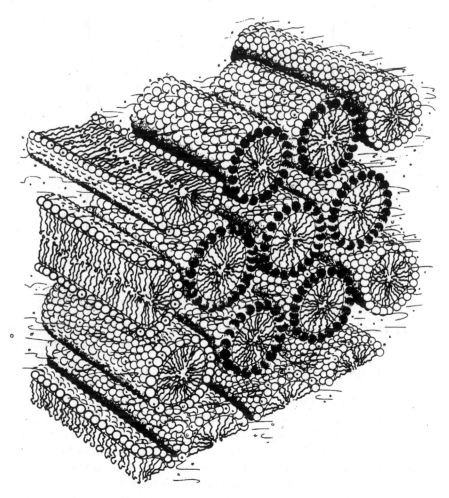

FIGURE 4.10 Schematic representation of the normal hexagonal phase.

surfactant $C_{12}E_5$ is given in Figure 4.13 (Strey et al., 1990b). At low temperatures the sequence of phases that occurs with increasing surfactant concentration is the same as described above for ionic surfactants and shown in Figure 4.12. However, the effect of temperature is much different. In the following paragraphs we consider the reasons for this behavior.

Both the ionic surfactant and the nonionic surfactant at low temperatures are quite hydrophilic. Owing to electrical repulsion between head groups for the ionic surfactant and the high degree of hydration of the EO chains for the nonionic surfactant, the natural tendency above the CMC is to form micelles in which the area a per head group on the micelle surface is significantly larger than the cross section of a straight hydrocarbon chain. That is, the packing parameter (Israelachvili et al., 1976) defined by v/al, where l is the length of the tail and v is the volume occupied by the tail, is small. With l the micelle radius, it is clear that

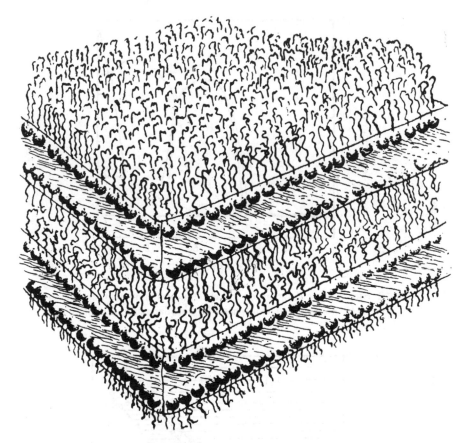

FIGURE 4.11 Schematic representation of the structure of the lamellar phase.

for these conditions, spheres, which have a $v/al = \frac{1}{3}$, are clearly preferred over cylinders or bilayers, where $v/al = \frac{1}{2}$ and 1. Thus spherical micelles are found in dilute solutions of hydrophilic surfactants. Clearly a model is needed to evaluate the individual terms in this dimensionless group. The one that is used is $l = l_c$, the fully stretched length of a surfactant molecule (hence closest to the largest radius possible), and $v = v_H$, the volume of the tail. Both are taken from crystallographic data (see Example 4.3). The surface area is an energy weighted value, as seen in Example 4.1, and is considered further at the end of this section. In the rest of the discussion, only v_H/al_c will be used.

As surfactant concentration increases, interaction between adjacent micelles becomes important. It can be shown that, for a fixed volume fraction of surfactant and length of surfactant molecule, the average spacing between micelles increases as the shape changes from spherical to cylindrical to planar. Moreover, an arrangement of uniform spheres is impossible above a volume fraction of about 74%, which corresponds to a close-packed hexagonal arrangement. The corresponding condition for cylinders is about 90 vol. %. Hence, if the micelles repel one another,

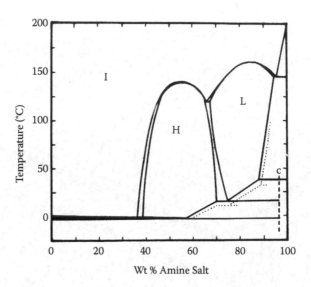

FIGURE 4.12 Binary phase diagram for dodecyl dimethylammonium chloride water. I: micellar solution, H: normal hexagonal phase, L: lamellar phase. From Broome et al. (1951) with permission.

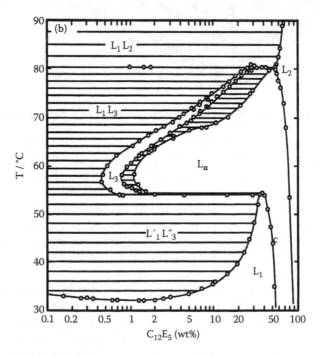

FIGURE 4.13 Phase diagram of the $C_{12}E_5$-water system over the temperature range 0°C to 100°C (H: normal hexagonal phase, L: lamellar phase, W: water, I_1, I_2, I_3: isotropic liquid phases, S: solid phase). From Strey et al. (1990b). Reproduced with permission of the Royal Chemical Society.

micelle interaction promotes a change from spherical to cylindrical micelles and eventually to bilayers with increasing surfactant concentration, even though these changes increase repulsion between head groups within a micelle by increasing the value of v_H/al_c above that which the individual surfactant molecules prefer. Micelle interaction is thus responsible for the appearance of cylindrical micelles and ultimately the hexagonal and lamellar phases with increasing concentration of a hydrophilic surfactant.

EXAMPLE 4.3 CYLINDRICAL MICELLES

Whereas communal entropy and micelle interaction energy play a role in the transformation of spheres to cylinders with increasing surfactant concentration, cylindrical micelles also form in dilute systems. One explanation has been offered by Debye and Anacker (1951). Crystallographic data show the volume of the hydrocarbon tail to be $v_H = [27.4 + 26.9(n_c - 1)] \times 10^{-3}$ nm^3 and a fully stretched length of $l_c = 0.15 + 0.1265(n_c - 1)$ nm (see Tanford, 1960). Calculate the maximum possible aggregation number allowable in a sphere for SDS, cetyl pyridium chloride, and sodium octyl sulfate.

SOLUTION

For $n_c = 12$, $v_H = 0.3233$ nm$^3 = 4/3(\pi R^3/N)$, where N is the aggregation number. The fully stretched length of the tail is 1.5415 nm. The radius of the micelle is $R = 0.4258N^{1/3}$ nm, but it is required that $R \leq l_c$ or $N^{1/3} \leq 3.621$ or $N \leq 58$ at 25°C, which is less than the reported value of 62 and the similar value found in Example 4.1. Hence relatively short cylindrical micelles are expected for SDS. For cetyl pyridium chloride, $n_c = 16$, $l_c = 2.68$ nm, and $N \leq 249$. Its aggregation number is 138, and such micelles are correctly predicted to be spherical. For sodium octyl sulfate, $N \leq 21.5$. Actual values lie in the range of 24 to 31 and experimental measurements indicate that it forms spherical micelles and not cylindrical ones, showing that the simple geometric criterion employed is not always adequate.

In contrast, the changes in phase behavior with temperature shown in Figure 4.13 for dilute nonionic surfactant-water mixtures are caused by changes in the preferred value of v_H/al_c of a surfactant molecule and associated changes in the micelle interaction. Basically the degree of hydration of the EO chains decreases with increasing temperature, causing a decrease in the effective head area a and hence an increase in v_H/al_c. At the same time, the reduced hydration reduces the steric repulsion between micelles, eventually allowing attractive forces to prevail.

A combination of these two effects is responsible for separation of the L$_1$ phase into micelle-rich and micelle-lean phases at the "cloud point" temperature. It is well known that entropic effects cause phase separation to occur in relatively dilute solutions of long, cylindrical particles, even in the absence of interparticle attraction. Hence phase separation is facilitated by micelle shape changes caused by higher values of v_H/al_c and, of course, also by a shift from a repulsive to an attractive interaction among micelles. The relative importance of these two factors appears to be different for different nonionic surfactants. Cylindrical micelles in the surfactant-rich phase frequently form a continuous network instead of remaining discrete entities. Note that the cloud point temperature varies with surfactant concentration and that a lower critical point exists at the minimum of the cloud point curve, where the two phases that separate are identical. For the $C_{12}E_5$- water system of Figure 4.13, the critical composition is about 1 wt. %.

At higher temperatures, v_H/al_c continues to increase and the lamellar liquid crystalline phase L_α and the L_3 phase are found. The latter is an isotropic liquid that is frequently termed the "sponge phase" because continuous, but tortuous water channels are separated by continuous surfactant bilayers whose large-scale morphology resembles, on electron micrographs, the solid portion of a sponge (Strey et al., 1990a). Locally the bilayers are saddle-shaped, with the two radii of curvature having opposite signs (see Chapter 1). It can be shown that under these conditions, v_H/al_c is slightly greater than one (Anderson et al., 1989), which is consistent with the occurrence of the L_3 phase at higher temperatures than the lamellar phase. As Figure 4.12 shows, the water content of the L_3 phase decreases with increasing temperature (i.e., the average cell volume in the sponge becomes smaller), the predicted effect of increasing v_H/al_c. The L_α and L_3 phases also occur widely in dilute ionic (Miller et al., 1986) and zwitterionic (Miller et al., 1990) surfactant systems as v_H/al_c decreases, i.e., as ionic strength increases for the case of ionic surfactants.

At the highest temperatures shown in Figure 4.13, water is in equilibrium with a surfactant-rich L_2 phase. In this phase, v_H/al_c is considerably greater than one, and the water is believed to be present as small drops with diameters on the order of 10 nm or less.

It may be seen from Figure 4.13 that the lamellar phase extends to surfactant concentrations as low as about 1 wt. %. Since the surfactant bilayers are some 3 nm in thickness, the water layers separating them must be very large indeed for these conditions. As there are no electrical double layers to cause long-range repulsion, the DLVO theory described in Chapter 3 cannot account for such large equilibrium distances between bilayers, and another repulsive force must be acting. This force stems from undulations of the bilayers produced by random thermal motion. Under these conditions, individual bilayers are very flexible and relatively little energy is required to deform them to a wavy configuration. Hence the amplitude of spontaneously generated undulations of an isolated bilayer would be large. However, nearby bilayers limit undulation amplitude, which is unfavorable from an entropic point of view. As a result, bilayer spacing increases to permit undulations of larger amplitude. This effect was first analyzed quantitatively by Helfrich (1978), who found a repulsive pressure proportional to $[3\pi^2(kT)^2/64\ k_b h^3]$, where k_b is the bending modulus (see Section 10) of a bilayer and h is the distance between bilayers. For flexible bilayers, k_b is small and the undulation pressure is large, leading to large equilibrium spacings. It should be noted that bilayers are very flexible and lamellar phases of high water content are also found in systems containing anionic surfactants and short-chain alcohols at moderate to high salinities.

Lamellar phases of high water content have also been observed in certain systems containing zwitterionic surfactants, such as tetradecyl dimethyl amine oxide, whose surfactant films have a very small positive charge at neutral pH. Indeed, lamellar phases exhibiting striking iridescent colors have been observed in systems containing this surfactant with some alcohol added to adjust the value of v_H/al_c (Hoffman and Ebert, 1988). The colors indicate that the spacing between bilayers is on the order of the wavelength of light. In this case, the long-range repulsion is electrical, since the colors disappear with the addition of a little sodium chloride.

The temperatures at which these phase changes occur for nonionic surfactants depend on the surfactant structure, increasing with increasing length of the EO chain and decreasing length of the hydrocarbon chain. The cloud point temperatures in dilute solutions, for example, are quite sensitive to EO number, being about 10°C for $C_{12}E_4$, 30°C for $C_{12}E_5$, and 50°C for $C_{12}E_6$. For $C_{12}E_4$, the other transitions also occur at significantly lower temperatures than those shown in Figure 4.13 for $C_{12}E_5$. On the other hand, $C_{12}E_6$ is sufficiently hydrophilic that the L_α and L_3 phases do not form in dilute systems, even at temperatures approaching 100°C. Mitchell et al. (1983) summarized extensive information on phase behavior of various nonionic surfactants.

Addition of a long-chain alcohol or other molecule with a small, neutral head group favors aggregates with denser packing of head groups, i.e., having larger overall values of v_H/al_c. For instance, the cloud point temperature of 1 wt. % $C_{12}E_5$ is lowered by more than 20°C by addition of 0.1 wt. % n-dodecanol (Raney and Miller, 1987). Similar effects occur in many other ternary systems. As Figure 4.14 shows, addition of n-decanol to fairly concentrated sodium octanoate-water mixtures causes transformation from a hexagonal to a lamellar phase. Although not shown in Figure 4.14, the L_3 phase is found in a small composition region at high water contents in this system (Benton and Miller, 1984). As would be expected, its alcohol:surfactant ratio is slightly higher than in the lamellar phase. A reverse hexagonal phase is found in the region denoted F in Figure 4.14, at low water content and high alcohol:surfactant ratios (large v_H/al_c). In this phase, water is inside the cylindrical micelles and the surfactant and alcohol tails are directed outward; just the opposite of the arrangement in the "normal" hexagonal phase of Figure 4.10.

"Cubic" phases with various structures constitute yet another type of liquid crystal known to exist in many systems containing surfactants (Fontell, 1981; Tiddy, 1980). Although they are isotropic, and hence do not exhibit birefringence like the hexagonal and lamellar phases, their microstructures, which cause them to be highly viscous, can be determined by x-ray diffraction. For instance, some are ordered arrangements of spherical micelles, others are bicontinuous, like the L_3 phase, but much more concentrated. Cubic and nematic phases frequently exist over only small composition ranges.

The dimensionless group v_H/al_c is seen to be very useful in classifying structures, a method that started very modestly with a model by Mukerjee (1974), where the surfactant molecule is visualized as a tapered peg. The inference is

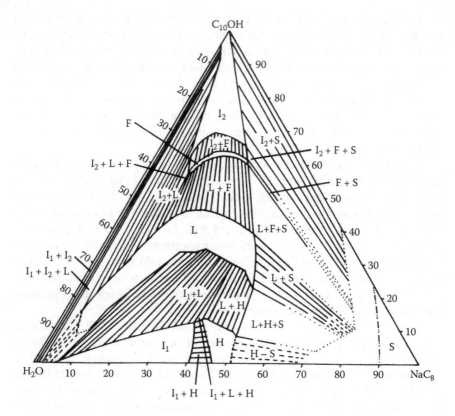

FIGURE 4.14 Phase diagram for the three component system sodium octanoate-water-decanol at 293°K. Symbols as in Figure 4.11, but with F: reverse hexagonal phase. After Friman et al. (1982) with permission.

that tapered pegs cannot be fitted next to one another without giving rise to a curved surface. Nonetheless, the method fails to give values beforehand to the three geometric parameters—volume, area per head group, and the length of the tail—particularly to a, the area per head group, although estimates for the others are available as indicated previously. Friberg et al. (1985) ask what relation, if any, exists between this a and the actual size of the head group by analyzing the phase equilibrium diagrams of a series of three homologous surfactants: tetradecyl trimethyl/triethyl/tri-n-propyl ammonium bromides (parts of which are shown in Figure 8.10). The most remarkable changes that occur can be explained by taking into account the increasing hydrophobicity of the head groups (i.e., by their interaction energies). More specifically, the term a should be found by minimizing the surface energies in Equation 4.E1.1 or in the similar expression given by Tanford (1980). As things stand, v_H/al_c is independent of chain length. However, when a is determined by minimizing the surface energy, the packing parameter does become weakly dependent on chain length, and the dependence does affect significantly the sphere-to-cylinder transition, as discussed by Nagarajan (2002);

that geometric considerations in predicting shapes may not be sufficient is indicated in Example 4.3.

Multiphase regions involving liquid crystalline phases are of some interest as well. The fluids proposed for injection in some surfactant processes for enhanced recovery of petroleum from underground reservoirs were stable dispersions of the lamellar phase (i.e., in the $L_1 + L_\alpha$ region) (Miller et al., 1986). In recent years, aqueous dispersions of liposomes or vesicles have attracted considerable attention as models of biological membranes and as possible vehicles for drug delivery. Although the former term includes particles of the lamellar phase, research has dealt primarily with small spherical particles — 10 to 100 nm — having aqueous interiors and surface regions consisting of a single bilayer or, in some cases, a few bilayers. Such vesicles can be formed in various ways, for instance, by ultrasonic treatment of a dispersion of the lamellar phase. Most dispersions of vesicles are thermodynamically unstable and ultimately revert to a dispersion of lamellar liquid crystalline particles, though the time scale may be very long. However, vesicles believed to be thermodynamically stable have been formed by mixing anionic and cationic surfactants in the proper proportions (Kaler et al., 1989).

Figures 4.12 through 4.14 and the above discussion make clear that, even in the absence of oil, the phase behavior of systems containing water and one or more surface-active compounds can be quite complex. Many phase diagrams similar to Figure 4.14 may be found in Ekwall (1975).

Alexandridis et al. (2000) identified all the structures discussed above on a water/oil/PEO-PPO-PEO phase diagram. Because the solutions in many parts of the phase diagram are very viscous for these polymeric surfactants, further lowering of the temperature makes it very difficult to equilibrate the system. As a result, very often full temperature dependence is not reported, and has not been reported by Alexandridis et al. (2000). However, for these nonionics it is important to consider temperature dependence because of the changes it brings to the EO groups. Thus an additional effect of cloud point should also exist at low temperatures in this system.

5. FORMATION OF COMPLEXES BETWEEN SURFACTANTS AND POLYMERS

Mixtures of surfactants and water-soluble polymers are of both fundamental and practical interest. If one excludes "hydrophobically modified" polymers, which have hydrocarbon chains attached periodically to rather hydrophilic backbones, polymer-surfactant complexes occur only for (a) ionic surfactants and nonionic polymers and (b) ionic surfactants and polyelectrolytes having opposite charges. In the former situation, for instance, mixtures of SDS and PEO, the EO segments can penetrate the micelle surface and separate the charged surfactant ions in much the same way as described above for ionic-nonionic surfactant mixtures. The EO segments also prevent water molecules from coming into direct contact with any

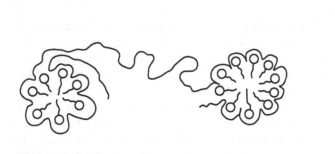

FIGURE 4.15 String of beads. Schematic representation from Cabane and Duplessix (1987) at right and Nagarajan and Kalpakcir (1982) at left.

CH_2 groups not shielded by the surfactant head groups alone. Hence micelles begin to form along the polymer chain at a concentration called the critical aggregation concentration (CAC), which is well below the CMC of the pure surfactant. The resulting arrangement has been likened to a string of beads (see Figure 4.15). These micelles, which generally contain slightly fewer surfactant molecules than ordinary surfactant micelles, continue to form until electrical repulsion between adjacent micelles along a polymer chain increases to the point that it becomes more favorable to form ordinary surfactant micelles in the bulk solution than to increase the number of micelles along a chain.

The dependence of surface tension on surfactant concentration when such behavior occurs is illustrated in Figure 4.16 for the SDS-polyvinylpyrrolidone (PVP) system. The symbols T_1 and T_2 denote the CAC and the surfactant concentration where the polymer chains are saturated with micelles. The plots indicate that the polymer itself has some surface activity (i.e., surface tension decreases with increasing polymer concentration when little surfactant is present). Moreover, T_2 is greater than the surfactant CMC and increases with increasing polymer concentration, as one would expect. It has been found that the surface tension curve does not vary significantly with polymer molecular weight except when the molecular weight is quite low (about 1500 for the SDS-PEO system).

In these systems, viscosity typically increases rapidly as surfactant concentration increases above the CAC. As more micelles form along the polymer chains, they become longer and straighter to increase micelle separation and reduce electrical repulsion. The greater length leads to a higher viscosity.

It should be noted that the interaction of nonionic polymers with cationic surfactants is much weaker than with anionic surfactants. The reason for this difference in behavior is not entirely clear but may stem from a small positive charge acquired by the EO chain in aqueous solution.

The presence of both polymers and surfactants affects phase behavior as well. For instance, the presence of SDS increases the cloud point temperature of PEO. The negatively charged micelles produce an electrical repulsion between polymer chains that impedes their aggregation to form a polymer-rich phase. (SDS

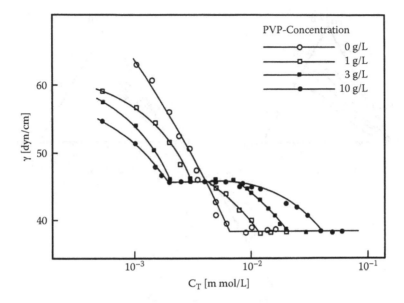

FIGURE 4.16 Surface tension as a function of SDS concentration in solutions of a charged polymer (PVP). The two "T" points can be seen. Reprinted from Lange (1971) with permission from Springer (copyright).

increases the cloud point temperatures of nonionic surfactants by a similar mechanism.)

Interaction between a surfactant molecule and a polymer segment is very strong when they are oppositely charged, and they readily form a complex. Since complexation has the effect of eliminating the head group charge, v_H/al_c increases, the CAC is low, and micelle shape may not be the same as for the pure surfactant. Indeed, formation of a second liquid phase or a precipitate is commonly seen when the composition is such that the number of surfactant molecules is approximately equal to the number of charged polymer segments.

In the absence of surfactant, electrical repulsion causes polyelectrolyte molecules in a dilute system to be rather extended and solution viscosity to be high. As complexation proceeds, the net charge of a polymer chain decreases, causing a decrease in chain length and a corresponding decrease in solution viscosity. At high surfactant concentrations, the net charge of the chain may be reversed, and its length may increase again, owing to repulsion between micelles. As a result, viscosity may again increase.

Figure 4.17 shows key aspects of phase behavior of the system water (H$_2$O)/C$_{14}$ trimethylammonium bromide/hyaluronan (an anionic polysaccharide) as a function of added sodium bromide (NaBr) (Lindman and Thalberg, 1993). With no added salt, a region with two liquid phases is seen for a range of surfactant:polymer ratios. One phase is rich in both polymer and surfactant. The phase separation stems from complexation with a low net charge on each polymer chain, as discussed above. As salt is added, the charges on both surfactant and

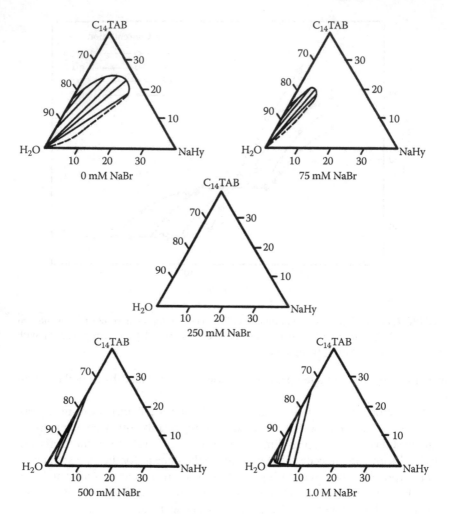

FIGURE 4.17 Phase equilibrium diagram of NaHy-Cl₄TAB-H₂O at different salt concentrations. Reproduced from Thalberg and Lindman (1991) with kind permission of Springer Science and Business Media.

polymer are screened, complexation decreases, and the two-phase region shrinks and ultimately disappears. With further salt addition another two-phase region appears with one phase rich in surfactant and the other rich in polymer. Under these conditions, screening of the charges is such that hardly any complexation occurs. The phase separation is like the well-known separation of "incompatible" uncharged polymers, the surfactant micelles here playing the role of one polymer.

Hydrophobically modified polymers provide another means for polymers and surfactants to interact because the hydrocarbon chains of the polymers may be incorporated into micelles. Various types of behavior are possible. At surfactant concentrations only slightly above the CMC, chains from more than one polymer

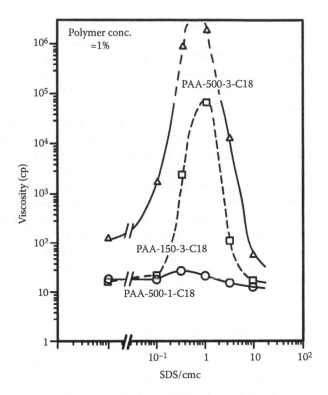

FIGURE 4.18 Viscosity of aqueous solutions of HM polyacrylates as a function of SDS concentration (expressed in CMC). Polymers are (circles) PAA-500-1-C18; (triangles) PAA-500-3-C18; (squares) PAA-150-3-C18, where the first number refers to PAA molecular weight, the second to the percentage of hydrophobic groups, and the third to the number of carbons in the alkyl chains. Reprinted with permission from Iliopoulos et al. (1991). Copyright 1991 American Chemical Society.

molecule may be present in the same micelle, in effect cross-linking the polymer molecules and increasing viscosity. At higher surfactant concentrations where there are more micelles, the probability of such cross-linking decreases, and viscosity reaches a maximum and decreases. An example of such behavior is shown in Figure 4.18 for SDS and hydrophobically modified acrylates. As both are negatively charged, no complexation will occur in this case for the unmodified polymer. Of course, the hydrocarbon chains of some hydrophobically modified polymer molecules form aggregates even in the absence of surfactant under suitable conditions.

Clearly, many possibilities exist for using polymers and surfactants to influence solution rheology and phase behavior. Reviews of systems containing polymers and surfactants may be found in Goddard and Ananthapadmanabhan (1993), Jönsson et al. (1998), and Kwak (1998).

One model for interactions between nonionic polymers and micelles is provided by Nagarajan (1989). The key feature of the model uses for the surface energy, in Equation 4.E1.1, a surface area of $(a - a_o - a_p)$. Here, a_o follows Tanford (1980), discussed earlier, as the area covered by the polar head group, and a_p is the area covered by the polymer. It is also possible to back out some key parameters of the surfactant structures from polymer-free surfactant solutions and to eventually use them in the model. It is possible to show that nonionic micelles may not bind to the polymer at all and that ionic micelles have two characteristic concentrations T_1 and T_2, the CAC and the concentration above which they do not attach any more. Finally, it is possible to extend the treatment to many other systems, including those with cationic and zwitterionic surfactants and with other surfactant structures such as vesicles and microemulsions.

Globular proteins can also be considered in this connection. Considerable data exist on complex formation between proteins and surfactants (Goddard and Ananthapadmanabhan, 1993). The trends are similar at different pH values, and hence different charges on protein. One common system studied is SDS and bovine serum albumin (BSA) (Reynolds et al., 1970). Binding is defined using υ = SDS:BSA, where υ increases with SDS concentration, which is maintained below the CMC. Complex formation follows in three stages: an almost linear increase at low SDS concentrations from hydrophobic interaction between surfactant tails and parts of the protein surfaces that are uncharged up to $\upsilon = 10$, followed by a more complex cooperative stage up to $\upsilon = 40$, and then a very large increase in complex formation. Here, irreversibility sets in. The protein unravels and at micellar concentrations it reaches a fixed loading that is uniquely determined by the molecular weight of the protein. Hence it can be used to measure protein molecular weight (Putnam, 1943). Very little appears to be known about complex formation between water soluble proteins and oil soluble two-tailed surfactants such as phospholipids.

Partition of proteins into water-in-oil microemulsions, which are easily formed from surfactants with two tails, has been considered for separation of proteins (Goklen and Hatton, 1987). It appears that the key to why proteins leave the aqueous phase to move into the aqueous interiors of the water-in-oil micro-emulsion drops is the charge effect. This partitioning is favorable when the charge on the protein is opposite of that on the surfactant molecules. Consequently, to effect separation of two proteins we can contact the protein solution at a pH such that one of them is positively charged and the other negatively charged. When contacted with a water-in-oil microemulsion, the protein with a charge opposite to that of the surfactant will move into the microemulsion phase. When subsequently this protein is moved out of the microemulsion phase it takes with it some of the two-tailed surfactant that has attached to it. This loss of surfactant can compromise the feasibility of the separation scheme. It is also interesting that these little pools of water inside the water-in-oil microemulsion droplets can be used as reactors and for reactions involving enzymes. The proximity of the enzyme to the reactant brings about a higher rate of reaction, which is called superactivity (Ruckenstein and Karpe, 1990). More on reactions in microemulsions is described in Section 13.

6. SURFACE FILMS OF INSOLUBLE SUBSTRATES

Let us turn away briefly from the solution properties and bulk phase behavior of surfactants. Much has been learned about the properties of surfactant molecules at interfaces by the study of molecules with hydrocarbon chains so long that they are virtually insoluble in water.

For many years, the main experimental tool for these studies was the film balance (see Figure 4.19). Various workers, including Pockels, Langmuir, and Adam, made major contributions to its development (see Gaines, 1966). A small, known quantity of the surfactant to be studied is dissolved in a volatile solvent and deposited carefully by pipette on the surface of a pool of water. The solvent is chosen so that it spreads rapidly over the water and then evaporates, leaving the surfactant uniformly distributed as a monomolecular layer (or monolayer) in the region between the two barriers. One of the barriers is movable, so that the area occupied by the surfactant film can be varied. A torsion balance is provided to measure the surface pressure (i.e., the difference between the surface tension of pure water and that of the film-covered surface). More commonly in modern instruments, a Wilhelmy plate is used to measure surface tension in the film region.

Film balance data are normally presented as plots of the surface pressure π as a function of the area σ per molecule. The latter is obtained by dividing the

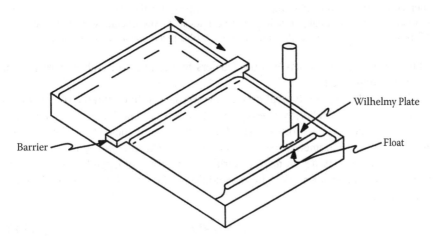

FIGURE 4.19 Schematic diagram of a Langmuir trough. The monolayer is deposited to the right of the barrier, and the barrier can be moved across the surface to change the area accessible to the monolayer. The surface pressure can be measured either by determining the force on a float that separates the monolayer from a clean water surface, or from the difference in the force exerted on the Wilhelmy plate when the plate is suspended in pure water and in water covered by the monolayer. From Knobler (1990), this material is used by permission of John Wiley & Sons, Inc.

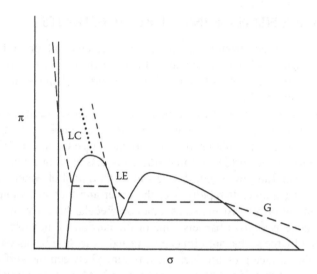

FIGURE 4.20 A schematic view of the $\pi - \sigma$ diagram adapted from Knobler (1990). The two phase region between gas (G) and liquid expanded (LE) phases ends in a critical point, as does the two phase region between LE and liquid condensed (LC) phases. The latter critical point is missing when the LC is an ordered solid, in which case the LE to LC is a second order transition shown by the oblique dashed line on the right that can back up towards the left until it reaches a point where it exhibits a special type of criticality. Such criticality is unrelated to the LE/LC critical point for the case of a liguid LC phase, and the associated critical points have not been located experimentally. Finally, the liquid LC phase shows a second order transition to a solid phase (dashed line on left).

total area of the region between the barriers by the known number of surfactant molecules deposited.

Figure 4.20 shows a typical $\pi - \sigma$ isotherm (dashed line) and the phase behavior inferred from isotherms determined at various temperatures for long-chain fatty acids and similar compounds. For very large values of σ (e.g., several hundred square nanometers per molecule), the surfactant molecules have their hydrocarbon tails lying along the surface and move independently, owing to random thermal motion. For these conditions, the two-dimensional ideal gas equation applies:

$$\pi \, \sigma = kT. \tag{4.16}$$

As indicated in Chapter 2, this equation can be modified to account for nonideal effects at smaller values of σ by methods similar to those used for three-dimensional equations of state for nonideal gases.

For the isotherm of Figure 4.20, the surface pressure is constant in a two-phase region where the gas G and a "liquid expanded" phase LE coexist. For pentadecanoic acid at 15°C, such behavior occurs for $\pi = 0.10$ mN/m. As the phase diagram of Figure 4.20 shows, there is a critical temperature (near 40°C for pentadecanoic acid) above which this two-phase region is not seen. Moreover, at low temperatures, the LE phase is not observed and the gas condenses to a phase of smaller σ discussed below.

More recently several additional techniques have been adapted to the study of monolayers, a topic which has been reviewed by Knobler (1990). The use of fluorescence microscopy and Brewster angle microscopy have been particularly important. In the former technique, a small quantity—less than 1%—of a fluorescent probe (i.e., an insoluble surfactant with an appropriate chromophore) is added to the long-chain fatty acid or other compound to be studied before it is spread on the surface of a small trough of water on the microscope stage. A very sensitive video camera is used to detect the fluorescence of the resulting monolayer. In the LE-G region, the morphology of the two phases can be observed because the LE phase, which has a much higher concentration of the probe than the G phase, appears brighter. This technique is useful not only for identifying two-phase regions, but also for investigating the kinetics of phase transformation.

The Brewster angle microscope developed by Henon and Meunier (1991) does not require addition of a probe and also yields information on the morphology of surface phases. Vollhardt (1996) reviewed results obtained with this technique.

The isotherm of Figure 4.20 indicates that the LE phase is rather compressible. Some evidence exists that the hydrocarbon chains lie along the water surface in the two-phase region but gradually tilt out of the surface and become straighter as the LE phase is compressed. At some point when they are still far from vertical, another first-order phase transition to the "liquid condensed" phase LC occurs. In the older literature, the isotherms are not horizontal in this two-phase region, probably due to the use of surfactants that contained some impurities, to leaching of impurities from the walls and barriers of the film balance, etc. However, fluorescence microscopy clearly confirms the transition (see Figure 4.21). In the LC phase, the chains are believed to be nearly straight and vertical (see the discussion by Knobler, 1990).

The morphology of phase domains when two phases are present in a monolayer depends on several factors. Among them is the line tension, which plays a role similar to that of interfacial tension in three dimensions. For instance, if line tension is small, the size of the critical nucleus for formation of a new phase decreases, and one expects to see more domains of smaller size. In mixtures of insoluble surfactants, one component may be "line active" and reside preferentially at the boundary between domains. Both these effects are illustrated by the behavior of the mixture of the (double-chain) phospholipid dipalmitoylphosphocholine (DPPC) and single-chain C_{18} or C_{20} lysophosphocholines (McConlogue and Vanderlick, 1998). Using fluorescence microscopy, these authors observed that smaller and more extended domains of the LC phase formed on compression of the monolayer when the mole fraction of the single-chain compound was larger. Their interpretation was that the single-chain compound was line active at the LC/LE boundary and decreased the line tension there.

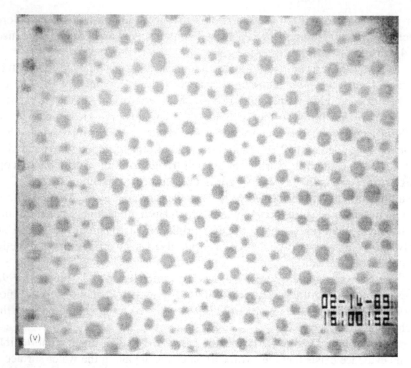

FIGURE 4.21 Two phase region near LC-LE boundary as seen under fluorescence. The system is pentadecanionic acid with 1 mole% fluorescent surface active probe at 20°C. From Knobler (1990); this material is used by permission of John Wiley & Sons, Inc.

Further compression typically leads to the formation of more compact and only slightly compressible liquid or solid phases still having vertically oriented chains, but with longer range order. For straight-chain uncharged compounds such as fatty acids and alcohols, σ has a value of about 0.20 nm²/molecule, the area of the hydrocarbon chain.

Continued compression of a compact film produces film "collapse," in which some molecules are ejected from the monolayer. From a thermodynamic point of view, a bulk phase of liquid or solid surfactant should begin to form at the "equilibrium spreading pressure" where it is in equilibrium with the compressed monolayer. While such behavior is usually found when the bulk phase is a liquid, the more typical behavior when the bulk phase is a solid is that the monolayer retains its integrity until pressures well beyond the equilibrium spreading pressure are reached. Ries and Kimball (1957) studied film collapse using electron microscopy. Harris (1964) developed a model of collapse based on the theory of buckling of an elastic plate. Smith and Berg (1980) treated collapse in terms of nucleation and growth of the bulk surfactant phase.

In recent years, interest has expanded rapidly in the formation of highly organized monolayers and multilayers of surface-active molecules on solid surfaces. Such materials are not new. Formation of a multilayer by successive deposition of monolayers on a solid was carried out several decades ago by Langmuir and Blodgett (1935). Basically they repeatedly dipped the solid into water covered by an insoluble monolayer while keeping the surface pressure of the monolayer constant. Successive monolayers of such a Langmuir-Blodgett film have opposite orientations; for example, a monolayer with its hydrocarbon tail directed toward the solid is deposited on top of a monolayer whose tail is directed away from the solid. The recent increase in research activity stems from possible uses of these films in microelectronics and other high technology applications.

Suitable monolayers, frequently referred to in the current literature as self-assembled monolayers (SAMs), are of interest in connection with corrosion and adhesion and as possible sensors that can selectively bind certain compounds present in the air or in aqueous solution. That is, a suitable group could be attached to the hydrocarbon chains of the surfactant molecules to effect the binding. Much of the work has employed model systems consisting of monolayers of various alkane thiols on gold. These can be characterized not only by traditional methods such as measuring the contact angle of a liquid drop on the surface, but by various spectroscopic techniques and by atomic force microscopy and its derivatives which provide detailed information on local structure. Further discussion and references to the original literature can be found in Bishop and Nuzzo (1996) and Whitesides and Laibinis (1990). The free ends of the hydrocarbon chains in SAMs can be functionalized to allow subsequent layers with different compositions to be deposited. For instance, if the free ends are given a negative charge, a second layer having positively charged ends or consisting of a positively charged polyelectrolyte can be deposited, then a third layer having negatively charged ends or consisting of a negatively charged polyelectrolyte can be deposited, etc. (i.e., a desired multilayer film structure can be built up using this layer-by-layer method).

Monolayers of other materials such as block copolymers and globular proteins at air-water interfaces have been studied. A diblock copolymer will have a block that is insoluble and spread on the interface. The other block will protrude into water. One key feature in expansion and contraction of the monolayer (below the CMC) is that of polymer entanglement. Only in some cases is a true equilibrium achieved. The experimental method where the $\pi - \sigma$ relation is obtained by compressing the monolayer at a slow and constant rate does not produce this equilibrium (Gragson et al., 1999). Where equilibrium exists, plateaus are seen in the $\pi - \sigma$ relations corresponding to transition to more favored configurations (Munoz et al., 2000).

Although proteins are surface active, they eventually denature at the air-water interface (Möbius and Miller, 1998). The resulting layer at higher concentrations can be almost like the skin that is formed on milk.

7. SOLUBILIZATION AND MICROEMULSIONS

Surfactant solutions with concentrations above the CMC can dissolve considerably larger quantities of organic materials than can pure water or surfactant solutions at concentrations below the CMC. This enhanced solubility is important in applications ranging from the formulation of pharmaceutical and personal care products, to detergency, to removal of organic contaminants from wastewaters, soils, and ground water aquifers.

Aliphatic hydrocarbons and other nonpolar compounds are thought to be incorporated or "solubilized" in the micelle interiors (Figure 4.22a). Molecules having some surface activity, such as alcohols, distribute themselves among the surfactant molecules, as shown in Figure 4.22b. This situation is closely related to formation of micelles from mixtures of surfactants, which was discussed previously. Rather polar substances may even occupy positions at the micelle surface (Figure 4.22c).

As would be expected, the capacity for solubilizing a compound is proportional to the number of micelles present and hence to the difference between the total surfactant concentration and the CMC. In this connection we note that the CMC decreases in the presence of dissolved hydrocarbon, though the effect is usually rather small.

Molecular theories of solubilization can be rather complex (Nagarajan and Ruckenstein, 1991). However, considerable insight can be gained from a relatively simple model that was suggested some years ago by Mukherjee (1970). Suppose that an aqueous micellar solution has reached its solubilization limit and is in equilibrium with an excess liquid phase of a pure hydrocarbon or other compound of low polarity. Equating the chemical potentials μ_b^o and μ_m of the contaminant in the bulk organic phase and the micelles, we have

$$\mu_b^o = \mu_m^o + RT \ln x_m ,\qquad (4.17)$$

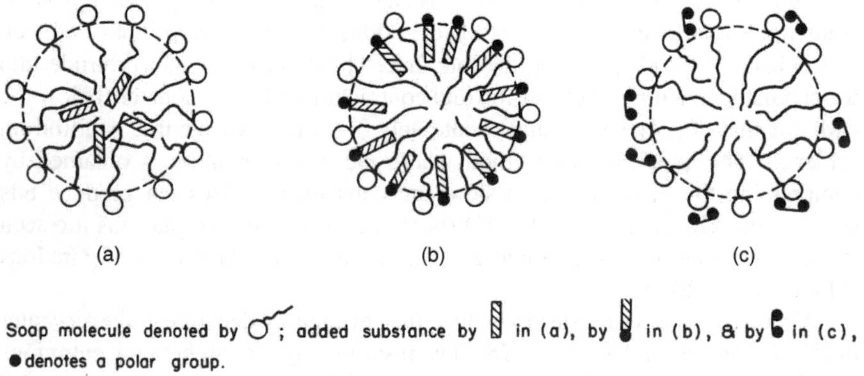

(a)   (b)   (c)

Soap molecule denoted by ◯⌒ ; added substance by ▯ in (a), by ▮ in (b), & by ● in (c), ● denotes a polar group.

FIGURE 4.22 Schematic view of solubilization. Reprinted from Alexander and Johnson (1949) with permission.

where x_m is the mole fraction of the solubilized species in the micelles and where ideal mixing in the micelles has been assumed. Now, according to the Young-Laplace equation, the pressure in the nonpolar interior of a spherical micelle is greater than that in the surrounding water by an amount $\gamma(2/r)$, where γ is the interfacial tension between the micelle and water and r is the micelle radius. Since μ_m^o must be evaluated at this higher pressure, we have

$$\mu_m^o = \mu_b^o + (2\gamma\, v_c\, /\, r)\,, \tag{4.18}$$

where v_c is the molar volume of the solubilized compound. Also pressure in the bulk oil phase has been taken equal to that in the water, a reasonable assumption here since the radii of curvature of the oil-water interfaces greatly exceed the micelle radius r, which is only some 1 to 3 nm. Substituting this equation into Equation 4.16, we obtain

$$x_m = \exp\left(-\frac{2\gamma\, v_c}{RTr}\right)\,. \tag{4.19}$$

According to this equation, compounds with small molar volumes v_c should be solubilized to a greater extent than those with large molar volumes, in agreement with existing experimental data for solubilization of hydrocarbons. Moreover, for a given compound, solubilization should increase with increasing micellar radius r (e.g., with increasing hydrocarbon chain length of the surfactant), a prediction also in agreement with experiments. Finally, solubilization should increase with decreasing interfacial tension between the micelles and water. It is known that interfacial tension decreases as a surfactant's hydrophilic and lipophilic properties become more nearly balanced, for instance, as salinity or hardness of the aqueous phase increases for the case of ionic surfactants (see Section 8). As before, this prediction is in agreement with experimental data.

Equation 4.19 is a plausible first approximation when micelles are small and spherical and solubilization is relatively low. Indeed, King (1995) showed that the dependence predicted by this model of x_m on both v_c and surfactant chain length, which determines r, is in quantitative agreement with experimental data for pure gases of low molecular weight solubilized in ionic surfactant micelles.

While the above derivation can apply both to nonpolar species solubilized in the micelle interior and to amphiphilic species such as alcohols, which basically form mixed micelles, caution should be used in the latter case. The reason is that alcohols less hydrophilic than the surfactant increase the value of v_H/al_c and can thereby cause the micelles to become rodlike, thus violating the assumption of spherical micelles employed in the analysis.

For solubilization of nonpolar compounds, the micelles can become larger in diameter as the surfactant becomes less hydrophilic and v_H/al_c increases. In this case the micelle interiors contain solubilized oil and the surfaces are covered by surfactant films. Thus l_c is not equal to the drop radius as in the earlier discussion of micelles. Indeed, if enough oil is present, drop radius increases until reaching a value where surfactant film geometry is compatible with the preferred value of v_H/al_c. When the diameters of such swollen micelles reach a few nanometers, the resulting phase is frequently called a "microemulsion" (see Figure 4.23). There is no universally accepted definition of a microemulsion. In this book we will use the term microemulsion to refer to a thermodynamically stable phase containing substantial amounts of both oil and water. Often solubilization increases greatly for small changes in system composition or temperature, so that a definite region where microemulsions exist can be identified. Use of the separate terms micellar solution and microemulsion to describe systems containing small and large drops, respectively, should not be allowed to obscure the fact that both can lie in different parts of the region of existence of a single thermodynamically stable phase.

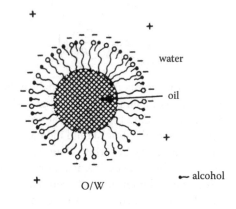

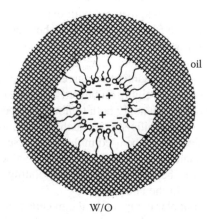

FIGURE 4.23 Schematic diagrams of oil-in-water and water-in-oil microemulsions. The small molecules shown are the cosurfactant.

Surfactant aggregation in oleic phases is more gradual than in water, formation of various small oligomers typically preceding that of larger aggregates. The presence of appreciable quantities of water generally promotes formation of "reverse micelles" with the polar groups directed inward, since water solubilized within the micelles is effectively removed from contact with the oil. As would be expected, solutions containing reverse micelles are capable of solubilizing various polar compounds. We note that a decrease in micelle curvature is favored by an increase in repulsion between head groups in this case. The packing parameter v_H/al_c discussed previously is greater than one, and the additional repulsion decreases its value toward one. Water-in-oil microemulsions also form under appropriate conditions (Figure 4.23). Indeed, most of the pioneering studies of microemulsions (Bowcott and Schulman, 1955; Hoar and Schulman, 1943; Schulman et al., 1959) dealt with oil-continuous phases.

Often the phase diagram of an oil-water-surfactant system is such that the domain of a single isotropic phase contains both regions with normal micelles and those with inverse micelles. Inversion between the two microstructures is continuous and does not involve the formation of the lamellar phase when v_H/al_c is approximately equal to 1.0. Such behavior is facilitated by the presence of amphiphilic compounds with short hydrocarbon chains, such as short-chain alcohols and sodium xylene sulfonate. The latter compound and others with similar properties are called hydrotropes and have many practical applications (Friberg, 1997).

Nagarajan and Ganesh (1989) extended the model for including polydispersity in detergent micelles [Section 2, Figure 4.6, Nagarajan and Ruckenstein (1977, 1991)] to both polydispersity in micelles from nonionic diblock copolymers as well as solubilization in such micelles. A very large range of results is obtained and are compared to their experimental results. One result is that for a given diblock and solubilizate, the aggregation number and the amount solubilized in the core do not vary much over variations in the length of the polar chains. Hurter et al. (1993) introduced more sophistication in their model by using more detailed aspects of the polymeric nature of the chains. This is expressed in the form of random walk in a lattice in a spherical system like that discussed in Chapter 3 (Section 6) (Scheutjens and Fleer, 1982). The polymer chains can interpenetrate each other and themselves, but for short range, a great deal of realism can be included. When these constraints are worked in, they result in a loss of close contact between PEO and the solvent (i.e., water) at higher temperatures. Although this is not equivalent to the breakdown of hydrogen bonding between the two that is expected at higher temperatures, the predicted temperature dependence of solubilization shows excellent agreement with the experiments.

Water-in-oil microemulsions have been reported in a water/oil/nonionic triblock polymer, where a bicontinuous phase was also seen on lowering the temperature up to 10°C (Mays et al., 1998). The problems that arise on lowering temperatures in polymeric systems, such as increases in viscosity, prevent the observation of a complete phase equilibrium diagram. A system where "water" and "oil" are replaced with polymers is a very important one. Often we would like to mix two polymers to get a better property. However, two polymers almost always turn out to be immiscible and the two-phase mixture often separates at the interface ("adhesive failure;" see Chapter 2, Section 3). To prevent this, the two-polymer blend ("polyblend") is required to have a very large surface area and a great deal of interpenetration of one phase in another. This can be achieved in melt form if a bicontinuous phase of polymer/block copolymer/polymer can be reached and then quenched to form a solid. Systems which at higher temperatures show microemulsions (hence they would be water-in-oil types) and at low temperatures show bicontinuous structures have been identified (Corvazier et al., 2001; Hillmeyer et al., 1999; Lee et al., 2003).

8. PHASE BEHAVIOR AND INTERFACIAL TENSION FOR OIL-WATER-SURFACTANT SYSTEMS

Let us consider mixtures containing a few percent of an anionic surfactant, approximately equal volumes of oil and NaCl brine, and usually some short-chain alcohol used as a "cosurfactant" or "cosolvent." If only the salinity is varied, a general pattern of phase behavior has been observed (Reed and Healy, 1977) (Figure 4.24).

At low salinities, an oil-in-water microemulsion coexists with nearly pure oil, as shown in Figure 4.24. Solubilization of oil and hence microemulsion drop size increase with increasing salinity (i.e., as repulsion between the charged head groups decreases). The interfacial tension between phases is low by ordinary standards, but is normally greater than 0.01 mN/m. This combination of phases is sometimes called Winsor type I behavior, after an early researcher on microemulsion phase behavior (Winsor, 1954).

At high salinities, the situation is reversed and an oil-continuous microemulsion coexists with excess brine, as the figure shows. Drop size increases with decreasing salinity (i.e., with increasing repulsion between head groups).

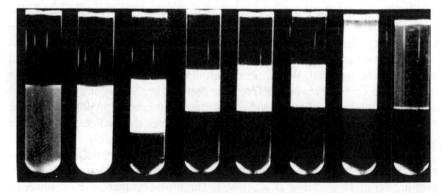

FIGURE 4.24 Phase behavior as a function of the salinity for a system containing a petroleum sulfonate surfactant, a short-chain alcohol, and approximately equal volumes of oil and brine.

Interfacial tensions in this case as well usually exceed 0.01 mN/m. This behavior is termed Winsor type II.

At intermediate salinities, three phases are present (Winsor type III), as Figure 4.24 indicates. Nearly pure oil and brine are in equilibrium with a microemulsion phase that contains almost all the surfactant in the system (except very near the transitions to the above-mentioned two-phase regions) as well as substantial quantities of oil and brine. This interesting and important phase is called the "surfactant" or "middle" phase. In the low salinity portion of its region of existence, its interfacial tension with brine is extremely low (see Figure 4.25). Conversely, in the high salinity portion of its region of existence, its interfacial tension with oil is ultralow. Between these extremes in suitably chosen systems is a region where the surfactant phase has the remarkable property of exhibiting ultralow tension with oil and brine simultaneously (Figure 4.25). The salinity where the two tensions are equal is called the "optimal" salinity and is of great importance in designing surfactant processes for enhanced oil recovery and remediation of groundwater aquifers, in detergency, and in controlling emulsion stability. For

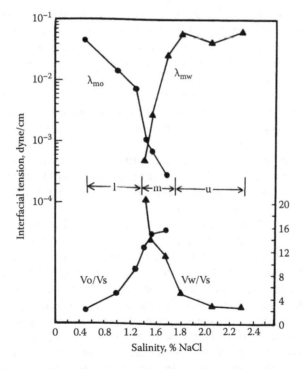

FIGURE 4.25 Interfacial tensions and solubilization parameters for a system containing 2% of a petroleum sulfonate surfactant and 1% of a short-chain alcohol. γ_{mo} is the interfacial tension between the microemulsion and an excess oil phase. V_o/V_s is the volumetric ratio of solubilized oil to surfactant in the microemulsion phase. γ_{mw} and V_w/V_s are defined in a similar manner. Reprinted from Reed and Healy (1977) with permission.

the system shown in Figure 4.23, the two tensions are about 0.0008 mN/m at 1.5% NaCl.

At optimal salinity, the volumes of oil and brine solubilized in the surfactant phase are approximately equal. These quantities can be obtained to a good approximation from measurements of phase volumes in samples such as those shown in Figure 4.24, provided that all surfactant is assumed to reside in the microemulsion phases. Oil solubilization increases with increasing salinity (Figure 4.25), while brine solubilization increases with decreasing salinity. Note that the ratios V_o/V_s and V_w/V_s of oil and brine volumes to surfactant volume are about 14 at optimal salinity in Figure 4.25—much greater than those found for the solubilization in small hydrophilic micelles discussed in Section 7 above, where the ratio is typically less than 1.0. High solubilization is accompanied by low interfacial tension. Indeed, correlations of wide applicability have been developed for relating microemulsion-oil interfacial tension to oil solubilization and microemulsion-brine interfacial tension to brine solubilization (Reed and Healy, 1977). With these correlations, interfacial tensions can be estimated from information on phase volumes as a function of salinity.

Various experiments indicate that properties of the microemulsion phase change continuously with increasing salinity as inversion from a water-continuous to an oil-continuous microstructure occurs. For instance, electrical conductivity decreases continuously with increasing salinity (Bennett et al., 1982). In addition, the self-diffusion coefficient of oil as measured by nuclear magnetic resonance (NMR) techniques increases from small values at low salinities where oil is the dispersed phase to a value comparable to that of the bulk oil phase near and above the optimal salinity. The self-diffusion coefficient of water, in contrast, decreases from a value comparable to that in pure NaCl brine below and near the optimal salinity to much smaller values at high salinities where water is the dispersed phase (Olsson et al., 1986). Thus the surfactant phase is bicontinuous near the optimal salinity, as originally proposed by Scriven (1976) and subsequently confirmed by electron microscopy (Jahn and Strey, 1988).

One additional feature of the phase behavior described above is significant. Variation of phase volumes and interfacial tension in the surfactant phase region suggests that both transitions to the three-phase region are associated with critical phenomena (Cazabat et al., 1982; Fleming and Vinatieri, 1979). Light scattering results near the transition confirm this interpretation (Cazabat et al., 1982; Huang and Kim, 1982). Thus, to the extent that the systems involved may be represented by ternary phase diagrams, the variation of phase behavior with salinity is as illustrated schematically in Figure 4.26. Note that critical end points where a microemulsion separates into two phases are associated with the appearance and disappearance of the three-phase region. For simplicity, other aspects of the phase behavior, including the existence of liquid crystalline phases, have been ignored in the ternary diagrams of Figure 4.24.

It is also useful to consider phase behavior as a function of surfactant concentration. With equal amounts of oil and water present, the plot for a pure surfactant takes the form of a "fish," as shown in Figure 4.27. Above a particular

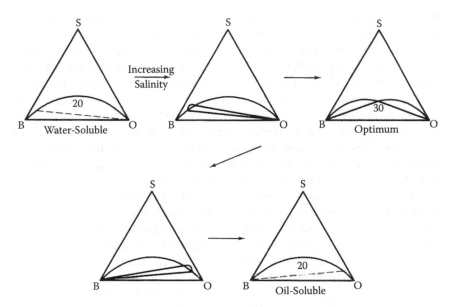

FIGURE 4.26 Schematic pseudo-ternary phase behavior. B, O, and S designate brine, oil, and surfactant, respectively.

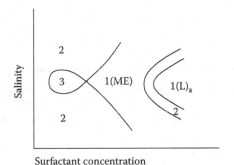

FIGURE 4.27 "Fish."

surfactant concentration, all the oil and water present can be solubilized near the optimal salinity and one finds a single-phase microemulsion instead of the three-phase region. At still higher concentrations, the lamellar phase is observed. While various workers had previously noted that a single-phase microemulsion appears at sufficiently high surfactant concentrations near the optimal salinity, Kahlweit et al. (1988, 1989) recognized the usefulness of the fish presentation and made extensive use of it in their systematic studies of phase behavior in oil-water-surfactant systems.

In view of the hydrocarbon-water interfacial tensions of some 35 to 50 mN/m given in Table 1.1, one may ask how ultralow tensions of 0.001 mN/m or even less (see Figure 4.25) can exist. The answer involves the existence of microstructure in surfactant-containing phases and the packing parameter v_H/al_c discussed earlier. If a planar interface between an oil phase and a water-continuous micellar solution or microemulsion is expanded, some surfactant must be transferred from the micelles or drops to the planar interface. The free energy change of this process involves both bending the transferred portion of the surfactant film from a curved to a flat configuration and a change in the free energy of mixing in the aqueous phase, owing to a decrease in the number of aggregates. As v_H/al_c increases toward unity, the diameter of the drops increases and less energy is required for the bending step. As a result, interfacial tension decreases. An example is the decrease of tension with increasing salinity in the Winsor I region of Figure 4.23. A similar argument involving the transfer of surfactant from inverse aggregates in the oil phase may be used to explain the decrease in interfacial tension with decreasing salinity in the Winsor II region of Figure 4.23. The magnitude of the tension can be understood in terms of the bending elastic coefficient k_b of the surfactant film, which was mentioned previously in connection with undulation forces. The energy required to bend a film from a sphere of radius R to the planar configuration is on the order of k_b/R^2. Experiments indicate that k_b is on the order of kT, where k is the Boltzmann constant (see Gradzielski, 1998). Thus if R is 10 nm, the bending energy at room temperature, and hence the interfacial tension between, say, an oil-in-water microemulsion and excess oil is on the order of 0.04 mN/m, in general agreement with measured tensions. Clearly, near optimal salinity, where v_H/al_c is approximately equal to unity, the bending energy is minimal and tensions can be very low indeed. This is a very important concept that will be utilized later to describe the surfactant phase formation.

These considerations apply to the tension between an oil-continuous and a water-continuous phase. The interface is covered by a surfactant monolayer and hence is relatively thin. However, attraction between micelles or droplets can cause separation into, for example, two water-continuous phases, one having a higher concentration of aggregates than the other. In this case the interface can be much thicker (e.g., on the order of a few droplet diameters) and interfacial tension can be low. In the limiting case of near criticality between the phases, the tension approaches zero and interfacial thickness becomes very large.

9. EFFECT OF COMPOSITION CHANGES

So far we have considered only the effect of salinity on microemulsion phase behavior. The effects of other compositional variables and temperature can be understood in terms of a single unifying principle. First, we summarize the results of numerous experiments by several workers in systems containing oil, brine, an anionic surfactant, and a short-chain alcohol.

   a. If the oil phase is a straight-chain hydrocarbon, increasing the oil chain length causes optimal salinity to increase (with temperature and other compositional variables fixed).

b. Changing from a single-chain to a double-chain surfactant with the same total number of carbon atoms causes optimal salinity to decrease.
c. Increasing the concentration of a relatively oil soluble alcohol such as n-pentanol or n-hexanol causes optimal salinity to decrease.
d. Increasing the chain length of a relatively oil soluble alcohol (e.g., from n-pentanol to n-hexanol) causes optimal salinity to decrease.
e. Increasing the concentration of a relatively water soluble alcohol such as isopropanol causes optimal salinity to increase.
f. Increasing temperature causes optimal salinity to increase. (It should be emphasized that the opposite occurs for nonionic surfactants and ionic surfactants with EO or PO groups.)

The key to understanding these results is to recognize that optimal salinity represents a balance between oil soluble (hydrophobic) and water soluble (hydrophilic) tendencies of the surfactant-alcohol films present at the surfaces of the microemulsion drops (i.e., the condition when the packing parameter v_H/al_c is approximately equal to unity). Here a and v_H must be interpreted as the area and volume of the film per surfactant molecule. Also, since the drops or other aggregates have oil or water cores, the length l_c of the surfactant molecule is no longer the same as the drop radius. If optimal salinity is to be maintained, any change (in a variable other than salinity) that decreases v_H/al_c and makes the surfactant-alcohol film more hydrophilic must be balanced by an increase in salinity, which increases v_H/al_c and makes the film more hydrophobic. An example is statement (a) above, where an increase in oil chain length decreases oil penetration into the hydrocarbon chain portion of the surfactant film, thereby decreasing both v_H and v_H/al_c). For statement (e), increasing temperature increases the thermal motion of ions, thereby reducing screening of the repulsion between adjacent ions in the surfactant film and increasing a while decreasing v_H/al_c). In contrast, any change that increases v_H/al_c and makes the film more oil soluble must be balanced by a decrease in salinity, which decreases v_H/al_c and makes the film more water soluble. This situation is represented, for example, by statement (b), since changing from a single to a double chain decreases l_c while keeping v_H constant, thus increasing v_H/al_c.

The statements involving alcohol deserve some comment. Salter (1977) showed that, with a given surfactant, adding one particular alcohol (e.g., isobutanol) might produce little or no change in optimal salinity. Adding longer chain alcohols caused decreases in optimal salinity; however, adding shorter chain alcohols caused increases in optimal salinity. Usually the surfactant chain is longer than the alcohol chain and thus determines l_c. Subject to this constraint, the longer chain alcohols produce a larger percentage change in v_H than a and so increase v_H/al_c, while the opposite is true for the shorter chain alcohols. It should be noted that while the surfactant molecules are probably almost all at the surfaces of the microemulsion drops, the alcohols partition between the drop surfaces and the oil and brine regions (Baviere et al., 1981). Partitioning behavior depends, of course, on such variables as salinity and temperature. In any case, the alcohol in the surfactant film at the drop surface is the important factor influencing phase behavior, as indicated above.

An empirical correlation describing how various compositional variables influence optimal salinity for petroleum sulfonate surfactants was developed by Salager et al. (1979):

$$\ln S_{opt} = 0.16(ACN) + f(A) - \sigma + 0.01(T - 25), \qquad (4.20)$$

where S_{opt} is optimal salinity (in g/dl), ACN is the equivalent alkane carbon number of the oil, for example, the chain length for straight-chain hydrocarbons, $f(A)$ is a function depending only on alcohol composition and concentration, σ is a parameter depending only on surfactant structure, and T is temperature (in °C). It is interesting that the oil, alcohol, surfactant, and temperature can be considered, for practical purposes, as making independent contributions to optimal salinity.

Wade, Schechter, and coworkers (Cash et al., 1977) proposed that mixtures of oils and surfactants can be described by Equation 4.20, provided that average values of ACN and σ are used.

$$(ACN)_{mix} = \sum_i x_i (ACN)_i \qquad (4.21)$$

$$\sigma_{mix} = \sum_i x_i \, \sigma_i \, . \qquad (4.22)$$

Here x_i is the mole fraction of component i. Further discussion of oil mixtures and of handling oils with naphthenic and aromatic groups can be found in Cayias et al. (1976) and Puerto and Reed (1983).

For certain classes of anionic surfactants, other effects are also important. For example, changes in pH affect the degree of ionization and hence the phase behavior when the surfactants include organic acids, amines, or other pH-sensitive compounds. An increase in the degree of ionization produces a more hydrophilic surfactant film and hence, other things being equal, increases optimal salinity (Qutubuddin et al., 1984).

To avoid complicating the discussion we have thus far dealt only with anionic surfactants. But Shinoda and Kunieda (1973) found that nonionic surfactants such as ethoxylated alcohols exhibit Winsor I-III-II phase transition with increasing temperature. Here the hydration of the EO chain decreases with increasing temperature, so that a decreases and v_H/al_c increases. Variation of interfacial tension and solubilization with temperature is similar to that described above for anionics with varying salinity. The temperature where the surfactant phase has equal interfacial tensions with oil and water is called the phase inversion temperature (PIT) or the hydrophilic-lipophilic balance (HLB) temperature.

The effects of surfactant and oil chain structure on phase behavior of nonionic surfactants are the same as given above for anionics (i.e., rules (a) and (b) above apply with PIT replacing optimal salinity). As might be expected, nonionics are much less sensitive to the addition of inorganic salts than anionics, but the direction of a shift in phase behavior is the same for most salts. The exceptions are salts containing ions such as thiocyanate (SCN), which disrupt water structure. Finally, we remark that an increase in the length of the EO chain increases a, thereby decreasing v_H/al_c and making the surfactant more hydrophilic. Accordingly, it increases the PIT.

In contrast to anionics, common nonionic surfactants have appreciable solubility in hydrocarbons. As a result, when mixtures of individual nonionic surfactants are present, as inevitably is the case for commercial products, or when anionic and nonionic surfactants are mixed, differential partitioning occurs and the composition of the surfactant films in the microemulsions formed depends on both the total surfactant concentration and the water:oil ratio. For instance, an increase in surfactant concentration while maintaining a fixed overall ratio of anionic to nonionic surfactant produces a reduced ratio of the two species in the surfactant films. The result is to increase v_H/al_c and reduce the PIT for the typical case where the nonionic surfactant has the higher value of v_H/al_c. Methods for relating the PIT or optimal conditions to surfactant concentration and the relative amounts of oil and water present have been developed (Kunieda and Ishikawa, 1985; Kunieda and Shinoda, 1985).

Example 4.4 Effect of Composition and Temperature on Optimal Salinity

A given anionic surfactant-alcohol formulation has an optimal salinity of 1.7 wt. % NaCl at 25°C with n-decane as the oil. Estimate the optimal salinity if (a) the oil is changed to n-tetradecane or (b) the temperature is raised to 40°C.

Solution

Equation 4.20 applies.

(a) If ACN changes from 10 to 14, $\ell n\, S_{opt}$ increases by $0.16 \times 4 = 0.64$ from 0.53 to 1.17. $S_{opt} = 3.2$ wt. % NaCl.

(b) If T increases from 25°C to 40°C, $\ell n\, S_{opt}$ increases by 0.15. $S_{opt} = 2.0\%$ NaCl.

Note that the temperature effect is relatively weak for anionic surfactants.

10. THERMODYNAMICS OF MICROEMULSIONS

As pointed out in Chapter 3, most colloidal dispersions, including emulsions, are unstable from a thermodynamic point of view, owing to their large interfacial area. How then can microemulsions be thermodynamically stable? Some years ago Rehbinder (1957) and Shchukin and Rehbinder (1958) suggested that a colloidal dispersion could be thermodynamically stable, provided that interfacial tension was low enough that the increase in interfacial energy accompanying dispersion of one phase in the other could be outweighed by the free energy decrease associated with the entropy of dispersion. Ruckenstein and Chi (1975) recognized the importance of this effect for microemulsions and developed a suitable analysis to describe it quantitatively. Subsequently Ruckenstein (1978) also pointed out that the free energy decrease accompanying adsorption of surfactant molecules from a bulk phase favors the existence of a large interfacial area and hence plays a major role in stabilizing microemulsions. In his actual calculations, attractive and repulsive forces between nearby drops were also included using the methods described in Chapter 3.

Figure 4.28 shows the general form of the calculated free energy ΔG_M of microemulsion formation as a function of drop radius (Ruckenstein and Chi, 1975) for a system with fixed amounts of oil, water, and surfactant. ΔG_M is lowest

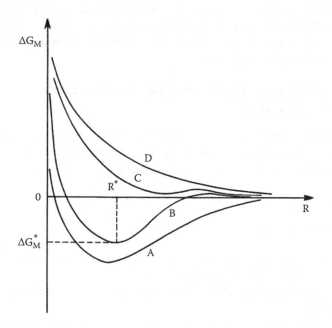

FIGURE 4.28 The Gibbs free energy of formation of microemulsions ΔG_M shown as a function of droplet radii R under various conditions. The curves A and B show that negative values of ΔG_M can be obtained (i.e., that microemulsions can form spontaneously). The actual radius is R^* corresponding to a minimum ΔG_M^*. The curve C shows kinetic stability and D is unstable. Reprinted from Ruckenstein and Chi (1975) with permission.

for curve A and highest for curve D. For curve B, the predicted radius is R^*, where the free energy is minimized. The negative value of the corresponding $\Delta G_M{}^*$ implies that the microemulsion is thermodynamically stable. The same is true for curve A, but the minimum for curve C occurs for a positive ΔG_M, indicating that any microemulsion that forms is metastable. Not even a metastable microemulsion is possible for curve D. For given amounts of oil, water, and surfactant, such curves can be calculated for both oil-in-water and water-in-oil microemulsions. The arrangement having the lower value of $\Delta G_M{}^*$ would be expected to occur.

An additional factor which frequently is important is the work required to bend a surfactant film (i.e., to change its mean curvature H) at a constant area (Murphy, 1966). As discussed in Problem 1.4, one more term is needed in Equation 1.9 to include the bending effects for interfaces which are of low tension or high curvature — conditions that are clearly fulfilled for microemulsions. Especially when a microemulsion is in equilibrium with an excess phase (e.g., an oil-in-water microemulsion with excess oil), bending effects play a major role in determining drop size. That is, the drop radius is typically not far from the "natural" radius the film assumes at an interface between oil and water. When head group repulsion is large, for example, the natural radius is small (or the "spontaneous curvature" large) with oil the interior phase.

Miller and Neogi (1980) and Mukherjee et al. (1983) analyzed the case of a microemulsion in equilibrium with excess dispersed phase (e.g., an oil-in-water microemulsion in equilibrium with excess oil) using concepts based on Hill's ([1963] 1992) thermodynamics of small systems and explicitly including bending effects. Their equations for equilibrium are given by

$$-4\pi r N \ [\gamma + (2(C/A)/r)] + (\partial G_{md}/\partial r)_{T,p,n_o,n_c,n_s} = 0 \qquad (4.23)$$

$$8\pi r N \ [\gamma - ((C/A)/r)] + (\partial G_{md}/\partial r)_{T,p,n_d,n_c,n_s} = 0 \qquad (4.24)$$

where N is the number of drops of radius r in a unit volume of microemulsion: γ is the interfacial tension of the droplets, A is the interfacial area, and C is the "bending stress" of the surfactant films of the droplets as defined in Problem 1.4. Also G_{md} is the contribution to the microemulsion free energy per unit volume resulting from the entropy of dispersion and the energy of interaction among droplets. For instance, in a simple case one could use the well-known Carnahan-Starling equation (Carnahan and Starling, 1969) based on a hard sphere model:

$$G_{md} = NkT\{\ln\varphi - 1 + \varphi[(4 - 3\varphi)/(1 - \varphi)^2] + \ln(3V_c/4\pi r^3)\} \qquad (4.25)$$

where ϕ is the volume fraction of droplets in the microemulsion and V_c is the volume of a molecule of the continuous phase of the microemulsion.

The partial derivative in Equation 4.23 is taken at constant temperature, pressure, and number of moles per unit volume n_c, n_d, and n_s of continuous phase, dispersed phase, and surfactant, respectively, in the microemulsion. This equation applies even when no excess phase is present and is used to find the equilibrium droplet radius in that case. The curves of Figure 4.28 were constructed using a similar equation, but with a somewhat different expression for G_{md} and without the term in C. In Equation 4.24, the partial derivative is taken at constant N, but with variable n_d. It is applicable only when excess disperse phase is present. When these equations were solved simultaneously, it was found that with a given amount of surfactant present, more droplets of smaller size are predicted than would be expected based on the relevant natural radius. That is, solubilization with an excess of the dispersed phase is somewhat less than if droplets assumed their natural radius. In this case, the energy required to bend the film beyond its natural radius is offset by the decrease in free energy associated with the increased entropy of the more numerous droplets.

It is also possible, as indicated previously, for a microemulsion containing spherical droplets to separate into a more concentrated microemulsion and excess continuous phase (e.g., an oil-in-water microemulsion in equilibrium with excess water), provided that attractive interaction among the droplets is sufficiently large. In other words, a microemulsion can have a limited capability to solubilize its continuous phase as well as its dispersed phase. Interfacial tension between the phases should be very low since both are continuous in the same component. Such a phase separation is similar to that which takes place at the cloud point of nonionic surfactants discussed previously. A simple theory of how it could occur for microemulsions was proposed by Miller et al. (1977). If excess continuous phase separates in this way and, at the same time, there is more dispersed phase present than can be solubilized, the microemulsion can coexist with both oil and water phases. While this situation of three-phase coexistence involving a micro-emulsion containing droplets probably exists for some compositions in some systems, in most situations the microemulsion in a three-phase region is bicon-tinuous. The above discussion emphasizes the early theoretical work on micro-emulsions with droplets, but numerous other developments have been reported since then.

Several theories of surfactant phase are available. Following Scriven (1976), this phase is assumed to be bicontinuous in oil and water, and the interface is assumed to have zero mean curvature, hence the pressure difference between oil and water is zero. Talmon and Prager (1978, 1982) divided up the medium into random polyhedra. The flat walls ensure no pressure difference between oil and water. They placed oil and water randomly into the polyhedra so that both oil and water were continuous when sufficient amounts of both phases were present. As in the earlier models of oil-in-water microemulsions, this randomness gave rise to an increased entropy which overcame the increased surface energy to yield a negative free energy of formation, reached only when the interfacial tension is ultralow. Such structures can form spontaneously. This random structure is char-acterized by a length scale. This led Jouffrey et al. (1982) to postulate that surfactant-laden surfaces between oil and water are characterized by a persistence length, which is the distance over which the surface can be approximated as flat. Pieces of surfaces less than this length scale are very difficult to bend. Hence random structures form with this length scale, which is directly related to the length scale of the polyhedra described earlier. Widom (1984) and Andelman et al. (1987) further developed this approach.

These theories also take into account the energy to deform the interface at a constant area. It is frequently expressed as (Safran, 1994)

$$(k_b/2)(2H - 2H_o)^2 + k_G K,$$

where H_o is the "spontaneous curvature" (i.e., that for which the bending stress C), defined as

$$C = -[\partial U^s / \partial(2H)]_{S,A,n_{is},K} ,$$

vanishes. This definition generalizes the expression for spheres given following Equation 4.24. Also, k_b is the bending modulus defined by

$$k_b = [(C/A)/(2H_o - 2H)].$$

It is often taken as constant in the absence of further information. K is the Gaussian curvature (the product of the two principal curvatures) and k_G is the "saddle splay" modulus defined as

$$k_G = [\partial U^s / \partial K]_{S,A,n_{is},H} .$$

It is a torsional modulus (Murphy, 1966); that is, it describes the work done in twisting the interface at a constant area and mean curvature. It is of higher order than C for small values of curvature.

Near optimal salinity k_b has been found to be small, so that the term in k_G can become significant. Neogi et al. (1987) show that such surfaces are unstable to infinitesimal disturbances at length scales above a critical value (critical wavelength). Thus surfaces of zero mean curvature will keep breaking up until they form structures at this length scale. Because of continued activity, the resulting structure may not have zero mean curvature on an average. Hofsäss and Kleinert (1987) introduced the effects of torsion in a discrete states model for microemulsions by Widom (1984) and found that the structures degrade to dimensions that are small solely due to this term, unless the bending modulus is quite high.

The common theme among these is that there is a small length scale involved that characterizes the structure, below which oil and water are segregated, but above which they are not. This scale is significantly larger than the size of the molecules involved.

11. APPLICATIONS OF SURFACTANTS: EMULSIONS

In contrast to microemulsions, ordinary emulsions are thermodynamically unstable, but they can be stable in a practical sense if the energy barrier to flocculation is sufficiently high. As with other colloidal dispersions, this energy barrier may be electrical in nature if the drops are charged, if water is the continuous phase, and if the ionic strength is not too high (cf. the discussion of DLVO theory in Chapter 3). Typically, stability is provided by adsorbed surfactants and polymers. However, it can also stem from small solid particles that are not completely wet by either phase and thus accumulate at the drop surfaces (Aveyard et al., 2003).

The phase behavior of oil-water-surfactant systems also has an important influence on emulsion stability. Friberg et al. (1969) show that the presence of the lamellar liquid crystalline phase instead of a simple surfactant monolayer at the drop surfaces can greatly enhance stability. Thus, shifting overall system composition from a two-phase region of the equilibrium phase diagram with aqueous and oleic phases present to a three-phase region where the lamellar phase is also found increases emulsion stability dramatically. In a similar manner, foams containing some liquid crystalline material are frequently quite stable. The liquid crystalline phase must spread spontaneously on the foam or emulsion films so that it covers their entire surfaces and protects them from breaking (Friberg and Ahmed, 1971). This mechanism of stabilization is particularly important for nonaqueous foams, such as those of liquid hydrocarbons, where electrical stabilization does not occur.

Even in portions of the phase diagram where microemulsions are found but not liquid crystals, emulsion stability is strongly correlated with phase behavior. As several groups have observed, emulsions break more rapidly in the region near optimal salinity where the surfactant phase coexists with excess oil and brine than in the regions at lower and higher salinities where a microemulsion and a single excess phase are present (Bourrel et al., 1979; Salager et al., 1982; Vinatieri, 1980). Salager et al. (1991) made use of this principle in their extensive studies of emulsion stability in various systems. Similarly, emulsions made with nonionic surfactants are least stable near PIT (Saito and Shinoda, 1970).

Generally speaking, surfactants preferentially soluble in water (i.e., $v_H/al_c <$ 1) are found to be best for making oil-in-water emulsions and surfactants preferentially soluble in oil (i.e., $v_H/al_c > 1$) are found to be best for making water-in-oil emulsions. For nonionic surfactants where phase behavior is a strong function of temperature, a general rule of thumb is that the temperature should be at least 20°C below (above) the PIT to obtain stable oil-in-water (water-in-oil) emulsions. Davies and Rideal (1963) suggest that both types of emulsions are formed during mixing, but only the type which flocculates and coalesces more slowly persists after agitation ceases. One might expect, for instance, oil-in-water emulsions to flocculate and coalesce slowly for water soluble surfactants because such surfactants are strongly hydrated or electrically charged, providing a substantial energy barrier for flocculation and for rupture of the thin aqueous films between oil drops approaching one another as a result of thermal motion or flow. Since any water-in-oil emulsion present coalesces rapidly when mixing stops in this case, the final dispersion is of the oil-in-water type.

Ivanov and Kralchevsky (1997) agree that the final emulsion type depends on the relative coalescence rates of oil-in-water and water-in-oil dispersions, but suggest that the rates are determined by the behavior of the thin liquid films formed when two drops approach. Consider, for example, a thin film of water between two oil drops. If the surfactant is preferentially soluble in water and has a low solubility in oil, it can remain in the film, slow film drainage, and stabilize the film at some equilibrium thickness or at least hinder film breakup, which leads to coalescence. In contrast, if the solubility of the surfactant in oil is high, it can diffuse into the oil, where it is not available to slow film drainage and hinder breakup. A similar argument can be made for the oil film between two water drops in a water-in-oil emulsion.

Kabalnov and Wennerström (1996) proposed yet another explanation for the difference in coalescence rates. It deals with nucleation of a small hole in a thin liquid film once it has reached its equilibrium thickness (note that some films become unstable without ever reaching an equilibrium thickness; see Chapter 5). If the hole is large enough to grow spontaneously, the film breaks and coalescence occurs. Consider again a water film between oil drops. A nucleus will have the shape shown in Figure 4.29. Note that the curvature in the view shown at point P is the same as that for water drops in oil. If the spontaneous or preferred curvature of the surfactant monolayer is that of oil drops in water (i.e., if v_H/al_c < 1), considerable energy must be supplied to bend the film to form a nucleus such as that shown in Figure 4.29, which has the opposite curvature. A similar argument indicates that only minimal bending energy must be supplied to form a nucleus in an oil film between water drops for the same surfactant monolayer. Consequently, thermal fluctuations, which have energy on the order of kT, are more likely to create a hole large enough to produce instability in the latter case. Coalescence is thus slower for the oil-in-water arrangement, and it will be the final state.

As we have seen, the optimal salinity or PIT concept based on the equilibrium phase behavior provides much useful insight in formulating emulsions and

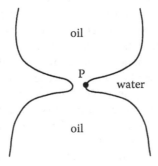

FIGURE 4.29 Sketch of the hole developing in the thin film between emulsion drops.

understanding their stability. Another widely used concept is the HLB value, originally proposed by Griffin (1949), which characterizes the relative oil and water solubility of surfactants. The smaller the HLB value, the more oil soluble the surfactant, with an HLB value of about 7 corresponding to a surfactant having equal oil and water solubilities. Methods of calculating HLB values from knowledge of surfactant structure are described by Davies and Rideal (1963). Two surfactants with different HLB values may be blended to obtain any desired intermediate value. Knowing the HLB value of a surfactant provides less information than knowing the PIT because the latter depends on properties of the oil as well as those of the surfactant, both of which are important in determining conditions for emulsion stability.

The microstructure of the continuous phase can also influence flocculation and coalescence. The water phase of the oil-in-water emulsion is typically a micellar solution containing solubilized oil or an oil-in-water microemulsion. It has been demonstrated that when surfactant concentration is sufficiently high, the micelles present can promote flocculation by the depletion mechanism discussed in Chapter 3 (Aronson, 1989). On the other hand, the presence of micelles can slow drainage of the film between approaching oil drops by means of the stepwise drainage process discussed more fully in Chapter 5 (Section 6). The result is a decrease in the rate of coalescence. In contrast, if the phase forming the film between approaching drops is bicontinuous, one might expect the barrier to coalescence to be minimal. Even if the film has only one continuous phase, coalescence can be rapid if there is frequent merging and breakup of drops in the microemulsion. For example, continuous water channels form transiently in some water-in-oil microemulsions, as shown by the observations that water has a low self-diffusion coefficient, confirming its presence as a dispersed phase, while at the same time electrical conductivity is relatively high because of the percolating behavior. In this case the microemulsion behaves much like a bicontinuous phase in promoting coalesence of water drops in a water-in-oil emulsion (Hazlett and Schecter, 1988).

Physical barriers to coalescence can increase emulsion stability. Crude oil is frequently produced as a water-in-oil emulsion. In some cases the water, or more accurately brine drops, are covered by rather rigid films that impede coalescence. Many such emulsions can be broken by adding small amounts of appropriate surfactants to the oil. These surfactants, which often are polymeric in nature with molecular weights of a few thousand, displace the rigid material at the drop surfaces, forming instead very mobile films, which readily permit drainage of the films between the drops, leading to coalescence. Similar behavior is seen for many food emulsions. In this case, proteins provide rigid films at the drop surfaces which can often be displaced when surfactants are added.

The relative amounts of oil and water employed also influence phase continuity. Other factors being equal, oil-in-water emulsions are formed when only small amounts of oil are present and water-in-oil emulsions are formed when only small amounts of water are present. Indeed, stable dilute emulsions can be formed without any surfactant at all if a procedure such as ultrasonic vibration is used, which generates very small drops. On the other hand, stable emulsions with high concentrations of the dispersed phase — even exceeding 99% — can be produced with a suitable choice of surfactant and mixing procedure (e.g., gradual addition of the dispersed phase). For such high volume fractions to be reached the drops must be distorted from spheres into polyhedra (Lissant, 1966), the arrangement becoming much the same as in a foam of high gas content. These emulsions are sometimes referred to as high internal phase ratio (HIPR) emulsions and can be made with either oil or water as the continuous phase (Kunieda et al., 1990). They exhibit non-Newtonian rheological behavior, including a yield stress (Princen, 1979), which makes them useful in certain applications.

The inversion of emulsions upon varying the water:oil ratio often exhibits hysteresis. As shown in Figure 4.30 (Silva et al., 1998), inversion of an oil-in-water emulsion does not take place until there is quite high oil content when the surfactant is hydrophilic (i.e., when v_H/al_c is small). Indeed, HIPR oil-in-water emulsions can be formed in this way, as indicated above. For the same surfactant and oil, inversion from water-in-oil to oil-in-water emulsions occurs at lower oil contents. Increasing surfactant concentration shifts both inversion points toward higher oil contents (Silva et al., 1998). A similar situation occurs for inversion when the surfactant is lipophilic, except that it occurs at high water contents, as Figure 4.30 indicates.

As mentioned in Chapter 3, development of methods to make solid particles with known composition and a narrow size distribution has contributed significantly to recent improvements in the fundamental understanding of colloidal dispersions. For stable emulsions, Bibette (1991) developed a method using successive fractionation to obtain emulsions with drops of nearly uniform diameter. It may have a similar effect on our understanding of emulsions.

It is, of course, not necessary to form an emulsion by mechanical dispersion of oil and water phases. One method that has been used to form oil-in-water emulsions with small and uniform drops in a nonionic surfactant system is to start with the surfactant phase near the PIT and cool it rapidly by perhaps 20°C to 30°C (Friberg and Solans, 1968; Förster et al., 1995; Sagitani, 1992). The capacity for solubilization of oil decreases dramatically upon cooling, and the excess oil nucleates as small drops from the supersaturated microemulsion. Provided that it solubilizes substantial oil, the lamellar liquid crystalline can also be cooled in this manner to form oil-in-water emulsions (Förster et al., 1995). Spontaneous emulsification can also be produced by diffusion, as discussed in Chapter 6.

Near the PIT, formation of multiple emulsions is frequently observed during mixing (e.g., one or more water drops within an oil drop, which is itself dispersed in water). This type of multiple emulsion is called water-in-oil-in-water. Of course, oil-in-water-in-oil emulsions also occur under these circumstances. One can also deliberately make multiple emulsions. For instance, a water-in-oil emulsion with relatively small drops is first formed by vigorous mixing with a suitable

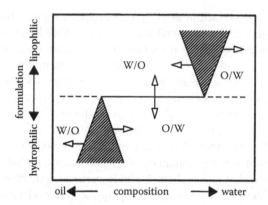

oil ◄──── composition ──► water

FIGURE 4.30 Emulsion inversion map. Adapted from Silva et al. (1998). Inversion can occur due to changes in hydrophilicity of the surfactant (vertical) or to changes in the water-to-oil ratio (horizontal).

(lipophilic) surfactant present. Then this emulsion is added to water and stirred gently with a hydrophilic surfactant to form a water-in-oil-in-water emulsion. A multiple emulsion made by using a micropipette to inject water droplets into an oil drop is shown in Figure 4.31 (Hou and Papadopoulos, 1997).

Multiple emulsions are the basis of so-called liquid membrane separation processes, where one solute is preferentially transferred from, say, the outer to the inner aqueous phase of a water-in-oil-in-water emulsion. The preferred solute is transferred faster because it is more soluble in the oil phase than other species dissolved in the inner aqueous phase. Transport across the oil phase may occur by diffusion of solute molecules or solute-containing reverse micelles or micro-emulsion drops. Sometimes a "carrier" is added to the oil to increase solute solubility in the oil by formation of a solute/carrier complex.

One of the challenges of liquid membrane separation is ensuring adequate multiple emulsion stability during transport while providing for relatively easy breakage after separation has occurred. Removal of soluble contaminants from wastewater is an example of where this process can be applied.

Multiple emulsions have also been considered as a possible mechanism for gradual release of drugs and cosmetic ingredients, which would have to diffuse from the interior drops of a water-in-oil-in-water emulsion to the external aqueous phase before becoming available. In this case, stability is desirable for both inner and outer emulsions. It can be increased by using proteins or other polymeric surfactants instead of or together with conventional surfactants. Since polymeric surfactants are not readily desorbed, there is, in contrast to the situation for ordinary nonionic surfactants, minimal transport of the lipophilic surfactant from the inner water-in-oil emulsion to the outer oil-in-water emulsion, and vice versa. Garti and Bisperink (1998) have reviewed the behavior and applications of multiple emulsions.

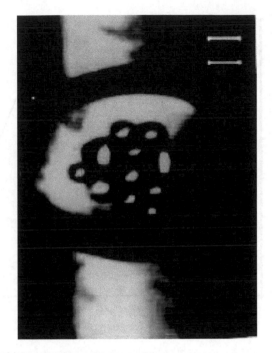

FIGURE 4.31 Photograph of a multiple emulsion.

12. APPLICATIONS OF SURFACTANTS: DETERGENCY

The detergent action of surfactants proceeds by different mechanisms under different conditions. Liquid soils of an oily nature are typically removed from hydrophilic solids such as cotton by the "rollback" mechanism. The soiled material is poorly wet by ordinary water. But if a suitable surfactant is present, it adsorbs at the water-solid interface, lowering the interfacial tension there. According to Young's equation (Equation 2.1), this reduction acts to lower the contact angle measured through the water. With sufficient reduction in contact angle, the aqueous solution spreads along the solid, causing the initial film of oil to contract and form a drop. Ultimately the drop is completely displaced from the solid or at least made large enough that most of it can be removed by agitation of the detergent solution.

Various aspects of the rollback mechanism have been reviewed by Carroll (1993). One interesting and important phenomenon is that the stable configuration for an oil drop on a fiber of uniform, circular cross section is axisymmetric when the contact angle, as measured through the water phase, is large but less than 180°. As the contact angle decreases, a point is reached when the axisymmetric configuration becomes unstable. The drop moves to one side of the fiber and becomes more spherical and is thus more readily removed by agitation in the washing bath.

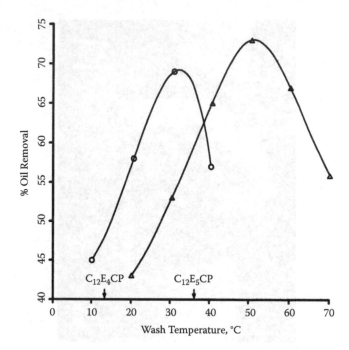

FIGURE 4.32 Detergency versus T for pure nonionic surfactants and hexadecane.

Because synthetic materials such as polyester are less hydrophilic than cotton, oily liquids adhere to them more strongly and the rollback mechanism has limited effectiveness. With the proper choice of surfactant and conditions, however, the soil can be removed by a solubilization-emulsification mechanism. With hydrocarbon soils, for instance, maximum soil removal for pure nonionic surfactants normally occurs near the PIT, as shown in Figure 4.32 for polyester/cotton fabric soiled with n-hexadecane and washed with pure $C_{12}E_4$ and $C_{12}E_5$. Observations using video microscopy show that near the PIT an "intermediate" middle phase microemulsion phase develops near the surface of contact between the water-surfactant mixture and the oil (Raney et al., 1987). As discussed previously, this microemulsion can solubilize considerable hydrocarbon. The very low interfacial tensions that exist under these conditions facilitate emulsification of the microemulsion into the agitated washing bath. As Figure 4.32 indicates, the PIT is well above the cloud point temperature in these systems, and indeed the surfactant is present in the initial washing bath as a dilute dispersion of the lamellar liquid crystalline phase in water.

At temperatures significantly below the PIT (e.g., in the Winsor I phase region), the microemulsion can solubilize much less hydrocarbon (see Figure 4.25). The rate of solubilization is also slower (Miller and Raney, 1993) and interfacial tension is higher. As a result, soil removal is considerably lower, as may be seen from Figure 4.32. Well below the cloud point temperature, no intermediate phase forms at all, and oil is solubilized very slowly into micelles in the surfactant solution.

Above the PIT in the Winsor II region the explanation for the poor detergency is different. In this case, both surfactant and water diffuse into the oil phase, so that it actually increases in volume instead of being solubilized. Moreover, video microscopy observations show that extensive spontaneous emulsification of water in the oil occurs. Here too the interfacial tension is higher than at the PIT. The results are that hydrocarbon removal is low (Figure 4.32) and that fabric weight actually increases during washing (Solans et al., 1988). It should be noted that the conditions for development of intermediate phases and for occurrence of spontaneous emulsification can be understood by the application of diffusion path theory (see Chapter 6).

When surfactant mixtures of practical interest containing multiple species were used (e.g., commercial nonionic surfactants or mixtures of anionic and nonionic surfactants), a maximum in hydrocarbon removal from polyester/cotton fabric similar to that in Figure 4.32 was again seen. For situations where the surfactant:oil ratio in the system is large, the typical case for household washing, the maximum occurred at the PIT of a system for which surfactant composition in the films separating oil and water domains of the microemulsion phase was the same as the initial surfactant composition (Raney and Miller, 1987). This result is reasonable since the small amount of hydrocarbon present can dissolve only a small portion of the total surfactant, leaving the remainder, which has nearly the initial composition, to make up the films. It should be noted that here too the PIT is well above the cloud point temperature of the mixed surfactant solution.

As indicated above, Figure 4.32 shows that maximum removal of n-hexadecane, which is widely used as a model compound representing mineral oils, occurred near the PIT. Similar results were found for 50/50 mixtures by weight of n-hexadecane and squalane, a highly branched hydrocarbon containing 30 carbon atoms (Raney et al., 1987). Thompson (1994) found similar behavior with n-decane. However, maximum removal of pure squalane with solutions of $C_{12}E_5$ occurred at approximately 40°C, well below the PIT of 63°C. Based on contact angle measurements, Thompson attributed this behavior to the fabric being rather oil wet at the PIT of squalane, which made the oil more difficult to remove.

Addition of a long-chain alcohol or fatty acid to nonionic surfactant-water-hydrocarbon systems depresses the PIT substantially. Indeed, in certain systems used as models for cleaning of skin sebum from fabrics, the PIT for a mixture of a hydrocarbon and such a long-chain amphiphilic compound is well below the surfactant cloud point temperature. For these systems, observations using video microscopy showed that for temperatures below the PIT, the soil was slowly solubilized into the surfactant solution with no intermediate phase formation (Lim and Miller, 1991). In contrast, lamellar liquid crystal formed as an intermediate phase for temperatures above the PIT. The liquid crystal grew as small, rather fluid filaments or myelinic figures that would presumably be broken off by agitation during actual washing processes. Washing experiments with polyester/cotton fabric showed that soil removal was considerably better above the PIT, where the oil was solubilized in the intermediate liquid crystalline phase, than at lower temperatures (Raney and Benson, 1990). Analysis of the diffusion process leading to formation of the intermediate liquid crystalline phase in such systems is provided in Chapter 6 (section 10).

Pure liquid triglycerides such as triolein are often used as models for soils consisting largely of cooking oils. Because of its high molecular volume, triolein is not readily incorporated into surfactant films and does not easily form water-continuous or bicontinuous microemulsions near ambient temperature (Mori et al., 1989). However, an intermediate microemulsion phase was observed during video microscopy contacting experiments in the $C_{12}E_5$-water-triolein system near 65°C, and the dependence of soil removal from polyester/cotton fabric on temperature had a sharp maximum at this temperature similar to that of Figure 4.32 (Mori et al., 1989). In contrast, when $C_{12}E_5$ was replaced by $C_{12}E_4$, triolein removal was moderate and nearly constant over a temperature range of some 20°C, falling off at lower and higher temperatures, although Thompson (1994) did observe a maximum in similar experiments. No intermediate phase was observed during the contacting experiments, and the triolein removal that did occur was attributed mainly to solubilization into the lamellar liquid crystal present initially in the washing bath. When a 50/50 mixture of triolein and n-hexadecane was used with the same surfactant, an intermediate microemulsion phase was observed at some temperatures and soil removal was increased significantly. All these results, which have been reviewed by Miller and Raney (1993), suggest that removal of oily liquid soils from synthetic fabrics is usually best accomplished by a solubilization-emulsification mechanism that is most effective when an intermediate phase capable of solubilizing considerable soil is formed and not when the soil is solubilized directly into micelles in the washing bath without intermediate phase formation. Such intermediate phases are most likely to occur when the surfactant/soil system is nearly balanced with respect to hydrophilic and lipophilic properties. This balance can be present initially as in the hydrocarbon and triglyceride systems discussed above, or it can develop during the washing process as a result of diffusion by surfactant into the soil, as shown more fully in Chapter 6 for hydrocarbon–long-chain alcohol soils.

Solid oily soils have received less attention than the corresponding liquids. At sufficiently high temperatures, solid fatty acids or alcohols swell in surfactant solutions, forming myelinic figures (Lawrence et al., 1964). This behavior can facilitate soil removal, as indicated above, if the myelinic figures can be broken away from the bulk of the soil by agitation and dispersed in the washing bath. At low temperatures, no swelling occurs. In this case the long-chain acid or alcohol dissolves slowly into the surfactant solution. Mass transfer studies have been carried out to clarify the mechanism of dissolution, which involves transport of surfactant micelles to the solid surface, where they temporarily adsorb, acquire some acid or alcohol molecules, then desorb and return to the bulk solution (Chan et al., 1976; Shaeiwitz et al., 1981).

Removal of layers of highly viscous abietic acid/alcohol mixtures from rotating silicon disks in nonionic surfactant solutions has been investigated (Beaudoin et al., 1995). As in the hydrocarbon–long-chain alcohol systems discussed above, both surfactant and water diffused into the acid layer, increasing its volume. More importantly, viscosity was substantially reduced, and the shear stresses exerted by the surfactant solution were able to detach much of the soil from the rotating disk and disperse it in the washing bath. Apparently no liquid crystalline phase was involved.

Solid inorganic soils (e.g., clay particles) must be removed mechanically by the washing process. One function of the anionic surfactants present is to adsorb on the clay and fabric surfaces, so that electrical repulsion aids removal of the particles and opposes their redeposition. Sometimes polymers that adsorb on the fabric are used to provide a steric component of the repulsion between particles and fabric by means of the mechanism discussed in Chapter 3 and thereby prevent redeposition.

13. CHEMICAL REACTIONS IN MICELLAR SOLUTIONS AND MICROEMULSIONS

Surfactant aggregates are a source of microstructure in solution that can provide a favorable environment for some chemical reactions. While, with the exception of emulsion polymerization, large-scale commercial use of micellar solutions or microemulsions as reaction media is not currently widespread, their potential is great and an area of active research (Grätzel and Kalyanasundararam, 1991; Holmberg, 1994; Pileni, 1993; Rathman, 1996).

Micelles can solubilize organic reactants having low solubilities in water, thereby allowing reactions to be carried out in aqueous media instead of organic solvents, whose use is undesirable for environmental reasons. Moreover, the concentration of such reactants in a micelle can be substantial. At the same time, surfactants can be chosen to increase the local concentration of water-soluble reactants at the micelle surface (e.g., cationic surfactants to attract anionic reactants). This combination of higher local concentrations of both organic and inorganic reactants near the micelle surface can greatly increase rates of a wide variety of reactions of commercial interest. In addition, both the specific heat and thermal conductivity of water are relatively large, making control of temperature easier.

A simple illustration of a reaction that proceeds faster in a micellar solution is the alkaline hydrolysis of trichlorotoluene to benzoate. As Table 4.2 shows (Menger et al., 1975), negligible conversion occurred in 1.5 hours in 20% sodium hydroxide (NaOH) at 80°C. However, the yield of benzoate was 98% in 1.5 hours when the solution also contained 0.01 M of the cationic surfactant cetyltrimethylammonium bromide (CTAB). The faster rate occurred because trichlorotoluene was solubilized in the surfactant micelles and hydroxide ions were attracted to the positively charged micelle surfaces. When commercial nonionic surfactant (Brij 35) having the average composition $C_{12}E_{23}$ was used instead, 97% conversion was achieved after 11 hours. In this case the micelles again solubulized trichlo-

TABLE 4.2
Effect of surfactants on conversion of trichlorotoluene to benzoic acid in 20% NaOH at 80°C (Menger et al., 1975)

Additive	Reaction time (hr)	Percent yield
0.01 M CTAB	1.5	98
None	1.5	0
None	60	97
0.006 M Brij 35	11	97

rotoluene and, as indicated in Section 5, may have a small positive charge as well. The same conversion was not achieved until 60 hours in the absence of surfactant.

Surfactant aggregates can also improve selectivity by having different effects on rates of different reactions. In aqueous nitric acid (HNO_3) solution, for example, nitration of phenol produces 35% o-nitrophenol and 65% p-nitrophenol. In an oil-in-water microemulsion, however, the corresponding results were 80% and 10% (Chhatre et al., 1993). Solubilized phenol molecules were oriented at the microemulsion droplet surfaces with the hydroxy (OH) group extending toward the aqueous solution. The ortho position was thus most accessible to nitronium ions (NO_2^+) in the aqueous solution.

Surfactants themselves can be synthesized in some cases by autocatalytic reactions using micelles. Kust and Rathman (1995), for instance, synthesized N,N-dimethyldodecylamine N-oxide by reaction of the amine with hydrogen peroxide (H_2O_2). The reaction rate increased greatly when enough surfactant had been produced to form micelles which solubilized the amine.

Microemulsions have also been used as reaction media for polymerization (Candau, 1992), the objective being to capture the small length scale of the initial microstructure in the polymer produced. Latex particles can be formed in either oil-in-water or water-in-oil microemulsions, depending on whether the monomer is preferentially soluble in oil or water. Particles of polymethyl methacrylate as small as 2 to 5 nm have been produced in water-in-oil microemulsions (Hammouda et al., 1995), although in most work the particles of various polymers investigated have been larger. Morgan et al. (1997) developed a model that yielded accurate predictions of conversion as a function of time for polymerization of hexyl methacrylate to form latex particles in a water-continuous microemulsion stabilized by a mixture of a single-chain and a double-chain cationic surfactant. Hermanson and Kaler (2003) extended the model to include the addition of monomer at specified points during polymerization for the same system. They also performed experiments confirming the theory and providing an understanding of how particle size distribution is affected by monomer addition.

Efforts are continuing to produce microporous solids for possible use as catalysts or adsorbents by polymerization of the oil or aqueous phase of bicontinuous microemulsions (Desai et al., 1996). In this case also, most of the work has yielded solids with characteristic sizes (here pore sizes) larger than those of the original microemulsions. A few experiments have been reported where different monomers in the oil and aqueous regions are polymerized simultaneously (Qutubuddin and Lin, 1994). Another approach to forming porous catalytic inorganic solids is to use the microstructure of surfactant aggregates as a template. The initial work used the hexagonal liquid crystalline phase of the cationic surfactant CTAB in a solution containing polysilicate ions. Coulombic attraction led to formation of mesoporous silica having long cylindrical channels in a hexagonal array (Beck et al., 1992; Kresge et al., 1992). Because the pores were on the order of 2 to 10 nm in diameter, and hence larger than in naturally occurring zeolites, these materials have great promise for catalysis involving relatively large molecules. Subsequent work, including efforts to synthesize oxides in this way, is summarized by Antonelli and Ying (1996).

Considerable research has also been directed toward using water-in-oil microemulsions to produce nanoscale solid particles (Pileni, 1993; Pillai et al., 1995). In one preparation method, two microemulsions are prepared, one containing the cation of the desired solid in its aqueous droplets, the other the anion. When they are mixed, nanoparticles are formed by precipitation. For example, titanium hydroxide was precipitated by mixing microemulsions containing titanium tetrachloride ($TiCl_4$) and ammonium hydroxide (NH_4OH). Subsequent calcination yielded titanium dioxide (TiO_2) particles (Pillai et al., 1995).

Metallic particles can be prepared by reducing salts dissolved in the droplets. For example, copper particles can be produced using hydrazine as a reducing agent (Pileni, 1993). Composite particles with a core of one metal surrounded by another have also been reported (see Belloni, 1996). Preparation of nanoparticles of various metals has been reviewed by Capek (2004). In some cases, supercritical carbon dioxide (CO_2) has been used instead of a hydrocarbon as the continuous phase in the microemulsion, which provides a system where the phase behavior can be relatively easily controlled by changing pressure and eliminates the use of volatile solvents. There is considerable interest in making nanoscale particles of semiconductors such as cadmium sulfide (CdS) using water-in-oil microemulsions because their electronic properties are different from those of bulk crystals (see Pileni, 1993).

The rates of some reactions catalyzed by enzymes are faster in water-in-oil microemulsions than in aqueous solutions, one reason being the higher concentration of organic reactants of low water solubility. A review with emphasis on lipase-catalyzed reactions is presented by Holmberg (1994).

REFERENCES

GENERAL REFERENCES ON SURFACTANTS AND THEIR BEHAVIOR

Becher, P. (ed.), *Encyclopedia of Emulsion Technology*, vols. 1–4, Marcel Dekker, New York, 1983–1996.

Bourrel, M. and Schechter, R.S., *Microemulsions and Related Systems*, Marcel Dekker, New York, 1988.

Gaines, G.L., *Insoluble Monolayers at Liquid-Gas Interfaces*, Wiley, New York, 1966.

Gelbart, W.M., Ben-Shaul, A., and Roux, D. (eds.), *Micelles, Microemulsions, and Monolayers*, Springer, New York, 1994.

Laughlin, R.G., *The Aqueous Phase Behavior of Surfactants*, Academic Press, New York, 1996.

Mittal, K.L. (ed.), *Micellization, Solubilization, and Microemulsions*, Plenum, New York, 1976. This was the first of a series of books coedited by Mittal containing papers presented at the biennial international conference on *Surfactants in Solution*.

Rosen, M.L., *Surfactants and Interfacial Phenomena*, 3rd ed., Wiley, New York, 2004.

Safran, S.A., *Statistical Thermodynamics of Surface, Interfaces, and Membranes*, Addison-Wesley, Reading, Massachusetts, 1994.

Schick, M.J. and Fowkes, F.M., consulting editors. Surfactant Science Series, Marcel Dekker, New York.

Tanford, C., *The Hydrophobic Effect: Formation of Micelles and Biological Membranes*, 2nd ed., Wiley, New York, 1980.

TEXT REFERENCES

Adderson, J.E. and Taylor, H., The effect of temperature on the critical micelle concentration of dodecylpyridinium bromide, *J. Colloid Sci.*, 19, 495, 1964.

Alexander, A.E. and Johnson, P., *Colloid Science*, vol. 2, Clarendon Press, Oxford, 1949, p. 686.

Alexandridis, P., Athanassiou, V., Fukuda, S., and Hatton, T.A., Surface activity of poly(ethylene oxide)-block-poly(propylene oxide)-block-poly(ethylene oxide) copolymers, *Langmuir*, 10, 2604, 1994.

Alexandridis, P. and Lindman, B. (ed.), *Amphiphilic Block Copolymers: Self-Assembly and Applications*, Elsevier, Amsterdam, 2000.

Anacker, E.W., Micelle formation of cationic surfactants in aqueous media, in *Cationic Surfactants*, Marcel Dekker, New York, 1970, p. 203.

Andelman, D., Cates, M., Roux, D., and Safran, S., Structure and phase equilibria of microemulsions, *J. Chem. Phys.*, 87, 7229, 1987.

Anderson, D., Wennerström, H., and Olsson, U., Isotropic bicontinuous solutions in surfactant-solvent systems: the L3 phase , *J. Phys. Chem.*, 93, 4243, 1989.

Antonelli, D.M. and Ying, J.Y., Mesoporous materials, *Curr. Opin. Colloid Interface Sci.*, 1, 523, 1996.

Aniansson, G. and Wall, S., Kinetics of step-wise micelle association, *J. Phys. Chem.*, 78, 1024, 1974.

Aniansson, G., and Wall, S., Kinetics of step-wise micelle association. Correction and improvement, *J. Phys. Chem.*, 79, 857, 1975.

Aronson, M.P., The role of free surfactant in destabilizing oil-in-water emulsions, *Langmuir*, 5, 494, 1989.

Ashakawa, T., Amada, K., and Miyagishi, S., Micellar immiscibility of lithium 1,1,2,2-tetrahydroheptadecafluorodecyl sulfate and lithium tetradecyl sulfate mixture, *Langmuir*, 13, 4569, 1997.

Aveyard, R., Binks, B.P., and Clint, J.H., Emulsions stabilised solely by colloidal particles, *Adv. Colloid Interface Sci.*, 100–102, 503, 2003.

Baviere, M., Wade, W.H., and Schechter, R.S., The effect of salt, alcohol and surfactant on optimum middle phase composition, in *Surface Phenomena in Enhanced Oil Recovery*, D.O. Shah (ed.), Plenum, New York, 1981, p. 117.

Beaudoin, S.P., Grant, C.S., and Carbonell, R.G., Removal of organic films from solid surfaces using aqueous solutions of nonionic surfactants. 1. Experiments, *Ind. Eng. Chem. Res.* 34, 3307, 3318, 1995.

Beck, J.S., Vartuli, J.C., Roth, W.J., Leonowicz, M.E., Krege, C.T., Schmitt, K.D., Chu, C.T.W., Olson, D.H. and Sheppard, E.W., A new family of mesoporous molecular sieves prepared with liquid crystal templates, *J. Amer. Chem. Soc.*, 114, 10834, 1992.

Belloni, J., Metal nanocolloids, *Curr. Opin. Colloid Interface Sci.*, 1, 184, 1996.

Ben-Naim, A., Statistical mechanical study of hydrophobic interaction. I. Interaction between two identical nonpolar solute particles, *J. Chem. Phys.*, 54, 1387, 1971.

Ben-Shaul, A. and Gelbart, W.M., Statistical thermodynamics of amphiphile self-assembly: Structure and phase transition in micellar solution, in *Micelles, Microemulsions, and Monolayers*, W.M. Gelbart, A. Ben-Shaul, and D. Roux (eds.), Springer, New York, 1994, p. 1.

Bennett, K.E., Hatfield, J.C., Davis, H.T., Macosko, C.W., and Scriven, L.E., Viscosity and conductivity of microemulsions, in *Microemulsions*, I.D. Robb (ed.), Plenum Press, New York, 1982, p. 65.

Benton, W.J. and Miller, C.A., A new optically isotropic phase in the dilute region of the sodium octanoate-decanol-water system, in *Surfactants in Solution*, K.L. Mittal and B. Lindman (eds.), Plenum Press, New York, 1984, p. 205.

Bibette, J., Depletion interactions and fractionated crystallization for polydisperse emulsion purification, *J. Colloid Interface Sci.* 147, 474, 1991.

Bishop, A.R. and Nuzzo, R.G., Self-assembled monolayers: recent developments and applications, *Curr. Opin. Colloid Interface Sci.*, 1, 127, 1966.

Blodgett, K., Films built by depositing successive monomolecular layers on a solid surface, *J. Am. Chem. Soc.*, 57, 1007, 1935.

Bowcott, J.E. and Schulman, J.H., Emulsions, Control of droplet size and phase continuity in transparent oil-water dispersions stabilized with soap and alcohol, *Z. Elektrochem.*, 59, 283, 1955.

Bourrel, M., Graciaa, A., Schechter, R.S., and Wade, W.H., The relation of emulsion stability to phase behavior and interfacial tension of surfactant systems, *J. Colloid Interface Sci.*, 72, 161, 1979.

Broome, F.K., Hoerr, C.W., and Harwood, H.J. (1951) *J. Am. Chem. Soc.* 73, 3350.

Cabane, B., and Duplessix, R., Decoration of semidilute polymer solutions with surfactant micelles, *J. Physique*, 48, 651, 1987.

Candau, F., Polymerization in microemulsions, in *Polymerization in Organized Media*, C.M. Paleos (ed.), Gordon and Breach, Philadelphia, 1992, p. 215 ff.

Capek, I., Preparation of metal nanoparticles in water-in-oil (w/o) microemulsions, *Adv. Colloid Interface Sci.*, 110, 49, 2004.

Carnahan, N.F. and Starling, K.E., Equation of state for nonattracting rigid spheres, *J. Chem. Phys.*, 51, 635, 1969.

Carroll, B.J., Physical aspects of detergency, *Colloids Surfaces A*, 74, 131, 1993.

Cash, R.L., Cayias, J.L., Fournier, G., MacAllister, D.J., Scharer, T., Schechter, R.S., and Wade, W.H., The application of low interfacial tension scaling rules to binary hydrocarbon mixtures, *J. Colloid Interface Sci.*, 59, 39, 1977.

Cayias, J.L., Schechter, R.S., and Wade, W.H., Modeling crude oils for low interfacial tension, *Soc. Petrol. Eng. J.*, 16, 351, 1976.

Cazabat, A.M., Langevin, D., Meunier, J., and Pouchelar, A., Critical behavior in microemulsions, *Adv. Colloid Interface Sci.*, 16, 175, 1982.

Chan, A.F., Evans, D.F., and Cussler, E.L., Explaining solubilization kinetics, *AIChE J.*, 22, 1006, 1976.

Chen, B.-H., Miller, C.A., Walsh, J.M., Warren, P.B., Ruddock, J.N., Garrett, P.R., Argoul, F., and Leger, C., Dissolution rates of pure nonionic surfactants, *Langmuir*, 16, 5276, 2000.

Chhatre, A.S., Joshi, R.A., and Kulkarni, B.D., Microemulsions as media for organic synthesis: selective nitration of phenol to *ortho*-nitrophenol using dilute nitric acid, *J. Colloid Interface Sci.*, 158, 183, 1993.

Corvazier, L., Messé, L., Salou, L.O., Young, R.N., Fairclough, J.P.A., and Ryan, A.J., Lamellar phases and microemulsions in model ternary blends containing amphiphilic block copolymers, *J. Mater. Chem.*, 11, 2864, 2001.

Davies, J.T. and Rideal, E.K., *Interfacial Phenomena*, 2nd ed., Academic Press, New York, 1963, chap. 8.

Debye, P. and Anacker, E.W., Micelle shape from dissymmetry measurements, *J. Phys. Colloid Chem.*, 55, 644, 1951.

Desai, S.D., Gordon, R.D., Gronda, A.M., and Cussler, E.L., Polymerized microemulsions, *Curr. Opin. Colloid Interface Sci.*, 1, 519, 1996.

Ekwall, P., Composition, properties, and structures of liquid crystalline phases in systems of amphiphilic compounds, in *Advances in Liquid Crystals*, vol. 1, Brown, G.H. (ed.), Academic Press, New York, 1975, p. 1.

Ekwall, P., Mandell, L., and Fontell, K., Solubilization in micelles and mesophases and the transition from normal to reversed structures, *Mol. Cryst. Liqd. Cryst.*, 8, 157, 1969.

Evans, D.F. and Wightman, P.J., Micelle formation above 100°C, *J. Colloid Interface Sci.*, 86, 515, 1982.

Fleming, P.D. and Vinatieri, J.E., Quantitative interpretation of phase volume behavior of multicomponent systems near critical points, *AIChE J.*, 25, 493, 1979.

Fontell, K., Liquid crystallinity in lipid-water systems, *Mol. Cryst. Liq. Cryst.*, 63, 59, 1981.

Förster, T., von Rybinski, W., and Wadle, A., Influence of microemulsion phases on the preparation of fine-disperse emulsions, *Adv. Colloid Interface Sci.*, 58, 119, 1995.

Franks, F., in *Water*, vol. 4, Franks, F., ed., Plenum Press, New York, 1975, p. 1.

Friberg, S.E., Hydrotropes, *Curr. Opin. Colloid Interface Sci.*, 2, 490, 1997.

Friberg, S. and Ahmed, S., Liquid crystals and the foaming capacity of an amine dissolved in water and *p*-xylene, *J. Colloid Interface Sci.*, 35, 175, 1971.

Friberg, S., Mandell, L., and Larsson, M., Mesomorphous phases, a factor of importance for the properties of emulsions, *J. Colloid Interface Sci.*, 29, 155, 1969.

Friberg, S. and Solans, C., Emulsification and the HLB-temperature, *J. Colloid Interface Sci.*, 66, 367, 1968.

Friberg, S.E., Venable, R.L., Kim, M., and Neogi, P., Phase equilibria in water, pentanol, tetradecyltrialkylammonium bromide systems, *Colloids Surfaces*, 15, 285, 1985.

Friman, R., Danielsson, I., and Stenius, P., Lamellar mesophase with high contents of water: X-ray investigations of the sodium octanoate-decanol-water system, *J. Colloid Interface Sci.*, 86, 501, 1982.

Gaines, G.L., *Insoluble Monolayers at Liquid-Gas Interfaces*, Wiley, New York, 1966.

Garti, N. and Bisperink, C., Double emulsions: progress and applications, *Curr. Opin. Colloid Interface Sci.*, 3, 657, 1998.

Gragson, D.E., Jensen, J.M., and Baker, S.M., Characterization of predominantly hydrophobic poly(styrene)-poly(ethylene oxide) copolymers at air/water and cyclohexane/water interfaces, *Langmuir*, 15, 6127, 1999.

Goddard, E.D. and Ananthapadmanabhan, K.P., *Interaction of Surfactants with Polymers and Proteins*, CRC Press, Boca Raton, Florida, 1993.

Goklen, K.E. and Hatton, T.A., Liquid-liquid extraction of low molecular weight proteins by selective solubilization in reverse micelles, *Sep. Sci. Tech.*, 22, 831, 1987.

Goldmints, I., Holzwarth, J.F., Smith, K.A., and Hatton, T.A., Micellar dynamics in aqueous solutions of PEO-PPO-PEO block copolymers, *Langmuir*, 13, 6130, 1997.

Gradzielski, M., Bending constants of surfactant layers, *Curr. Opin. Colloid Interface Sci.*, 3, 478, 1998.

Grätzel, M. and Kalyanasundaram, K., *Kinetics and Catalysis in Microheterogeneous Systems*, Marcel Dekker, New York, 1991.

Griffin, J., Classification of Surface-active Agents by "HLB", *J. Soc. Cosmet. Chem.*, 1, 311, 1949.

Hammouda, A., Gulik, T., and Pileni, M., Synthesis of nanosize latexes by reverse micelle polymerization, *Langmuir*, 11, 3656, 1995.

Harris, W.F., A study of the mechanism of collapse of monomolecular films at fluid/fluid interfaces, Master's thesis, University of Minnesota, 1964.

Hazlett, R.D. and Schechter, R.S., Stability of macroemulsions, *Colloids Surfaces*, 29, 53, 1988.

Helfrich, W., Steric interaction of fluid membranes in multilayer systems, *Z. Naturforsch*, 33a, 305, 1978.

Henon, S. and Meunier, J., Microscope at the Brewster angle: Direct observation of first-order phase transitions in monolayers, *Rev. Sci. Instrum.*, 62, 936, 1991.

Hermanson, K.D. and Kaler, E.W., Kinetics and mechanism of the multiple addition microemulsion polymerization of hexyl methacrylate, *Macromolecules*, 36, 1836, 2003.

Hill, T.L. (1992) *Thermodynamics of Small Systems*, Dover, New York; originally published by W.A. Benjamin, New York as Vol. 1 (1963) and Vol. 2 (1964).

Hillmeyer, M.A., Maurer, W.W., Lodge, T.P., Bates, F.S., and Almdal, K., Model bicontinuous microemulsions in ternary homopolymer/block copolymer blends, *J. Phys. Chem. B*, 103, 4814, 1999.

Hoar, T.P. and Schulman, J.H., Transparent water-in-oil dispersions: the oleopathic hydromicelle, *Nature*, 152, 102, 1943.

Hoffmann, H. and Ebert, G., Surfactants, micelles and fascinating phenomena, *Angew. Chem. Int. Ed. Engl.*, 27, 902, 1988.

Hofsäss, T. and Kleinert, H., Gaussian curvature in an Ising model of microemulsions, *J. Chem. Phys.*, 86, 3565, 1987.

Holland, P.M., and Rubingh, D. (eds.), *Mixed Surfactant Systems*, ACS Symposium Series 501, American Chemical Society, Washington, D.C., 1992.

Holmberg, K., Organic and bioorganic reactions in microemulsions, *Adv. Colloid Interface Sci.*, 51, 137, 1994.

Hou, W. and Papadopoulos, K.D., W1/O/W2 and O1/W/O2 globules stabilized with Span 80 and Tween 80, *Colloids Surfaces A*, 125, 181, 1997.

Hsiano, L., Dunning, H.N., and Lorenz, P.B., Critical micelle concentrations of polyoxyethylated non-ionic detergents, *J. Phys. Chem.*, 60, 657, 1956.

Huang, J.S. and Kim, M.W., Critical scaling behavior of microemulsions, SPE/DOE preprint 10787, presented at the Symposium on Improved Oil Recovery, Tulsa, Oklahoma, 1982.

Hurter, P.N., Scheutjens, J.M.H.M., and Hatton, T.A., Molecular modeling of micelle formation and solubilization in block copolymer micelles. 1. A self-consistent mean-field lattice theory, *Macromolecules*, 26, 5592, 1993.

Iliopoulos, I., Wang, T.K., and Audebert, R., Viscometric evidence of interactions between hydrophobically modified poly(sodium acrylate) and sodium dodecyl sulfate, *Langmuir*, 7, 617, 1991.

Israelachvili, J.N., Mitchell, D.J., and Ninham, B.W., Theory of self-assembly of hydrocarbon amphiphiles into micelles and bilayers, *J. Chem. Soc. Faraday Trans. II*, 72, 1525, 1976.

Ivanov, I.B. and Kralchevsky, P.A., Stability of emulsions under equilibrium and dynamic conditions, *Colloids Surfaces A*, 128, 155, 1997.

Jahn, W. and Strey, R., Microstructure of microemulsions by freeze fracture electron microscopy, *J. Phys. Chem.*, 92, 2294, 1988.

Jeng, J.-F. and Miller, C.A., Theory of microemulsions with spherical drops I. Phase diagrams and interfacial tensions in gravity-free systems, *Colloids Surfaces*, 28, 247, 1987.

Jönsson, B., Lindman, B., Holmberg, K., and Kronberg, B., *Surfactants and Polymers in Aqueous Solutions*, Wiley, New York, 1998.

Jouffroy, J., Levinson, P., and de Gennes, P., Phase equilibriums involving microemulsions. (Remarks on the Talmon-Prager model) *J. Physique*, 43, 1241, 1982.

Kabalnov, A. and Wennerström, H., Macroemulsion stability: The oriented wedge theroy revisited. *Langmuir* 12, 276, 1996.

Kahlweit, M., Kinetics of formation of association colloids, *J. Colloid Interface Sci.*, 90, 92, 1982.

Kahlweit, M., Strey, R., Firman, P., Haase, D., Jen, J., Schomäcker, R., General patterns of the phase behavior of mixtures of water, nonpolar solvents, amphiphiles, and electrolytes. 1, *Langmuir*, 4, 499, 1988.

Kahlweit, M., Strey, R., Schomäcker, R., and Haase, D., General patterns of the phase behavior of mixtures of water, nonpolar solvents, amphiphiles, and electrolytes. 2, *Langmuir*, 5, 305, 1989.

Kaler, E.W., Murthy, A.K., Rodrigues, B.E., and Zasadzinski, J.A.N., Spontaneous vesicle formation in aqueous mixtures of single-tailed surfactants, *Science*, 245, 1371, 1989.

King, A.D., Solubilization of Gases, in *Solubilization in Surfactant Aggregates*, Christian, S.D. and Scamehorn, J.F. (eds.), Marcel Dekker, New York, 1995, p.35.

Knobler, C.M., Recent developments in the study of monolayers at the air-water interface, *Adv. Chem. Phys.*, 77, 397, 1990.

Kresge, C.T., Leonowicz, M.E., Roth, W.J., Vartuli, J.C., and Beck, J.S., Ordered meso-porous molecular sieves synthesized by a liquid-crystal template mechanism, *Nature*, 359, 710, 1992.

Kunieda, H., Evans, D.F., Solans, C., and Yoshida, M., The structure of gel-emulsions in a water/ nonionic surfactant/oil system, *Colloids Surfaces*, 47, 35, 1990.

Kunieda, H. and Shinoda, K., Evaluation of the hydrophile-lipophile balance (HLB) of nonionic surfactants. I. Multisurfactant systems, *J. Colloid Interface Sci.*, 107, 107, 1985.

Kunieda, H., and Ishikawa, I., Evaluation of the hydrophile-lipophile balance (HLB) of nonionic surfactants. II. Commercial-surfactant sysems, *J. Colloid Interface Sci.*, 107, 122, 1985.

Kust, P.R. and Rathman, J.F., Synthesis of surfactants by micellar autocatalysis: N,N-dimethyldodecylamine N-oxide, *Langmuir*, 11, 3007, 1995.

Kwak, J.C.T. (ed.), *Surfactant-Polymer Systems*, Marcel Dekker, New York, 1998.

Lang, J. and Zana, R., Chemical relaxation methods, in *Surfactant Solutions*, R. Zana (ed.), Marcel Dekker, New York, 1987, p. 405.

Lange, H., Interactions of sodium alkyl sufate and poly(vinylpyrrolidinone) in aqueous solutions, *Kolloid Z.Z. Polym.*, 243, 101, 1971.

Lawrence, A.S.C., Bingham, A., Capper, C.B., and Hume, K., The penetration of water and aqueous soap solutions into fatty substances containing one or two polar groups, *J. Phys. Chem.*, 68, 3470, 1964.

Lee, J.H., Ruegg, M.L., Balsara, N.P., Zhu, Y., Gido, S.P., Krishnamoorti, R., and Kim, M.-H., Phase behavior of highly immiscible polymer blends stabilized by a balanced block copolymer surfactant, *Macromolecules*, 36, 6537, 2003.

Lessner, E., Teubner, M., and Kahlweit, M., Relaxation experiments in aqueous solutions of ionic micelles. 2. Experiments on the system water-sodium dodecyl sulfate-sodium perchlorate and their theoretical interpretation, *J. Phys. Chem.*, 85, 3167, 1981.

Lim, J.-C. and Miller, C.A., Dynamic behavior and detergency in systems containing nonionic surfactants and mixtures of polar and nonpolar oils, *Langmuir*, 7, 2021, 1991.

Lindman, B. and Thalberg, K., Polymer-surfactant interacions: recent developments, in *Interaction of Surfactants with Polymers and Proteins*, CRC Press, Boca Raton, Florida, 1993, p. 205.

Lissant, K.J., The geometry of high-internal-phase-ratio emulsions, *J. Colloid Interface Sci.*, 22, 462, 1966.

Mays, H., Almgren, M., Brown, W., and Alexandridis, P., Cluster and network formation toward percolation in the microemulsion L_2 phase formed by an amphiphilic triblock copolymer and water in *p*-xylene, *Langmuir*, 14, 723, 1998.

McBain, M.E.L. and Hutchinson, E., *Solubilization*, Academic Press, New York, 1955.

McConlogue, C.W. and Vanderlick, T.K., Monolayers with one component of variable solubility: studies of lysophosphocholine/DPPC mixtures, *Langmuir*, 14, 6556, 1998.

Menger, F.M., Rhee, J.U., and Rhee, H.K., Applications of surfactants to synthetic organic chemistry, *J. Org. Chem.*, 40, 3803, 1975.

Miller, C.A. and Neogi, P., Thermodynamics of microemulsions: combined effects of dispersion entropy of drops and bending energy of surfactant films, *AIChE J.*, 26, 212, 1980.

Miller, C.A. and Raney, K.H., Solubilization-emulsification mechanisms of detergency, *Colloids Surfaces A*, 74, 169, 1993.

Miller, C.A., Ghosh, O., and Benton, W.J., Behavior of dilute lamellar liquid-crystalline phases, *Colloids Surfaces*, 19, 197, 1986.

Miller, C.A., Hwan, R.W., Benton, W.J., and Fort, T., Jr., Ultralow interfacial tensions and their relation to phase separation in micellar solutions, *J. Colloid Interface Sci.*, 61, 554, 1977.

Miller, C.A., Gradzielski, M., Hoffmann, H., Kramer, U., and Thunig, C., Experimental results for the L3 phase in a zwitterionic surfactant system and their implications regarding structures, *Colloid Polymer Sci.*, 268, 1066, 1990.

Missel, P.J., Mazer, N.A., Carey, M.C., and Benedeck, G.B., Thermodynamics of the sphere-to-rod transition in alkyl sulfate micelles, in *Solution Behavior of Surfactants*, vol. 1, Plenum Press, New York, 1982, p. 373.

Mitchell, D.J., Tiddy, G.J.T., Waring, L., Bostock, T., and Macdonald, M.P., Phase behaviour of polyoxy ethylene surfactnats with water, *J. Chem. Soc. Faraday Trans. I*, 79, 975, 1983.

Möbius, D. and Miller, R. (eds.), *Proteins at Liquid Interfaces*, Elsevier, New York, 1998.

Morgan, J.D., Lusvardi, K.M., and Kaler, E.W., Kinetics and mechanism of microemulsion polymerization of hexyl methacrylate, *Macromolecules*, 30, 1897, 1997.

Mori, F., Lim, J.-C., Raney, O.G., Elsik, C.M., and Miller, C.A., Phase behavior, dynamic contacting and detergency in systems containing triolein and nonionic surfactants, *Colloids Surfaces*, 40, 323, 1989.

Mukerjee, P., Micellar properties of drugs: micellar and nonmicellar patterns of self-association of hydrophobic solutes of different molecular structures—monomer fraction, availablity, and misuses of micellar hypothesis, *J. Pharm. Sci.*, 63, 972, 1974.

Mukerjee, P., Odd-even alternation in the chain length variation of micellar properties. Evidence of some solid-like character of the micelle core, *Kolloid-Z.-Z. Polymere*, 236, 76, 1970.

Mukerjee, S., Miller, C.A., and Fort, T., Jr., Theory of drop size and phase continuity in microemulsions I. Bending effects with uncharged surfactants, *J. Colloid Interface Sci.*, 91, 223, 1983.

Murphy, C.L., Thermodynamics of low tension and highly curved surfaces, Ph.D. dissertation, University of Minnesota, 1966.

Munch, M.R. and Gast, A.P., Block copolymers at interfaces. 1. Micelle formation, *Macromolecules*, 21, 1360, 1988.

Munoz, M.G., Monroy, F., Ortega, F., Rubino, R.G., and Langevin, D., Monolayers of symmetric triblock copolymers at the air-water interface. 1. Equilibrium properties, *Langmuir*, 16, 1083, 2000.

Nagarajan, R., Association of nonionic polymers with micelles, bilayers, and microemulsions, *J. Chem. Phys.*, 90, 1980, 1989.

Nagarajan, R., Molecular packing parameter and surfactant self-assembly: the neglected role of the surfactant tail, *Langmuir*, 18, 31, 2002.

Nagarajan, R. and Ganesh, K., Block copolymer self-assembly in selective solvents: theory of solubilization in spherical micelles, *Macromolecules*, 22, 4312, 1989.

Nagarajan, R., and Kalpakcir, B., Viscosimetric investigation of complexes between polyethylene oxide and surfactant micelles, *Polym. Prepr. Am. Chem. Soc. Div. Polym. Chem.*, 23, 41, 1982.

Nagarajan, R. and Ruckenstein, E., Critical micelle concentration: A transition point for micellar size distribution: A statistical thermodynamical approach, *J. Colloid Interface Sci.*, 60, 221, 1977.

Nagarajan, R. and Ruckenstein, E., Theory of surfactant self-assembly: a predictive molecular thermodynamic approach, *Langmuir*, 7, 2934, 1991.

Neogi, P., Kim, K., and Friberg, S.E., Micromechanics of surfactant microstructures, *J. Phys. Chem.*, 91, 605, 1987.

O'Connell, J.P. and Brugman, R.J., Some thermodynamic aspects and models of micelles, microemulsions, and liquid crystals, in *Improved Oil Recovery by Surfactant and Polymer Flooding*, Shah, D.O. and Schechter, R.S. (ed.), Academic Press, New York, 1977, p. 339 ff.

Oetter, G. and Hoffmann, H., Correlation between interfacial tensions and micellar structures, *J. Disp. Sci. Technol.*, 9, 459, 1988.

Olsson, U., Shinoda, K., and Lindman, B., Change of the structure of microemulsions with the hydrophile-lipophile balance of nonionic surfactant as revealed by NMR self-diffusion studies, *J. Phys. Chem.*, 90, 4083, 1986.

Pileni, M.P., Reverse micelles as microreactors, *J. Phys. Chem.*, 97, 6961, 1993.

Pillai, V., Kumar, P., Hou, M.J., Ayyub, P., and Shah, D.O., Preparation of nanoparticles of silver halides, superconductors and magnetic materials using water-in-oil microemulsions as nano-reactors, *Adv. Colloid Interface Sci.*, 55, 241, 1995.

Princen, H.M., Highly concentrated emulsions. I. Cylindrical systems, *J. Colloid Interface Sci.*, 71, 55, 1979.

Puerto, M.C. and Reed, R.L., Three-parameter representation of surfactant/oil/brine interaction, *Soc. Petrol. Eng. J.*, 23, 669, 1983.

Putnam, F.W., *The Proteins v IB*, Academic Press, New York, 1943, p. 807.

Puvvada, S. and Blankschtein, D., Thermodynamic description of micellization, phase behavior, and phase separation of aqueous solutions of surfactant mixtures, *J. Phys. Chem.*, 96, 5567, 1992.

Qutubuddin, S., Lin, C.S. and Tajuddin, Y., Novel polymeric composites from microemulsions, *Polymer*, 35, 4606, 1994.

Qutubuddin, S., Miller, C.A., and Fort, T., Jr., Phase behavior of pH-dependent microemulsions, *J. Colloid Interface Sci.*, 101, 46, 1984.

Ramadan, M., Evans, D.F., and Lumry, R., Why micelles form in water and hydrazine. A reexamination of the origins of hydrophobicity, *J. Phys. Chem.*, 87, 4538, 1983.

Raney, K.H. and Miller, C.A., Optimum detergency conditions with nonionic surfactants: II. Effect of hydrophobic additives, *J. Colloid Interface Sci.*, 119, 539, 1987.

Raney, K.H. and Benson, H.L., The effect of polar soil components on the phase inversion temperature and optimum detergency conditions, *J. Am. Oil Chem. Soc.*, 67, 722, 1990.

Raney, K.H., Benton, W.J., and Miller, C.A., Optimum detergency conditions with nonionic surfactants: I. Ternary water-surfactant-hydrocarbon systems, *J. Colloid Interface Sci.*, 117, 282, 1987.

Rathman, J.F., Micellar catalysis, *Curr. Opin. Colloid Interface Sci.*, 1, 514, 1996.

Reed, R.L. and Healy, R.N., Some physico-chemical aspects of microemulsion flooding: a review, in *Improved Oil Recovery by Surfactant and Polymer Flooding*, Shah, D.O. and Schechter, R.S. (eds.), Academic Press, New York. 1977, p. 383.

Rehbinder, P.A., *Proc. 2nd Intl. Cong. Surface Activity*, 1, 476, 1957.

Reynolds, J.A., Gallagher, J.P., and Steinhardt, J., Effect of pH on the binding of N-alkyl sulphates to bovine serum albumin, *Biochemistry*, 9, 1232, 1970.

Ries, H.E. and Kimball, W.A., Structure of fatty-acid monolayers and a mechanism for collapse, *Proc. 2nd Intl Cong. Surface Activity*, 1, 75, 1957.

Roe, C.P. and Brass, P.D., The adsorption of ionic surfactants and their gegenions at the air-water interface of aqueous solutions, *J. Am. Chem. Soc.*, 76, 4703, 1954.

Rosen, M.J., *Surfactants and Interfacial Phenomena*, 3rd ed., Wiley-Interscience, New York, 2004.

Ruckenstein, E., The origin of thermodynamic stability of microemulsions, *Chem. Phys. Lett.*, 57, 517, 1978.

Ruckenstein, E. and Chi, J.C., Stability of microemulsions, *J. Chem. Soc. Faraday Trans. II*, 71, 1690, 1975.

Ruckenstein, E. and Karpe, P., On the enzymatic superactivity in ionic reverse micelles, *J. Colloid Interface Sci.*, 139, 408, 1990.

Safran, S.A., *Statistical Thermodynamics of Surface, Interfaces, and Membranes*, Addison-Wesley, Reading, Massachusetts, 1994.

Sagitani, H., , in *Organized Solutions*, Friberg, S. and Lindman, B. (eds.), Marcel Dekker, New York, 1992, p. 259 ff.

Saito, H. and Shinoda, K., The stability of W/O type emulsions as a function of temperature and of the hydrophilic chain length of the emulsifier, *J. Colloid Interface Sci.*, 32, 647, 1970.

Salager, J.L., Bourrel, M., Schechter, R.S., and Wade, W.H., Mixing rules for optimum phase-behavior formulations of surfactant/oil/water systems, *Soc. Petrol. Eng. J.*, 19, 271, 1979.

Salager, J.L., Loaiza-Maldonado, I., Minana-Perez, M., and Silva, F., Surfactant-oil-water systems near the affinity inversion. Part I: Relationship between equilbrium phase behavior and emulsion type and stability, *J. Disp. Sci. Technol.*, 3, 279, 1982.

Salager, J.L., Lopez-Castellanos, G., Minana-Perez, M., Parra, C., Cucuphat, C., Graciaa, A., and Lachaise, J., Surfactant-oil-water systems near the affinity inversion: Part VII: phase behavior and emulsions with polar oils, *J. Disp. Sci. Technol.*, 12, 59, 1991.

Salter, S.J., , The influence of type and amount of alcohol on surfactant-oil-brine phase behavior and properties, SPE preprint 6843, presented at the annual fall meeting of the Society of Petroleum Engineers, Denver, Colorado, 1977.

Scamehorn, J.F. (ed.), *Phenomena in Mixed Surfactant Systems*, ACS Symposium Series 311, American Chemical Society, Washington, D.C., 1986.

Scamehorn, J.F. and Harwell, J.H. (eds.), *Surfactant-Based Separation Processes*, Marcel Dekker, New York, 1989.

Schulman, J.H., Stoeckenius, W., and Prince, L.M., Mechanism of formation and structure of micro emulsions by electron microscopy, *J. Phys. Chem.*, 63, 1677, 1959.

Scheutjens, J.M.H.M. and Fleer, G.J., Effect of polymer adsorption and depletion on the interaction between two parallel surfaces, *Adv. Colloid Interface Sci.*, 16, 361, 1982.

Scriven, L.E., Equilibrium bicontinuous structure, *Nature*, 263, 123, 1976.

Shaeiwitz, J.A., Chan, A.F., Evans, D.F., and Cussler, E.L., The mechanism of solubilization in detergent solutions, *J. Colloid Interface Sci.*, 84, 47, 1981.

Shah, D.O. and Schechter, R.S. (ed.), *Improved Oil Recovery by Surfactant and Polymer Flooding*, Academic Press, New York, 1977.

Shchukin, E.D. and Rehbinder, P.A., Formation of new surfaces during the deformation and rupture of a solid in a surface-active medium, *Colloid J. USSR*, 20, 645, 1958.

Shinoda, K., Kobayashi, M., and Yamaguchi, N., Effect of "iceberg" formation of water on the enthalpy and entropy of solution of paraffin chain compounds: the effect of temperature on the critical micelle concentration of lithium perfluorooctane sulfonate, *J. Phys. Chem.*, 91, 5292, 1987.

Shinoda, K. and Kunieda, H., Conditions to produce so-called microemulsions: Factors to increase the mutual solubility of oil and water by solubilizers, *J. Colloid Interface Sci.*, 42, 381, 1973.

Shinoda, K., Nakagawa, T., Tamamushi, B.I., and Isemura T., *Colloid Surfactants*, Academic Press, New York, 1963.

Silva, F., Pena, A., Minana-Perez, M., and Salager, J.L., Dynamic inversion hysteresis of emulsions containing anionic surfactants, *Colloids Surfaces A*, 132, 221, 1998.

Smith, R.D. and Berg, J.C., The collapse of surfactant monolayers at the air-water interface, *J. Colloid Interface Sci.*, 74, 273, 1980.

Solans, C., Azemar, N., Parra, J., and Calbet, J., Phase behavior and detergency in water/nonionic surfactant/hydrocarbon systems, *Proceedings CESIO 2nd World Surfactants Congress*, vol. 2, Paris, 1988, p. 421 ff.

Strey, R., Jahn, W., Portze, G., and Bassereau, P., Freeze fracture electron microscopy of dilute lamellar and anomalous isotropic (L3) phases, *Langmuir*, 6, 1635, 1990a.

Strey, R., Schomächer, R., Roux, D., Nallet, F., and Olsson, U., Dilute lamellar and L3 phases in the binary water-C12E5 system, *J. Chem. Soc. Faraday Trans.*, 86, 2253, 1990b.

Talmon, Y. and Prager, S., Statistical thermodynamics of phase equilibria in microemulsions, *J. Chem. Phys.*, 69, 2984, 1978.

Talmon, Y. and Prager, S., The statistical thermodynamics of microemulsions. II. The interfacial region, *J. Chem. Phys.*, 76, 1535, 1982.

Tanford, C., *The Hydrophobic Effect: Formation of Micelles and Biological Membranes*, Wiley, New York, 1980.

Thalberg, K. and Lindman, B., Polyelectrolyte-ionic surfactant systems: phase behavior and interactions, in *Surfactants in Solution*, Mittal, K. and Shah, D.O. (eds.), Plenum Press, New York, 1991, pp. 11, 243.

Texter, J. and Tirrell, M., Chemical processing by self-assembly, *AIChE J.*, 47, 1706, 2001.

Thompson, L., The role of oil detachment mechanisms in determining optimum detergency conditions, *J. Colloid Interface Sci.*, 163, 61, 1994.

Tiddy, G.J.T., Surfactant-water liquid crystal phases, *Phys. Rep.*, 57, 1, 1980.

Vinatieri, J.E., Correlation of emulsion stability with phase behavior in surfactant systems for tertiary oil recovery, *Soc. Petrol. Eng. J.*, 20, 402, 1980.

Vollhardt, D., Morphology and phase behavior of monolayers, *Adv. Colloid Interface Sci.*, 64, 143, 1996.

Whitesides, G.M. and Laibinis, P.E., Wet chemical approaches to the characterization of organic surfaces: self-assembled monolayers, wetting, and the physical-organic chemistry of the solid-liquid interface, *Langmuir*, 6, 87, 1990.

Widom, B., A model microemulsion, *J. Chem. Phys.*, 81, 1030, 1984.

Winsor, P.A., *Solvent Properties of Amphiphilic Compounds*, Butterworth's, London, 1954.

Yu, G., Deng, Y., Dalton, S., Wang, Q., Attwood, D., Price, C., and Booth, C., Micellization and gelation of triblock copoly (oxyethylene oxypropylene oxyethylene), F127, *J. Chem. Soc. Faraday Trans.*, 88, 2537, 1992.

PROBLEMS

4.1   IF $\Delta G°$ IS KNOWN AS A FUNCTION OF TEMPERATURE, SHOW THAT
$\Delta H° = -T^2 \, \partial(\Delta G°/T)/\partial T$ AND $\Delta S° = (\Delta H° - \Delta G°)/T$. THE HIGH
TEMPERATURE DATA OF EVANS AND WIGHTMAN (1982) ARE

T (°C)	10^3 CMC (mole/l)	$-\Delta \bar{G}°$ (kcal/mole)	$-\Delta \bar{H}°$ (kcal/mole)	$-\Delta \bar{S}°$ (cal/deg-mole)
25.2	3.79	5.79	4.99	2.7
25.2	3.82	5.78	4.99	2.6
40.3	4.22	5.86	6.21	−1.1
54.2	4.94	5.78	7.50	−5.3
76.3	6.69	5.61	10.25	−13.6
95.5	9.83	5.35	11.34	−16.3
95.5	9.91	−	−	−
114.0	13.56	4.96	12.42	−19.2
134.9	21.38	4.52	13.23	−21.4
166.0	39.35	3.81	13.47	−22.0

where the actual values of α (see Equation 4.3) have been used for the calculations and the overbars denote that these quantities have been divided by $\bar{N}$. Calculate $\Delta \bar{G}°$, $\Delta \bar{H}°$, and $\Delta \bar{S}°$ using $\alpha = 0$ and a modified form of Equation 4.6 with concentrations instead of mole fractions. Note that enthalpic effects are responsible for micelle formation in this case.

More conventional data from Adderson and Taylor (1964) for dodecylpyridinium bromide in water are

T (°C)	10^3 CMC (mole/ℓ)
5	11.5
10	11.2
15	11.0
20	11.2
25	11.4
30	11.8
35	12.2
40	12.8
45	13.5
50	14.0
55	14.8
60	15.4
65	16.3
70	17.2

Repeat the previous calculations. Note that CMCs pass through a minimum, which is not an uncommon feature.

4.2   Obtain an expression for $\Delta \bar{G}_N^o$ for the data in the accompanying figure
from Shinoda et al. (1963). What is the significance of the fact that the
lines have almost the same slope? (Obtain first $\Delta \bar{G}_N^o / \bar{N}$ as a function
of the carbon number n.)

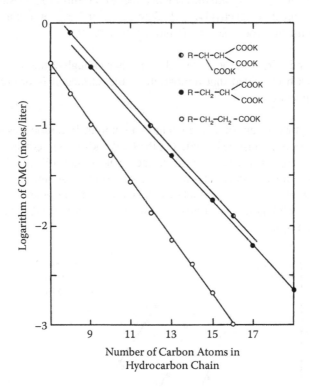

PROBLEM 4.2 Data for Problem 4.2. Reprinted from Shinoda et al. (1983) with permission.

4.3   Since micelle aggregation numbers are large and initial micelle formation occurs over a narrow concentration range, micelle formation,
including that with mixed surfactants, can be modeled with reasonable
success by a phase separation model. Assume that the following expressions give the chemical potential of a surfactant species in the aqueous
and micellar phases, respectively:

$$\mu_i = \mu_i^o + RT \ln x_i$$

$$\mu_{im} = \mu_{im}^o + RT \ln x_{im}.$$

(a) Consider a surfactant mixture with a ratio of species 2 to species
1 of n on a mole basis. Show that if this mixture is added to water,
micelles first form when

$$\frac{x_1}{x_{1c}} = \frac{x_{2c}/x_{1c}}{n+(x_{2c}/x_{1c})},$$

where x_{1c} and x_{2c} are the critical micelle concentrations of surfactants 1 and 2, respectively. Calculate x_1/x_{1c} and x_{1m} at initial micelle formation for $x_{2c}/x_{1c} = 10$ and $n = 0.1, 1, 10$.

(b) Find x_2/x_1 in the aqueous solution when enough of the same mixture has been added that almost all of the surfactant is present in the micelles.

4.4   Show that for an ionic surfactant that completely dissociates in water, a factor of 2 must be included in the Gibbs adsorption equation (Equation 1.26). Then calculate the surface concentration Γ_2 just below the critical micelle concentration (i.e., where the discontinuity in slope occurs) for the three surfactants shown in the figure (Oetter and Hoffman, 1988). DMAO is dimethylamine oxide; DMPO is the corresponding phosphine oxide.

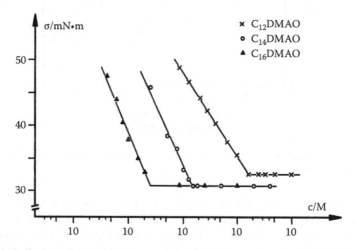

PROBLEM 4.4 Reprinted from Oetter and Hoffman (1988) with permission. Copyright 1988 Marcel Dekker.

4.5   Ashakawa et al. (1997) report CMC data as shown in the figure for mixtures of a partially fluorinated lithium decyl sulfate (1) and lithium tetradecyl sulfate (2). Analyze assuming ideal solution theory in the bulk solution and regular solution theory in the micelles. Show that an azeotrope is formed in this system. Plot x_{1m} as a function of $x_1/(x_1 +$

x_2) in the bulk solution and show that an azeotrope exists where the two quantities are equal. Find β/RT and the azeotrope composition.

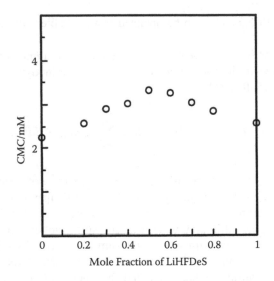

PROBLEM 4.5 Reprinted from Ashakawa et al. (1997). Copyright 1997 American Chemical Society, with permission.

4.6 Use the thermodynamic approach of Chapter 1 (Section 8) for an insoluble monolayer to show that the mole fraction of surfactant at the surface is given by

$$x_s^s = 1 - \exp\left[-\frac{\pi\sigma}{RT}\right],$$

where σ is the area per molecule of both surfactant and water at the surface.

4.7 Consider a lamellar liquid crystalline phase (Figure 4.9) whose surfactant bilayers and water layers have thicknesses h_s and h_w, respectively. The repeat spacing ($h_s + h_w$) can be obtained experimentally using low-angle x-ray diffraction techniques. Suppose that the spacing is measured over a range of surfactant:water ratio values in the liquid crystal.

(a) If h_s is constant, show that a straight line is obtained when the spacing is plotted as a function of $(1/\Phi_s)$, where Φ_s is the volume fraction of surfactant in the liquid crystal. Also explain how h_s can be obtained from the slope of this line or by extrapolating the line to $1/\Phi_s = 1$.

(b) If v_s is the volume of a surfactant molecule, derive an expression for area per surfactant molecule in the bilayers.

Thus both h_s and σ_s can be obtained from suitable x-ray data.

4.8 The differential dG_m of the Gibbs free energy of a microemulsion can be written in the following form using the thermodynamics of small systems (Hill, 1992):

$$dG_m = -S_m dT + V_m dp + \sum_i \mu_{im} dn_{im} + E dN \ ,$$

where p is the overall pressure, N is the number of drops, and E is $(\partial G_m / \partial N)_{T,p,ni}$. For simplicity we suppose that temperature and external pressure are maintained constant and that the microemulsion contains drops of pure oil in pure water with the surfactant entirely at the drop surfaces. Then we write dG_m as the sum of contributions dG_{mw}, dG_{mo}, and dG_{mi} from the water, oil, and interface and dG_{md} from the configurational entropy of the drops and the energy of interaction among them. With this framework, we have for the equilibrium condition for a single microemulsion phase

$$E = 0 = \left(\frac{\partial G_{mi}}{\partial N} \right)_{T,p,n_{im}} + \left(\frac{\partial G_{md}}{\partial N} \right)_{T,p,n_{im}} .$$

Note that there is no term in G_{mo} in this equation even though the pressure p_d of oil inside the drops may vary since G_m is given by $U_m + pV_m - TS_m)$, that is, the external pressure p must be used in the pV term.

(a) Use Equation 1.9 supplemented by a term $Cd(2/a)$ for dG_{mi} (see Problem 1.4) to show that the above equilibrium condition simplifies to

$$-\frac{3}{a^2} V_d \ \gamma - \frac{2C}{a^2} + \left(\frac{\partial G_{md}}{\partial a} \right)_{T,p,n_{im}} = 0 \ ,$$

where V_d is the (constant) total volume of the microemulsion drops. Derive an expression for the last term when G_{md} is given by Equation 4.25.

(b) Now suppose that enough oil is present that an oil-in-water micro-emulsion is in equilibrium with an excess oil phase. Show that minimization of total system free energy leads to a second equilibrium equation:

$$8\pi a N \gamma - \frac{2C}{a^2} + \left(\frac{\partial G_{md}}{\partial a}\right)_{T,p,n_{wm},n_{sm},N} = 0 \, .$$

Again evaluate the last term using the same expression for G_{md}.

(c) To demonstrate that bending effects are important when an excess oil phase is present, evaluate γ, C/A, $2\gamma/a$, and $2(C/A)/a^2$ for $a = 3$, 10, and 50 nm and $\Phi = 0.01$, 0.05, and 0.30. The significance of the last two terms may be seen from Equation 1.4.i of Problem 1.4. Take $V_w = 0.015$ nm^3 and $T = 300°$K in your calculations.

4.9   (a) Consider an oil-in-water emulsion whose drop surfaces are negatively charged due to adsorption of an anionic surfactant. Will local pH at the surface be larger or smaller than in the bulk aqueous phase? By how much? This effect can be important when the surfactant is an organic acid or amine whose degree of dissociation depends on pH.

(b) Consider a system containing oil, the sodium salt of an anionic surfactant, and a NaCl solution. Suppose that it separates into oil-in-water microemulsion and an excess NaCl brine phase. If the microemulsion is sufficiently concentrated that much of its aqueous material is part of the double-layer regions between adjacent droplets, explain why the NaCl concentration in the excess brine phase is greater than that of the original NaCl solution.

5 Interfaces in Motion: Stability and Wave Motion

1. BACKGROUND

So far the discussion of interfaces has dealt mainly with their equilibrium properties. Such topics as interfacial thermodynamics, equilibrium shapes of fluid interfaces, equilibrium contact angles, and equilibrium phase behavior in systems containing surfactants are central to the understanding of interfacial phenomena and are discussed to a greater or a lesser extent in most textbooks on the subject.

Interfaces in motion or otherwise not in equilibrium occur frequently in situations of practical importance. Nonequilibrium interfaces are common in mass transfer processes such as distillation, gas desorption, and liquid-liquid extraction, which are ubiquitous and of great importance in the chemical processing industry. Nonequilibrium behavior in the form of interfacial motion is also decisive in both the formation and destruction of dispersed systems. Whether energy is supplied by mechanical stirring, ultrasonic vibration, electric fields, or other means, processes for forming emulsions, foams, or aerosols starting from the two bulk phases necessarily involve interfacial motion, instability, and breakup. On the other hand, destruction of a disperse fluid system is the result of coalescence of many pairs of drops (or bubbles). Before coalescence occurs, the film between two drops must thin by drainage. Coalescence itself is initiated by an instability of the draining film.

In the remainder of this book, we present information on phenomena where dynamic interfacial effects are important. We begin in this chapter with interfacial stability and the closely related subject of interfacial oscillation or wave motion. It is frequently of great interest to know the conditions for interfacial instability. We may ask, for instance, how far a fluid jet leaving a circular orifice travels before it breaks up into drops. Or when we can expect spontaneous convection to arise near an interface across which one or more species diffuse.

Suitable stability analyses can provide answers to these and similar questions. They basically establish whether perturbations from some initial state grow or diminish with time. In the former, case the system is unstable and never returns to its initial state. In the latter case, the system is stable and the initial state is regained, although sometimes via oscillations of ever decreasing amplitude.

If only small perturbations are considered, terms involving the squares of or products of the various perturbed quantities may be neglected. Under these conditions the stability analysis is linear in nature and hence much simplified from the general case.

Nonlinear analysis is required for larger perturbations. Though more complex mathematically, it provides additional information, e.g., whether or not an unstable system eventually approaches some new steady state, and if so, the nature of that steady state. It also shows whether a system stable with respect to small perturbations can be unstable with respect to some large perturbations. In this book we shall consider only small perturbations and the techniques of linear stability analysis.

2. LINEAR ANALYSIS OF INTERFACIAL STABILITY

2.1 DIFFERENTIAL EQUATIONS

We consider stability of the interface between two incompressible fluids A and B, as illustrated in Figure 5.1. When the upper fluid is the denser of the two, a gravitationally produced instability arises. It is the prototype of various instabilities that are produced by acceleration of fluids in contact and which are employed in forming emulsions, aerosols, and foams by various shaking processes and ultrasonic vibration. While important in its own right, it also serves here as a vehicle for introducing the techniques of linear stability analysis which will be employed throughout this chapter and the next.

Initially the interface in Figure 5.1 occupies the horizontal plane $z = 0$, and there is no motion in either fluid. At time $t = 0$, the interface is deformed slightly to a new configuration $\bar{z}(x)$. Whatever the functional form of $\bar{z}(x)$, it can be expressed as a Fourier integral:

$$\bar{z}(x) = \int_0^\infty [B_1(\alpha) \cos \alpha x + B_2(\alpha) \sin \alpha x] \, d\alpha . \tag{5.1}$$

The analysis given below focuses on determining the time-dependent interfacial position $z_{1\alpha}^*(x,t)$ (or $z_{2\alpha}^*(x,t)$) when the initial deformation is that produced

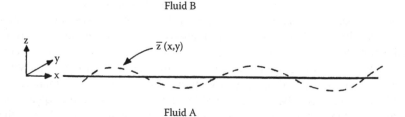

Fluid B

$\bar{z}(x,y)$

Fluid A

FIGURE 5.1 Wavy perturbation of the interface between superposed fluids. Gravity acts in the negative z direction.

by a single Fourier component $\overline{z}_{1\alpha} = \cos \alpha x$ (or $\overline{z}_{2\alpha} = \sin \alpha x$). The linear nature of the governing equations for small perturbations implies that the interfacial position $z^*(x,t)$ for a general deformation $\overline{z}(x)$ may be obtained by superposition of these results:

$$z^*(x,t) = \int_0^\infty [B_1(\alpha)\, z^*_{1\alpha}(x,t) + B_2(\alpha)\, z^*_{2\alpha}(x,t)]\, d\alpha \ . \tag{5.2}$$

The underlying idea behind the stability analysis is that all possible initial perturbations $\overline{z}(x)$ consistent with system boundary conditions may occur. If all of these perturbations diminish in amplitude with time, the system is stable. But if any perturbation grows, instability occurs and the interface never returns to its initial flat configuration. It is clear from Equation 5.2 that if the system is stable with respect to all initial interfacial deformations having the forms $\overline{z}_{1\alpha}$ and $\overline{z}_{2\alpha}$ of individual Fourier components, it is stable with respect to a general deformation. But if it is unstable with respect to even one Fourier component, interfacial deformation can be expected to increase continuously with time, and there is no return to the initial state.

Accordingly, we concentrate on solving the equations of change for an initial deformation given by the single Fourier component $B \cos \alpha x$ (or $B \sin \alpha x$), which has a wavelength λ equal to $(2\pi/\alpha)$. Actually, for reasons explained below, we choose to write the deformation in the form $e^{i\alpha x}$, it being understood that only the real (or imaginary) part of this expression is of physical significance. For simplicity two-dimensional initial deformations of the form $\overline{z}(x,y)$ are not considered here. The basic method of solution is the same, however, and is described elsewhere (Miller, 1978).

For small interfacial deformations, the velocity components v_x and v_z are small. If we neglect terms involving the squares or products of these components, a valid procedure when disturbance amplitude is much smaller than the wavelength, the equations of motion in either fluid A or fluid B simplify to

$$\rho \frac{\partial v_x}{\partial t} = -\frac{\partial p}{\partial x} + \mu \nabla^2 v_x \tag{5.3}$$

$$\rho \frac{\partial v_z}{\partial t} = -\frac{\partial p}{\partial z} - \rho g + \mu \nabla^2 v_z \ . \tag{5.4}$$

Here p is the pressure, ρ and μ are fluid density and viscosity, t is time, g is the gravitational acceleration, and ∇^2 is the operator $(\partial^2/\partial x^2 + \partial^2/\partial z^2)$. The pressure may be eliminated by differentiating Equation 5.3 with respect to z and Equation 5.4 with respect to x and subtracting:

$$\rho \frac{\partial \omega}{\partial t} = \mu \nabla^2 \omega. \tag{5.5}$$

$$\omega = \frac{\partial v_x}{\partial z} - \frac{\partial v_z}{\partial x}. \tag{5.6}$$

We note that ω in these equations is the y component of the vorticity $\nabla \times v$.

If Equation 5.5 is differentiated with respect to z, the result can be written as

$$\rho \frac{\partial}{\partial t} \left[\frac{\partial^2 v_x}{\partial z^2} - \frac{\partial}{\partial x} \left(\frac{\partial v_z}{\partial z} \right) \right] = \mu \nabla^2 \left[\frac{\partial^2 v_x}{\partial z^2} - \frac{\partial}{\partial x} \left(\frac{\partial v_z}{\partial z} \right) \right]. \tag{5.7}$$

Moreover, local continuity for an incompressible fluid requires that

$$\frac{\partial v_x}{\partial x} + \frac{\partial v_z}{\partial z} = 0. \tag{5.8}$$

Substituting Equation 5.8 into Equation 5.7, we obtain

$$\left(\frac{\partial}{\partial t} - \nu \nabla^2 \right) \nabla^2 v_x = 0, \tag{5.9}$$

where ν is the kinematic viscosity (μ/ρ). A similar manipulation leads to

$$\left(\frac{\partial}{\partial t} - \nu \nabla^2 \right) \nabla^2 v_z = 0. \tag{5.10}$$

Of course, these equations can also be readily obtained using vector calculus (see Miller, 1978).

Let us suppose that v_z has the same periodic behavior as the interfacial deformation itself, so that

$$v_z (x,z,t) = u (z,t) e^{i\alpha x}. \tag{5.11}$$

The validity of this procedure will be confirmed if we are successful in finding a solution that satisfies the governing equations and boundary conditions. Substitution of Equation 5.11 into Equation 5.10 yields

$$\left(\frac{\partial}{\partial t} - \nu \frac{\partial^2}{\partial z^2} - \nu \alpha^2 \right) \left(\frac{\partial^2}{\partial z^2} - \alpha^2 \right) u = 0. \tag{5.12}$$

Next, we try a further separation of variables and suppose that

$$u\ (z,t) = W\ (z)\ \delta\ (t).$$ (5.13)

When this expression for u is substituted into Equation 5.12, the result may be written as

$$\frac{\delta'(t)}{\delta(t)} = \nu \frac{\left(\dfrac{\partial^2}{\partial z^2} - \alpha^2\right)\left(\dfrac{\partial^2}{\partial z^2} - \alpha^2\right) W(z)}{\left(\dfrac{\partial^2}{\partial z^2} - \alpha^2\right) W(z)} = \beta\ .$$ (5.14)

Here β is a separation constant. Since the first two terms in Equation 5.14 are functions of t and z, respectively, and since they are equal, both terms must be equal to the same constant β.

Equation 5.14 yields two ordinary differential equations for $\delta(t)$ and $W(z)$. The solutions in fluids A and B, subject to the condition that v_z becomes vanishingly small at large distances from the interface arc

$$\delta_A(t) = \delta_{oA} e^{\beta_A t}$$ (5.15)

$$\delta_B(t) = \delta_{oB} e^{\beta_B t}$$ (5.16)

$$W_A(z) = D_1 e^{\alpha z} + D_2 e^{\alpha r_A z}$$ (5.17)

$$W_B(z) = D_3 e^{-\alpha z} + D_4 e^{-\alpha r_B z}$$ (5.18)

$$r_A^2 = 1 + \frac{\beta_A}{\nu_A \alpha^2}; \quad r_B^2 = 1 + \frac{\beta_B}{\nu_B \alpha^2}\ .$$ (5.19)

The tangential velocity v_x is given by the real (or imaginary) part of

$$v_x = \frac{i}{\alpha} W'\ (z)\ e^{i\alpha x}\ \delta(t)\ .$$ (5.20)

As is easily shown, this expression satisfies the differential equation (Equation 5.9).

In employing certain boundary conditions, we also require an expression for pressure. Differentiating Equations 5.3 and 5.4 with respect to x and z, respectively, adding the resulting expressions, and invoking Equation 5.8, we find

$$\frac{\partial^2 p}{\partial x^2} + \frac{\partial^2 p}{\partial z^2} = 0 . \tag{5.21}$$

Before perturbation the initial pressure distribution is hydrostatic,

$$p_{iA} = p_o - \rho_A g z$$

$$p_{iB} = p_o - \rho_B g z, \tag{5.22}$$

where p_o is the initial pressure at the interface. Let us assume that the perturbation p_p from this initial pressure distribution has the following form in each fluid:

$$p_p = P(z)e^{i\alpha x}\delta(t). \tag{5.23}$$

Combining this equation with Equation 5.21, we find that

$$\frac{d^2 P}{dz^2} = \alpha^2 P . \tag{5.24}$$

Similarly, the derivative of Equation 5.4 with respect to z yields

$$\frac{d^2 P}{dz^2} = -\rho\beta\frac{dW}{dz} + \mu\left(\frac{d^3 W}{dz^3} - \alpha^2\frac{dW}{dz}\right) . \tag{5.25}$$

Finally, equating the right-hand members of Equations 5.24 and 5.25 leads to the following expression for P:

$$P(z) = \frac{\mu}{\alpha^2}\frac{d^3 W}{dz^3} - \mu r^2\frac{dW}{dz} . \tag{5.26}$$

Hence the pressure distributions in the two fluids are readily found using Equations 5.17 and 5.18.

The basic procedure employed here, that is, writing each dependent variable as the sum of an initial contribution and a small perturbation, substituting these expressions into the governing differential equations, neglecting squares and products of perturbations, and solving the resulting partial differential equations by separation of variables, is applicable in most cases where a linear stability analysis is to be performed.

2.2 BOUNDARY CONDITIONS

The above solution of the equations of motion for a given value of the wavenumber α contains several unknown constants : D_1, D_2, D_3, D_4, β, β_A, and β_B (the constants δ_{oA} and δ_{oB} can be incorporated into the constants D_i when the overall expressions for v_{zA} and v_{zB} are written in accordance with Equations 5.11 and 5.13). Suitable boundary conditions along the wavy interface and the initial condition must be used to determine these constants. It is through these conditions that interfacial properties such as interfacial tension enter the analysis.

Some boundary conditions are basically requirements that quantities such as overall mass, momentum, energy, and electrical charge be conserved. We shall derive expressions for the first two of these in this chapter, although in forms more general than required for solution of the simple problem of superposed fluids considered in the preceding section. Other conservation equations, including individual component material balances, are treated in the following chapter which deals with transport processes.

We begin with conservation of mass for a "pillbox" control volume such as that shown in Figure 1.3. We take the control volume itself as fixed in space but permit the reference surface S to move normal to itself with velocity $\dot{a}_n$. Employing a procedure similar to that used in Chapter 1, we first write the mass balance for the overall control volume:

$$\int_{S_A} \rho_A \, v_A \cdot n \, dS - \int_{S_B} \rho_B v_B \cdot n \, dS - \int_{S_o} \rho v \cdot M \, dS \; = \frac{d}{dt} \int_v \rho \, dV \, . \quad (5.27)$$

We then subtract the following expressions which would apply if the regions below and above S were occupied by bulk phases A and B, respectively:

$$\int_{S_A} \rho_A v_A \cdot n \, dS - \int_S \rho_A (v_A \cdot n - a_n) \, dS - \int_{S_{oA}} \rho_A v_A \cdot M \, dS \; = \frac{d}{dt} \int_{V_A} \rho_A \, dV \quad (5.28)$$

$$\int_S \rho_B (v_B \cdot n - a_n) \, dS - \int_{S_B} \rho_B v_B \cdot n \, dS - \int_{S_{oB}} \rho_B v_B \cdot M \, dS \; = \frac{d}{dt} \int_{V_B} \rho_B \, dV \quad (5.29)$$

The result is

$$\int_S [\rho_A (v_{An} - a_n) - \rho_B (v_{Bn} - a_n)] \, dS - \int_{S_o} \Delta \, (\rho v) \cdot M \, dS \; = \frac{d}{dt} \int_v \Delta\rho \, dV \, , \quad (5.30)$$

where

$$\rho v - \rho_A v_A \ in \ V_A$$

$$\Delta (\rho v) =$$

$$\rho v - \rho_B v_B \ in \ V_B.$$

Note that use of these equations implies that even in the nonequilibrium situation, a method can be found for extrapolating the bulk phase densities and velocities into the interfacial region. In many situations of interest, this extrapolation presents no difficulties.

If we make the same small curvature approximation as in Chapter 1, the second integral in Equation 5.30 becomes

$$\int_C \left[\int_{\lambda_A}^{\lambda_B} \Delta(\rho v) \, d\lambda \right] \cdot m \, ds = \int_C \Gamma v \cdot m \, ds = \int_C \Gamma v_s \cdot m \, ds = \int_S (\nabla_s \cdot \Gamma v_s) \, dS \ ,$$

(5.30a)

with m the outward pointing normal to C (see Figure 1.3). In writing this equation we have made use of the fact that m has no component in the normal direction to obtain $(v \cdot m) = (v_s \cdot m)$ and then invoked Equation 1.36. The third integral in Equation 5.30 becomes

$$\frac{d}{dt} \int_S \Gamma dS = \int_S \frac{\partial \Gamma}{\partial t} dS + \int_S (-2H) \dot{a}_n \Gamma \, dS \ .$$

(5.30b)

Because S_o, and hence C, are fixed, there is no term here involving the line integral around C. The last term in this equation represents the rate of change of the area of S due to translation of S perpendicular to itself (cf. Equation 1.19). Substituting Equations 5.30a and 5.30b into Equation 5.29, we obtain

$$\int_S \left[\rho_A(v_{An} - a_n) - \rho_B(v_{Bn} - a_n) - \frac{\partial \Gamma}{\partial t} + 2H\Gamma a_n - \nabla_s \cdot \Gamma v_s \right] dS \ .$$ (5.31)

As the extent of S is arbitrary, the integrand itself must vanish, and we obtain the differential interfacial mass balance:

$$\frac{\partial \Gamma}{\partial t} = -\nabla_s \cdot \Gamma v_s + 2H\Gamma a_n + \rho_A(v_{An} - a_n) - \rho_B(v_{Bn} - a_n) \ .$$ (5.32)

According to this general equation, the local surface excess mass per unit area Γ can change as a result of (a) the inflow or outflow of mass by surface convection, (b) an increase or decrease in surface area due to surface motion, or (c) an inflow from bulk fluids.

For the case of superposed fluids, considered in Section 2.1, we neglect the terms in Equation 5.32 involving the surface excess mass Γ. Moreover, we note that the term $\rho_A(v_{An} - \dot{a}_n)$ represents the rate at which material from phase A crosses the interface. Since no transfer of either A or B across the interface occurs in this case, we conclude that

$$v_{An} = v_{Bn} = \dot{a}_n \, . \tag{5.33}$$

In terms of Equations 5.15 through 5.18, we have

$$W_A(0) \, \delta_A(t) = W_B(0) \, \delta_B(t). \tag{5.34}$$

This equation implies that the two exponential time factors β_A and β_B are equal. We shall therefore drop the subscripts and simply use β. Note that W_A and W_B are evaluated at $z = 0$. Although the interface is slightly perturbed from this position, the small correction is of higher order and hence neglected in our linear analysis because the quantities W_A and W_B are themselves small perturbations.

A similar procedure can be used to derive a general interfacial momentum balance. The result, a generalization of the static momentum balance derived in Chapter 1 (Section 4), is

$$\frac{\partial Y}{\partial t} = -\nabla_S \cdot Y v + \Gamma \hat{F} + (p_A + \tau_{An} - p_B - \tau_{Bn})n + 2H\gamma n + \, . \tag{5.35}$$

$$\nabla_S \gamma + (\tau_{At} - \tau_{Bt}) + \rho_A v_A (v_{An} - \dot{a}_n) - \rho_B v_B (v_{Bn} - \dot{a}_n)$$

Here, Y is the surface excess momentum defined by

$$Y = \int_{\lambda_A}^{\lambda_B} \Delta(\rho v) \, d\lambda \, , \tag{5.36}$$

with $\Delta(\rho v)$ defined as in Equation 5.30. Also, τ_{An} and τ_{Bn} are the normal viscous stresses at the interface and τ_{At} and τ_{Bt} are the shear stresses at the interface in fluids A and B, respectively. The sign convention used for the normal viscous stresses and shear stresses is that employed by Bird et al. (2002). We note that no excess stresses along the interface caused by interfacial flow itself have been considered. Some discussion of such stresses produced by an effective interfacial viscosity is provided in Chapter 7.

The last two terms in Equation 5.35 represent the net momentum input to the interface by material crossing the interface (e.g., material which changes from liquid to vapor during a boiling process). Normally these terms are small and their effect on the interfacial momentum balance is negligible. We shall consider briefly, however, in Chapter 6 an interesting exception where such momentum effects are the source of one type of interfacial instability.

For the superposed fluids of Section 2.1, we neglect terms in Equation 5.35 involving surface excess mass and momentum, assume that interfacial tension is uniform, and take the bulk fluids to be Newtonian with viscosities μ_A and μ_B in evaluating the normal viscous and shear stresses. The general vector equation then simplifies to the following two scalar equations reflecting the normal and tangential components:

$$\left(p_A - 2\mu_A \frac{\partial v_{zA}}{\partial z} \right) - \left(p_B - 2\mu_B \frac{\partial v_{zB}}{\partial z} \right) + 2H\gamma = 0 . \tag{5.37}$$

$$\mu_A \left(\frac{\partial v_{xA}}{\partial z} + \frac{\partial v_{zA}}{\partial x} \right) = \mu_B \left(\frac{\partial v_{xB}}{\partial z} + \frac{\partial v_{zB}}{\partial x} \right) \tag{5.38}$$

The pressures p_A and p_B include both the initial pressures of Equation 5.22 and the perturbations given by Equation 5.23. They must be evaluated at the actual position z^* of the interface, but with all terms of second and higher order in small perturbation quantities neglected. The result for the pressure in fluid A is, for instance,

$$p_A = p_o - \rho_A g z^* + P_A(0) \, e^{i\alpha x} \delta(t). \tag{5.39}$$

Equation 5.37 also contains a term involving curvature. There is only one nonzero radius of curvature in this case. It can be evaluated using Equation 1.52, neglecting terms of second and higher order as before. The result is that the mean curvature H is given by $1/2 \, (\partial^2 z^*/\partial x^2)$.

In addition to the boundary conditions based on conservation of mass and momentum, another condition required for the case of superposed fluids is continuity of tangential velocity at the interface:

$$v_{xA} = v_{xB} \text{ at } z = 0. \tag{5.40}$$

Finally, the initial condition that the interface is at position $B \cos \alpha x$ (or $B \sin \alpha x$) at $t = 0$ must be imposed. Since the interface moves with the fluids, the time derivative (dz^*/dt) of interfacial position is given by v_{zA} at the interface

$$\frac{dz^*}{dt} = W_A(o)\, e^{i\alpha x}\delta(t) = (D_1 + D_2)\, e^{i\alpha x}\, e^{\beta t}. \qquad (5.41)$$

It is clear from this expression that the initial condition is satisfied, provided that

$$B = \frac{D_1 + D_2}{\beta}. \qquad (5.42)$$

2.3 STABILITY CONDITION AND WAVE MOTION FOR SUPERPOSED FLUIDS

Let us summarize our progress to this point in analyzing the stability of the interface between the superposed fluids of Figure 5.1. We solved the equations of continuity and motion in Section 2.1 to obtain the perturbations in the velocity and pressure distributions in the two fluids (Equations 5.17 through 5.20 and Equation 5.26). With Equation 5.42 and the equality of β_A and β_B implied by Equation 5.34, the list of unknown constants in these solutions given at the beginning of Section 2b has been reduced to D_1, D_2, D_3, D_4, and β. Moreover, we derived four applicable boundary conditions in Section 2.2 (Equations 5.34, 5.37, 5.38, and 5.40). When the velocity and pressure distributions are substituted into these equations, we find the following relationships among the unknown constants:

$$D_1 + D_2 - D_3 - D_4 = 0 \qquad (5.43)$$

$$D_1\mu_A\left(1 + r_A^2 + \frac{\gamma\alpha^2 + (\rho_A - \rho_B)\,g}{\alpha\beta\mu_A}\right) +$$

$$D_2\mu_A\left(2r_A + \frac{\gamma\alpha^2 + (\rho_A - \rho_B)\,g}{\alpha\beta\mu_A}\right) + D_3\mu_B(1 + r_B^2) + 2D_4\mu_B r_B = 0 \qquad (5.44)$$

$$2\mu_A D_1 + \mu_A(1 + r_A^2)\, D_2 - 2\mu_B D_3 - \mu_B(1 + r_B^2)\, D_4 = 0 \qquad (5.45)$$

$$D_1 + r_A D_2 + D_3 + r_B D_4 = 0. \qquad (5.46)$$

Note that these are four linear, homogeneous equations in the four unknowns D_1 through D_4. An obvious solution, but an uninteresting one because it implies no interfacial deformation and no flow, is the trivial solution $D_1 = D_2 = D_3 = D_4 = 0$. According to the theory of linear equations, the condition for a nontrivial solution to exist is that the determinant of coefficients vanishes for Equations 5.43 through 5.46. This relationship, which of course does not contain any of the D_is, may be solved to obtain the time factor β as a function of the wavenumber α and the physical properties of the fluids. If the real part of β is positive, the perturbation grows and the interface is unstable. If the real part of β is negative, the perturbation diminishes and the interface ultimately returns to the flat configuration. A nonzero imaginary part of β implies that oscillation can be expected.

It is instructive to consider some special cases. First let fluid A be a gas of negligible density and viscosity. Then Equations 5.43 and 5.46, which require continuity of normal and tangential velocity at the interface, are not needed. If the derivation leading to Equation 5.42 is repeated for fluid B, the quantity $D_1 + D_2$ in that equation is replaced by $D_3 + D_4$. Thus, if A is a gas, Equation 5.44 becomes

$$D_3\left(1+r_B^2+\frac{\gamma\alpha^2-\rho_B g}{\alpha\beta\mu_B}\right)+D_4\left(2r_B+\frac{\gamma\alpha^2-\rho_B g}{\alpha\beta\mu_B}\right)=0 . \qquad (5.47)$$

Noting that terms in D_1 and D_2 also disappear from Equation 5.45, we find, after some manipulation, that the determinant of coefficients of Equations 5.45 and 5.47 simplifies to

$$r_B^4 + 2r_B^2 - 4r_B + 1 = \frac{-\beta_B^{*2}}{v_B^2\alpha^4} , \qquad (5.48)$$

where

$$\beta_B^{*2} = \frac{1}{\rho_B}(\gamma\alpha^3 - \rho_B g\alpha) . \qquad (5.49)$$

When liquid B has a low viscosity, we anticipate that the dimensionless group $|\beta/v_B\alpha^2| \gg 1$. In this case, if we retain only the leading term of the left side of Equation 5.48, we obtain

$$\beta^2 = -\beta_B^{*2} = g\alpha - \frac{\gamma\alpha^3}{\rho_B} . \qquad (5.50)$$

Thus for small wavenumbers α, that is, long wavelengths $(2\pi/\alpha)$, β is positive and the perturbation grows. Such gravitationally produced instability is often called the Rayleigh-Taylor instability and is expected in this case where the liquid is placed above the gas. However, for sufficiently large α (i.e., for short wavelengths), interfacial tension is able to stabilize the interface in spite of the unfavorable positions of the fluids relative to gravity. Then we have

$$\beta = \pm i\,\beta_B^*, \quad \beta_B^* > 0 , \qquad (5.51)$$

and the interface oscillates with frequency β_B^*. (There is no damping of the oscillations since viscous dissipation has been neglected in going from Equation 5.48 to Equation 5.50.)

The critical wavenumber α_c for which β vanishes is given by

$$\alpha_c = [\rho_B g/\gamma]^{1/2}. \tag{5.52}$$

For the air-water system, this critical wavelength $(2\pi/\alpha_c)$ is about 1.7 cm. Only perturbations having wavelengths greater than this value will grow. Hence the instability can be suppressed by limiting the lateral extent of the interface between the fluids.

We expect the rate of perturbation growth to be zero for $\alpha = \alpha_c$ at the transition between stable and unstable regions and also to be small for long wavelengths (small α), where interfacial deformation and hence the gravitational driving force for the instability vary slowly along the interface. Between these extremes there should be a fastest growing perturbation. By differentiating Equation 5.50 with respect to α we can find the wavenumber α_M that has the fastest rate of growth. The result is

$$\alpha_M = [\rho_B g/3\gamma]^{1/2} \tag{5.53}$$

$$\beta_M = [4\rho_B g^3/27\gamma]^{1/4}. \tag{5.54}$$

For the air-water system, the fastest growing disturbance has a wavelength $2\pi/\alpha_M$ of about 2.9 cm and its time constant β_M^{-1} for growth is about 0.027 seconds.

We expect disturbances with wavenumbers near α_M to dominate the early stages of the instability where the linear theory applies. In practice, the dominant disturbance observed experimentally in many cases of instability frequently has a wavenumber near α_M even though the linear analysis is not applicable for the rather large perturbations that are usually required to detect the instability.

As the liquid viscosity μ_B increases, the condition of Equation 5.52 for the critical wavenumber remains unchanged, as it is basically a thermodynamic condition (see Section 7). But the rates of growth of unstable disturbances decrease. The slowing of growth is greatest for short wavelengths where velocity gradients are greatest. As a result, the fastest growing disturbance shifts to longer wavelengths (smaller α).

Another limiting case is reached when μ_B becomes very large. Now we have $|\beta/\nu_B\alpha^2| \ll 1$ and expansion of Equation 5.48 in terms of this quantity leads to

$$\beta = \frac{-\beta_B^{*2}}{2\nu_B\alpha^2}. \tag{5.55}$$

Under these conditions there is no oscillation and the disturbance either grows or decays exponentially, depending on the sign of β_B^{*2}. The critical wavenumber α_c is still given by Equation 5.52, but there is no finite α_M. The smaller the value of α (consistent with $|\beta/\upsilon_B\alpha^2| \ll 1$), the greater the rate of growth in the unstable region.

If we consider instead the situation where A is a liquid and B a gas, we obtain instead of Equation 5.48:

$$r_A^4 + 2r_A^2 - 4r_A + 1 = \frac{-\beta_A^{*2}}{v_A^2\alpha^4} \tag{5.56}$$

$$\beta_A^{*2} = \frac{1}{\rho_A}(\gamma\alpha^3 + \rho_A g\alpha). \tag{5.57}$$

We expand as before for the low viscosity limit, but this time we retain the two leading terms of the left side of Equation 5.56. The result is

$$\beta^2\left(1 + \frac{4v_A\alpha^2}{\beta}\right) = -\beta_A^{*2}. \tag{5.58}$$

In the case of an inviscid liquid ($v_A = 0$), we obtain $\beta = \pm i\,\beta_A^*$ as before, and the interface oscillates without damping. There is no instability because, with the liquid below the gas, both gravity and interfacial tension act to restore the deformed interface to its initial planar configuration. But with no viscous resistance to flow, the momentum developed during the return carries the interface beyond the flat configuration to a new deformation of opposite sign to the original one.

When v_A is small, we anticipate that β has the form $(\pm\, i\,\beta_A^* + \beta_1)$, where β_1 is a small damping factor produced by viscous dissipation such that $(\beta_1/\beta^*) \ll 1$. Substituting this expression into Equation 5.58 and keeping only the leading terms, we find

$$\beta = \pm i\beta_A^* - 2v_A\alpha^2. \tag{5.59}$$

The second term in this expression is the rate at which the oscillations are damped. As might be expected, higher viscosities lead to more rapid damping in this case where the flow is primarily determined by the oscillatory motion.

A second limiting case of Equation 5.56 is the high viscosity situation for which $|\beta/v_A\alpha^2| \ll 1$. If Equation 5.56 is expanded in powers of $\beta/v_A\alpha^2$ and only the leading nonzero term is kept, we find in a manner analogous to that used in deriving Equation 5.55 that

$$\beta = \frac{-\beta_A^{*2}}{2v_A\alpha^2} . \qquad (5.60)$$

Here the resistance to flow is so great that the interface does not continually overshoot its equilibrium position as above (i.e., no oscillatory motion develops). Instead, the deformed interface simply returns gradually to the flat configuration. Higher viscosities produce slower damping because they reduce the velocities and velocity gradients.

In the liquid-liquid case, it can be shown by similar procedures (Chandrasekhar, 1961) that in the limit of low fluid viscosities

$$\beta^2 = -\beta^{*2} = -\left[\frac{\gamma\alpha^3 + (\rho_A - \rho_B)g\alpha}{\rho_A + \rho_B}\right]. \qquad (5.61)$$

Thus β is real and positive and the perturbation grows whenever β^{*2} is negative. The expressions for the critical and fastest growing wavenumbers α_c and α_M can be obtained by replacing ρ_B by $\rho_B - \rho_A$ in Equations 5.52 and 5.53. But the growth factor β_M for the fastest growing disturbance is given by

$$\beta_M = \left[\frac{4(\rho_B - \rho_A)^3 g^3}{27\gamma(\rho_A + \rho_B)^2}\right]^{1/4}. \qquad (5.62)$$

When the interface is stable, the time factor β in the low viscosity limit has the form (Wehausen and Laitone, 1960)

$$\beta = \pm i\beta^* - (1 \pm i)\left[\frac{(2\beta^*\mu_A\mu_B\rho_A\rho_B)^{1/2}}{(\rho_A + \rho_B)\left[(\mu_A\rho_A)^{1/2} + (\mu_B\rho_B)^{1/2}\right]}\right]. \qquad (5.63)$$

Thus the frequency of oscillation is slowed slightly from the inviscid value β^* by viscous effects and the damping factor has an entirely different form than in Equation 5.59, being dependent, for instance, on the frequency β^*. The reason is that the requirement of Equation 5.40 for continuity of tangential velocity at the interface leads to formation of an oscillating boundary layer near the interface in which viscous dissipation is much greater than that for the liquid-gas case.

In the high viscosity limit, the result analogous to Equations 5.55 and 5.60 is

$$\beta = \frac{-\beta^{*2}(\rho_A + \rho_B)}{2\alpha^2(\mu_A + \mu_B)} . \qquad (5.64)$$

As before, this equation applies to both stable and unstable interfaces.

EXAMPLE 5.1 CHARACTERISTICS OF WAVE MOTION FOR
FREE INTERFACES

Calculate the frequency of oscillation and the rate of damping at an air-water interface for wavelengths $\lambda\, (= 2\pi/\alpha)$ of 100, 10, 1, and 0.1 cm. Repeat for an oil-water interface with $\rho_B = 0.85$ g/cm^3, $\mu_B = 1$ cp for oil, and with interfacial tension $\gamma_{AB} = 50$ mN/m.

SOLUTION

For the air-water interface, we take $\rho_A = 1$ g/cm^3, $\mu_A = 1$ cp, and $\gamma = 72$ mN/m. The frequency β_A^* can be calculated from Equation 5.57 and the damping factor β_1 from Equation 5.59 in the low viscosity limit. Results are

λ (cm)	α (per cm)	β_A^* (per second)	$-\beta_1$ (per second)
100	0.0628	7.85	7.9×10^{-5}
10	0.628	25.2	7.9×10^{-3}
1	6.28	155	0.79
0.1	62.8	4230	79

Similarly for the oil-water interface, Equations 5.61 and 5.63 can be used. The corrected frequency is given by β_C^*.

λ (cm)	β^* (per second)	β_c^* (per second)	$-\beta_1$ (per second)
100	2.24	2.23	4.14×10^{-3}
10	7.52	7.45	7.59×10^{-2}
1	84.9	82.3	2.55
0.1	2590	2450	141

In both cases, damping increases with decreasing wavelength. This behavior is due partly to the increased frequency and partly to the increased velocity gradient for short wavelengths.

3. DAMPING OF CAPILLARY WAVE MOTION BY INSOLUBLE SURFACTANTS

Mass transfer across a fluid interface is enhanced by convection in the vicinity of the interface. One source of such convection is wave motion. An increase in the rate of damping of waves can thus be expected to reduce mass transfer rates. As we shall now show, surfactants can cause a significant increase in the damping of capillary waves at a liquid-gas interface.

We consider the simplest case of an insoluble surfactant. In the initial motion-less state with a flat interface, the surfactant is uniformly distributed and inter-facial tension is uniform. During wave motion, the local concentration of surfac-tant varies with position along the interface, with the result that interfacial tension gradients arise. Because these gradients influence the interfacial momentum balance (through the term $\nabla_s \gamma$ in Equation 5.35), the velocity distribution is altered and with it the damping rate.

The solution of the equations of motion and application of the boundary conditions proceeds as before except for an additional term $(d\gamma/dx)$ on the right side of Equation 5.38. If fluid B is a gas, we obtain instead of Equations 5.44 and 5.45:

$$D_1\left(1 + r_A^2 + \frac{\beta_A^{*2}}{\beta\alpha^2 v_A}\right) + D_2\left(2r_A + \frac{\beta_A^{*2}}{\beta\alpha^2 v_A}\right) = 0 \qquad (5.65)$$

$$2\mu_A D_1 + \mu_A(1 + r_A^2)D_2 + \frac{(d\gamma/d\Gamma_s)(\partial\Gamma_s/\partial x)}{(-i\alpha)e^{i\alpha x}e^{\beta t}} = 0 , \qquad (5.66)$$

where Γ_s is the surface concentration of surfactant. Now $d\gamma/d\Gamma_s$ is a property of the surfactant film related to its compressibility. Here we shall assume that its value, which normally has a negative sign, is constant at least over the small range of Γ_s involved in the perturbation.

The general boundary condition for conservation of mass of some species in the interfacial region is derived in Chapter 6. For the present case of an insoluble surfactant, and in the absence of surface diffusion, we anticipate that the first two terms of Equation 5.32 should suffice with Γ replaced by Γ_s:

$$\frac{\partial\Gamma_s}{\partial t} = -\frac{\partial(\Gamma_s v_x)}{\partial x} . \qquad (5.67)$$

If we assume that Γ_s is the sum of an initial concentration Γ_{os} and a perturbation term, we may write

$$\Gamma_s = \Gamma_{os} + \Gamma_{1s}e^{i\alpha x}e^{\beta t}. \qquad (5.68)$$

Substituting this expression into Equation 5.67, we find

$$\Gamma_{1s} = \frac{\Gamma_{os}}{\beta}W_A'(0) = \frac{\Gamma_{os}\alpha}{\beta}(D_1 + r_A D_2) . \qquad (5.69)$$

Thus Equation 5.66 becomes

$$D_1\left[2 + \frac{\Gamma_{os}\alpha(-d\gamma/d\Gamma_s)}{\beta\mu_A}\right]$$

$$+ D_2\left[1 + r_A^2 + r_A\frac{\Gamma_{os}\alpha(-d\gamma/d\Gamma_s)}{\beta\mu_A}\right] = 0 \qquad (5.70)$$

The determinant of coefficients of Equations 5.65 and 5.70 yields the following dispersion equation for periodic motion of a liquid pool covered by an insoluble monolayer:

$$-(r_A - 1)^2 + \left[r_A^2 + r_A + \frac{\beta_A^{*2}}{\beta v_A \alpha^2} \right]\left[1 + r_A + \right.$$

(5.71)

$$\left. \left(\frac{v_A \alpha^2}{\beta} \right) \cdot \left(\frac{\Gamma_{os}(-d\gamma / d\Gamma_s)}{\mu_A v_A \alpha} \right) \right] = 0$$

Two limiting cases are of interest. In one, the expression in the first set of brackets in this equation is very large (e.g., owing to a large value of interfacial tension and hence β_A^{*2}). As a result, the expression in the second set of brackets is nearly zero. Under these conditions, "longitudinal" wave motion ensues (Lucassen, 1968) where the flow near the surface is dominated by the effects of surface tension variation rather than by surface deflection. Longitudinal waves, which are an important tool for measuring elastic properties of soluble surfactants, are considered further in Problem 5.3 and in Chapter 6.

In the second limiting case, the surface elasticity $\Gamma_{os}(-d\gamma/d\Gamma_s)$, or more precisely the dimensionless surface elasticity number $[\Gamma_{os}(-d\gamma/d\Gamma_s)/\mu_A v_A \alpha]$, becomes so large that the first expression in brackets must be nearly zero. Lateral flow along the interface is virtually precluded and effects of surface deflection are important in this "inextensible" case which might be expected for a concentrated, nearly incompressible monolayer.

Setting the first expression in brackets in Equation 5.71 equal to zero and taking the low viscosity limit in the usual way, we find

$$\beta = \pm i\beta_A^* - (1+i)\frac{\alpha(\beta_A^* v_A)^{1/2}}{2^{3/2}} .$$

(5.72)

The expression for the damping factor here is very different from that found at Equation 5.59 for the free surface case, and indeed resembles that of Equation 5.63 for a free liquid-liquid interface. We conclude that a surfactant film causes an oscillating boundary layer to develop and thus increases the damping rate. The boundary layer arises because the film prevents all lateral motion at the surface in this inextensible limit. Indeed, Equation 5.70 reduces to a requirement that $D_1 + r_A D_2$, and hence v_x must vanish at the surface. That the damping rate is higher with the surfactant film is indicated by its proportionality to $\mu_A^{1/2}$ in Equation 5.72, in contrast to μ_A in Equation 5.59, which of course, is the smaller quantity for low viscosities. For intermediate surfactant concentrations, the time factor β is still calculated by requiring that the determinant of coefficients of Equations 5.65 and 5.70 vanish, but with all terms of the latter equation retained. It can be shown that for low fluid viscosities, the damping factor passes through a maximum with increasing surfactant concentration Γ_{os} [see Dorrestein (1954) and Problem 5.1]. The maximum damping factor is twice that given by Equation 5.72. Evidently velocity gradients in the oscillating boundary layer can achieve values greater than those of the inextensible case because tangential velocities at the surface and near the edge of the boundary layer are out of phase and thus have opposite directions during some parts of the oscillation cycle.

For an inextensible liquid-liquid interface, the time factor β is given by the following expression (Miller and Scriven, 1968):

$$\beta = \pm i\beta^* - (1 \pm i)\frac{\alpha\beta^{*1/2}[(\mu_A\rho_A)^{1/2} + (\mu_B\rho_B)^{1/2}]}{2^{3/2}(\rho_A + \rho_B)}. \qquad (5.73)$$

Damping is predominantly due to an oscillating boundary layer, as for a surfactant-free interface (cf. Equation 5.63). Because surfactants do not change the basic nature of the flow, their damping effect is smaller at a liquid-liquid than at a liquid-gas interface.

Experimentally it is not convenient to measure damping with time of standing waves of a fixed wavelength, the situation considered so far. Instead, oscillations of a fixed frequency β^* and fixed amplitude are imposed on a surface by, for instance, vibration of a bar placed on the surface or application of an oscillating electric field to a thin metal plate located within about 1 mm of the surface. For liquids of low viscosity in air, waves having a definite wavelength develop that move away from the bar or plate, their amplitude diminishing with increasing distance of travel. Both the wavelength and the variation of amplitude with position can be measured by optical or acousto-mechanical methods (see Langevin, 1999; Lucassen-Reynders and Lucassen, 1970).

In our previous discussion, the "dispersion equation" obtained by setting the determinant of coefficients equal to zero was used to solve for the time factor β, which could be a complex number, in terms of a known real wavenumber α. For the situation of the experiments just described, the same equation is solved for α, which may be complex, for a given imaginary value of β corresponding to the imposed frequency. Since interfacial configuration varies as $e^{i\alpha x}$, according to Equation 5.41, we see that the real part α_R is the wavenumber (i.e., the wavelength λ is $2\pi/\alpha_R$). The imaginary part α_I is a damping factor characterizing the exponential decay of wave amplitude with distance from the source of vibrations. We note that it was this situation involving a complex α that was the motivation for using $e^{i\alpha x}$ instead of $\cos \alpha x$ or $\sin \alpha x$ in equations such as Equation 5.11.

In the low viscosity limit, the analysis shows that α_R is given by Equation 5.57, with β_A^* the imposed frequency. The waves travel outward at a speed of β_A^*/α_R or $(\beta_A^* \gamma/\rho_A)^{1/3}$ for short wavelengths where the effect of gravity is unimportant. For a free surface, the damping factor α_I is found under the same circumstances to be

$$\alpha_I = \frac{4\mu_A\beta_A^*}{3\gamma}. \qquad (5.74)$$

For high surfactant concentrations, the interface is inextensible, an oscillating boundary layer develops, and we obtain

$$\alpha_I = \frac{2^{1/2}}{6}(\beta_A^{*5}\rho_A\mu_A^3 / \gamma^4)^{1/6} \ . \tag{5.75}$$

At intermediate surfactant concentrations a maximum in α_I is predicted, just as was a maximum in the damping factor β_1 of the earlier discussion.

Figure 5.2 shows experimental damping factors α_I obtained by the experimental technique described previously. The data are for monolayers of various fatty acids on water and the frequency of oscillations is 200 Hz (Lucassen and Hansen, 1967). As the theory predicts, adding surfactant produces a significant increase in damping. Moreover, there is a maximum damping rate at an intermediate surfactant concentration. We note that only the long-chain acids are described by the present theory. The short-chain acids have appreciable solubility in water and adsorption-desorption and bulk diffusion effects must be considered. Because diffusion provides a means of surfactant transport from regions of high to low surfactant concentration along the interface, interfacial tension gradients are smaller and less damping occurs, as the figure shows. This effect is considered in Chapter 6.

Experiments of this type can be used to determine the surface tension γ and surface compressibility $\Gamma_{os}(- d\gamma/d\Gamma_s)$. The former is obtained from the measured value of wavelength using Equation 5.57, the latter from the damping factor and the general dispersion equation found by setting equal to zero the determinant of coefficients of Equations 5.65 and 5.70. A comparison of surface properties obtained in this way with those obtained independently from film balance measurements is shown in Figure 5.3 for monolayers of a particular anionic surfactant. Agreement is good, a confirmation of the validity of the capillary wave motion analysis presented above.

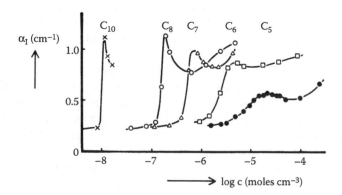

FIGURE 5.2 Damping on monolayers of fatty acids at 200 Hz. Reprinted with permission from Lucassen and Hansen (1967).

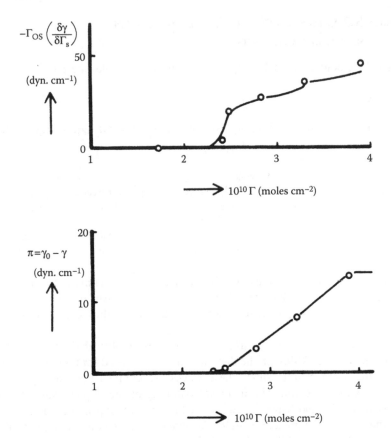

FIGURE 5.3 Comparison of measured surface pressure and elasticity coefficient with values calculated from ripple properties. Monolayers of dodecyl-p-toluenesulfonate: (- -) measured; (o) calculated. Reprinted with permission from Lucassen-Reynders and Lucassen (1970).

In a similar manner, the frequency of oscillation and the rate of damping of surface waves on an oscillating drop or bubble can be used to determine interfacial tension and compressibility. A derivation of pertinent equations and tabulation of key results can be found in the paper by Miller and Scriven (1968). Some useful formulas are given in Problem 5.4.

The above discussion has centered on wave motion imposed on a surface by, for instance, an oscillating bar. But thermal fluctuations cause wave motion of small amplitude even on interfaces that are not disturbed by external means. With laser light scattering techniques it is possible to measure interfacial tension from analysis of surface fluctuations. This method has been applied to the measurement of ultralow interfacial tensions between liquid phases (Bouchiat and Meunier, 1972; Cazabat et al., 1983; Zollweg et al., 1972). Presumably it could also be used to determine surface compressibility or other rheological properties.

EXAMPLE 5.2 CHARACTERISTICS OF WAVE MOTION FOR INEXTENSIBLE INTERFACES

Repeat the calculations of Example 5.1 for an inextensible interface.

SOLUTION

Equations 5.72 and 5.73 apply to the liquid-gas and liquid-liquid cases, respectively.

	Air-water		Oil-water	
λ (cm)	β_c^* (per second)	$-\beta_1$ (per second)	β_c^* (per second)	$-\beta_1$ (per second)
100	7.84	6.22×10^{-3}	2.23	4.24×10^{-3}
10	25.1	0.112	7.45	7.77×10^{-2}
1	152	2.77	82.3	2.61
0.1	4080	144	2450	144

As expected, the high surface elasticity greatly increases damping for the liquid-gas case over the free surface values of Example 5.1. But it has only a small effect for the liquid-liquid case, where boundary layer flow exists even with a free surface.

Surface elasticity is very effective in damping ripples of short wavelengths on a liquid surface. Although the damping rate is increased for long wavelengths, too, as the above results show, these waves are damped very slowly with or without a surfactant film. The increased damping of the ripples causes them to disappear rapidly and leave the surface with a smooth appearance. The same is true when a second immiscible liquid is added. A combination of these two effects is responsible for the ability of added oil to produce an apparent calming of the sea, a phenomenon that stimulated Benjamin Franklin to carry out experiments foreshadowing more recent monolayer studies (Tanford, 1989).

4. INSTABILITY OF FLUID CYLINDERS OR JETS

A slow-moving, thin cylindrical stream of water issuing from a faucet can often be seen to become nonuniform in diameter at some distance below the faucet and eventually to break up into drops. In a similar manner the beads on a spider's web are formed by instability of a cylinder of sticky liquid surrounding the thread leaving the spider's body. This instability is sometimes used in industrial processes to disperse one fluid in another by injecting it in the form of a jet.

We consider a cylinder of an incompressible, inviscid liquid A in gas B (see Figure 5.4). Initially there is no flow and the cross section is everywhere circular with radius R. At time $t = 0$, the fluid interface is deformed to the position given by the real part of the following expression:

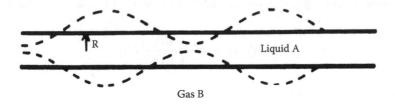

FIGURE 5.4 Wavy deformation of a liquid jet.

$$\bar{r} = R + Be^{i(m\theta + \alpha z)}$$

$$m = 0,1,2,\ldots\ldots$$

(5.76)

If effects of gravity are neglected, the equation of motion simplifies to

$$\rho \frac{\partial v}{\partial t} = -\nabla p \ .$$

(5.77)

Taking the divergence of this equation and invoking the continuity equation $\nabla \cdot v = 0$, we find

$$-\rho \frac{\partial}{\partial t}(\nabla \cdot v) = 0 = \nabla^2 p \ .$$

(5.78)

We look for a solution to this equation which has the form

$$p(r, \theta, z, t) = P(r)e^{i(m\theta + \alpha z)}\delta(t).$$

(5.79)

Substitution of Equation 5.79 into Equation 5.78 yields

$$r^2 \frac{d^2 P}{dr^2} + r\frac{dP}{dr} - (m^2 + \alpha^2 r^2)P = 0 \ .$$

(5.80)

The solution to this equation, which remains finite at $r = 0$, is given by

$$P(r) = DI_m(\alpha r) = Di^{-m}J_m(i\alpha r),$$

(5.81)

where I_m is a modified Bessel function of the first kind.

In a similar manner we take the radial velocity v_r to have the form

$$v_r = W(r)e^{i(m\theta + \alpha z)}\delta(t).$$

(5.82)

Substituting this expression into the radial component of Equation 5.77 we obtain

$$W(r)\delta'(t) = -\frac{1}{\rho}\frac{dP(r)}{dr}\delta(t) .$$

(5.83)

Separation of variables yields

$$\frac{\delta'(t)}{\delta(t)} = -\frac{dP(r)/dr}{\rho W(r)} = \beta ,$$

(5.84)

where β is a separation constant. The solutions are

$$\delta(t) = \delta_o e^{\beta t}$$

(5.85)

$$W(r) = -\frac{1}{\rho\beta}\frac{dP(r)}{dr} = -\frac{D\alpha}{\rho\beta}\left(\frac{m}{\alpha r}I_m(\alpha r) + I_{m+1}(\alpha r)\right) .$$

(5.86)

We next invoke the requirement that the time derivative of the interfacial perturbation r^* must equal the radial velocity v_r at the interface. Clearly this condition and the initial condition of Equation 5.76 can be satisfied if the following relationships hold:

$$r^* - R = Be^{i(m\theta + \alpha z)}e^{\beta t}$$

(5.87)

$$B = W(R)/\beta.$$

(5.88)

The normal component of the interfacial momentum balance of Equation 5.35 must be satisfied at the fluid interface. The tangential component cannot be imposed because neglect of viscous effects has reduced the order of the equation of motion and thus diminished the number of boundary conditions that can be satisfied. Since there are no viscous stresses here, Equation 5.35 simplifies to the Young-Laplace equation:

$$p_A - p_B = -2H\gamma.$$

(5.89)

Before the interface is perturbed, this equation simply states that p_A exceeds p_B by an amount γ/R. For the case of an axisymmetric perturbation ($m = 0$), the curvature of the interface in a plane passing through the axis can be calculated from Equation 1.42 by a procedure similar to that outlined following Equation 5.39 for a nearly plane interface. The second radius of curvature is the local radius of the cylinder. Hence

$$-2H = -\frac{\partial^2 (r^* - R)}{\partial z^2} + \frac{1}{r^*} \tag{5.90}$$

$$= \frac{1}{R} + \left(\alpha^2 - \frac{1}{R^2}\right)(r^* - R) \quad ; \quad m = 0 . \tag{5.91}$$

For $m \neq 0$, it can be shown that this expression must be augmented as follows:

$$-2H = \frac{1}{R} + \left(\alpha^2 + \frac{m^2 - 1}{R^2}\right)(r^* - R) . \tag{5.92}$$

Substituting Equations 5.81, 5.87, 5.88, and 5.92 into Equation 5.89, we obtain the following equation for the time factor β:

$$\beta^2 = -\frac{\gamma}{\rho}\left(\alpha^3 + \frac{\alpha(m^2 - 1)}{R^2}\right)\left(\frac{m}{\alpha R} + \frac{I_{m+1}(\alpha R)}{I_m(\alpha R)}\right) . \tag{5.93}$$

It is clear that for nonaxisymmetric disturbances ($m \neq 0$), the right side of Equation 5.93 is always negative. That is, β is an imaginary number, and oscillations exist without growth or decay, the expected behavior for a stable, inviscid system. But for the axisymmetric case ($m = 0$), instability does occur whenever the wavenumber is less than a critical value α_c given by

$$\alpha_c = \frac{1}{R} . \tag{5.94}$$

That is, a fluid cylinder is unstable with respect to disturbances having wavelengths greater than $2\pi/\alpha_c$ or the circumference $2\pi R$. The reason is that interfacial area and hence interfacial free energy actually decrease for such disturbances.

As before, the wavenumber α_M of the fastest growing disturbance can be obtained by differentiating Equation 5.93 with respect to α (with $m = 0$) and requiring the resulting expression to vanish. This calculation, which was first carried out by Rayleigh, yields the following numerical results:

$$\alpha_M = 0.697/R \tag{5.95}$$

$$\beta_M = 0.3433(\gamma/\rho R^3)^{1/2}. \tag{5.96}$$

Thus the fastest growing disturbance has a wavelength λ_M approximately equal to $9.0R$. For a jet of water in air with a radius R of 0.2 cm, the corresponding time factor β_M is about 32.6 per second^{-1}.

Comparing Equation 5.96 with Equation 5.54, we see that β_M increases with increasing surface tension for instability of a cylinder but that the opposite is true for superposed fluids. This result is not surprising since the change in interfacial free energy is the source of the instability here, whereas it is a stabilizing effect for superposed fluids.

While the above analysis applies to breakup of a stationary cylinder, the results can be used to obtain a rough estimate of the length L_B required for breakup of a jet injected with uniform velocity V. In particular, $L_B = Vt_B$, where t_B is the time required for the amplitude of a perturbation with wavelength λ_M to increase from its initial value B to a value comparable to the radius R. Unfortunately the initial amplitude is unknown. However, the exponential nature of the growth causes t_B to be somewhat insensitive to the value of B. For instance, t_B for the water jet considered above is 0.14 seconds if the amplitude must increase by a factor of 100 before breakage and 0.21 seconds if it must increase by a factor of 1000. Using these values, we would estimate a jet length L_B of several centimeters for a jet velocity V of 50 cm/sec.

Another procedure for determining L_B is to recast the stability analysis to recognize explicitly that the instability of a jet is a spatial rather than a temporal one, i.e., that perturbation amplitude increases with distance from the point of injection rather than with time at a fixed location, as in the stability analysis developed so far. For the spatial analysis, one proceeds much as in Section 3 and assumes that a perturbation with frequency β^* is imposed on the moving jet as it leaves the orifice. The time factor β is taken as $i\beta^*$ and α is obtained from the stability analysis. Its real part is related to the perturbation wavelength and its imaginary part to the rate of growth (or decay) of perturbation amplitude with distance from the orifice. Perturbations of all frequencies would be deemed possible, and L_B would be the breakup length for the frequency whose amplitude increases most rapidly with distance from the orifice. Such an analysis has been given by Keller et al. (1973).

5. OSCILLATING JET

As indicated above, all perturbations of a fluid cylinder for which $m \neq 0$ are stable. For $m = 1$, consideration of Equation 5.87 shows that the cross section remains circular with radius R_o at all points but that the axis becomes sinuous. For $m = 2$, the axis remains straight but the cross section is perturbed to a somewhat elliptical shape. It is this latter case that is of interest in this section.

If a jet issues from an orifice of circular cross section, we suppose that it can experience perturbations of the type specified by Equation 5.76 with all possible values of m and α. As shown in the previous section, an instability develops characterized by wavenumber α_M and growth factor β_M. But if the orifice has a rather elliptical cross section, a perturbation characterized by $m = 2$ and $\alpha = 0$ is imposed. Since Equation 5.93 is not convenient to use for $\alpha = 0$, we return to

Equation 5.80 and solve directly for $P(r)$ in this limit. The solution, which remains finite for $r = 0$, is given by

$$P(r) = E \cdot r^m. \tag{5.97}$$

The remainder of the analysis proceeds in the same manner as before, the final result being

$$\beta^2 = -\frac{\gamma m(m+1)(m-1)}{\rho R^3}. \tag{5.98}$$

For $m = 2$, the behavior is a sustained oscillation with frequency β^*:

$$\beta = \pm i \beta^* = \pm i (6\gamma/\rho R^3)^{1/2}. \tag{5.99}$$

If the liquid moves along the jet with velocity V, the distance Λ traveled during one cycle of oscillation is given by

$$\Lambda = \frac{2\pi V}{\beta^*} = \pi V (2R^3\rho / 3\gamma)^{1/2}. \tag{5.100}$$

If the wavelength Λ of jet oscillation is measured, for instance, from a photograph of the jet, and if ρ, R, and V are known, the surface tension γ can be calculated from Equation 5.100. In practice, the oscillating jet is used to measure the variation of surface tension with time due to diffusion of surfactant from the bulk liquid to the surface and adsorption there, a subject considered further in Chapter 6. The procedure is simply to measure Λ at various distances from the orifice corresponding to different times available for surfactant diffusion and adsorption.

The above equation is based on a linear analysis that applies only for infinitesimal perturbations. Perturbations of finite amplitude are required, however, for a practical experiment. An early paper of Niels Bohr (1909) dealt with the necessary extension. He also included the effect of liquid viscosity. A corrected equation suitable for perturbations whose amplitude b is finite but still smaller than the mean radius R is

$$\gamma = \frac{2\rho V^2 \pi^2 R^3}{3\Lambda^2} \left[\frac{1 + \dfrac{37b^2}{24R^2}}{1 + \dfrac{5}{3}\dfrac{\pi R^2}{\Lambda^2}} \right]. \tag{5.101}$$

$$\left[1 + 2\left(\frac{\mu\Lambda}{\rho V \pi R^2}\right)^{3/2} + 3\left(\frac{\mu\Lambda}{\rho V \pi R^2}\right)^2 \right]$$

Other corrections that are sometimes required are discussed by Defay and Petre (1971).

EXAMPLE 5.3 SURFACE TENSION OF OSCILLATING JETS

A jet of an aqueous solution has a mean radius R of 0.2 cm and travels with a velocity V of 60 cm/sec. Density and viscosity may be taken as 1 g/cm³ and 1 cp, respectively. Calculate the surface tension using the simple and corrected Equations 5.100 and 5.101 if the wavelength Λ for an oscillation is 2.40 cm and if b/R is 0.25.

SOLUTION

Solving Equation 5.100 for γ and substituting the above values, we find $\gamma = 32.9$ mN/m. With Equation 5.101 the corresponding result is $\gamma = 34.8$ mN/m. In this particular case the viscosity correction is negligible, so that the chief correction is for the finite amplitude.

6. STABILITY AND WAVE MOTION OF THIN LIQUID FILMS: FOAMS AND WETTABILITY

Coalescence of drops or bubbles is a phenomenon of fundamental importance to the understanding of dispersed fluid systems such as foams and emulsions. As two drops or bubbles approach, a film forms between them. It drains, and if coalescence occurs, ultimately breaks. We are concerned here with this last step, which involves film instability.

Careful experiments by Sheludko (1967) on thin liquid films showed that instability often occurs when film thickness is a few tens of nanometers (i.e., a few hundred Angstrom units). For these very thin films, the effects of London–van der Waals forces and of electrical double-layer interaction, discussed in Chapter 3, must be included in the analysis of film stability. Each of them contributes a local "body force" that acts on each volume element of liquid within the film and influences flow much as does the familiar gravitational body force per unit volume ρg. As indicated in Chapter 3 (Equation 3.23), the electrical body force is $-\rho_e \nabla \psi$, where ρ_e is the local free charge density and ψ is the electrical potential. It is well known that the gravitational body force can be written as $-\nabla \varphi_g$, where φ_g is the usual gravitational potential energy. Similarly the electrical body force has the form $-\nabla \varphi_e$, provided that flow is sufficiently slow that diffusional equilibrium of the ions is maintained at all times (Miller and Scriven, 1970). The potential is given by

$$\varphi_e = -\varepsilon \varepsilon_o \kappa^2 \left(\frac{kT}{\nu e_o} \right)^2 (\cosh u - 1) . \qquad (5.102)$$

This expression is similar to those of Equations 3.31 and 3.32, as must be true since integrals of the pressure and body forces are equal in magnitude but opposite in sign, according to Equation 3.23. Finally, the van der Waals body force can be written as $-\nabla\varphi_{vw}$, where the van der Waals energy φ_{vw} can be calculated, for instance, by integration as in Chapter 3 (Section 1).

If we neglect gravity in a thin film of liquid A and assume that it is of uniform thickness h_o with no flow (see Figure 5.5), the equation of motion simplifies to

$$\frac{d\Phi}{dz} = 0 ,\qquad (5.103)$$

with

$$\Phi = p_A + \varphi_e + \varphi_{vw}.\qquad (5.104)$$

Thus the total potential Φ is uniform in such a film, but the initial pressure p_{Ai} varies.

If the film surfaces are given a small wavy perturbation, the equations of motion now become Equations 5.3 and 5.4, with p replaced by Φ. Clearly Φ can be eliminated from these equations in the same manner as was p in Equation 5.5. Hence the basic differential equations (Equations 5.9 and 5.10) for the velocity components, and naturally also the solutions to these equations, such as Equation 5.17, show no explicit dependence on van der Waals and electrical double-layer forces. As we shall see shortly, these forces do make their appearance in the boundary conditions. Note that the use of Equations 5.3 and 5.4 involves an assumption that the film is at rest before perturbation, i.e., flow due to film drainage (see Chapter 7) is neglected. This assumption amounts to requiring that the instability develop rapidly in comparison with changes in film thickness due to drainage.

One feature of the thin film analysis is that two interfaces exist, each of which can receive a general deformation such as that given by Equation 5.1. For each wavenumber α, let us consider a general perturbation in which the two interfaces are displaced to $\bar{z}_1 = B_{11}\, e^{i\alpha x}$ and $\bar{z}_2 = B_{12}\, e^{i\alpha x}$. Such a perturbation can be expressed as a linear combination of two special perturbations: a symmetric one about the film centerline for which $\bar{z}_{1s} = -\bar{z}_{2s} = B_s\, e^{i\alpha x}$, and an antisymmetric

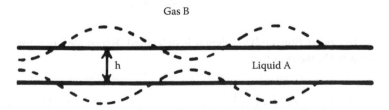

FIGURE 5.5 Wavy deformation of a thin liquid film.

one for which $\bar{z}_{1a} = \bar{z}_{2a} = B_a\,e^{i\alpha x}$. In particular, B_{11} can be written as the sum of B_a and B_s, and B_{12} as their difference. Thus it suffices to analyze symmetric and antisymmetric perturbations for all values of α.

Of primary interest here are symmetric perturbations, because these lead to variations in film thickness and eventual film breakup if they are unstable. Since the resulting velocity distribution must be antisymmetric, we obtain instead of Equation 5.17

$$W_A(z) = D_1 \sinh \alpha{\cdot}z + D_2 \sinh \alpha r_A z. \qquad (5.105)$$

The tangential velocity and the perturbation Φ_p in the total potential Φ defined in Equation 5.103 can be found from Equations 5.20 and 5.26, respectively, using this expression for W_A. If we assume that the film surfaces contain sufficient surfactant to prevent lateral motion, a common situation in practice, we have the inextensible case considered previously. Hence the tangential component of the momentum balance becomes a requirement that v_x vanish. A derivation similar to that of Section 2.2 leads to the following counterpart of Equation 5.42:

$$\beta\,B_s = D_1 \sinh \frac{\alpha h_o}{2} + D_2 \sinh \frac{\alpha r_A h_o}{2} . \qquad (5.106)$$

The key to the analysis is the normal component of the momentum balance. Because the liquid film is surrounded by gas, the terms for fluid B in Equation 5.37 may be neglected. But another term must be added, as in Equation 3.29, to account for electrical stresses acting on the interface (Miller and Scriven, 1970; Sanfeld, 1968). We thus obtain the following form for the boundary condition:

$$p_A - \varepsilon \frac{\varepsilon_o}{2}\left(\frac{\partial\psi}{\partial z}\right)^2 - 2\mu_A \frac{\partial v_{Az}}{\partial z} + 2H\gamma = 0 . \qquad (5.107)$$

It remains to evaluate the first two terms of this equation, or more precisely the perturbation,

$$\Delta\left(p_A - \frac{\varepsilon\varepsilon_o}{2}\left(\frac{\partial\psi}{\partial z}\right)^2 \right)$$

which occurs at the film surface when it is deformed. Expressing in terms of the total potential Φ, we find that

$$\Delta\left(p_A - \frac{\varepsilon\varepsilon_o}{2}\left(\frac{\partial\psi}{\partial z}\right)^2 \right) = \Delta\Phi - \Delta\varphi_{vw} - \Delta\left(\varphi_e + \frac{\varepsilon\varepsilon_o}{2}\left(\frac{\partial\psi}{\partial z}\right)^2 \right) . \qquad (5.108)$$

The quantity $\Delta\Phi$ in Equation 5.108 can be evaluated by making use of Equation 5.26 for the initial and perturbation contributions to Φ:

$$\Delta\Phi = \frac{d\Phi_i}{dz}\Big|_{\frac{h_o}{2}} B_s + \Phi_p\left(\frac{h_o}{2}\right)e^{i\alpha x}e^{\beta t}$$

$$= \left(D_1\,\mu\alpha\cosh\frac{\alpha h_o}{2} + D_2\,\mu\alpha r^2\cosh\frac{\alpha r h_o}{2}\right)e^{i\alpha x}e^{\beta t}$$

(5.109)

Also, $\Delta\varphi_{vw}$ can be readily evaluated if we limit consideration to perturbations for which $(\alpha\,h_o) \ll 1$ (i.e., those having wavelengths much greater than initial film thickness h_o). Then the potential energy at the surface of a film with local thickness h relative to that at the interface of a bulk fluid can be found by integration:

$$\varphi_{vwo} = \int_0^\infty \int_h^\infty \frac{2\pi n^2\beta r\,dr\,dz}{(r^2 + z^2)^3}.$$

(5.110)

The integration can be carried out using the procedure described following Equation 3.2. In this manner we obtain

$$-\Delta\varphi_{vw} = \frac{d}{dh}\left(\frac{-A_H}{6\pi h^3}\right)2\,B_s\,e^{i\alpha x}e^{\beta t}.$$

(5.111)

The quantity in parentheses in this equation has the dimensions of pressure and has been termed the van der Waals contribution Π_{vw} to the film "disjoining pressure" by Derjaguin. We note that $-\Pi_{vw}$ is the derivative with respect to h of the film potential energy given by Equation 3.14. This definition of Π_{vw} is used throughout this chapter and Chapter 7. Widely used by workers on thin liquid films, it differs in sign from that given at Equation 2.26. Either definition gives correct results if used consistently.

Again, restricting consideration to the case $\alpha\,h_o \ll 1$, we can use Equation 5.102 to write the last term of Equation 5.108 in terms of an electrical contribution Π_{el} to the disjoining pressure:

$$-\Delta\left[\varphi_e + \frac{\varepsilon\varepsilon_o}{2}\left(\frac{\partial\psi}{\partial z}\right)^2\right] = \frac{d\Pi_{el}}{dh}2B_s\,e^{i\alpha x}\,e^{\beta t}$$

(5.112)

$$\Pi_{el} = \varepsilon\varepsilon_o\kappa^2\left[\left(\frac{kT}{\nu e_o}\right)^2(\cosh u_o - 1) - \frac{\varepsilon\varepsilon_o}{2}\left(\frac{\partial\psi}{\partial z}\right)^2\right].$$

(5.113)

Comparing with Equations 3.32 and 3.33, we see that Π_{el} is equal to the electrical force per unit area F_e which acts on the surface of a film of uniform thickness. Substituting Equations 5.108, 5.109, 5.111, and 5.112 into Equation 5.107 and evaluating the mean curvature H as in Section 2.2, we obtain

$$D_1 \left[\mu_A \alpha \cosh \frac{\alpha h_o}{2} + \frac{1}{\beta} \sinh \frac{\alpha h_o}{2} \left(2 \frac{d}{dh} (\Pi_{vw} + \Pi_{el}) - \gamma \alpha^2 \right) \right]$$

$$+ D_2 \left[\mu_A \alpha r_A^3 \cosh \frac{\alpha r_A h_o}{2} + \frac{1}{\beta} \sinh \frac{\alpha r_A h_o}{2} \left(2 \frac{d}{dh} (\Pi_{vw} + \Pi_{el}) \right. \right. \quad (5.114)$$

$$\left. \left. - \gamma \alpha^2 \right) \right] = 0$$

The condition for no tangential flow discussed above implies that

$$D_1 \cosh \frac{\alpha h_o}{2} + r_A D_2 \cosh \frac{\alpha r_A h_o}{2} = 0 \ . \quad (5.115)$$

The determinant of coefficients of these two equations yields, as before, a relation that can be used to calculate the time factor β. We have already assumed $\alpha h_o \ll 1$ in writing Equation 5.110. If we further assume $|\alpha r_A h_o| \ll 1$, we can expand the hyperbolic functions and obtain a result first given by Vrij et al. (1970):

$$\beta = \frac{-\left[\dfrac{\gamma \alpha^3}{\rho_A} - \dfrac{2\alpha}{\rho_A} \dfrac{d(\Pi_{vw} + \Pi_{el})}{dh} \right] (\alpha h_o)^3}{24 \nu_A \alpha^2} \ . \quad (5.116)$$

That is, the perturbation either grows or decays without oscillation. The form of Equation 5.116 is similar to that found above for a single interface in the high viscosity limit (e.g., Equation 5.60), yet no assumption of high viscosity has been made here. Basically the inextensible interfaces so inhibit flow that a high viscosity is not required to prevent oscillatory motion in a thin film.

For the hypothetical case of an inviscid liquid film with free surfaces, it is readily shown that for small values of αh_o, oscillatory motion is predicted by analysis with an angular frequency β_F^* given by

$$\beta_F^{*2} = (\alpha h_o) \left[\frac{\gamma \alpha^3}{\rho_A} - \frac{2\alpha}{\rho_A} \frac{d(\Pi_{vw} + \Pi_{el})}{dh} \right] . \quad (5.117)$$

Hence we can write Equation 5.116 in a form even more similar to Equation 5.60:

$$\beta = -\frac{\beta_F^{*2}(\alpha h_o)^2}{24 v_A \alpha^2} \,. \tag{5.118}$$

Since $(d\Pi_{vw}/dh) > 0$, according to Equation 5.111, for a liquid film in air where $A_H > 0$, we see from Equation 5.116 that van der Waals forces favor instability. With film thickness less than the effective range of intermolecular forces, the potential energy of a molecule in a thin region of the film is greater than that of a molecule in a thick region that is surrounded by more molecules with which it can interact. Hence the body force due to van der Waals interactions promotes flow from thin to thick regions of the film. This flow increases perturbation amplitude and leads eventually to film breakage. We note that the destabilizing effect of van der Waals forces is entirely consistent with Equation 3.14, which states that total film energy per unit area (as opposed to the energy per molecule of the above argument) decreases with decreasing film thickness.

As indicated above, the electrical disjoining pressure Π_{el} is equal to the electrical force F_e discussed in Chapter 3. Examining the various expressions for F_e given there, we conclude that dF_e/dh, and hence $d\Pi_{el}/dh$ are negative. With this information and Equation 5.116, we see that repulsion between the double layers of the two film surfaces has a stabilizing influence on the film, the expected result. Also stabilizing is the tension γ of the film surfaces, especially for short wavelengths. Thus perturbations having short wavelengths comparable to film thickness, which are not adequately described by the present analysis, are of little interest for film stability.

For situations where the film is unstable, Equation 5.116 can be solved for the critical wavenumber α_c for which β vanishes.

$$\alpha_c^2 = \frac{2}{\gamma} \frac{d}{dh} (\Pi_{vw} + \Pi_{el}) \tag{5.119}$$

Only wavelengths exceeding $2\pi/\alpha_c$ are unstable. Similarly the fastest growing disturbance can be found. It is characterized by

$$\alpha_M^2 = \frac{1}{\gamma} \frac{d}{dh} (\Pi_{vw} + \Pi_{el}) \tag{5.120}$$

$$\beta_M = \frac{(\alpha h_o)^3}{24 \mu_A \alpha} \frac{d}{dh} (\Pi_{vw} + \Pi_{el}) \,. \tag{5.121}$$

In terms of disjoining pressure, the film is always stable when the slope of the disjoining pressure curve, that is, $d(\Pi_{vw} + \Pi_{el})/dh$, is negative and unstable except at short wavelengths when the slope is positive. Figure 5.6 shows a typical disjoining pressure curve where both van der Waals attractive and electrical

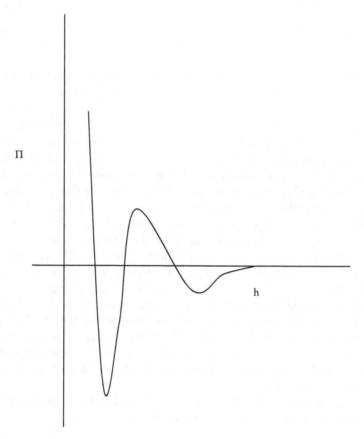

FIGURE 5.6 Typical disjoining pressure curve.

double-layer contributions are important. Such a film is stable at relatively large thicknesses and unstable at small thicknesses.

Of course, not all thin films are unstable. If the stabilizing effect of electrical forces is sufficiently large it can outweigh the destabilizing influence of van der Waals forces. In this case β_F^*, as calculated from Equation 5.117 is positive, and the film is stable with respect to disturbances of all wavelengths.

Also of significance is that initial instability of a thin film in accordance with the above mechanism does not inevitably lead to film rupture. The analysis, like all others in this chapter, is based on linear stability theory, and hence is valid only for small amplitude perturbations. It has been observed experimentally that at low surfactant concentrations instability of a film some tens of nanometers in thickness does produce rupture. But for many surfactants it is found that above a critical concentration, the instability leads to formation of "black" films which are only slightly thicker than the total length of two surfactant molecules (Sheludko, 1967). These black films can be very stable and are a major factor in foam stability.

When surfactant concentration is above the CMC, variation of disjoining pressure with film thickness can be more complex. In particular, stepwise drainage is sometimes observed. That is, after the film or some substantial portion of it reaches a nearly uniform thickness, a small region of lesser thickness nucleates and grows until none of the thicker film remains. The decrease in film thickness as a result of this process is approximately the diameter of a micelle in the surfactant solution from which the film was formed (Kralchevsky et al., 1990; Nikolov et al., 1989). As surfactant concentration increases, more such steps are observed. The explanation for this behavior is that micelles form layers within the film parallel to its surfaces, so that metastable states exist corresponding to situations with one layer, two layers, etc. The disjoining pressure curve in this case has one or more intermediate metastable states, as shown in Figure 5.7.

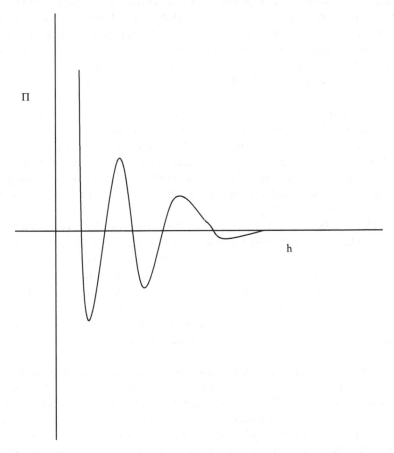

FIGURE 5.7 Disjoining pressure curve with oscillations due to the layering of micelles.

Nucleation of a thinner layer stems from a thermal fluctuation requiring an input of energy. Since the energy needed to form a nucleus of critical size which will grow spontaneously decreases as the number of micelle layers increases, films with one layer persist longer than films with more layers before nucleation takes place. Short-range oscillatory forces in pure liquids due to packing effects, called solvation forces, were also discussed in Chapter 3, at the end of Section 4.

The portions of the disjoining pressure curves that have negative slope and thus correspond to stable films can be measured experimentally. One technique involves formation of a film in a hole in a porous plate. At equilibrium, disjoining pressure is equal to the capillary pressure applied to the film, which can be measured. Film thickness is determined by interferometry. Further information may be found elsewhere (Bergeron and Radke, 1992; Exerowa and Kruglyakov, 1998).

An important application of thin film behavior is foams. Consumers prefer detergents and shampoos that produce some foam during use (but not too much). Formulators must design their products accordingly. Foams are also used in firefighting, mineral flotation, food products, and "foam fractionation" processes which can separate materials having different degrees of surface activity. And, of course, as with emulsions, foams often develop where they are unwanted and hinder process performance. In this case they must be broken or steps taken to prevent their initial formation (Garrett, 1993).

Many foams of practical interest contain large proportions of the gas phase (more than 90% by volume). The gas cells are polyhedral since such high volume fractions are impossible in dispersions of spheres having moderate drop size distributions. The cells are separated by thin liquid films of uniform thickness. The junctions of the films are called Plateau borders. A balance of forces requires that three films meet at a plateau border and that the angle between adjacent films be 120°, as illustrated in Figure 5.8a.

In view of the interfacial curvature in the Plateau border region (see Figure 5.8a), it is clear from the Young-Laplace equation (Equation 1.22) that the pressure there is less than that in the film (i.e., a driving force exists for film drainage). Thus, when a foam forms above a bulk liquid phase, liquid drains from the individual films into the plateau borders and from the latter into the bulk liquid. Note that since hydrostatics requires liquid pressure to decrease with increasing height, the radius of curvature of the plateau borders must decrease as well. As a result, the liquid holdup in the plateau borders is less at greater elevations above the liquid phase.

An interesting feature of Plateau borders is their nonzero contact angles with the individual films. Owing to the effects of disjoining pressure, the effective tension γ_F of a thin film differs from the sum 2γ of the tensions of its two interfaces with the surrounding gas. A balance of forces at the junction between the Plateau border and thin film thus requires the existence of a finite contact angle λ (see Figure 5.8b). Indeed, γ_F exceeds by 2γ the excess energy E_T of the film per unit area, which is given by Equation 3.58. Hence, in view of the geometry of Figure 5.8b, we have

$$\gamma_F = 2\gamma + E_T = 2\gamma \cos \lambda. \qquad (5.122)$$

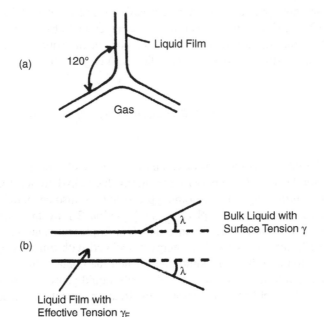

FIGURE 5.8 (a) Plateau border in a foam. (b) Detailed view showing the finite contact angle between the thin film and bulk liquid in the plateau border region.

Values of such contact angles are typically small, ranging from somewhat less than one degree to a few degrees (Huisman and Mysels, 1969; Princen and Frankel, 1971). Because $\Pi = -(dE_T/dh)$, simultaneous measurement of the contact angle and film thickness provides a method for determining disjoining pressure as a function of thickness and hence of obtaining information on the magnitude of intermolecular forces, e.g., values of the Hamaker constant A_H.

It should be noted that while the above discussion of thin film stability and disjoining pressure has emphasized foam films, much of it is readily adapted to emulsion films and to thin liquid films on solid surfaces. Of course, the appropriate expression for the Hamaker constant must be used (see Equation 3.14). Instability of a thin liquid film on a solid surface (Ruckenstein and Jain, 1974; Sharma and Jameel, 1993) is important in various situations, including flotation, film boiling, and condensation, and continuity of the tear film on the cornea of the eye. If the bulk fluid phase bounding the film is a gas or vapor, instability can lead to the development of "dry spots" on the solid, which is usually undesirable. Such an instability can arise when the Hamaker constant is positive (i.e., the film is nonwetting) (see below). In oil reservoirs the bulk phase may be crude oil and the thin film an aqueous salt solution. If the film is unstable, crude oil can contact the pore surfaces. Typically crude oils contain compounds such as asphaltenes which can then adsorb on the solid and make it more hydrophobic or oil wet. The change in wettability alters the locations of oil and water during subsequent flow and can significantly affect oil recovery. Of special interest is the relationship between disjoining pressure and wetting behavior [see the detailed discussion by Hirasaki (1991)]. Consider, for example, the contact angle between a liquid and a solid where the liquid forms a stable thin film on the solid. In this case the generalization of Young's equation (Equation 2.1) is given by

$$(\gamma_{SL} + \gamma_{LV} + \int_{h_{eq}}^{\infty} \Pi \, dh) + \Pi_{eq} \, h_{eq} = \gamma_{SL} + \gamma_{LV} \, \cos \lambda . \qquad (5.123)$$

Note that the integral is the sum of E_e and E_{vw} if only electrical and dispersion forces are significant at h_{eq} (otherwise the energy due to hydration forces or other interactions contributing to the disjoining pressure must also be included). In any case it seems reasonable to replace γ_{sv} in Equation 2.1 by the expression in parentheses in Equation 5.123 when a thin film is present. The term $\Pi_{eq} h_{eq}$ in Equation 5.123 represents the free energy associated with transferring material between the film and the bulk liquid phase when the contact line moves. Since $\Pi_{eq} = 0$ for a thin film in equilibrium with a bulk liquid phase on a solid surface, Equation 5.123 simplifies to Equation 2.26. Rearrangement of Equation 5.123 also yields

$$\cos \lambda = 1 + \left(\int_{0}^{\Pi_{eq}} h \, d\Pi \right) / \gamma_{LV} . \qquad (5.124)$$

Inspection of Figure 5.6 shows that a finite contact angle can exist if the electrical contribution to the disjoining pressure is not too large. The disjoining pressure curve of Figure 5.6 is for the case of a positive Hamaker constant. If the Hamaker constant is negative, which is possible if the materials in regions A, B, and C of Figure 3.1 are all different, Π is positive for all values of film thickness h, and no contact angle exists (i.e., the liquid spreads on the solid). As Equation 3.14 indicates, a negative Hamaker constant can occur when a strong attractive interaction exists between the solid and the liquid (corresponding to regions A and C in Figure 3.1). As indicated in Chapter 2, this is precisely when spontaneous spreading is expected. For a complete description of a foam film it is also necessary to use an overall force balance equation:

$$\Pi_{eq} = p_c, \qquad (5.125)$$

where p_c is the capillary pressure calculated for the curved region in the plateau border using Laplace pressure. It is also seen that if the contact angle formed by Equation 5.124 is too large, the curvatures in the vicinity of the contact lines become large and the system becomes unstable to infinitesimal perturbations (Zhang and Neogi, 2002). Thus only small contact angles can be sustained where the contact line ends in a film of constant thickness, or else only a film of constant thickness h_{eq} will result at equilibrium. That large curvatures can give rise to a destabilizing effect is discussed in the next section.

Finally, returning briefly to foams, we note that the number of cells in a polyhedral foam can decrease by a mechanism other than film rupture by the mechanisms discussed above. As the pressure is largest for gas in the smallest cells, a driving force exists for transfer of gas from small to large cells by diffusion across the thin liquid films separating them. Ultimately this process leads to the disappearance of small cells and growth of large cells. Lemlich (1978) and Ranadive and Lemlich (1979) present a model of gas diffusion in foams.

Drainage and instability of individual films are very important in foam behavior, but other factors must be considered as well. For example, the lateral extent of individual films may increase during deformation of a foam, the additional liquid being drawn from the surrounding plateau borders (Lucassen, 1981; Mysels et al., 1959). The new portions of the film are normally thicker than the old. By such behavior a foam may remain stable in spite of mechanical perturbations.

EXAMPLE 5.4 STABILITY OF A THIN LIQUID FILM

Consider a thin liquid film in air for which the Hamaker constant A_H is 5×10^{-13} erg and electrical double-layer effects are negligible. If viscosity is 1 cp, density is 1 g/cm³, film thickness is 50 nm, and surface tension is 40 mN/m, find the wavenumbers α_c and α_M of the critical and fastest growing disturbances and the time factor β_M for growth of the latter using the inextensible interface results given above. Compare with the corresponding results for a free interface in the inviscid approximation.

SOLUTION

The disjoining pressure is $-A_H/6\pi h^3$, according to Equation 5.111. Hence, from Equations 5.119 through 5.121, we have, for inextensible film surfaces,

$$\alpha_c = \left(\frac{A_H}{\pi \gamma h_o^4} \right)^{1/2} = 2.52 \times 10^3 \ cm^{-1}; \ \lambda_c = (2\pi / \alpha_c) = 2.49 \times 10^{-3} \ cm$$

$$\alpha_M = (\alpha_c/2^{1/2}) = 1.78 \times 10^3/cm^{-1}, \ \lambda_M = 3.53 \times 10^{-3} \ cm.$$

$$\beta_M = \frac{\alpha_M^2 A_H}{48\pi\mu_A h_o} = 0.422 \ sec^{-1}; \ \text{time constant} \ \beta_M^{-1} = 2.37 \ sec ,$$

The inviscid result is simply

$$\beta^2 = -\beta_F^{*2}$$

where β_F^{*2} is given by Equation 5.117. It is clear from inspection of this equation and Equation 5.117 that α_c is the same as before, the expected result since α_c is determined by thermodynamics alone. Upon differentiating Equation 5.117, we find that α_M is again given by Equation 5.120, so that it also has the same value as above. But β_M now has a value of 4.50×10^4/sec, so that the time constant for growth is 2.22×10^{-5} sec. Hence a surfactant free film breaks very rapidly because of the rapid development of the instability. As a result, stable foams cannot be produced in the absence of surfactants.

7. ENERGY AND FORCE METHODS FOR THERMODYNAMIC STABILITY OF INTERFACES

In the examples of interfacial stability considered thus far, the systems have been at rest in their initial states. Hence the predictions of when instability can be expected are, in fact, conditions for thermodynamic stability. We have chosen not to emphasize this point and to carry out the analyses in terms of perturbations of the general equations of change because we obtain in this way not only the stability condition but also the rates of growth of unstable perturbations and the appropriate frequencies of oscillation and damping factors for stable perturbations. Also, the basic method of analysis used above is applicable to systems not initially in equilibrium, as we shall see later in this chapter and in Chapter 6.

Nevertheless, there may be situations where information on growth rates and the characteristics of the fastest growing disturbance is not of prime importance. In this case the energy and force methods, when they are applicable, can be used to determine the condition for thermodynamic stability with less effort than required for the more general analysis presented above (Miller and Scriven, 1970).

The energy method can be used whenever an energy function such as the free energy exists that is known to have a local minimum at any state of stable equilibrium. Basically the change in this function is calculated for all possible perturbations from the equilibrium state. If all such changes are positive, the initial state is stable. But if any possible perturbation produces a decrease in the energy function, instability can be expected.

The force method is often the easiest method of stability analysis to use because it requires only that the local normal force acting on the interface after deformation be calculated. If, for all possible deformations, this force acts to return the interface to its initial configuration, the system is stable. But if the force for any possible deformation is in a direction to increase deformation amplitude, the interface is unstable. The force method applies to fewer situations than the other methods, however. It requires not only that an energy function exist, but also that body forces be irrotational, that fluids be incompressible, and that normal displacement at the system boundaries vanish except for the interface being analyzed (Miller and Scriven, 1970).

Tyuptsov (1966) and Huh (1969) independently derived a general thermodynamic stability condition for an interface between fluids A and B using the energy method. If the fluids are of infinite extent or are bounded by rigid solid surfaces except at their interface, if gravity is the only external force acting on the system, and if any contact lines at the periphery of the interface are either fixed or move with no change in contact angle from the initial equilibrium values, the stability condition simplifies to

$$\int_S \{\gamma[-\nabla_s^2 \, \eta + (2K - 4H^2) \, \eta] - \eta \, \boldsymbol{n}.\nabla(p_A - p_B)\} \frac{\eta}{2} dS > 0 \, . \quad (5.126)$$

Here η is the local displacement of the interface in the normal direction (toward fluid B) and K is the local Gaussian curvature (product of the principal curvatures $1/r_1$ and $1/r_2$). Hydrostatics require

$$-\nabla p_A + \rho_A g = 0$$

$$-\nabla p_B + \rho_B g = 0,$$

which lead to

$$\int_S \{\gamma[-\nabla_s^2 \, \eta + (2K - 4H^2) \, \eta] - (\rho_A - \rho_B)g_n\eta\} \frac{\eta}{2} dS > 0 \, , \quad (5.127)$$

where g_n is the local normal component of the gravitational acceleration and the integration is over the entire fluid interface.

Equation 5.127 simply states that stability requires the system's free energy to increase for all possible perturbations. The last term of the integrand clearly represents the local change in gravitational potential energy due to interfacial deformation. The first term, however, is only part of the local change in interfacial free energy, that is, the part due to changes in curvature during the deformation. For were the interface deformed at constant curvature, the change in interfacial free energy would, in view of the Young-Laplace equation (Equation 1.22), exactly balance the work done at the interface by the pressures in the two fluids.

According to Equation 5.127, interfacial curvature changes for two reasons. The first, represented by the term ($-\nabla_s^2\eta$) is the nonuniformity (e.g., waviness) of the perturbation itself. The second, represented by the term $(2K - 4H^2) \, \eta$, is displacement of an initially curved surface normal to itself. For instance, uniform outward displacement of a spherical surface produces an increase in the radii of curvature and hence a decrease in the magnitude in mean curvature H. Whereas the first term tends to stabilize the system, the second term tends to destabilize it. However, in the case of a spherical liquid drop, the critical wavelength turns out to be so large that disturbances of such large wavelengths cannot be accommodated and drops are stable to infinitesimal perturbations. This constraint does not arise in liquid cylinders, which are seen to be unstable. Thus the fact that curved surfaces tend to be more unstable should be kept in mind.

Equation 5.127 can be applied to some of the situations we have already analyzed, such as gravitationally produced instability of superposed fluids and capillary instability of a fluid cylinder. It has also been applied to more complex situations, such as stability of a pendant drop (Huh, 1969; Pitts, 1974). We remark that the expression in braces in Equation 5.127 is the local force acting in the normal direction on the deformed interface to restore it to its initial configuration. Hence application of the force method amounts to a requirement that this expression be positive for stability.

Energy methods can also be applied to systems not initially at equilibrium. A Liapunov function must be found which (a) vanishes for the initial state, (b) is positive for all other states, and (c) decreases in value as perturbation amplitude decreases. The initial state is stable if the function's value decreases continuously during system response to all possible perturbations. As with the thermodynamic energy method, the Liapunov method is generally easier to use than perturbation of the governing equations, especially for perturbations of large amplitude. However, a Liapunov function must first be found. Further information on this approach may be found in Denn (1975), Dussan (1975), and Joseph (1976).

EXAMPLE 5.5 ENERGY METHOD FOR STABILITY OF SUPERPOSED FLUIDS

Use Equation 5.127 to determine the stability condition for superposed fluids.

SOLUTION

Suppose that the interface in Figure 5.1 is displaced from the plane $z = 0$ to the position $\bar{z} = B \cos \alpha x$. In this case, $\eta = \bar{z}$, $H = K = 0$ and Equation 5.127 becomes

$$\int_S \left[\gamma \alpha^2 \frac{\bar{z}^2}{2} + (\rho_A - \rho_B) g \frac{\bar{z}^2}{2} \right] dS > 0 .$$

Hence, at marginal stability,

$$\alpha_c^2 = \frac{(\rho_B - \rho_A) g}{\gamma} .$$

This result agrees with that obtained by the general analysis above (see remark following Equation 5.61).

In systems where A is a thin liquid film in gas, the effect of gravity in Equation 5.126 is replaced by that of disjoining pressure. As Equation 5.104 indicates for an equilibrium film having a uniform value of Φ, taken as zero for convenience, ρ_a is given by $-(\phi_{vw} + \phi_e)$, which, in turn, is $\nabla(\Pi_{vw} + \Pi_e)$. Recognizing that interfacial curvature is determined by local deformation η of a single film surface but that disjoining pressure depends on total change in film thickness, one obtains

$$\int_S \left\{ \gamma[-\nabla_s^2\eta + (2K - 4H^2)\eta] - 2\eta\frac{d\Pi}{dh}\right\} \frac{\eta}{2}\,dS > 0 \ , \qquad (5.128)$$

Manipulations similar to those above for a single interface yield the same expression for α_c as in Equation 5.119. For a thin liquid film on a solid surface, only the fluid interface is deformed and the factor of 2 disappears from the term multiplying $(d\Pi/dh)$ in Equation 5.128.

As discussed previously, petroleum oil reservoirs contain, besides oil, other fluids, and as the pores in the reservoirs are very narrow, thin films of brine or water are formed next to the solid walls. The wettability of the rock changes depending on whether such films are stable or unstable. A relation exists between the capillary pressure and wettability in reservoirs, which has been explored with a simplified version of Equation 5.128 by Basu and Sharma (1996).

8. INTERFACIAL STABILITY FOR FLUIDS IN MOTION: KELVIN-HELMHOLTZ INSTABILITY

We now turn to interfaces in systems not initially at rest. From the manifold possible situations of this type we choose two for detailed study. One is the so-called Kelvin-Helmholtz instability at the interface between two fluids initially moving in a direction parallel to the interface, but at different velocities. The other is wave motion on a falling liquid film, a situation of great practical interest.

We begin with an analysis of the stability of the interface between two inviscid, incompressible fluids which initially have uniform velocities V_A and V_B, respectively, in the x direction (see Figure 5.9). Kelvin's interest in this problem

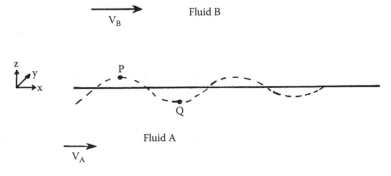

FIGURE 5.9 Wavy perturbation of the interface between fluids moving with different velocities in a direction parallel to the interface.

was as a possible source of wind-generated wave motion on the sea. But the instability also occurs in various cases of layered or annular flow of fluids, for instance, with high-speed liquid jets. In the atmosphere it causes cloud surfaces to be irregular. Large vertical gradients in wind velocity may even be the source of much "clear air turbulence" (Dutton and Panofsky, 1970).

After initial deformation of the interface from its initial planar state to the position $\bar{z} = Be^{i\alpha x}$, the equation of motion, to the first order in perturbation terms, takes the form

$$\rho\left(\frac{\partial \mathbf{v}}{\partial t} + V \frac{\partial \mathbf{v}}{\partial x}\right) = -\nabla p + \rho \mathbf{g} . \qquad (5.129)$$

Taking the divergence of this equation, we obtain

$$\nabla^2 p = 0. \qquad (5.130)$$

If we take the perturbation in pressure to have the form $P(z)e^{(i\alpha x + \beta t)}$, we can easily solve Equation 5.130 using procedures similar to those of Section 2.1. The solutions that vanish far from the interface are

$$P_A(z) = A_1 e^{\alpha z} \qquad (5.131)$$

$$P_B(z) = A_2 e^{-\alpha z}. \qquad (5.132)$$

Now, with the z component v_z of velocity given by $W(z)e^{(i\alpha x + \beta t)}$, we find from the z component of Equation 5.129 that

$$W(z) = -\frac{1}{\rho(\beta + i\alpha V)} \frac{dP}{dz} . \qquad (5.133)$$

As before, the perturbation in the tangential velocity v_x can be found from Equation 5.20.

Now the rate of interfacial motion is equal to the normal component of velocity at the interface. The latter includes not only v_z at the interface as before, but also the component normal to the deformed interface of the initial velocity V. This condition requires that

$$\beta B = W_A(0) - i\alpha V_A B \qquad (5.134)$$

$$\beta B = W_B(0) - i\alpha V_B B. \qquad (5.135)$$

These equations can be solved for A_1 and A_2 in terms of B:

$$A_1 = -\frac{\rho_A}{\alpha}(\beta + i\alpha V_A)^2 B \qquad (5.136)$$

$$A_2 = \frac{\rho_B}{\alpha}(\beta + i\alpha V_B)^2 B . \qquad (5.137)$$

As before, the tangential component of the interfacial momentum balance cannot be imposed for inviscid fluids, nor can continuity of tangential velocity at the interface. But the normal component of the momentum balance is, as usual, a key equation. For the present situation, Equation 5.37 simplifies to

$$A_1 - A_2 = (\gamma\alpha^2 + (\rho_A - \rho_B)g)B. \qquad (5.138)$$

Substituting Equations 5.136 and 5.137 into this expression and solving for β we obtain

$$\beta = \frac{-i\alpha(\rho_A V_A + \rho_B V_B)}{\rho_A + \rho_B}$$

$$\pm i\left[\frac{\gamma\alpha^3 + (\rho_A - \rho_B)\,g\alpha}{\rho_A + \rho_B} - \frac{\alpha^2\rho_A\rho_B(V_A - V_B)^2}{(\rho_A + \rho_B)^2}\right]^{1/2} . \qquad (5.139)$$

According to Equation 5.139, instability occurs whenever the expression inside the brackets is negative. That β always has an imaginary part implies that wave motion exists, whether perturbation amplitude grows or not. The wave travels in the x direction, its speed being $(\rho_A V_A + \rho_B V_B)/(\rho_A + \rho_B)$, and thus intermediate between V_A and V_B for unstable disturbances. The terms involving gravity and interfacial tension indicate that these effects act to stabilize the interface just as they do in the absence of tangential flow. As the destabilizing term inside the brackets is proportional to $(V_A - V_B)^2$, occurrence of the instability does not depend on which fluid has the greater velocity. When the two tangential velocities are equal, Equation 5.139 reduces to the result Equation 5.61 found previously for superposed inviscid fluids.

One way of viewing the basic instability mechanism is to recognize that for the situation illustrated in Figure 5.9 where $V_B > V_A$, tangential velocity increases for the portion of fluid A displaced into the region near point P which was initially occupied by B. Similarly tangential velocity decreases for the portion of fluid B displaced into the region near point Q. Because kinetic energy is proportional to the square of the velocity, more kinetic energy is extracted from the flow during the latter process than is added during the former. Thus energy is made available to overcome gravity and interfacial tension and deform the interface.

The Kelvin-Helmholtz instability has been confirmed experimentally by Francis (1954) for a situation where air was blown over a viscous oil. At a critical air speed, small ripples appeared and grew rapidly. The critical speed was in good agreement with that predicted by Equation 5.139.

EXAMPLE 5.6 KELVIN-HELMHOLTZ INSTABILITY FOR AIR-WATER SYSTEM

Calculate the critical velocity and wavenumber for air at 1 atm and 25°C blowing over water.

SOLUTION

We have $\gamma = 72$ mN/m, $\rho_A = 1$ g/cm³, $\rho_B = pM/RT$ by the ideal gas law equals 0.00119 g/cm³. At marginal stability, the expression in brackets in Equation 5.139 vanishes. Inspection of Equation 5.139 shows that for a given value of $|V_A - V_B|$, this condition occurs whenever the stabilizing quantity $\gamma \, \alpha + (\rho_A - \rho_B)g/\alpha$ is minimized. Thus at marginal stability we have

$$\alpha^2 = \alpha_c^2 = \frac{(\rho_A - \rho_B)g}{\gamma} .$$

Hence $\alpha_c = 3.69/\text{cm}^{-1}$ ($\lambda_c = 1.70$ cm) when the instability first develops. The critical value of the velocity difference is given by

$$|V_A - V_B| = \left[\frac{(\nu\alpha_c + (\rho_A - \rho_B)g/\alpha_c)(\rho_A + \rho_B)}{\rho_A \rho_B} \right]^{1/2} = 669 \text{ cm / sec or about 15 mph.}$$

The speed c of the traveling wave can be found from the first term of Equation 5.139:

$$c = -\frac{\beta}{i\alpha} = \frac{\rho_A V_A + \rho_B V_B}{\rho_A + \rho_B} = 0.794 \ cm/sec .$$

Thus air blowing over a lake or the sea should be able to generate waves when wind velocity exceeds about 15 mph (about 13 nautical mph).

EXAMPLE 5.7 PEAK HEAT FLUX

In pool boiling, a saturated liquid at temperature T_s lies on a horizontal solid surface at temperature T_w. Boiling starts and gives rise to large heat fluxes q. A schematic plot of q versus $\Delta T = T_w - T_s$ is shown in Figure 5.10. To the left of the peak is the nucleate boiling regime and to the right is the film boiling regime. In the nucleate boiling regime, bubbles grow on the surface and detach. As ΔT

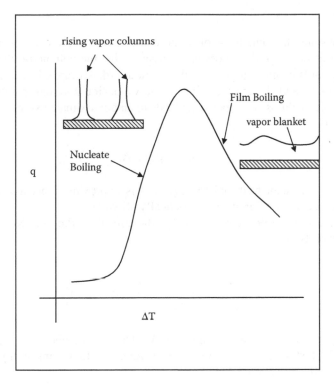

FIGURE 5.10 Schematic view of heat flux as a function of ΔT. On the left of the maximum (the peak heat flux) is the nucleate boiling regime where cylinders of vapor form as the peak is approached. On the right of the peak is the region of film boiling.

is increased, the bubbling becomes vigorous and it appears that columns of vapor form that keep breaking off because of an instability. In the film boiling regime, a blanket of vapor covers the solid surface. Since vapors have poorer heat transfer properties than liquids, the heat flux decreases.

Zuber (1958) postulated that as the peak heat flux (or the critical heat flux) is reached, the columns become unstable at the peak when it is approached from the left. When liquid evaporates at the solid surface, the vapor rises under gravity and the liquid falls to replenish the rising vapor. Since the vapor and the liquid move tangentially past the interface, it gives rise to Kelvin-Helmholtz instability. On the other hand, in the film boiling regime, the vapor has a lower density but is covered by the higher density liquid. This gives rise to Rayleigh-Taylor instability.

Assume that the peak is characterized by the critical condition for Kelvin-Helmholtz instability and for this critical wavelength the fastest growing wavelength from Rayleigh-Taylor instability can be used. Show that it can be used to obtain a correlation for the peak heat flux. Also assume where needed that $\rho_L \gg \rho_v$.

SOLUTION

At steady state, the same mass of liquid comes down as the mass of vapor that rises. But as $\rho_L \gg \rho_V$, it is possible to assume that the volumetric flow rate of vapor (upward) is much larger than that of liquid (downward). Hence the velocities $V_V \gg V_L$ and the cross section of the vapor-liquid system parallel to the solid surface is almost all vapor. This leads to an approximate expression for the heat flux as

$$q \simeq \rho_V V_V \Lambda,$$

where Λ is the latent heat and the equation is approximate because the cross section has been considered to be practically all vapor.

For Equation 5.139 and Example 5.6, the critical condition in Kelvin-Helmholtz instability is

$$V_V \simeq \left[\frac{\alpha_c \gamma + g(\rho_L - \rho_V)/\alpha_c}{\rho_V \rho_L} (\rho_L + \rho_V) \right]^{1/2},$$

where V_L has been neglected compared to V_V. The wavenumber α_c is replaced by that for the fastest growing disturbance in Rayleigh-Taylor instability (see the comments after Equation 5.61):

$$\alpha_M = \left[\frac{g(\rho_L - \rho_V)}{3\gamma} \right]^{1/2}.$$

Combining the three equations one has

$$q_{max} \simeq 1.52 \rho_V \Lambda \left(\frac{\gamma(\rho_L - \rho_V)g}{\rho_V^2} \right)^{1/4},$$

where the approximation $\rho_L \gg \rho_V$ has been made. Zuber, in his detailed analysis, obtained a numerical factor of 0.18 as well as a correction factor when ρ_V is not negligible (say in the region close to critical). Zuber's equation was arrived at empirically by Kutateladze (1951), who also provided the experimental verification for the correlation.

9.   WAVES ON A FALLING LIQUID FILM

It has been found that when a film of liquid flows down an inclined surface, as in Figure 5.11, waves often develop along the fluid interface. The enhanced

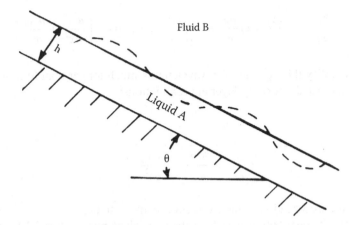

FIGURE 5.11 Flow of a liquid film down an inclined plane.

convection near the interface can lead to appreciable increases in heat and mass transfer rates between the liquid and gas phases. We note that the films of interest here are much thicker than those considered in Section 6, so that effects of disjoining pressure are negligible.

The velocity distribution in a film of uniform thickness h, where the flow is laminar, is well known (Bird et al., 2002):

$$v_x = \frac{\rho \, g \, \sin \theta}{2\mu}(2zh - z^2) \, . \tag{5.140}$$

It can be written in terms of the average velocity $\bar{v}_x$ in the film as

$$v_x = \frac{3\bar{v}_x}{2}\left(2\,\frac{z}{h} - \frac{z^2}{h^2}\right) \, . \tag{5.141}$$

We seek information about characteristics of the traveling waves that develop when the interface is deformed. Because of the initial velocity, a direct stability analysis based on the differential equations of change is more complex mathematically than the analyses presented above, although it has been carried out (Benjamin, 1957; Krantz and Goren, 1970; Yih, 1963). We present here a simpler but slightly less accurate analysis based on the integral momentum equation. It follows for the most part the procedure of Kapitsa (1948) which was described by Levich (1962).

The x component of the equation of motion for the falling film of Figure 5.11 has the general form

$$\frac{\partial v_x}{\partial t} + v_x \frac{\partial v_x}{\partial x} + v_z \frac{\partial v_x}{\partial z} = -\frac{1}{\rho}\frac{\partial p}{\partial x} + g \sin\theta + \nu\left(\frac{\partial^2 v_x}{\partial x^2} + \frac{\partial^2 v_x}{\partial z^2}\right). \quad (5.142)$$

We can simplify this equation for wavelengths much longer than film thickness by using the usual boundary layer approximations:

(a)

$$\frac{\partial^2 v_x}{\partial x^2} \ll \frac{\partial^2 v_x}{\partial z^2},$$

(b) pressure variation in the z direction is hydrostatic,
(c) velocity distribution at each position x is given by Equation 5.141 using local values of $\bar{v}_x$ and h. From condition (b) and the Young-Laplace equation at the fluid interface we obtain

$$p = \rho g(h-z)\cos\theta - \gamma\frac{\partial^2 h}{\partial x^2}. \quad (5.143)$$

With this equation and condition (a), Equation 5.142 becomes

$$\frac{\partial v_x}{\partial t} + v_x \frac{\partial v_x}{\partial x} + v_z \frac{\partial v_x}{\partial z} = \frac{\gamma}{\rho}\frac{\partial^3 h}{\partial x^3} - g\cos\theta\frac{\partial h}{\partial x} + g\sin\theta + \nu\frac{\partial^2 v_x}{\partial z^2}. \quad (5.144)$$

We now integrate Equation 5.144 over the entire film thickness for a given value of x using condition (c). Integrating by parts and invoking continuity, we write for the integral of the third term

$$\int_0^h v_z \frac{\partial v_x}{\partial z}\,dz = v_x v_z\Big|_0^h - \int_0^h v_x \frac{\partial v_z}{\partial z}\,dz$$

$$= -v_x(h)\int_0^h \frac{\partial v_x}{\partial x}\,dz + \int_0^h v_x \frac{\partial v_x}{\partial x}\,dz \qquad (5.145)$$

With this expression and with Equation 5.140 for v_x and its derivatives, the integral momentum equation becomes, to first order in perturbation quantities,

$$\frac{3}{2}\frac{v_o}{h_o}\frac{\partial h}{\partial t} + \frac{3}{2}\frac{v_o^2}{h_o}\frac{\partial h}{\partial x} = \frac{\gamma}{\rho}\frac{\partial^3 h}{\partial x^3} - g\cos\theta\frac{\partial h}{\partial x}, \quad (5.146)$$

where v_o and h_o are the average velocity and thickness of the initial unperturbed film of uniform thickness. An overall mass balance at each position x is also required. Noting from Equations 5.140 and 5.141 that $\bar{v}_x$ is proportional to h^2, and again keeping only first-order terms, we find

$$\frac{\partial h}{\partial t} = -\frac{\partial}{\partial x}(\bar{v}_x h) = -3\, v_o\, \frac{\partial h}{\partial x}. \tag{5.147}$$

Let us now suppose that the waves travel down the film with a velocity c and with no increase or decrease in amplitude. That is, we focus on the condition of marginal stability at the boundary between stable and unstable regions. Then the thickness h should have the form

$$h = h_o(1 + h_1 e^{i\alpha(x-ct)}), \tag{5.148}$$

where α and c are both real. Substitution of this expression into Equation 5.147 yields

$$c = 3v_o. \tag{5.149}$$

The speed of the traveling waves is thus three times the average velocity or twice the surface velocity for a uniform film.

Similarly substitution of Equations 5.148 and 5.149 into Equation 5.146 provides an equation for the wavenumber α:

$$\alpha^2 = \frac{\rho}{\gamma}\left(3\frac{v_o^2}{h_o} - g\,\cos\,\theta\right)$$

$$= \frac{\rho}{\gamma}\left(\frac{\rho^2 g^2 h_o^3\,\sin^2\theta}{3\mu^2} - g\,\cos\,\theta\right) \tag{5.150}$$

For a film with a given inclination θ and with a fluid of given properties, a real solution for α exists only when h_o exceeds some critical value. We interpret this result to mean that the film is stable for sufficiently small values of h_o (i.e., for sufficiently small velocities v_o).

If we define a film Reynolds number N_{Re} by $v_o h_o \rho/\mu$, we find that the condition for $\alpha = 0$ in Equation 5.150 becomes

$$N_{Re} = \cot\,\theta, \tag{5.151}$$

that is, disturbances of all wavelengths are stable for $N_{Re} < \cot\,\theta$. The more rigorous analysis (Kapitsa, 1948) predicts stability for $N_{Re} < (5\cot\,\theta)/6$. In any case, a film flowing vertically ($\theta = \pi/2$) is always unstable with respect to perturbations with sufficiently long wavelengths, no matter how small the Reynolds number. In practical situations, however, the rate of growth for small Reynolds numbers is so slow that the instability is not usually detected.

For fixed values of film inclination and Reynolds number, Equation 5.150 gives the value of α for which wave amplitude remains constant. We might expect that disturbances of shorter wavelength (smaller α) would be stabilized by interfacial tension while disturbances of longer wavelength (larger α) would be unstable. The more detailed analysis confirms this expectation, although the value of α at marginal stability differs slightly from that given by Equation 5.150.

EXAMPLE 5.8 WAVE MOTION ON FALLING WATER FILM

A film of water flows down a surface with an inclination θ from the horizontal of 60°.

(a) Find the Reynolds number, average velocity, and film thickness when the film first becomes unstable with respect to perturbations having very long wavelengths.

(b) For a Reynolds number of 5.0, find the speed of propagation of the waves and their wavelength for the situation where no growth or damping occurs.

SOLUTION

(a) Equation 5.151 gives a critical Reynolds number of 0.577. Using Equations 5.140 and 5.141 to obtain the average velocity $\bar{v}_x$, we find

$$N_{Re} = 0.577 = \frac{\rho \bar{v}_x h}{\mu} = \frac{\rho^2 g \sin \theta \, h^3}{3\mu^2}.$$

Solving for h (with $\mu = 1$ cp and $\rho = 1$ g/cm³), we find $h = 5.89 \times 10^{-3}$ cm. The average velocity in the film is $\rho g h^2 \sin \theta / 3\mu$ or 0.981 cm/sec. Note that the more rigorous analysis gives a critical Reynolds number of 0.480 and even smaller values of film thickness and velocity.

(b) For $N_{Re} = 5$, we find by a similar calculation that $h = 1.21 \times 10^{-2}$ cm and $\bar{v}_x = 4.14$ cm/sec. From Equations 5.150 and 5.151 we obtain a speed of wave propagation of 3×4.14, or 12.4 cm/sec, and a wavenumber α of 7.22/cm. The corresponding wavelength is $2\pi/\alpha$, or 0.869 cm.

REFERENCES

GENERAL REFERENCES ON INTERFACIAL STABILITY AND
WAVE MOTION

Chandrasekhar, S., *Hydrodynamic and Hydromagnetic Stability*, Clarendon Press, Oxford, 1961.

Ivanov, I.B. (ed.), *Thin Liquid Films*, Marcel Dekker, New York, 1988.

Levich, V.G., *Physicochemical Hydrodynamics*, Prentice-Hall, Englewood Cliffs, N.J., 1962.

Lucassen-Reynders, E.H. and Lucassen, J., Properties of capillary waves, *Adv. Colloid Interface Sci.*, 2, 347, 1970.

Miller, C.A., Stability of interfaces, in *Surface and Colloid Science*, vol. 10, Matijevic E. (ed.), Plenum Press, New York, 1978, p. 227.

Wehausen, J.V. and Laitone, E.V., Surface waves, in *Handbuch der Physik*, vol. 9, Springer-Verlag, Berlin, 1960, p. 446.

TEXT REFERENCES

Basu, S. and Sharma, M.M., Measurement of critical disjoining pressure for dewetting of solid surfaces, *J. Colloid Interface Sci.*, 181, 443, 1996.

Benjamin, T.B., Wave formation in laminar flow down an inclined plane, *J. Fluid Mech.*, 2, 554, 1957.

Bergeron, V. and Radke, C.J., Equilibrium measurements of oscillatory disjoining pressures in aqueous foam films, *Langmuir*, 8, 3020, 1992.

Bird, R.B., Stewart, W.E., and Lightfoot, E.N., *Transport Phenomena*, 2nd ed., Wiley, New York, 2002.

Bohr, N., Determinaion of the surface-tension of water by the method of jet vibration, *Philos. Trans. R. Soc. (Lond.)*, A209, 281, 1909.

Bouchiat, M.A. and Meunier, J., Light scattering from surface waves on carbon dioxide near its critical point, *J. Physique*, Colloque Cl, Suppl. no. 2-3, 33, Cl-141, 1972.

Cazabat, A.M., Langevin, D., Meunier, J., and Pouchelar, A., Critical behavior in microemulsions, *Adv. Colloid Interface Sci.*, 16, 175, 1983.

Chandrasekhar, S., *Hydrodynamic and Hydromagnetic Stability*, Clarendon Press, Oxford, 1961.

Defay, R. and Petre, G., Dyanmic surface tension, in *Surface and Colloid Science*, vol. 3, Matijevic, E. (ed.), Wiley, New York, 1971, p. 27.

Denn, M.M., *Stability of Reaction and Transport Processes*, Prentice-Hall, Englewood Cliffs, N.J., 1975.

Dorrestein, R., General linearized theory of the effect of surface films on water ripples. I and II, *Proc. Konikl. Ned. Akad. Wetenschap.*, B54, 260, 350, 1954.

Dussan V., Hydrodynamic stability and instability of fluid systems and interfaces, *Arch. Ratl. Mech. Anal.*, 57, 363, 1975.

Dutton, J.A. and Panofsky, H.A., Clear air turbulence: a mystery. may be unfolding, *Science*, 167, 937, 1970.

Exerowa, D. and Kruglyakov, P.M., *Foams and Foam Films*, Elsevier, Amsterdam, 1998.

Francis, J.R.D., Wave motions and aerodynamic drag on a free oil surface, *Philos. Mag. Ser. 7*, 45, 695, 1954.

Garrett, P.R. (ed.), *Defoaming*, Marcel Dekker, New York, 1993.

Hirasaki, G.J., Thermodynamics of thin films and the three-phase regions, in *Interfacial Phenomena in Petroleum Recovery*, Morrow, N.R. (ed.), Marcel Dekker, New York, 1991, p. 23.

Huh, C., Capillary hydrodynamics: interfacial instability and the solid/fluid/fluid contact line, Ph.D. dissertation, University of Minnesota, 1969.

Huisman, H.F. and Mysels, K.J., Contact angle and the depth of the free-energy minimum in thin liquid films. Their measurement and interpretation, *J. Phys. Chem.*, 73, 489, 1969.

Joseph, D.D., *Stability of Fluid Motions*, vol. 2, Springer-Verlag, Berlin, 1976, chap. 14.

Kapitsa, P.L., Wave flow of thin viscous liquid layers, *Zh. Ekper. Teoret. Fiz.*, 18, 3, 1948.

Keller, J.B., Rubinow, S.I., and Tu, Y.O., Spatial instability of a jet, *Phys. Fluids*, 16, 2052, 1973.

Kralchevsky, P.A., Nikolov, A.D., Wasan, D.T., and Ivanov, I., Formation and expansion of dark spots in stratifying foam films, *Langmuir*, 6, 1180, 1990.

Krantz, W.B. and Goren, S.L., Finite-amplitude, long waves on liquid films flowing down a plane, *Ind. Eng. Chem. Fundam.*, 9, 107, 1970.

Kutateladze, S.S., A hydrodynamic theory of changes in a boiling process under free convection, *Izv. Akad. Nauk. SSR, Otd. Tekhn. Nauk*, 4, 529, 1951.

Langevin, D., Optical techniques for the characterization of monolayers and thin liquid films, in *Modern Characterization Methods of Surfactant Systems*, Binks, B.P. (ed.), Marcel Dekker, New York, 1999, 181.

Lemlich, R., Prediction of changes in bubble size distribution due to interbubble gas diffusion in foam, *Ind. Eng. Chem. Fundam.*, 17, 89, 1978.

Levich, V.G., *Physicochemical Hydrodynamics*, Prentice-Hall, Englewood Cliffs, N.J., 1962.

Lucassen, J., Longitudinal capillary waves, *Trans. Faraday Soc.*, 64, 2221, 2230, 1968.

Lucassen, J., Dynamic properties of free liquid films and foams, in *Anionic Surfactants*, Lucassen-Reynders, E.H. (ed.), Marcel Dekker, New York, 1981, p. 217.

Lucassen, J. and Hansen, R.S., Damping of waves on monolayer-covered surfaces: II. Influence of bulk-to-surface diffusional interchange on ripple characteristics, *J. Colloid Interface Sci.*, 23. 319, 1967.

Lucassen-Reynders, E.H. and Lucassen, J., Properties of capillary waves, *Adv. Colloid Interface Sci.*, 2, 347, 1970.

Miller, C.A., *AIChE J.*, 23, 959, 1977.

Miller, C.A., Stability of interfaces, in *Surface and Colloid Science*, vol. 10, Matijevic, E. (ed.), Plenum Press, New York, 1978, p. 227.

Miller, C.A. and Scriven, L.E., The oscillations of a droplet immersed in another fluid, *J. Fluid Mech.*, 32, 417, 1968.

Miller, C.A. and Scriven, L.E., Interfacial instability due to electrical forces in double layers: I. General considerations, *J. Colloid Interface Sci.*, 33, 360, 1970.

Mysels, K.J., Shinoda, K., and Frankel, S., *Soap Films; Studies of Their Thinning and Bibliography*, Pergamon, New York, 1959.

Nikolov, A.D., Kralchevsky, P.A., Ivanov, I., and Wasan, D.T., Ordered micelle structuring in thin films formed from anionic surfactant solutions : II. Model development, *J. Colloid Interface Sci.*, 133, 13, 1989.

Pitts, E., The stability of pendant liquid drops, Part 2. Axial symmetry, *J. Fluid Mech.*, 63, 487, 1974.

Princen, H.M. and Frankel, S., Contact angles in soap films from diffraction of light traversing a plateau border, *J. Colloid Interface Sci.*, 35, 386, 1971.

Ranadive, A.Y., and Lemlich, R. (1979), The effect of initial bubble size distribution on the interbubble gas diffusion in foam, *J. Colloid Interface Sci.* 70, 392.

Ruckenstein, E. and Jain, R.K., Spontaneous rupture of thin liquid films, *J. Chem. Soc. Faraday Trans. II*, 70, 132, 1974.

Sanfeld, A., *Introduction to the Thermodynamics of Charged and Polarized Layers*, Wiley, New York, 1968.

Sharma, A. and Jameel, A.T., Nonlinear stability, rupture, and morphological phase separation of thin fluid films on apolar and polar substrates, *J. Colloid Interface Sci.*, 161, 190, 1993.

Sheludko, A., Thin liquid films, *Adv. Colloid Interface Sci.*, 1, 391, 1967.

Tanford, C., *Ben Franklin Stilled the Waves*, Duke University Press, Durham, N.C., 1989.

Tyuptsov, A.D., Hydrostatics in weak force fields, *Fluid Dynam.*, 1(2), 51, 1966.

Vrij, A., Hesselink, F. Th., Lucassen, J., and van den Tempel, M., Waves in thin liquid films. II. Symmetrical modes in very thin films and film rupture, *Proc. Konikl. Ned. Akad. Wetenschap.*, B73, 124, 1970.

Wehausen, J.V. and Laitone, E.V., Surface waves, in *Handbuch der Physik,* vol. 9, Springer-Verlag, Berlin, 1960, p. 446.

Yih, C.S., Stability of liquid flow down an inclined plane, *Phys. Fluids*, 6, 321, 1963.

Zhang, X. and Neogi, P., Stable drop shapes under disjoining pressure: II. Multiplicity and stability , *J. Colloid Interface Sci.*, 249, 141, 2002.

Zollweg, J., Hawkins, G., Smith, I.W., Giglio, M., and Benedek, G.B., , *J. Physique,* Colloque Cl, Suppl. no. 2-3, 33, p. Cl-135, 1972.

Zuber, N., On the stability of boiling heat transfer, *Trans. ASME*, 80, 711, 1958.

PROBLEMS

5.1 Show that for a liquid of low viscosity in contact with a gas the damping factor (i.e., the real part of the time factor β) passes through a maximum with increasing surfactant concentration and that it approaches the value given by Equation 5.72 for large surfactant concentrations.

(a) Form the determinant of coefficients of Equations 5.65 and 5.70 and show that the dispersion equation has the form

$$r_A^4 + 2r_A^2 - 4r_A + 1 + r_A(r_A^2 - 1)N = -\frac{\beta_A^{*2}}{v_A^2\alpha^4}\left[1 + \frac{v_A\alpha^2}{\beta}(r_A - 1)N\right],$$

where

$$N = \frac{\Gamma_{os}\alpha\left(-\dfrac{d\gamma}{d\Gamma_s}\right)}{\beta\,\mu_A}.$$

(b) In the low viscosity limit (i.e., for $|\beta/v_A\alpha^2| \gg 1$), show that the above equation may be approximated by its leading terms to obtain

$$\beta^2\left[1 + N\left(\frac{v_A\alpha^2}{\beta}\right)^{1/2} + 4\frac{v_A\alpha^2}{\beta}\right] = -\beta^{*2}\left[1 + N\left(\frac{v_A\alpha^2}{\beta}\right)1/2 - N\left(\frac{v_A\alpha^2}{\beta}\right)\right].$$

(c) Assume that the order of magnitude of N is at least as large as $(\beta/v_A\alpha^2)^{1/2}$. Then set β equal to $(i\beta^* + \beta_1)$ with $|\beta_1/\beta^*| \ll 1$. Neglecting term β_1^2, show that the result obtained in (b) becomes

$$\beta_1 \simeq \frac{\alpha(\beta^* v_A)^{1/2}}{2i} \left(\frac{\beta^*}{\beta}\right)^{1/2} \frac{1}{1 + \dfrac{1}{N} \left(\dfrac{\beta}{v_A \alpha^2}\right)^{1/2}} \cdot$$

(d) Show that $\beta^{1/2} \simeq \beta^{*1/2}(1 + i)/(2)^{1/2}$. Substitute this expression into the result of (c) and show that

$$\beta_1 \simeq -\frac{\alpha(\beta^* v_A)^{1/2}}{2^{3/2}} \times \frac{1 + i(1 - 2E)}{1 - 2E + 2E^2},$$

where

$$E = \frac{\beta^{*3/2}(\mu_A \rho_A)^{1/2}}{2^{1/2}\, \Gamma_{os}\, \alpha^2 \left(\dfrac{-d\gamma}{d\Gamma_s}\right)} \cdot$$

(e) Show that in the limit $E \to 0$, the inextensible result of Equation 5.72 is obtained. In addition, show that the real part of β_1 has its greatest magnitude for $E = 1/2$ and that the damping factor in this case is twice that for an inextensible surface. Finally, show that if the wavelength is short enough for gravitational effects to be negligible, the surface compressibility $\Gamma_{os}(-d\gamma/d\Gamma_s)$ at maximum damping is given by

$$\Gamma_{os}\left(\frac{-d\gamma}{d\Gamma_s}\right) = 2^{1/2} \gamma^{3/4} \alpha^{1/4} v_A^{1/2} \rho_A^{1/4} \cdot$$

5.2   Use Equation 5.71 to derive the expressions for damping of capillary waves given by Equations 5.74 and 5.75. Also, find the speed at which waves travel away from the source of vibrations.

(a) Show that in the free surface case the dispersion equation is given by

$$r^4 + 2r^2 - 4r + 1 + \frac{\beta^{*2}}{v^2 \alpha^4} = 0.$$

(b) Show that if $\beta = i\bar{\beta}$, where $\bar{\beta}$ is the imposed frequency, the speed c of the traveling waves is ($\bar{\beta}/\alpha_R$), where α_R is the real part of α (i.e., $\alpha = \alpha_R + i\alpha_I$). If $|\alpha_I/\alpha_R| \ll 1$, substitute these expressions for β and α into the dispersion equation and keep only the lead terms of the real and imaginary parts of this equation in the low viscosity approximation $|\beta/v\alpha^2| \gg 1$. Then solve for α_R and α_I. Show that for short wavelengths where gravity is negligible, α_I is given by Equation 5.74 and

$$\alpha_R = \left(\frac{\rho\bar{\beta}^2}{\gamma}\right)^{1/3}$$

$$c = \left(\frac{\bar{\beta}\gamma}{\rho}\right)^{1/3}.$$

(c) Repeat for the inextensible surface case. At one point in this derivation you will need to prove that the lead term of r in the low viscosity approximation is

$$(\bar{\beta}^{1/2}/\alpha_R \nu^{1/2})(1 - i)/2^{1/2}.$$

5.3 In the experiments described in Section 3, "surface wave" motion was generated by small oscillations of a bar in a direction perpendicular to the surface. In another type of experiment for surfactant-containing interfaces, the bar receives small oscillations in a direction parallel to the surface and "longitudinal wave" motion ensues (Lucassen, 1968). Develop an analysis of this situation for the case of an insoluble surfactant film at a liquid-gas interface that remains flat during longitudinal wave motion.

(a) Show that the dispersion equation in this case is given by

$$r^2 - 1 + N(r - 1) = 0,$$

where N is a defined in Problem 5.1.

(b) With $\beta = -i\bar{\beta}$ and $\bar{\beta}$ the imposed frequency, simplify this equation to show that in the low viscosity limit

$$i\bar{\beta}^3 \mu\rho = \alpha^4 \Gamma_{os}^2 \left(\frac{-d\gamma}{d\Gamma_s}\right)^2.$$

(c) Let $\alpha/\alpha_o = \alpha_R + i\alpha_I$, where $\alpha_o = [\bar{\beta}^3\mu\rho/\Gamma_{os}^2 (-d\gamma/d\Gamma_s)^2]^{1/4}$. Show that

$$(\alpha/\alpha_o) = e^{i\pi/8}$$

and hence that $\alpha_R = \cos \pi/8$ and $\alpha_I = \sin \pi/8$.

(d) Calculate the real and imaginary parts of α for a surfactant film on an aqueous surface with $\beta = 300/\text{sec}^{-1}$, $\mu = 1$ cp, $\rho = 1$ g/cm³, and $[\Gamma_{os}(-d\gamma/d\Gamma_s)] = 5$ dyne/cm. What are the wavelength and the speed of travel of the longitudinal waves generated in this case?

5.4 For a drop of a viscous liquid having a radius R oscillating in a gas, the time factor β for the case of a free surface is given by

$$\beta = \pm i\,\beta_D^* - \frac{v(\ell-1)\,(2\ell+1)}{R^2}\,,$$

where

$$\beta_D^{*2} = \frac{\gamma\ell(\ell-1)\,(\ell+2)}{\rho R^3}$$

and ℓ is a positive integer characterizing the mode of oscillation. For example, the drop is deformed into approximately an ellipsoidal shape for $\ell = 2$.

Similarly, for an inextensible interface, we have

$$\beta = \pm i\,\beta_D^*\,(1\mp i)\,\frac{(\beta_D^* v)^{1/2}(\ell-1)^2}{2^{3/2}\,R\,(\ell+1)}\,.$$

If $\ell = 2$, $R = 1.5$ cm, $\mu = 1$ cp, $\gamma = 72$ dyne/cm, and $\rho = 1$ g/cm³, calculate the oscillation frequency and the damping rates. Repeat for a bubble. The formulas are

$$\beta = \pm i\,\beta_B^* - \frac{v(2\ell+1)\,(\ell+2)}{r^2}$$

$$\beta = \pm i\,\beta_B^* - (1\mp i)\,\frac{(\beta_B^* v)^{1/2}(\ell+2)^2}{2^{3/2}\,\ell\,R}\,,$$

where

$$\beta_B^{*2} = \frac{\gamma(\ell+1)\,(\ell-1)\,(\ell+2)}{\rho R^3}\,.$$

Why does the surfactant have a much larger effect on damping for a bubble? Hint: Consider the tangential velocities in the bulk fluids and at the interface as the drop or bubble becomes more ellipsoidal. See Problem 5.5.

5.5 As shown in the text, there is no lateral motion along an inextensible plane interface subjected to small wavy deformations, i.e., for a surface where $\Gamma_{os}(- d\gamma/d\Gamma_s)$ is very large. The same is *not* true for a spherical interface because, in view of Equation 1.19, local expansion occurs in regions where the interface moves outward and local compression occurs in regions where it moves inward. If the surfactant film is nearly incompressible, tangential flow develops from the latter to the former regions. This situation is the one described by the formulas given in Problem 5.4.

It is also conceivable that a film might be rather compressible but exhibits a very large resistance to shear. In this case, where the so-called shear elasticity is large, there is little lateral motion along the interface and the following expressions can be derived (Miller, 1977):

Drop

$$\beta = \pm\, i\, \beta_D^* - (1 \mp i)\, \frac{(\beta_D^* \nu)^{1/2}(\ell + 1)}{2^{3/2}\, R}$$

Bubble

$$\beta = \pm\, i\, \beta_B^* - (1 \mp i)\, \frac{(\beta_B^* \nu)^{1/2}\, \ell}{2^{3/2}\, R}$$

Using the same values of physical properties as in Problem 5.4, calculate damping factors using these formulas. Comment on the results in comparison with those of Problem 5.4.

5.6 Develop an analysis for surface wave motion of a liquid having a finite depth H. Show that for wavelengths comparable in magnitude to H or larger, an oscillating boundary layer develops at the solid surface in the low viscosity limit and the time factor β is given for a free surface by

$$\beta = i\, \beta^* - (1 \mp i)\, \frac{\alpha(\beta^* \nu)^{1/2}}{2^{3/2}}\, \text{sech } \alpha H \text{ csch } \alpha H \,,$$

with

$$\beta^{*2} = \frac{\gamma \, \alpha^3 + \rho_A g \alpha}{\rho_A \, \coth \, \alpha \, H} \, .$$

5.7   Find α_c, α_M, and β_M for the instability with respect to axisymmetric perturbations of a hollow fluid cylinder (i.e., a cylinder of gas in an infinite expanse of liquid).

5.8   Use the force method to investigate the stability of a cylinder of A in B when both fluids rotate at an angular velocity ω around the axis of the cylinder. Assume that interfacial tension exerts a local restoring force equal to $-\gamma d(2H)$, where H is the local mean curvature. If $\rho_B > \rho_A$, how large must ω be to overcome the basic capillary instability described in Section 4.

5.9   For air blowing over water under the conditions given in Example 5.6, find the range of unstable wavenumbers α, the wavenumber of the fastest growing disturbance α_M, its rate of growth, and its speed of wave propagation along the interface when wind velocity is 20 mph (892 cm/sec).

5.10 Repeat the analysis of wave motion on a falling film for the case where the fluid surface is immobile (i.e., it can deflect in the z direction, but tangential velocity v_x is zero). Such a flow might exist in the event of surfactant buildup along the film because some barrier far downstream prevents surfactant from leaving the film surface. Find the critical Reynolds number for instability as a function of θ and find c and α at marginal stability as a function of film thickness h.

6 Transport Effects on Interfacial Phenomena

1. INTERFACIAL TENSION VARIATION

Experiments demonstrate that interfacial tension varies with both interfacial composition and temperature. At equilibrium, both these quantities are uniform along an interface and hence so is interfacial tension. If transport processes (e.g., diffusion of surfactant to a newly created interface) cause interfacial tension to be time dependent while remaining uniform at each time, changes in interfacial shape may occur. Such processes may be followed by, for instance, monitoring the dimensions of a sessile or pendant drop as a function of time.

Flow or transport may also produce spatial variations in interfacial temperature or concentration and hence in interfacial tension. Indeed, we have already seen in Chapter 5 that flow associated with surface wave motion causes surfactant concentration gradients to develop at an interface with an insoluble monolayer. The resulting interfacial tension gradients were found to significantly enhance the damping of wave motion at liquid-gas interfaces, with resulting adverse effects on transfer of heat or mass across the interface.

In this chapter we deal first with flow generated by interfacial tension gradients, the so-called Marangoni effect. The development of "tears" on the inner surface of a glass of strong wine is a common example. Such flow is a useful way to produce fluid motion in the gravity-free environment of space and in certain microfluidic devices. We shall be particularly interested in cases of interfacial stability where flow develops by this mechanism for converting thermal or chemical energy to mechanical energy. Such instabilities can bring about severalfold increases in heat or mass transfer rates across interfaces. Later in the chapter we discuss spontaneous emulsification, instability of solid surfaces leading to dendritic growth, and other interfacial effects involving transport such as dynamic surface tension and formation and growth of "intermediate" phases at an interface.

2. INTERFACIAL SPECIES MASS BALANCE AND ENERGY BALANCE

General mass and momentum balances for an interfacial region were derived in Chapter 5 and used in the analyses of interfacial stability presented there. In dealing with heat and mass transport near interfaces we require additional balances for energy and individual species.

We consider the latter first because it can be derived in a manner similar to that employed previously for the overall mass balance, but with additional terms to account for diffusion and chemical reaction. The starting point is the mass balance for species i for the pillbox control volume of Figure 1.3:

$$\int_{s_A} (\rho_{iA} v_A + j_{iA}) \cdot n \, dS - \int_{s_B} (\rho_{iB} v_B + j_{iB}) \cdot n \, dS - \int_{s_o} (\rho_i v + j_i) \cdot M \, dS$$

$$+ \int_V r_i dV = \frac{d}{dt} \int_V \rho_i dV. \tag{6.1}$$

Here ρ_i is the local mass density of species i, j is the local diffusional flux of i relative to the mass average velocity v, and r_i is the local rate of production of i per unit volume as the result of one or more chemical reactions. As before, the corresponding equations for the extrapolated bulk phases must be subtracted from this equation, various manipulations performed to convert the resulting equation to a statement that an integral over the reference surface S must vanish, and the arbitrary extent of S invoked to argue that the integrand itself must be zero (compare Equations 5.28 through 5.32).

The resulting differential interfacial mass balance is given by

$$\frac{\partial \Gamma_i}{\partial t} + \nabla_S \cdot \Gamma_i v - \rho_{iA}(v_{An} - a_n) + \rho_{iB}(v_{Bn} - a_n) - j_{iAn} + j_{iBn} - R_i + \nabla_s \cdot j_{is} = 0 \ , \tag{6.2}$$

where $\dot{a}_n$ is the normal velocity of S as before and

$$\Gamma_i = \int_{\lambda_A}^{\lambda_B} \Delta \rho_i d\lambda \tag{6.3}$$

$$R_i = \int_{\lambda_A}^{\lambda_B} \Delta r_i d\lambda \tag{6.4}$$

$$j_{is} = \int_{\lambda_A}^{\lambda_B} \Delta j_i d\lambda \cdot (I - nn) \ . \tag{6.5}$$

The quantities $\Delta \rho_i$, Δr_i, and Δj_i are defined in a manner analogous to $\Delta \rho$ and Δp_T in Equation 1.33. The term involving the unit dyadic (or tensor) I employed in Equation 6.5 ensures that j_{is} includes only the tangential portion of the surface excess flux. Also, the convective flux in the integral over S_0 has been neglected, as in the previous derivation of the overall mass balance, because S_0 can be made small in comparison to S by reducing pillbox thickness.

Equation 6.2 states that the local interfacial concentration Γ_i changes as a result of interfacial convection, interchange with the bulk phases due to convection and diffusion, interfacial chemical reaction, and diffusion within the interface. When the last effect is important, some relation is required between j_{is} and Γ_i. Usually j_{is} is taken as $-D_{is}\nabla_s\Gamma_i$, where D_{is} is an interfacial diffusion coefficient.

The interfacial energy balance can be derived in a similar manner, again beginning with a balance over the pillbox control volume of Figure 1.3:

$$\frac{d}{dt}\int_v \rho\hat{E}dV = \int_{S_A}[q_A + \rho_A\hat{E}_A v_A + p_A v_A + \tau_A \cdot v_A]\cdot n\ dS$$

$$-\int_{S_B}[q_B + \rho_B\hat{E}_B v_B + p_B v_B + \tau_B \cdot v_B]\cdot n\ dS .\qquad(6.6)$$

$$-\int_{S_A}[q + \rho\hat{E}v + p_T v + \tau\cdot v]\cdot M\ dS$$

In this equation, q represents a conductive heat flux, $\hat{E}$ the total energy per unit mass including internal, kinetic, and gravitational potential energy contributions, and τ is the viscous stress dyadic. The reference states for the various internal energies are chosen in such a way that energy effects of phase changes and chemical reactions are accounted for in this formulation.

By the same basic procedure as used before, we arrive at the following energy balance:

$$\frac{\partial E^s}{\partial t} + \nabla_s \cdot E^s v - (\rho_A\hat{E}_A + p_A + \tau_{An})(v_{An} - a_n) - \tau_{At}v_{At}$$

$$+ (\rho_B\hat{E}_B + p_B + \tau_{Bn})(v_{Bn} - a_n) + \tau_{Bt}v_{Bt} - q_{An} + q_{Bn}\qquad . \qquad(6.7)$$

$$+ \nabla_s \cdot q_s - \nabla_s \cdot \gamma v_s = 0$$

Here the surface excess energy per unit area and the surface excess heat flux are given by

$$E^s = \int_{\lambda_B}^{\lambda_A} \Delta(\rho\hat{E})\,d\lambda\qquad(6.8)$$

$$q_s = \int_{\lambda_A}^{\lambda_B} \Delta q d\lambda \cdot (I - nn) \ . \tag{6.9}$$

Convective transport and work done by viscous stresses have been neglected in the integral over S_0 because of the small pillbox thickness. Finally, tangential velocity v_s has been assumed to be uniform in the interior region so that

$$\int_{\lambda_A}^{\lambda_B} \Delta(p_T v_s)\, d\lambda = v_s \int_{\lambda_B}^{\lambda_B} \Delta p_T \, d\lambda = -\gamma v_s \ . \tag{6.10}$$

This assumption should be acceptable for thin pillbox control volumes.

The general equation (Equation 6.7) is rather lengthy, but frequently many of the terms are unimportant. Usually the terms involving E_s and q_s are small, for instance. The last term in Equation 6.7 is of interest because it indicates that interconversion between thermal or chemical and mechanical energy is possible during interfacial flow. As indicated above, such interconversion is precisely what happens during Marangoni flow, when interfacial tension varies with position owing to gradients in temperature or concentration. Mechanical energy release owing to contractive interfacial flow in regions of high interfacial tension exceeds mechanical energy removal by expansive flow in regions of low tension. The excess is available to replace mechanical energy lost by viscous dissipation in the bulk fluids and thus maintain the flow. In most realistic situations, both these effects are small in comparison to the normal heat fluxes q_{An} and q_{Bn}.

3.   INTERFACIAL INSTABILITY FOR A LIQUID HEATED FROM BELOW OR COOLED FROM ABOVE

More than a century ago, Benard (1900) observed development of a regular hexagonal pattern of convection cells in an initially stagnant thin layer of liquid heated from below. Figure 6.1 is an example of this striking phenomenon. For many years it was assumed that the flow was generated by natural convection effects, since low-density fluid was situated beneath high-density fluid in the stagnant film and hence had a tendency rise. Pearson's analysis (1958) demonstrated that interfacial tension gradients could produce cellular convection independent of any natural convection. His analysis explained why cellular convection was sometimes observed in thin paint films of various orientations, including those on the underside of horizontal surfaces where the effect of natural convection should be stabilizing. We note that such instabilities are undesirable in paint films because they impart a cellular appearance, popularly known as "orange

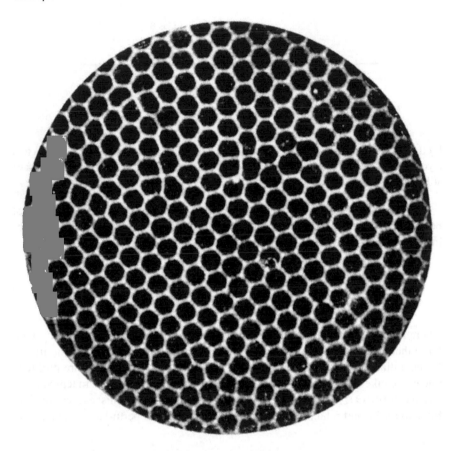

FIGURE 6.1 Cellular convection produced by surface tension gradients in a thin liquid layer heated from below. Reproduced from Avsec (1939) with permission from the French Ministry of Defense.

peel," to the finished surface. Convective instabilities in paint films have been reviewed by Hansen and Pierce (1973).

We shall apply linear stability analysis to the thin layer illustrated in Figure 6.2. In order to appreciate the basic mechanism, let us suppose that, due to a small disturbance, some warm liquid from the interior of the layer reaches the interface at point P. Since interfacial tension decreases with increasing temperature, this disturbance produces a local lowering of interfacial tension at P and hence an outward directed lateral force along the interface. As a result, an outward flow arises which, in turn, draws more warm liquid to the surface, reinforcing the initial flow (see Figure 6.2). Thus the initial stagnant layer is unstable, interfacial convection develops, and heat transfer rates increase substantially over those in the absence of flow. Note that if the layer were cooled from below, the interfacial tension gradients would oppose the initial flow and the system would be stable.

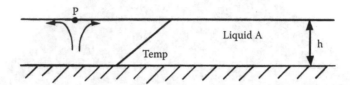

FIGURE 6.2 Origin of cellular convection driven by interfacial tension gradients in a thin layer of liquid heated from below or cooled from above. Warm fluid reaching the interface at P reduces the interfacial tension locally.

The stability analysis is similar in principle to those of Chapter 5, but must be extended to account for heat transport effects. The differential equation describing heat transport in the liquid layer is

$$\frac{\partial T}{\partial t} + v \cdot \nabla T = D_T \nabla^2 T , \qquad (6.11)$$

where T is the temperature and D_T is the thermal diffusivity given by $k/\rho c_p$, with k the thermal conductivity, ρ the density, and c_p the specific heat of the liquid. In the initial stagnant film, Equation 6.11 implies that the temperature profile is linear. We take its slope to be ζ. If we assume that the perturbation in temperature has the form $\Theta (z) \, e^{i\alpha x} e^{\beta t}$, we find that upon substituting this expression into Equation 6.11 and retaining only terms of first order in perturbation quantities:

$$\frac{d^2\Theta}{dz^2} - \alpha^2 q^2 \Theta = \frac{\zeta}{D_T} W , \qquad (6.12)$$

where $W(z)$ is a function describing the z dependence of the normal component of velocity as before and

$$q^2 = 1 + \frac{\beta}{D_T \alpha^2} . \qquad (6.13)$$

Now $W(z)$ must satisfy Equation 5.12 if the variation of density and viscosity with temperature is neglected. Thus if we apply the operator of Equation 5.12 to both sides of Equation 6.12, we obtain

$$\left(\frac{d^2}{dz^2} - \alpha^2 r^2 \right) \left(\frac{d^2}{dz^2} - \alpha^2 \right) \left(\frac{d^2}{dz^2} - \alpha^2 q^2 \right) \Theta(z) = 0 . \qquad (6.14)$$

This equation has the general solution

$$\Theta(z) = A_1' \sinh \alpha rz + A_2' \cosh \alpha rz + A_3' \sinh \alpha z + A_4' \cosh \alpha z$$
$$+ A_5' \sinh \alpha qz + A_6' \cosh \alpha qz \qquad (6.15)$$

The normal velocity $W(z)$ may be obtained from this equation using Equation 6.12 and the tangential velocity and pressure distribution can be found using Equations 5.20 and 5.26.

To simplify the algebra in the following discussion, we shall restrict our attention to states of marginal stability where the time factor β vanishes. As a result, we shall be able to identify conditions where the system is stable or unstable, the information of prime interest, but we shall not be able to compute growth rates. With this scheme, Equations 6.12, 6.14, and 6.15 are replaced by

$$\frac{d^2\Theta}{dz^2} - \alpha^2 \Theta = \frac{\zeta}{D_T} W \qquad (6.16)$$

$$\left(\frac{d^2}{dz^2} - \alpha^2 \right)^3 \Theta = 0 \qquad (6.17)$$

$$\Theta(z) = A_1 \sinh \alpha z + A_2 \cosh \alpha z + A_3 z \sinh \alpha z + A_4 z \cosh \alpha z$$
$$+ A_5 z^2 \sinh \alpha z + A_6 z^2 \cosh \alpha z. \qquad (6.18)$$

It is easily verified by direct substitution that Equation 6.18 satisfies Equation 6.17. The normal velocity distribution can be obtained from Equations 6.16 and 6.18:

$$W(z) = \frac{D_T}{\zeta} [2\alpha A_3 \cosh \alpha z + 2\alpha A_4 \sinh \alpha z + A_5(2 \sinh \alpha z + 4\alpha z \cosh \alpha z)$$
$$+ A_6(2 \cosh \alpha z + 4\alpha z \sinh \alpha z)] \qquad (6.19)$$

The constants A_i can be found by imposition of suitable boundary conditions. At the solid surface $z = 0$, the velocity components v_z and v_x vanish and, if the solid is an excellent conductor such as a metal, the temperature remains uniform at its initial value. Using Equations 6.19 and 6.20 to evaluate v_z and v_x, we find that these three conditions lead to the following equations:

$$A_2 = 0 \qquad (6.20)$$

$$A_5 = -\alpha A_4/3 \tag{6.21}$$

$$A_6 = -\alpha A_3. \tag{6.22}$$

At marginal stability, interfacial deflection remains constant so that the normal velocity at $v_z = H$ is zero. Making use of Equations 6.21 and 6.22, we find from this condition that

$$A_3 = \frac{A_4}{3\alpha^*}(1 - \alpha^* \coth\alpha^*), \tag{6.23}$$

where α^* is a dimensionless wavenumber αH.

The normal component of the interfacial momentum balance equation takes the form

$$\left(p - 2\mu\frac{\partial v_z}{\partial z}\right)\Bigg|_{H+z^*} = -2H\gamma . \tag{6.24}$$

The interfacial tension γ in this equation is that of the initial stagnant layer, with interfacial tension variations providing a second-order correction that is neglected in our linear analysis. With the pressure given by Equations 5.39 and 5.26, z^* given by $Be^{i\alpha x}$, and A_3, A_5, and A_6 eliminated by Equations 6.21 through 6.23, Equation 6.24 becomes

$$0 = B + \frac{4\nu D_T\alpha^3}{3\zeta\beta^{*2}}A_4\left[\frac{(\alpha^*-1)}{\alpha^*}\cosh\alpha^* - 2\alpha^* \sinh\alpha^* + 2\alpha^* \cosh\alpha^* \coth\alpha^*\right], \tag{6.25}$$

where β^* is as defined at Equation 5.57.

The key boundary condition of the analysis is the tangential component of the interfacial momentum balance, which becomes

$$\tau_{zx}\big|_H + \frac{d\gamma}{dT}\frac{\partial T}{\partial x}\bigg|_H = 0 . \tag{6.26}$$

The coefficient $d\gamma/dT$ of interfacial tension variation with temperature is assumed to have a constant value. It is virtually always negative, a typical value being about -0.1 mN/m·K. According to Equation 6.26, development of a temperature gradient along an interface causes flow to arise having shear stresses that balance the lateral force produced by the interfacial tension gradient. In terms of the coefficients A_i, Equation 6.26 becomes

$$0 = -A_1 N_{Ma} \sinh \alpha^* + \frac{A_4 H}{3} [8(-\cosh \alpha^* + \alpha^* \cosh \alpha^* \coth \alpha^*$$

$$-\alpha^* \sinh \alpha^*) - N_{Ma}(\cosh \alpha^* - \alpha^* \sinh \alpha^* + \frac{\sinh \alpha^*}{\alpha^*} \qquad , \quad (6.27)$$

$$+ \alpha^* \cosh \alpha^* \coth \alpha^*]$$

where the dimensionless "Marangoni number" N_{Ma} is defined by

$$N_{Ma} = \frac{\varsigma \left(\frac{d\gamma}{dT} \right) H^2}{D_T \mu} .$$

The final boundary condition is the energy balance at the interface. We use a very simple form of this balance that employs a heat transfer coefficient h to describe transfer into the gas phase. The latter is presumed to have some bulk temperature T_b:

$$-k \frac{\partial T}{\partial z} \bigg|_{H+z^*} = h(T - T_b) \big|_{H+z^*} . \qquad (6.28)$$

Substitution of the temperature profile into this equation yields after some manipulation

$$0 = B N_{Nu} \varsigma + A_1 (\alpha^* \cosh \alpha^* + N_{Nu} \sinh \alpha^*) + \frac{A_4 H}{3} [\cosh \alpha^* + \frac{\sinh \alpha^*}{\alpha^*}$$

$$+ \alpha^* \cosh \alpha^* \coth \alpha^* + N_{Nu}(\cosh \alpha^* + \frac{\sinh \alpha^*}{\alpha^*} - \alpha^* \sinh \alpha^* \qquad , \quad (6.29)$$

$$+ \alpha^* \cosh \alpha^* \coth \alpha^*)]$$

with the Nusselt number N_{Nu} given by

$$N_{Nu} = \frac{hH}{k} . \qquad (6.30)$$

Equations 6.25, 6.27, and 6.29 are three linear, homogeneous equations in A_1, A_4, and B. For a nontrivial solution to exist, the determinant of coefficients for these equations must vanish, a condition which leads to

$$N_{Ma} = \frac{8\alpha^*(\alpha^*\cosh\alpha^* + N_{Nu}\sinh\alpha^*)\,(\alpha^* - \sinh\alpha^*\cosh\alpha^*)}{\alpha^{*3}\cosh\alpha^* - \sinh^3\alpha^* - \dfrac{8N_{Cr}\alpha^{*5}\cosh\alpha^*}{N_B + \alpha^{*2}}}. \quad (6.31)$$

Here the "crispation number" N_{Cr} and the Bond number N_B are given by

$$N_{Cr} = \frac{\mu D_T}{\gamma H} \qquad (6.32)$$

$$N_B = \frac{\rho g H^2}{\gamma}. \qquad (6.33)$$

Figure 6.3 shows plots of this marginal stability relationship for selected values of the dimensionless parameters. For Marangoni numbers N_{Ma} below those on a curve corresponding to a particular set of parameters, the system is stable. For Marangoni numbers above the curve, instability occurs. Thus, in view of the definition of N_{Ma} following Equation 6.27, instability is favored by larger initial temperature gradients, larger variations of interfacial tension with temperature, and larger film thicknesses (because of lower resistance to flow). It is opposed

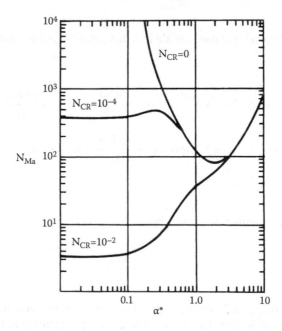

FIGURE 6.3 Marginal stability condition for thin layer heated from below. From Smith (1966) with permission.

by higher fluid viscosities and higher fluid thermal conductivities, the latter because temperature gradients along the interface are diminished by conduction through the liquid.

Let us first examine the limit $N_{Cr} \rightarrow 0$, corresponding to high interfacial tensions or thick layers. In this case, which was the one considered in the original analysis (Pearson, 1958), interfacial deflection is small. We see from Figure 6.3 that for low values of N_{Ma}, the system is stable (i.e., the small interfacial temperature gradient is insufficient to overcome the viscous resistance to flow in the thin layer). As N_{Ma} increases owing to, for instance, application of higher temperature gradients ζ, instability first sets in at critical values for N_{Ma} and α_c^* of about 80 and 2, respectively. Higher values of N_{Mac} are required to produce instability for wavelengths (cell sizes) both shorter and longer than $2\pi/\alpha_c^*$, the former owing to the higher velocity gradients and viscous dissipation rates in small convection cells, the latter because the interfacial temperature gradient that promotes the flow decreases faster with increasing cell size than does the rate of viscous dissipation, which is determined largely by layer thickness H for long wavelengths.

Calculations show that with increasing values of the Nusselt number N_{Nu} the system becomes more stable (i.e., larger values of N_{Ma} are needed to produce instability). For instance, N_{Mac} increases from about 80 to 150 as N_{Nu} increases from near 0 to 2. This result can be understood by recognizing that the enhanced heat loss to the gas in hot regions along the interface acts to diminish the interfacial temperature gradient. Larger values of the heat transfer coefficient h, and hence of N_{Nu}, increase the magnitude of this effect.

When N_{Cr} has a nonzero value (i.e., when surface deflection is permitted), the system becomes, according to Figure 6.3, much less stable to perturbations having long wavelengths (small values of α^*). The reason is that deflection promotes instability by depressing high temperature regions such as that near point P of Figure 6.4, moving them closer to the hot solid surface at the bottom of the layer (Scriven and Sternling, 1964). Similarly, low temperature regions such as those near point Q are elevated, increasing their distance from the hot solid surface. This effect is especially important when wavelengths are long, as first recognized by Smith (1966), because then both flow and lateral conduction are slow and deflection is the dominant factor determining the interfacial temperature gradient. For sufficiently long wavelengths, the stabilizing effect of

FIGURE 6.4 Marangoni flow with surface deflection.

gravity becomes important and the Marangoni number required to produce instability becomes nearly constant, as shown in Figure 6.3.

As fluid depth H increases, α^* becomes large and Equation 6.31 simplifies to

$$N_{Ma} = 8\alpha^{*2}\left(1 + \frac{N_{Nu}}{\alpha^*}\right). \tag{6.34}$$

Rearrangement of this equation yields

$$N'_{Ma} = \frac{\zeta\dfrac{d\gamma}{dT}}{\mu D_T\alpha^2} = 8\left(1 + \frac{h}{\alpha k}\right). \tag{6.35}$$

That is, stability depends on modified Marangoni and Nusselt numbers in which
α^{-1} instead of H is the characteristic length. The small magnitude of N_{Cr} for large
H suggests that very small interfacial deflections can be expected. If the layer is
of infinite depth, it is unstable for any nonzero value of the initial temperature
gradient ζ at sufficiently long wavelengths. The reason is that viscous resistance
to flow is much reduced by removal of the solid surface. Of course, wavelengths
exceeding the lateral dimensions of the experimental apparatus are impossible,
so that in practical cases the temperature gradient must reach some critical finite
value for instability to be observed.

Several extensions of the above analysis are possible. Frequently instability
is not seen until Marangoni numbers significantly higher than the critical value
predicted by the above analysis are reached (Berg et al., 1966). The enhanced
stability can be attributed to small amounts of surface active contaminant. Flow
produced by interfacial temperature gradients sweeps surfactant from warm to
cool regions, causing an increase in interfacial tension at the former and a decrease
at the latter. This effect opposes the original interfacial tension gradient produced
by temperature variations and hence acts to diminish the flow. Relatively small
amounts of an insoluble surfactant can prevent the instability from occurring
altogether (Berg and Acrivos, 1965). Palmer and Berg (1971) analyzed the situation for a soluble surfactant where diffusion somewhat reduces this strong
stabilizing effect. They also confirmed some of their predictions experimentally.

In Benard's original experiments and in the diagram of Figure 6.2, the liquid
layer is heated from below and the temperature profile is linear. If convection is
generated by the cooling effect of evaporation at the fluid interface, the temperature falls rapidly near the interface but more slowly in the remainder of the layer.
Flow from within the layer does not, therefore, raise the surface temperature at
a point such as P of Figure 6.2 as much as would be the case if the temperature
gradient near the interface were present throughout the entire layer. Accordingly,
the interface is much more stable than would be expected from the above analysis,
as shown by Vidal and Acrivos (1968).

While interfacial tension gradients are responsible for instability in thin liquid layers, natural convection due to adverse density gradients becomes an important destabilizing influence with increasing layer thickness. Nield (1964) analyzed this situation, finding, as expected, that in layers of intermediate thickness, instability occurs more readily with both effects present than with either alone. His predictions regarding dependence of the stability condition on layer thickness were confirmed experimentally by Palmer and Berg (1971). Generally speaking, Marangoni flow is the major destabilizing factor in layers of organic liquid less than about 1 mm thick, while natural convection is more important in thicker layers.

The linear stability analysis we have considered predicts when the instability will occur and frequently also the wavelength to be expected. However, it is unable to predict whether the instability will lead to a steady state with cellular convection, as Benard observed, or to ultimate breakup of the interface, which is seen in many hydrodynamic instabilities such as the Rayleigh–Taylor instability for superposed fluids discussed in Chapter 5. Moreover, even assuming that steady convection does develop, linear analysis yields no information on preferred cell shape. While, for simplicity, we have considered here only two-dimensional disturbances ("roll cells"), the corresponding three-dimensional linear analysis yields only a critical wavenumber which is consistent with various cell shapes, such as roll cells, hexagons, and rectangles. Nonlinear analysis has led to some progress in understanding the development of cellular convection and the conditions when certain cell shapes are preferred (Colinet et al., 2001; Hinkbein and Berg, 1970; Wollkind and Segel, 1970).

EXAMPLE 6.1 CONDITIONS FOR DEVELOPMENT OF MARANGONI INSTABILITY

Determine the conditions for development of instability when N_{Nu} is very small, the initial temperature gradient $(-\zeta)$ is $1°K/cm$, and the liquid has the following properties:

$$\rho = 0.8 \text{ g/cm}^3$$
$$\mu = 10 \text{ cp}$$
$$\gamma = 30 \text{ mN/m (at initial temperature)}$$
$$D_T = 1.3 \times 10^{-3} \text{ cm}^2/\text{sec}$$

$$\frac{d\gamma}{dT} = -0.1 \; mN \, / \, (mK) \; .$$

SOLUTION

Since the thickness H of the liquid layer is unknown, we first estimate H assuming that $N_{Cr} = 0$. According to Figure 6.2, the critical value of the Marangoni number N_{Ma} is about 80. From Equation 6.27, we find $H = 0.32$ cm. From Equation 6.32, we find $N_{Cr} = 1.3 \times 10^{-5}$, which is indeed negligible. The critical wavenumber satisfies $\alpha_c^* = \alpha_c H = 2$, which corresponds to a critical wavelength of λ_c $(= 2\pi/\alpha_c)$ of 1.0 cm.

4. INTERFACIAL INSTABILITY DURING MASS TRANSFER

Spontaneous flow, often known as interfacial turbulence, can develop near an interface across which one or more species is being transferred. Frequently the flow patterns are less regular than the Benard cells of Figure 6.1, though a basic periodicity is usually discernable. Its source is interfacial tension gradients produced by interfacial concentration gradients. Variation of interfacial tension with concentration is typically rather strong. Hence when both temperature and concentration vary along an interface, as in evaporating multicomponent liquids, the concentration effect usually dominates in determining whether and under what conditions interfacial turbulence develops. When it does, mass transfer rates normally increase several fold.

Absorption of a solute from a gas phase into a thin liquid layer and the reverse process of desorption provide the closest mass transfer analogies to the analysis of heat transfer given in Section 3. Indeed, with one important exception, the analysis can be used directly. When the interfacial energy balance of Equation 6.28 is converted to a solute mass balance, an additional term must be added in the frequently encountered case of a solute having some surface activity. This term corresponds to the second term of Equation 6.2 and represents solute transport along the interface by convection. As might be expected from the discussion of surfactant effects in Section 3, it provides a significant stabilizing effect in the desorption case, which is the mass transfer analog of a liquid layer heated from below. That is, values of N_{Ma} required for instability are much higher than predicted by the analysis of Section 3 (Brian, 1971), a conclusion that is in agreement with experimental findings. Absorption of a solute with some surface activity is the analog of a liquid cooled from below, and hence no instability arises.

Instability accompanying solute transfer across a liquid-liquid interface was first analyzed by Sternling and Scriven (1959). Let us consider the situation shown in Figure 6.5 where solute diffuses from phase A to phase B, and further suppose

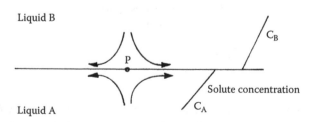

FIGURE 6.5 Convection due to interfacial tension gradients during solute transfer between immiscible liquid phases.

that interfacial tension decreases with increasing solute concentration. If a small disturbance at point P of Figure 6.5 brings some solute-rich liquid from the interior of phase A to the interface, the resulting interfacial tension gradient generates a flow that draws more liquid in A to the interface as shown, thus promoting instability. But the flow also brings solute-lean fluid from B to the interface, which acts to increase interfacial tension at P and oppose the instability. Thus, whether instability develops depends on the relative magnitudes of these opposing effects.

The analysis is very similar to that of Section 3 except, of course, flow and transport in both phases must be considered. When $\alpha^* \gg 1$ (i.e., when both fluids are of great depth), and when solute convection along the interface can be ignored, the marginal stability condition is that obtained by Sternling and Scriven (1959):

$$(N_{Ma})_A = \frac{8\left(m_{AB}\dfrac{D_B}{D_A} + 1\right)\left(\dfrac{\mu_B}{\mu_A} + 1\right)}{\left(1 - \dfrac{D_A}{D_B}\right)}, \qquad (6.36)$$

where the Marangoni number is given by

$$(N_{Ma})_A = \frac{\left(\dfrac{d\gamma}{dC_A}\right)\left(\dfrac{dC_{Ai}}{dz}\right)}{\mu_A D_A \alpha^2}. \qquad (6.37)$$

Also, m_{AB} is the distribution coefficient C_B/C_A for the solute at the interface, D_A and D_B are the solute diffusivities in A and B, $d\gamma/dC_A$ is the rate of change of interfacial tension with respect to the interfacial concentration in phase A, and dC_{Ai}/dz is the initial (uniform) concentration gradient in A.

Equation 6.36 is closely related to the heat transfer result in Equation 6.35 in the limit of large α^*. Indeed, setting $\mu_B \ll \mu_A$ and $D_A \ll D_B$, as might be expected in a liquid-gas system, Equation 6.36 becomes

$$(N_{Ma})_A = 8\left(1 + m_{AB}\dfrac{D_B}{D_A}\right). \qquad (6.38)$$

This equation still differs from Equation [6.35]. But if one recognizes that (a) the analog of the "distribution coefficient" in heat transfer is unity, (b) D_B/D_A is equal to the ratio of the concentration gradients $(dC_{Ai}/dz)/(dC_{Bi}/dz)$, and (c) $h/k\alpha$ in Equation 6.35 is the ratio of the temperature gradient in the liquid to the quantity $\alpha \Delta T_{gas}$, which has the dimensions of a temperature gradient, the relationship between the two results becomes clear.

Since $(N_{Ma})_A > 0$ for the case of a surface active solute diffusing from A to B, we see from Equation 6.36 that instability can occur only when $D_A < D_B$ and that, moreover, the value of $(N_{Ma})_A$ required for instability becomes larger as D_A becomes more nearly equal to D_B. These results can be explained in terms of the steeper concentration gradient in A than B for $D_A < D_B$. Since solute concentration increases rapidly in A with distance from the interface, the flow transports liquid quite rich in solute to point P of Figure 6.5. Thus the destabilizing effect due to the flow in A outweighs the stabilizing effect of the flow in B. We note that no matter how small the mass transfer rate, Equation 6.36 predicts that instability can occur for sufficiently long wavelengths (small α).

Equation 6.36 also indicates that the critical value of $(N_{Ma})_A$ decreases when the viscosity ratio μ_B/μ_A decreases. We might expect high values of μ_A to be associated with low velocity gradients in A, i.e., a slow falloff in tangential velocity with distance from the interface. Under these conditions the flow extends deeply into the bulk of liquid A and can bring to the interface material that is quite rich in solute, a destabilizing effect.

Finally, Equation 6.36 shows that low values of the distribution coefficient m_{AB} promote instability. The reason is that solute concentration in B is low under these conditions and the stabilizing influence of the flow in B is reduced.

For a solute that increases interfacial tension, $(N_{Ma})_A$ is negative for solute diffusion from A to B, and instability can occur only if $D_B < D_A$, according to Equation 6.36. This result is entirely reasonable since in this case it is the flow in B that promotes instability and the flow in A that opposes it, just the opposite of that found for surface active solutes.

The stability condition given by Equation 6.36 is, unfortunately, not the whole story. As we saw in Chapter 5, β is in general a complex number with its real part β_R describing disturbance growth or decay and its imaginary part β_I an oscillatory motion. It is therefore possible to pass from a stable to an unstable region via a state where oscillatory motion of constant amplitude occurs with $\beta_R = 0$ and $\beta_I \neq 0$. Under these conditions the system is said to be in a state of "marginal oscillatory stability," whereas the condition of Equation 6.36 with $\beta_R = \beta_I = 0$ describes states of "marginal stationary stability" (i.e., without oscillation).

Both types of marginal stability were encountered in Chapter 5. For gravitationally produced instability of superposed fluids (Section 2.3) and the other cases where the interface was unstable from a thermodynamic point of view, no oscillatory motion was found at marginal stability. But the state of marginal stability was an oscillatory one for the Kelvin–Helmholtz instability (Section 8), with traveling waves of constant amplitude moving along the interface.

In the present situation, both types of marginal stability are possible, as Sternling and Scriven (1959) demonstrated in their original paper. To find the states of marginal oscillatory stability, one must use the governing equations in their general forms, such as Equations 6.12 (adapted to the case of mass transport) and 6.14. The basic analysis is as before, but the equation for β obtained by setting the determinant of coefficients of the boundary conditions equal to zero is rather complicated and must be solved numerically. We shall omit the details of this procedure and simply quote some key results (Sternling and Scriven, 1959).

For the case of a solute that lowers interfacial tension, we have seen already that instability occurs when transport is from the phase of lower diffusivity to one of higher diffusivity. With the general analysis we can draw the following conclusions:

(a) If $D_A/D_B < 1$, but $v_A/v_B > 1$, instability occurs only when solute transfer is from A to B. Both these conditions favor instability for transfer from A to B and stability for transfer from B to A by the physical arguments given above.

(b) If $D_A/D_B < 1$ and $v_A/v_B < 1$, instability occurs for transport in both directions. The diffusivity ratio is favorable to instability when transfer is from A to B, while the kinematic viscosity ratio is favorable to instability for transfer in the opposite direction. We remark that systems of type (a) are more common than those of type (b).

(c) If $D_A = D_B$, instability occurs only when transfer is from the phase of higher kinematic viscosity to that of lower kinematic viscosity, as might be expected from the above results.

Generally speaking, experimental results are in agreement with these predictions, although exceptions are known. As Berg (1972) has pointed out, some of the exceptions may be due to factors not considered in the analysis, such as natural convection effects, nonuniform transfer rates due to system geometry, and solute transport along the interface by convection. The latter effect, which can greatly enhance interfacial stability, has been incorporated in an analysis for liquid-liquid systems by Gouda and Joos (1975).

The discussion thus far has, for simplicity, been limited to plane interfaces. Burkholder and Berg (1974) studied the instability of fluid cylinders and low velocity jets when there is mass transfer between phases. They found that tangential flow produced by variations in interfacial tension modifies the stability condition from that given in Chapter 5 (Section 4) for a fluid cylinder in the absence of mass transfer. In particular, transfer of solute from a liquid cylinder into a surrounding gas phase has a stabilizing effect when the solute is surface active. The reason can be seen from Figure 6.6. When the cylinder begins to deform, the usual capillary instability causes liquid rich in solute to flow from thin to thick portions of the cylinder. But the resulting increase in

FIGURE 6.6 Tangential flow during axisymmetric deformation of a fluid cylinder. In the absence of interfacial tension gradients flow is from Q to P as shown.

(a)   (b)

FIGURE 6.7 Schematic illustration of surface tension gradients on the stability of the neck in drop formation. Shaded areas denote a liquid of higher surface tension, and the arrows indicate the direction of interface movement. In (a), Marangoni flow hinders and in (b) it promotes drop break-off. From Bainbridge and Sawistowski (1964) with permission.

solute concentration at P reduces interfacial tension there and causes a reverse flow that opposes the instability. A similar argument shows that transfer of a surface active solute from the surrounding gas into the jet has a destabilizing effect, i.e., the distance from where the jet is formed to where it breaks up is reduced. These effects were confirmed in a qualitative manner by simple experiments.

These effects of Marangoni flow on stability apply, of course, not only to jets deliberately injected into a fluid but also to fluid cylinders formed during mixing or agitation. At high vapor rates during distillation with sieve plate columns, for instance, agitation of the liquid causes more or less cylindrical protrusions to develop, as illustrated in Figure 6.7. If the more volatile component has the lower surface tension, the argument given above indicates that Marangoni flow has a stabilizing effect when a neck starts to form, thus hindering drop break-off (Figure 6.7a). But if the less volatile component has the lower surface tension, its transfer from vapor to liquid generates a Marangoni flow that promotes neck thinning and drop break-off (Figure 6.7b). Since the latter behavior increases interfacial area, plate efficiencies are, other factors being equal, higher in the latter case (Bainbridge and Sawistowski, 1964).

For a liquid jet in an immiscible liquid, the situation is more complicated because the interfacial concentration is influenced by transport in both phases and, as discussed above for plane interfaces, the effects are in opposite directions. The original paper (Burkholder and Berg, 1974) and an extension incorporating relative motion between the phases due to overall jet motion (Coyle et al., 1981) should be consulted for details of the theoretical analysis. Figures 6.8 and 6.9 show theoretical predictions and experimental measurements of jet length for benzene jets in water with acetone transferring into the aqueous phase (Berg, 1982). The major prediction of the analysis, viz., a strong destabilizing effect of the Marangoni flow is confirmed by the experimental results.

Various other extensions of the analysis of Marangoni instability during mass transfer have been made. They incorporate such phenomena as chemical reactions in the bulk fluids (Nelson and Berg, 1982; Ruckenstein and Berbente, 1964) and at the interface (Sanfeld et al., 1979; Vedove and Sanfeld, 1981; von Gottberg et al., 1995), nonlinear initial concentration gradients (Sorensen, 1979), heat of solu-

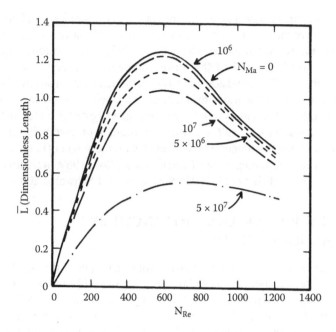

FIGURE 6.8 Computed linear stability results for normalized length of a benzene jet in water (undergoing outward transfer of acetone) as a function of the jet Reynolds number is predicted to decrease the jet length considerably, particularly in the region of jet length maximum. Reprinted from Coyle et al. (1981) with permission.

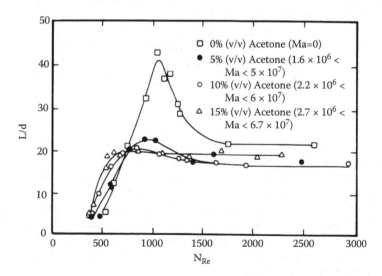

FIGURE 6.9 Jet breakup length for the system of Figure 6.8. L/d is the ratio of jet length to the mean diameter. Reprinted from Coyle et al. (1981) with permission.

tion effects in binary systems where interfacial concentration gradients are not possible at constant temperature (Perez de Ortiz and Sawistowski, 1973), and spherical geometry (Sorensen and Hennenberg, 1979). Also, careful experimental observations have been made of the convection patterns generated in various systems (Linde et al., 1979; Sawistowski, 1971), although many features of the observations as yet lack explanation. In addition, the stability of contact lines in the presence of both heat and mass transfer has been investigated (Charles and Cazabat, 1993; Kataoka and Troian, 1997). Further information on instability produced by Marangoni flow may be found in review articles on the subject (Berg, 1972, 1982; Berg et al., 1966; Kenning, 1968; Levich and Krylov, 1969; Rabinovich, 1991; Sawistowski, 1971; Birikh et al, 2003; Narayanan and Schwabe, 2003).

5.   OTHER PHENOMENA INFLUENCED BY MARANGONI FLOW

Interfacial tension gradients are responsible for other types of flow besides cellular convection. For instance, a gas bubble in a liquid subjected to a temperature gradient moves toward the high temperature region because by so doing it reduces its interfacial free energy. The flow that produces this motion is generated by interfacial tension gradients and, for the case of low Reynolds numbers, can be analyzed using the basic equations for flow near a drop or bubble given in Chapter 7. In a similar manner Marangoni flow can cause motion of drops or bubbles in a fluid having a concentration gradient. The Marangoni effect has been used to produce flow in space, where gravity is absent (Subramanian and Balasubramaniam, 2001), and in microfludic devices.

The time required for coalescence to occur when two drops or bubbles approach one another is usually determined mainly by the time required for drainage of the fluid film between them. Marangoni flow can, under suitable conditions, either enhance or retard film drainage. Suppose, for example, that a solute that lowers interfacial tension transfers from the dispersed to the continuous phase. We expect solute concentration to build up faster in the film than in the bulk of the continuous phase, with a resulting decrease in the interfacial tension of film surfaces. The resulting Marangoni flow speeds drainage and promotes coalescence. Such behavior can have a significant effect in mass transfer processes, being unfavorable when it is desirable to maximize the area for mass transfer and hence to minimize coalescence. Note that for transfer of the same solute from the continuous to the dispersed phase, the interfacial tension gradient is reversed and its effect on flow is to retard drainage and coalescence. For a solute that increases interfacial tension, a similar argument shows that transfer into the continuous phase slows and transfer into the dispersed phase speeds up film drainage and coalescence.

When distillation or gas absorption is carried out in a packed column, it is important for maintaining a large area for transfer that the liquid flows as a film covering the entire solid surface and that it not break up into rivulets. Let us consider binary distillation for a system where the more volatile component has

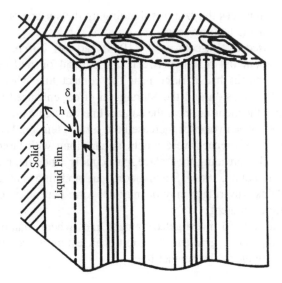

FIGURE 6.10 Falling film subject to disturbances leading to the formation of rivulets. From Johnson and Berg (1983).

the higher surface tension. If a transverse wavy perturbation develops along a falling liquid film, as illustrated in Figure 6.10, we would expect the concentration of the more volatile species to be greater in the thick than in the thin portions of the film at a given height. The resulting interfacial tension gradient draws liquid from thin to thick regions and promotes film breakup into rivulets. In contrast, Marangoni flow opposes ripple formation when the more volatile component has the lower surface tension. Zuiderweg and Harmens (1958) found that mass transfer rates for similar conditions were about half as large in the benzene-n-heptane system, where rivulet flow is favored, than in the n-heptane-toluene system, where it is not. Wang et al. (1971) presented an analysis of this situation. Johnson and Berg (1983) extended the analysis to include the presence of a third component which is highly surface active, since such materials have been found in some cases to reduce rivulet formation.

As a final example we mention the effect of interfacial tension gradients on the rate of spreading of a liquid drop over a solid surface. We consider a drop of two components of which the more volatile has the lower surface tension. Near the drop periphery, where a thin advancing layer of liquid forms (see Chapter 7 for an extensive discussion of the spreading process), evaporation causes liquid composition to be leaner in the more volatile component than in the bulk drop and surface tension to be higher. An interfacial tension gradient arises that enhances flow from the bulk drop to the advancing layer and hence increases the rate of spreading, an effect that has been observed by Bascom et al. (1964). This mechanism is responsible for the formation of wine tears. To a first approximation, wine may be considered a mixture of ethanol and water, with the former being more volatile and having a lower surface tension.

Marangoni flow enhances the rate of spreading up the surface of the glass until enough liquid collects to form tears.

The liquids used by Bascom et al. (1964) had low differences in surface tension values, about 0.4 mN/m between the original oil and oil stripped of volatiles. The meniscus had a ridge near the contact line when the volatiles lowered the surface tension and the Marangoni effect enhanced the spreading. On the other hand, the meniscus near the contact line was a thin film that thickened somewhat abruptly when volatiles increased the surface tension. When the values of surface tension differences were much larger, along with larger Marangoni effects, catastrophic spreading took place, particularly when the liquid was non-wetting (Cottington et al., 1964). Some theoretical modeling is available (Neogi, 1985) that explains why ridges may or may not form depending on the sign of the surface tension gradient.

Cazabat et al. (1990) investigated contact lines where Marangoni effect is induced by a temperature gradient. An additional feature is seen in the form of a periodic structure along the contact line.

6.   NONEQUILIBRIUM INTERFACIAL TENSIONS

When a system contains a soluble surfactant, diffusion, adsorption, and desorption of surfactant may cause interfacial tension to vary with both position and time. If, for instance, a fresh interface is formed on a stagnant pool of a surfactant solution, interfacial tension is found to decrease with time as surfactant diffuses to the interface and adsorbs.

Let us consider such a situation with a plane interface. If there is no convection, the diffusion equation in the bulk liquid ($z > 0$) is given by

$$\frac{\partial c}{\partial t} = D \frac{\partial^2 c}{\partial z^2} . \tag{6.39}$$

Initially c has its bulk value c_∞ for all $z > 0$. Also, at all times, $c \to c_\infty$ far from the interface (i.e., as $z \to \infty$). Were the solution concentration $c(0,t)$ at the interface known, as a function of time, Equation 6.39 could be solved. Instead, it is known from the interfacial mass balance that

$$\frac{\partial \Gamma}{\partial t} = D \frac{\partial c}{\partial z}\bigg|_{z=0} . \tag{6.40}$$

With the additional condition that $\Gamma(0) = 0$, the solution for $\Gamma(t)$ is given by

$$\Gamma(t) = 2\left(\frac{D}{\Pi}\right)^{1/2}\left[c_\infty t^{1/2} - \int_0^{t^{1/2}} c(0, \, t - \tau) \, d(\tau^{1/2})\right]. \tag{6.41}$$

This equation was first derived by Ward and Tordai (1946) and later by Hansen (1961), who used the method of Laplace transforms.

If local equilibrium prevails at the interface, i.e., if adsorption is rapid, $\Gamma(t)$ and $c(0,t)$ are related by an adsorption isotherm. Moreover, interfacial tension $\gamma(t)$ can be obtained by combining the isotherm with the Gibbs adsorption equation to derive a surface equation of state, as shown in Chapter 2 (Section 7). Hence, if the isotherm is known, both $\Gamma(t)$ and $\gamma(t)$ can be found for given values of c_∞ and D. Alternatively, the measured behavior of interfacial tension $\gamma(t)$ can be used with the known isotherm and equation of state to find the diffusion coefficient D.

Equations 6.39 and 6.40 can be made dimensionless by scaling c and Γ to c_∞ and Γ_∞ and defining the characteristic length L and characteristic time τ as Γ_∞/c_∞ and $\Gamma_\infty^2/Dc_\infty^2$, respectively. Here Γ_∞ is the interfacial concentration in equilibrium with c_∞. As D does not vary greatly from surfactant to surfactant, one sees that L and τ are larger for more surface active species (i.e., those with large values of Γ_∞/c_∞). That is, equilibration takes longer because saturation of the interface requires that molecules diffuse to it from a greater distance L. For surfactants with small values of Γ_∞ (i.e., large areas per molecule), L and τ are small, and surface tension decreases more rapidly toward an equilibrium value.

When the adsorption isotherm is linear, so that $\Gamma(t) = Hc(0,t)$, with H the Henry's law constant (see Chapter 2), an analytical solution exists (Sutherland, 1952):

$$\Gamma(t)/\Gamma_\infty = \{1 - \exp(Dt/H^2)\mathrm{erfc}[(Dt)^{1/2}/H]\} = \{1 - \exp(t/\tau)\mathrm{erfc}[(t/\tau)]^{1/2}\}. \qquad (6.42)$$

Here $\Gamma_\infty = Hc_\infty$ has been invoked. Equation 6.42 confirms that τ is the characteristic time for equilibration for diffusion-controlled processes, a conclusion more widely applicable than the linear isotherm itself, which holds only for low surface coverages (see Ferri and Stebe, 2000). Variation of surface tension with time can be found by combining Equations 2.48 and 6.42.

A more realistic isotherm is the Langmuir isotherm:

$$\Gamma = \Gamma_o[c/(c + a)], \qquad (6.43)$$

where Γ_o is the limiting value of Γ for large concentrations c and a is a constant that decreases as the surface activity increases. Miller and Kretzschmar (1991) tabulated coefficients for an approximate polynomial solution for Γ based on a numerical solution of the problem.

Another possible isotherm is the Frumkin isotherm:

$$\Gamma = \Gamma_o\{c/[c + a \exp(B\Gamma/\Gamma_o)]\}. \qquad (6.44)$$

Here B is a parameter proportional to the interaction energy between adsorbed molecules. When $B = 0$, Equation 6.44 simplifies to the Langmuir isotherm

(Equation 6.43). Of interest here is that for $B < -4$, a surface phase separation occurs into liquid and gaseous phases. As long as both phases coexist, surface tension remains unchanged even though adsorption continues. In this way, long "induction times" with constant tension observed in some systems can be explained (Ferri and Stebe, 1999; Lin et al., 1991). Once the surface concentration reaches the value for which the gaseous phase disappears, tension decreases as further adsorption occurs.

Although the assumptions of rapid adsorption and local equilibrium at the interface are justified in many situations of interest, sometimes the rates of adsorption and desorption must be considered. Equation 6.41 still applies, but the analysis must be modified. The terms "adsorption barrier" and "desorption barrier" are sometimes used when kinetic limitations exist for the respective processes. If a surface active solute diffuses between phases under conditions where there is an appreciable desorption barrier, for example, interfacial concentration Γ will attain higher values than in the absence of the barrier, and interfacial tension will be lower. England and Berg (1971) and Rubin and Radke (1980) have studied such situations. Figure 6.11 shows an example of predicted interfacial tension as a function of time for various values of a dimensionless rate constant. The low transient interfacial tension is evident.

Another interfacial rate process that has been invoked to explain dynamic surface tension data for some pure nonionic surfactants is reorientation of adsorbed surfactant molecules between a state in which they lie nearly flat along the surface, which is favored at low values of Γ, and a state in which their orientation is nearly vertical, which is favored at high values of Γ (Fainerman et al., 1996; Miller et al., 1999). Neutron reflection data are consistent with such an interpretation (Lu et al., 1993). Further discussion and references dealing with

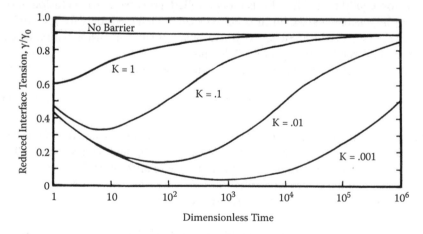

FIGURE 6.11 Time dependence of the interfacial tension in the presence of a desorption barrier. Smaller values of K correspond to larger desorption barriers. Reprinted from England and Berg (1971) with permission.

interfacial tension variation when rates of adsorption, desorption, or reorientation must be considered can be found in Dukhin et al. (1995) and Joos (1999).

One important case is where the solute is charged, which occurs very frequently with surfactants and almost always with proteins. In this case, the charge on the surface where the adsorption is taking place hinders the movement of the solute, which has a charge of the same sign. In addition, the hindrance increases with time as more adsorption gives rise to a larger surface charge and larger repulsion. The governing equations are nonlinear and difficult to solve. An approximate solution for adsorption of sodium n-dodecyl sulfate (SDS) at the water-oil interface and below the critical micelle concentration (CMC) has been given for situations close to equilibrium (Bonfillon et al., 1994; Datwani and Stebe, 2001). Narsimhan and Uraizee (1992) provide a more complete, but nevertheless an approximate solution for adsorption of proteins. The solution takes into account the configuration of the adsorbent at the interface, such as that discussed in the preceding paragraph. Agreement with experiments in both cases appears to be very good. With exceptions such as these, the developments given below are generally confined to nonionic solutes.

While the above equations were derived for a plane interface, they may also be used for curved interfaces provided that the radii of curvature are much greater than the thickness of the layer in the bulk solution where the concentration deviates significantly from c_∞. This thickness is, to an order of magnitude, $(Dt)^{1/2}$ and hence only about 0.1 mm for $D = 5 \times 10^{-6}$ cm²/sec and $t = 20$ sec. Thus interfacial tension data $\gamma(t)$ obtained with the oscillating jet technique described in Chapter 5 may be used with Equation 6.41 in many cases. Another method of measuring dynamic interfacial tension is the falling meniscus method discussed in Problem 6.4.

In the preceding examples there was no convection, and transport of the surfactant was by diffusion alone. A situation of considerable practical interest where convection must be considered is transport of a surface active solute in the vicinity of a growing drop or bubble with radius $R(t)$. If the flow is radial, continuity requires that

$$v_r = (R^2/r^2)(dR/dt). \tag{6.45}$$

This expression may be substituted into the equation for the conservation of solute in each phase, which is given by

$$\frac{\partial c}{\partial t} + v_r \frac{\partial c}{\partial r} = D\left[\frac{\partial^2 c}{\partial r^2} + \frac{2}{r}\frac{\partial c}{\partial r}\right]. \tag{6.46}$$

The surfactant mass balance at the interface takes the form

$$\frac{\partial \Gamma}{\partial t} = -D_1 \left. \frac{\partial c_1}{\partial r} \right|_R + D_2 \left. \frac{\partial c_2}{\partial r} \right|_R - \Gamma \frac{d \ln A}{dt} \, , \qquad (6.47)$$

where subscripts 1 and 2 denote the interior and exterior phases.

MacLeod and Radke (1994) presented a solution to the above equations applicable when both phases have uniform compositions initially, local equilibrium is maintained at the interface, drop volume increases at a constant rate, and the thickness δ_1 of each diffusion boundary layer, which is on the order $(D_1 t)^{1/2}$, is small in comparison with R. An interesting feature of this analysis is that it predicts that a steady-state interfacial tension will be reached. Since R is large for these conditions, the last term of Equation 6.47 may be neglected and one finds

$$D_1 \frac{(c_{1\infty} - c_{1s})}{\delta_1} = D_2 \frac{(c_{2\infty} - c_{2s})}{\delta_2} \, , \qquad (6.48)$$

where $c_{i\infty}$ and c_{is} are the bulk and steady-state interfacial concentrations in phase i. Making use of the above relationship between δ_i and D_i, this equation may be written as

$$\frac{c_{1s}}{c_{1\infty}} = \left[1 + \frac{c_{2s}}{c_{2\infty}} \cdot \frac{D_1}{D_2} \right] \bigg/ \left[1 + K \frac{D_1}{D_2} \right] , \qquad (6.49)$$

where K is the partition coefficient c_{2s}/c_{1s}. This equation was also obtained by van Hunsel and Joos (1987).

According to Equation 6.49, the interfacial and bulk concentrations are not equal at steady state unless either $c_{2\infty}/c_{1\infty} = K$ (i.e., the bulk phases are in equilibrium) or $c_{2\infty} = K = 0$ (i.e., the surfactant is insoluble in phase 2). Hence, with the same exceptions, the steady-state interfacial tension is not the ultimate equilibrium value that would be measured after equilibration of initial phases. Figure 6.12 compares predictions of the theory with experimental results for drops consisting of aqueous solutions of decanol growing at different rates in air. Initially interfacial tension increases with time because, for a fixed rate of volume increase, the fractional rate of increase in interfacial area is large when the radius is small. Later the tension reaches a maximum and decreases as the fractional rate of increase in interfacial area decreases. The steady-state tension is about 45 mN/m, which is considerably higher than the equilibrium value of 35 mN/m for the given alcohol concentration $c_{1\infty}$ in the bulk aqueous phase.

So far we have used the term "surface active" in an all-encompassing way, with the only stipulation that such species be soluble in the bulk fluid. Needless to say, exceptions exist. For example, when the solute is a surfactant above the CMC, many investigators consider that micelles do not adsorb as a whole on the surface, but instead that only the singly dispersed amphiphiles adsorb. In that

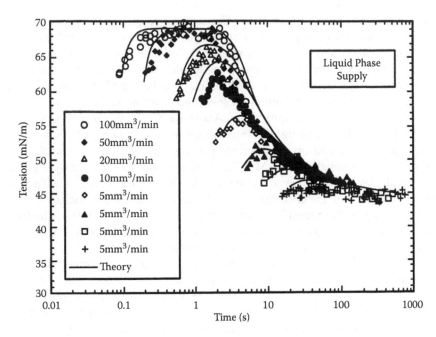

FIGURE 6.12 Model predictions and experimental dynamic surface tension data for 1.7×10^{-4} kmol/m^3 aqueous decanol drops growing into decanol-free air. The equilibrium value is 35 mN/m. Reprinted from MacLeod and Radke (1994) with permission from Elsevier.

case, once adsorption starts, the adsorption process can become limited by the rate at which micelles demicellize to give rise to singly dispersed amphiphiles. This rate is connected to the demicellization time constant τ_2 (see Chapter 4, Section 2). It is possible to write the conservation equations of both micelles and singly dispersed amphiphiles to include diffusion and micellization-demicellization. Eventually the results can be cast into a generalized Sutherland form (Equation 6.42) containing τ_2 as an additional parameter, thus allowing the extraction of τ_2 from the dynamic surface tension data (Noskov, 2002). Different experimental techniques lead to values of τ_2 separated by many orders of magnitude for sodium dodecyl sulfate. Others have obtained a better fit for a nonionic surfactant (Vlahovska et al., 1997). It should be noted that the basic theory of micellization-demicellization kinetics is actually available for nonionics only, although it continues to be applied to ionic surfactants as well. Clearly more needs to be done.

It is possible to include in a model a small but nonzero volume of the interfacial region in analyzing transport phenomena such as mass transfer across or to an interface. Mathematical construction can now be performed to divide the domain into a small interfacial region, called the inner region, where all surface effects apply. The bulk is the outer region where usual convective-diffusive mass transfer (and possibly reactions as in micellization-demicellization) dominate. This is exactly the method of matched asymptotic expansions

discussed in Chapter 7 (Section 9). The separate results in these two regions
have to match. This analysis given by Brenner and Leal (1978, 1982) helps us
understand how boundary conditions to Equation 6.46, such as Equation 6.47,
arise. In fact, this is no more than a specific and a more detailed treatment of
how conservation equations for the "surface" are separated from those in the
"bulk" described at the beginning of this chapter for dynamic systems and in
Chapter 1 for systems at equilibrium. Above the CMC, this analysis provides
a direction for investigating dynamic surface tension in two parts. Since only
single amphiphiles adsorb, the surface part can be assumed to be unchanged
from the situation below the CMC. The bulk part was investigated by Weinheimer et al. (1981), who showed that it is possible to define an effective diffusivity
where unspecified micellization-demicellization kinetics are included. This
holds as long as the system is not far from equilibrium (Neogi, 1994). The form
for effective diffusivity has been experimentally verified in both flow (Weinheimer et al., 1981) and nonflow (Leaist, 1986a,b) systems.

Generally the inventory at a liquid-liquid interface as a function of time is
measured only indirectly by measuring the interfacial tension. This cannot be
done at a solid surface. The displacement of a less surface active species by a
more surface active species is of importance in underground remediation operations, where an adsorbed pollutant is released to the flowing water by displacing
it with a surfactant. Such steps are also of importance in physiological processes.
To monitor the process in a laboratory, one of the species can carry a tracer
(optical in the case below). Consider a solid-liquid interface with a weakly
adsorbed polymer A (polystyrene), where the liquid and the interface are in
equilibrium. The liquid containing A is now removed and replaced with a liquid
containing a polymer B (polyisoprene or polymethylmethacrylate), which is more
surface active. Subsequently A is lost from the interface and if A contains a tracer,
its concentration at the interface can be monitored. The rate can be written as

$$\frac{d\Gamma_A}{dt} = k_a(\Gamma_{A\infty} - \Gamma_A)c_A\big|_{x=0} - k_d\Gamma_A = -D_A\frac{\partial c_A}{\partial x}\bigg|_{x=0}, \qquad (6.50)$$

where $x = 0$ is the location of the interface and k_a and k_d are the adsorption and
desorption rate constants. $\Gamma_{A\infty}$ is the saturation value of Γ_A in the adsorption term.
Compare Equation 6.50 to Equation 6.47 and the equation above Equation 2.30.
Douglas et al. (1993) found that at high temperature, 45°C, the decay of the signal
was exponential, indicating desorption control. At low temperature, 0°C, it was
proportional to the square root of time t, which made it diffusion controlled,
although these authors provide a different explanation (see Problem 6.11). The
issues of competitive transport and adsorption, among others, are of importance
in the study of dynamic surface tensions of surfactant mixtures at liquid-fluid
interfaces (Daniel and Berg, 2002; Mulqueen et al., 2001).

7.   EFFECT OF SURFACTANT TRANSPORT ON
WAVE MOTION

Diffusion also influences the damping of capillary wave motion by surfactants. It can be shown (see Problem 6.6) that when surfactant diffusion is included in the analysis, the surface elasticity term $\Gamma_{os}(-d\gamma/d\Gamma_s)$ in Equation 5.71 must be divided by the quantity $(1 + (\alpha q D/\beta(d\Gamma_s/dc))$, where D is the diffusion coefficient, $d\Gamma_s/dc$ the slope of the adsorption isotherm for the initial surface concentration, and $q^2 = [1 + (\beta/D\alpha^2)]$. If capillary waves with an angular frequency $\bar{\beta}$ are imposed by vertical oscillation of a bar at the surface, as discussed in Chapter 5, and if the usual low viscosity assumption is made along with its diffusional counterpart $|\beta/D\alpha^2| \gg 1$, manipulation of the above expression leads to the following generalized elasticity coefficient E which governs the surfactant effect on wave damping (see Problem 6.6):

$$E = \Gamma_{os}\left(-\frac{d\gamma}{d\Gamma_s}\right)\left(1 + \frac{\alpha q D}{\beta(d\Gamma_s / dc)}\right)^{-1}$$

$$(6.51)$$

$$= \Gamma_{os}\left(-\frac{d\gamma}{d\Gamma_s}\right)\left[\frac{1 + B + iB}{1 + 2B + 2B^2}\right]$$

$$B = \left(\frac{d\Gamma_s}{dc}\right)^{-1}\left(\frac{D}{2\bar{\beta}}\right)^{1/2}. \qquad (6.52)$$

We see from these equations that for high frequencies $\bar{\beta}$, B is small, the elasticity is $\Gamma_{os}(-d\gamma/d\Gamma_s)$, and the behavior is that found in Chapter 5 for an insoluble monolayer. In this case, oscillation is so fast that bulk diffusion is not able to appreciably diminish the interfacial concentration gradients produced by the oscillation. On the other hand, at low frequencies $B \gg 1$ and we find

$$E = \Gamma_{os}\left(-\frac{d\gamma}{d\Gamma_s}\right)\frac{1 + i}{2B}. \qquad (6.53)$$

Clearly the real part of the elasticity is substantially reduced by diffusion effects from its value for an insoluble monolayer, an effect that was already noted for the lower molecular weight compounds of Figure 5.2.

The ratio of the imaginary part of E to $\bar{\beta}$ is often called surface viscosity, although in this case the complex nature of E arises naturally from the diffusion problem and is unrelated to any relationship between interfacial stress and the rate of strain. The reason for this terminology is that the analysis of wave motion in Chapter 5 is carried out without any explicit consideration of surfactants, but including appropriate terms in a surface elasticity Λ and a surface viscosity (η + ε). (η and ε are defined and discussed in Chapter 7, Section 6.) Equation 5.71 is obtained with $\Gamma_{os}(-d\gamma/d\Gamma_s)$ replaced by a complex elasticity E given by

$$E = \Lambda + \beta(\eta + \varepsilon)$$

$$= \Lambda + i\bar{\beta}(\eta + \varepsilon) \tag{6.54}$$

with $\bar{\beta}$ again the imposed frequency. Thus the imaginary part of E in Equation 6.52 does indeed play the same role in the analysis as would the product of frequency and an overall surface viscosity ($\eta + \varepsilon$) defined in terms of interfacial stresses, as in Chapter 7. Of course, it is conceivable that in some cases there would exist both an actual surface viscosity and the apparent surface viscosity involving diffusion effects, which is of concern here.

In any case, we see that measurement of the damping rate for capillary waves at high frequencies yields information on $\Gamma_{os}(-d\gamma/d\Gamma_s)$. This value can be used with similar damping measurements at low frequencies to calculate the diffusivity D if the adsorption isotherm is known. Or if D is known from separate experiments, the damping rate can be used to obtain information about $d\Gamma_s/dc$ as a function of surfactant concentration, and hence to determine the adsorption isotherm.

Another method for measuring surfactant properties by their influence on fluid motion is the use of longitudinal waves. In the basic concept, a bar lying along the interface is subjected to lateral oscillations instead of vertical ones as for the capillary waves considered so far. Variations in surface concentration cause variations in interfacial tension that strongly influence the flow. Longitudinal waves are thus closely related to the Marangoni flows discussed above. As indicated in Chapter 5, the dispersion relation for longitudinal waves is simply a requirement that the second term in brackets in Equation 5.71 vanish. Diffusion is incorporated by employing the complex elasticity $\underline{E}$ given by Equation 6.51. If oscillations are imposed with an angular frequency β , the wavenumber α is a complex number, as for capillary waves, with its real part related to disturbance wavelength and its imaginary part to the rate of damping of wave amplitude with distance from the oscillating bar. Longitudinal waves have some advantages over capillary waves for measuring surfactant properties since both wavelength and damping factor are influenced by elastic and diffusional phenomena, in contrast to the capillary wave case where only the damping factor varies significantly. Moreover, damping of longitudinal waves is greater than for capillary waves.

Two experimental techniques have been employed in studies of longitudinal waves. One follows closely the basic concept described above, with the motion of small test particles placed at various positions along the interface being used to determine the wavelength and damping rate (Maru and Wasan, 1979; Maru et al., 1979). Lucassen and van den Tempel (1972) and their coworkers, who originated the study of longitudinal waves, chose instead the arrangement illustrated

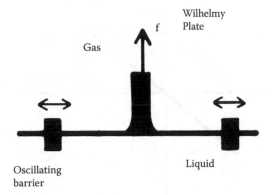

FIGURE 6.13 Schematic of an apparatus for studying longitudinal waves, after Lucassen and Barnes (1972).

in Figure 6.13 (Lucassen and Barnes, 1972). Here, two parallel barriers at the surface oscillate out of phase so that the area enclosed between them undergoes a sinusoidal oscillation. Thus at each instant of time the surface experiences a uniform expansion or compression. A Wilhelmy plate measures the interfacial tension as a function of interfacial area during the oscillation.

It can be shown that when frequency is high enough that the usual low viscosity approximation applies, but low enough that disturbance wavelength is much greater than the lateral dimensions of interest, the flow described by the longitudinal wave analysis does indeed correspond to a uniform expansion or contraction of the interface at each time. Moreover, the change in interfacial tension from its initial value is simply the product of the fractional area change ($\Delta A/A_o$) with the complex elasticity E. If the former is given by $A^* \sin \bar{\beta} t$ and the latter by Equation 6.51, we find

$$\Delta\gamma = \Gamma_{os}\left(-\frac{d\gamma}{d\Gamma_s}\right) A^* \left[\frac{1+B}{1+2B+B^2}\sin \bar{\beta} t + \frac{B}{1+2B+B^2}\cos \bar{\beta} t\right], \qquad (6.55)$$

with

$$\frac{\Delta A}{A_o} = A^*\sin \bar{\beta} t . \qquad (6.56)$$

In general, a plot of $\Delta\gamma$ as a function of $\Delta A/A_o$ is an ellipse, as illustrated in Figure 6.14, although it degenerates to a straight line in the high frequency limit where $B \to 0$ and effects of diffusion are negligible.

Figure 6.15 shows data obtained with this latter technique for two concentrations of decanoic acid in water (Lucassen and van den Temple, 1972). The magnitude $|E|$ of the elasticity is found to increase with increasing frequency, as

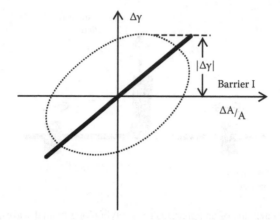

FIGURE 6.14 Change of surface tension with movement of the oscillating barrier at a given distance from the starting position of the barrier. Solid line: purely elastic behavior; dotted line: viscoelastic behavior. The eccentricity and tilt of the ellipse vary with distance from the barrier unless multiple reflections render the surface deformation uniform. From Lucassen-Reynders (1981) with permission.

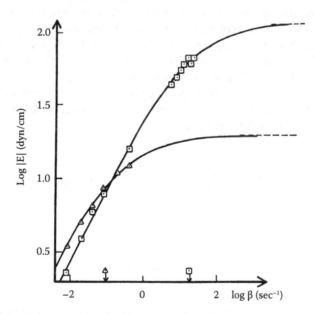

FIGURE 6.15 Experimental and theoretical values of $|E|$ as a function of wave frequency $\bar{\beta}$ for decanoic acid solutions. $\Delta{:}c = 2 \times 10^{-5}$ M; $\square{:}c = 10^{-4}$ M. From Lucassen-Reynders (1981) with permission.

expected from Equation 6.51. In the high frequency limit, $|E|$ is $\Gamma_{os}(-d\gamma/d\Gamma_s)$ and, as would be expected, is larger for the more concentrated solution. As indicated previously, the slope of the linear portion of the curve at low frequencies can be used to calculate $d\Gamma_s/dc$ at each concentration.

Significant deviations are seen when the solute is charged. The surface elasticity experiments deal with small perturbations from equilibrium. As a result, the nonlinear conservation equation for charged species alluded to in the previous section can be linearized and solved. Excellent comparison between theory and experiments is reported by Bonfillon and Langevin (1994) for SDS and double-chained sodium bis(2-ethylhexyl) sulfosuccinate (AOT) below the CMC at the water-oil interface.

The effects of soluble surfactants above the CMC are better explained using a model where the dissolution of micelles (accompanied by a loss in numbers) plays a strong role. This step, represented by the slow rate time constant τ_2, determines how much singly dispersed surfactant is available for adsorption at the interface. At frequencies higher than τ_2^{-1}, there is no effect on surface elasticity. But at lower frequencies there is a significant effect on the shapes of elasticity (real or imaginary) versus frequency curves, although qualitative shapes remain the same. As a function of concentration, the CMC is marked by an abrupt change. The values of τ_2 calculated from these curves are very close to those measured by bulk methods over a range of surfactant concentrations for three ionic surfactants where both kinds of data are known (Noskov, 2002).

8. STABILITY OF MOVING INTERFACES WITH PHASE TRANSFORMATION

We turn now to an entirely different mechanism of transport-related instability. It is important at solid-fluid interfaces where phase transformations or chemical reactions take place and is, for instance, the basic instability giving rise to dendritic growth during solidification. We illustrate it using the simple case of pure component solidification. Figure 6.16a shows the initial moving planar interface with A and B the solid and liquid phases, respectively. Solidification is presumed to be caused by subcooling the bulk liquid below its equilibrium melting temperature T_0. The initial temperature profile is then as shown. The rate of solidification is limited by the rate at which the heat of fusion released at the interface can be transported into the bulk liquid. If the interface is perturbed, as in Figure 6.16b, points such as P are exposed to more of the cooler liquid than before, the local rate of heat loss increases there, and hence so does the local solidification rate. A similar argument indicates a decrease in local solidification at points such as Q. Both these changes cause perturbation amplitude to grow (i.e., they are destabilizing).

Suppose instead that the temperature is uniform and equal to T_0 in the liquid B, but that the solid A is cooled (Figure 6.16c). Then when the wavy perturbation arises, P is exposed to less of the cooler solid than before and the local

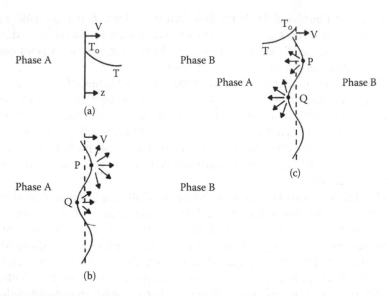

FIGURE 6.16 Transport effect on interfacial stability during phase transformation. (a) Phase transformation with transport into shrinking phase B; initial situation. (b) Effect of wavy perturbation of the interface. Transport into phase B is enhanced at point P and diminishes at Q. (c) Interfacial perturbation with transport into the growing phase. Transport into phase A is enhanced at point Q and diminished at point P.

solidification rate decreases. Similarly Q is exposed to more of the cooler solid and the local solidification rate increases. Both these effects are stabilizing since they cause perturbation amplitude to decrease.

We see from this example that heat transport is a key factor influencing stability. In solidification of binary mixtures or alloys both heat and mass transport must be considered. If the solute is preferentially soluble in the liquid phase, for example, solute must diffuse from the interface to the bulk liquid as solidification proceeds. Since diffusion coefficients are often much smaller than thermal diffusivities in liquids, the solidification rate is often limited primarily by solute diffusion. By an argument similar to that given above for heat transport, we conclude that solute diffusion in the liquid is destabilizing and can lead to dendritic growth.

Let us consider solidification of a pure component in detail. For the planar interface situation of Figure 6.16a, conservation of energy in the liquid (phase B) requires that the following equation be satisfied in a coordinate system moving with interfacial velocity V:

$$-V\frac{dT_i}{dz} = D_T\frac{d^2T_i}{dz^2}. \qquad (6.57)$$

This equation, which assumes steady state in the moving coordinate system, has the solution

$$T_i = T_I + \frac{GD_T}{V}\left[1 - \exp\left(-\frac{Vz}{D_T}\right)\right],$$ (6.58)

where T_I and G are the temperature and temperature gradient at the interface. These quantities and the velocity V are related by the boundary conditions, which are discussed next.

Conservation of energy at the interface implies that

$$kG + V\rho\lambda = 0,$$ (6.59)

where k and λ are the liquid thermal conductivity and the latent heat of fusion. In this simplified derivation, the solid and liquid densities are assumed to have the same value ρ.

For the remaining boundary condition we use an empirical expression that allows for the possibility that the kinetics of crystal structure formation at the interface may significantly influence the solidification rate:

$$V = b(T_o - T_i).$$ (6.60)

Here T_o is the equilibrium melting temperature of a flat interface and b is an empirical coefficient. In the limiting case of large b, we have $T_i = T_o$. In this special case, crystal formation kinetics are unimportant and the solidification rate is limited solely by heat transport in the liquid.

We now proceed as in the stability analyses of Chapter 5 and suppose that perturbation shifts the interface from the plane $z = 0$ in the moving coordinate system to the wavy surface z^* given by $Be^{i\alpha x}e^{\beta t}$. We further suppose that the temperature distribution after perturbation is given by

$$T(x,y,z) = T_i(z) + T_p(z)e^{i\alpha x}e^{\beta t}.$$ (6.61)

In this case, the differential energy equation becomes

$$\beta T_p - V\frac{dT_p}{dz} = D_T\left(\frac{d^2T_p}{dz^2} - \alpha^2 T_p\right).$$ (6.62)

The solution to this equation, which vanishes far from the interface, is

$$T_p = a_1 e^{-\omega z}$$ (6.63)

$$\omega = \frac{V}{2D_T} + \left(\frac{V^2}{4D_T^2} + \alpha^2 + \frac{\beta}{D_T} \right)^{1/2} . \qquad (6.64)$$

This solution is valid only when $\alpha^2 + \beta/D_T > 0$ (i.e., for unstable situations and for stable situations where the decay rate is not too rapid). Because it turns out that the unstable wavenumbers are typically fairly large (about 1000/cm), we can usually ignore β/D_T in comparison with α^2, a simplification that is employed in the remainder of the derivation.

Local energy conservation along the wavy interface requires that

$$k\left(-\omega a_1 - \frac{VGB}{D_T} \right) + \beta\lambda\rho B = 0 . \qquad (6.65)$$

Similarly, local application of Equation 6.60 leads to

$$\beta B + b\left(a_1 + GB + \frac{\gamma\alpha^2 T_o}{\lambda\rho} B \right) = 0 . \qquad (6.66)$$

The last term in parentheses in Equation 6.66 represents the effect of interfacial tension and curvature on the equilibrium freezing temperature. The basic idea is that curvature produces a difference in pressure between solid and melt. Since the chemical potentials in the two phases are functions of pressure, and since they must be equal at equilibrium, the equilibrium freezing temperature depends on curvature (see Problem 6.7).

Equations 6.65 and 6.66 are linear, homogeneous equations in a_1 and B. For a nontrivial solution to exist, their determinant of coefficients must vanish, a condition which leads to

$$\beta = \frac{kG\left(-\omega + \dfrac{V}{D_T} \right) - (\omega k\gamma T_o\alpha^2)/(\lambda\rho)}{\lambda\rho\left(1 + \dfrac{\omega k}{b\lambda\rho} \right)} . \qquad (6.67)$$

The first term in the numerator of this equation represents the basic effect of heat transport on interfacial stability. Since $G < 0$ and $\omega > V/D_T$, this term makes a positive contribution to β (i.e., it is destabilizing). The second term represents a stabilizing effect produced by interfacial tension. It is most important for short wavelengths (large α), and indeed outweighs the destabilizing effect of transport for sufficiently large α. The term $\omega k/b\rho\lambda$ in the denominator of Equation 6.67 represents the ratio of the rates of heat transport and interface crystallization processes. When this ratio is small, owing to, for instance, a large value of b, crystallization is rapid and the solidification rate is limited by heat transport. But when the ratio is large, the kinetics of interfacial crystallization control the solidification. Note that, when b has a finite value, kinetic effects are always rate controlling for perturbations with sufficiently short wavelengths since the time required for heat transport over this length scale becomes very small. When kinetic effects dominate, Equation 6.67 shows that β is small in magnitude. Thus interfacial kinetic effects cannot remove the basic instability produced by transport, but they can slow drastically the rate of growth of the unstable perturbations.

It should be noted that instability leading to dendrite formation can also occur in insoluble surfactant monolayers of the type discussed in Chapter 4. In this case, supersaturation of a two-dimensional liquid phase occurs during rapid compression of the monolayer, and the instability is the result of mass transfer to the growing solid phase. Vollhardt (1996) provides examples.

When an initially plane fluid interface is deformed, interfacial tension initiates flow which acts to restore the planar configuration, as we saw in the discussion of wave motion in Chapter 5. This stabilizing effect is normally strong enough during ordinary condensation or vaporization to overwhelm transport effects on interfacial stability, such as those in the above analysis of solidification (Miller, 1973). However, transport effects on stability are important for certain moving "fronts" in fluid systems, which are not true interfaces, but instead are regions of rapid composition changes. Examples are moving fronts where fog forms or vaporizes (Miller and Jain, 1973) (see Section 10) and moving condensation or combustion fronts in porous media (Armento and Miller, 1977; Gunn and Krantz, 1982; Miller, 1975; Yortsos, 1982).

An intriguing phenomenon important in the vaporization of organic liquids under high vacuum conditions is instability due to the "vapor recoil" mechanism (Hickman, 1952; Palmer, 1976) (see Figure 6.17). In this case, the difference in momentum between liquid entering and vapor leaving the interface becomes important (i.e., the last two terms of Equation 5.35 must be included in the interfacial momentum balance). Since vapor density is very low, the interfacial mass balance of Equation 5.32 indicates that vapor velocity normal to the interface must be high. Accordingly, vapor momentum must be high as well and liquid pressure must exceed vapor pressure in order for the normal component of the interfacial momentum balance of Equation 5.35 to be satisfied. Hydrostatic effects thus dictate that surface elevation be lowest in locations where the vaporization rate is highest.

Suppose that some bulk liquid reaches the surface at point P of Figure 6.18, producing a local increase in temperature. The result is a local increase in vaporization rate and, according to the above argument, a slight additional depression of the surface. Analysis (Palmer, 1976) shows that the shear stress produced by vapor flow along the wavy surface near P drags liquid outward from P, which in turn brings more warm liquid from the bulk liquid to P. The initial perturbation is thus enhanced and an instability involving appreciable convection near the

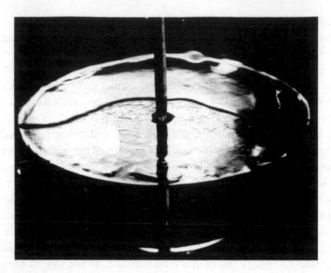

FIGURE 6.17 The surface of a rapidly evaporating mineral oil: evaporation is much more rapid in the foreground because of vapor recoil instability. From Palmer (1976) with permission.

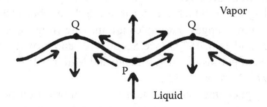

FIGURE 6.18 Schematic of convection in liquid produced by the vapor recoil mechanism. The local evaporation rate increases at point P and decreases at point Q.

interface develops. As a result, the vaporization rate increases by several-fold over that for a stagnant fluid surface.

EXAMPLE 6.2 CHARACTERISTICS OF INTERFACIAL INSTABILITY DURING SOLIDIFICATION

Find the wavelength of the fastest growing disturbance and its rate of growth for solidification of water when (a) the rate of crystal formation is very fast ($b \to \infty$) and (b) $b = 10^{-5}$ cm/sec·K. Assume that the temperature gradient ($-G$) in the water at the interface is 10°C/cm and use the following values:

$\rho = 1$ g/cm³ $\lambda = 80$ cal/g

$k = 1.4 \times 10^{-3}$ cal/(sec·cm·K) $\gamma = 100$ mN/m

$D_T = 1.4 \times 10^{-3}$ cm²/sec $T_o = 273°$K

SOLUTION

(a) For $b \to \infty$ we have, according to Equation 6.60, $T_I = T_o$ at the interface. From Equation 6.59 we can calculate that interfacial velocity V is 1.75×10^{-4} cm/sec. If $V/D_T\alpha \ll 1$, $\omega = \alpha$ and Equation 6.67 becomes

$$\beta = \frac{-kG\alpha - k\gamma T_o\alpha^3 / \lambda\rho}{\lambda\rho}. \qquad (6.E2.1)$$

By differentiating this equation with respect to α and setting the derivative equal to zero, we find

$$\alpha_m = \left(\frac{-G\lambda\rho}{3\gamma T_o}\right)^{1/2}. \qquad (6.E2.2)$$

With the above values, α_m is 639/cm, i.e., the wavelength $2\pi/\alpha_m$ of the fastest growing perturbation is 98.3 μm. Note that $V/D_T\alpha_m = 1.9 \times 10^{-4}$, so that the assumption made above is justified. The time factor β_m for growth of this disturbance is 0.069/sec (i.e., the time constant β_m^{-1} is 14.7 sec).

(b) When b is very small, the rate of crystal formation controls the solidification rate. If $(\alpha k/b\lambda\rho) \gg 1$ and $V/D_T\alpha \gg 1$, Equation 6.60 simplifies to

$$\beta = b\left(-G - \frac{\gamma T_o\alpha^2}{\lambda\rho}\right). \qquad (6.E2.3)$$

Comparing this equation with Equation 6.E2.1, we see that the wavenumber α_c at marginal stability (i.e., where $\beta = 0$) is the same as in (a) for a given temperature gradient G. But the time factor β now increases with decreasing α and reaches its maximum value $\beta = -bG$ in the limit $\alpha \to 0$. Thus β_m in the present case is 10^{-4}/sec, corresponding to a time constant of 10^4 seconds, even though G and V have the same values as in (a). We conclude that the development of instability is greatly slowed when solidification kinetics are important.

Note that for $\alpha k/b\lambda\rho$ to have a magnitude of order unity, α must be about 0.57/cm, or the wavelength $2\pi/\alpha$ must be about 11 cm. As seen in (a), the wavelengths of interest are much shorter than this value, so that the assumption made in deriving Equation 6.E2.3 is valid. Note further that with $V = 1.75 \times 10^{-4}$ cm/sec, the temperature decrease at the interface is 17.5°C, according to Equation 6.60.

9. STABILITY OF MOVING INTERFACES WITH CHEMICAL REACTION

Moving interfaces where one or more chemical reactions occur are subject to transport-related instabilities similar to those discussed in Section 8 which lead to dendritic growth during solidification. However, the wide variety of possible reaction schemes and kinetic relationships makes possible different types of behavior.

Let us consider a solid S that is being consumed by reaction with a species A that diffuses from the bulk fluid to the interface (Figure 6.19). Based on the qualitative argument given in Section 8, we would expect enhanced diffusion of A to points such as P when the interface is deformed, with a resulting increase in local reaction rate and interfacial velocity. This effect is clearly a stabilizing one.

But suppose now that the reaction is exothermic, so that heat must be transported from the interface into the bulk liquid. Now the perturbation also produces a local increase in heat flux at P, which favors a reduction in local temperature and hence reaction rate, a destabilizing effect. If it exceeds the stabilizing effect of reactant diffusion, instability can result (Cannon and Denbigh, 1957a,b; Knapp and Aris, 1972; Miller, 1978).

The stability analysis is very similar to that of Section 8 except that there are equations analogous to Equations 6.61 and 6.62 for both heat and mass transport in the liquid. Moreover, interfacial conservation equations both for energy and for the reactant A must be invoked.

The key equation in the analysis, however, is the boundary condition giving the local reaction rate r. For simplicity we assume that the reaction proceeds by formation of an activated complex C^* in equilibrium with the reactants A and S. The equilibrium constant for this reaction is taken as K_e. The complex then decays irreversibly with a rate constant k_1 to form the product B in the fluid phase: thus we have

$$A + S \underset{}{\overset{K_e}{\rightleftharpoons}} C^* \xrightarrow{k_1} B .$$ (6.68)

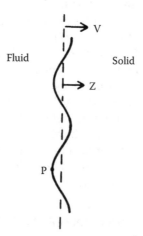

FIGURE 6.19 Schematic perturbation of reacting solid surface.

With this scheme the reaction rate r can be written in terms of K_e and k_1 as follows:

$$r = k_1 c_{c*} \cdot \exp\left(-\frac{E_a}{RT}\right)$$

$$= k_1 K_e c_A (f_s / f_s^o) \exp\left(-\frac{E_a}{RT}\right)$$

(6.69)

where c_{c*} and c_A are concentrations of C^* and A at the interface, E_a is the activation energy for the reaction, and f_s/f_s^o is the ratio of the fugacity f_s of the reacting solid to its standard state value f_s^o.

Now f_s^o is evaluated at the pressure p_1 exerted by the fluid at the initial flat interface, which is also the pressure in the solid under these conditions if the Young-Laplace equation is assumed to apply. But once the interface is deformed, as in Figure 6.19, the pressure in the solid near the interface becomes a function of position. Indeed, if pressure in the fluid phase is presumed to be uniform, we can use the Young-Laplace equation and the well-known dependence of fugacity upon pressure to write

$$\frac{f_s}{f_s^o} = \exp\left(-\frac{v_s\gamma}{RT}\alpha^2 z^*\right).$$

(6.70)

with v_s the molar volume of the solid and z^* the interfacial displacement. Substituting this expression into Equation 6.68, we obtain

$$r = k_1 K_e c_A \exp\left(-\frac{\gamma v_s}{RT}\alpha^2 z^*\right)\exp\left(-\frac{E_a}{RT}\right).$$

(6.71)

According to Equation 6.70, solid fugacity increases over the standard state value at points such as P of Figure 6.19, where $z^* < 0$. An increase in the local reaction rate at P results. We see therefore that, as before, interfacial tension has a stabilizing effect that enters the analysis through the equation describing interface kinetics. The origin of the effect is made plain here, however, because the kinetic equation is based on a physical model of reaction and is not simply an empirical expression like Equation 6.60.

When the stability analysis is carried out in the usual way, the time factor β for short wavelengths where $V/\alpha D$ and $V/\alpha D_T$ are small is found to be given by

$$\beta = \frac{V\left(\dfrac{G_A}{c_{Ao}} + \dfrac{E_a G k}{RT_o^2(k + k_s)} - \dfrac{\gamma M_s \alpha^2}{\rho_s RT_o}\right)}{1 + \dfrac{v\rho_s}{c_{Ao}DM_s\alpha} + \dfrac{E_a V \Delta H_R \rho_s}{RT_o^2 \alpha(k + k_s)}}.$$

(6.72)

Here c_{Ao} and T_o are reactant concentration and temperature at the interface before perturbation, G_A and G are the initial concentration and temperature gradients, ΔH_R is the heat of reaction per unit mass of solid S consumed, ρ_s and M_s are the density and molecular weight of S, and k and k_s are the thermal conductivities of the fluid and solid, respectively. The initial temperature in the solid phase has been presumed uniform, and $\beta/D_T\alpha^2$ and $\beta/D\alpha^2$ are taken as small quantities (cf. discussion following Equation 6.64).

Since $G_A < 0$, it is clear from the numerator of Equation 6.72 that reactant diffusion has a stabilizing effect, as expected. If the reaction is exothermic, heat transport in the fluid is away from the interface, $G > 0$, and heat transport is destabilizing. Finally, the effect of interfacial tension is stabilizing and of greatest importance for short wavelengths (small α), the same as found in Section 8 for solidification. Thus instability is possible if the heat transport effect outweighs the combined effect of reactant diffusion and interfacial tension. Of course, $G < 0$ for an endothermic reaction, all three effects are stabilizing, and no instability is possible.

An exothermic reaction has autocatalytic properties in the sense that heat generated by the reaction increases temperature and hence speeds up the reaction. It might be expected that other reaction schemes having autocatalytic features could cause interfacial instability. Suppose, for instance, that the following two reactions occur:

$$S + aA + b_1B \rightarrow cC + \text{other products} \qquad (6.73)$$

$$cC + dD + eE \rightarrow b_2B + \text{other products.} \qquad (6.74)$$

Here species A, B, C, D, and E exist only in the fluid phase. Further suppose that $b_2 > b_1$ so that some B is produced by the overall reaction obtained by adding Equations 6.73 and 6.74. Finally, suppose that the rate of Equation 6.73 is much slower than that of Equation 6.74. Then the rate of the overall reaction is, for all practical purposes, equal to the rate of Equation 6.73, and hence proportional to $c_B^{b_1}$, where c_B is the concentration of B at the solid-fluid interface. In other words, the rate of the overall reaction is proportional to the concentration of one of its products, B. The reaction is thus of an autocatalytic nature and instability is possible (see Problem 6.8).

A possible example of a reaction scheme of this type involves leaching of the metal ore chalcopyrite ($CuFeS_2$). In this case, Equation 6.73 is the basic dissolution reaction, with the reactants A and B being dissolved oxygen and hydrogen ion, respectively, while C is ferric ion Fe^{+3}. The second reaction involves precipitation of Fe^{+3} by sulfate ions D and water E to form, for example, hydrogen jarosite ($Fe_3(SO_4)_2(OH)_5\cdot2H_2O$). Hydrogen ions are also produced. A reaction scheme of this type is reported by Braun et al. (1974), who also describe certain observations in which ore particles are leached nonuniformly (i.e., the amount of leaching is different at different positions along the particle surface). Such behavior was associated with an increased overall leaching rate for the entire particle, a desirable effect. A possible explanation of the nonuniform leaching is instability of the reacting surface as a result of the mechanism just described in accordance with the analysis outlined in Problem 6.8.

So far we have dealt with reactions in which a solid has been consumed. When a solid surface is built up by a reaction, transport of reactants in the fluid phase is destabilizing, just as is the case during solidification (Seshan, 1975). An example of considerable practical importance is electrodeposition. In this case, instability is normally undesirable because it leads to an irregular surface that does not have the preferred bright appearance. It has been found that instability can sometimes be prevented by including small quantities of certain surface active organic additives in the electrodeposition bath (Edwards, 1964). The additives diffuse to the solid-liquid interface and are either incorporated into the developing deposit or consumed by an electrochemical reaction. Since they are surface active, they adsorb at the interface and produce a decrease in the rate of the electrodeposition reaction, possibly because they block some sites where metal ions would otherwise deposit.

Although details of the action of organic additives remain poorly understood, their qualitative effect can be described empirically by an equation of the following type for the electrodeposition rate r:

$$ r = kc_M / c_A^m , \tag{6.75} $$

where c_M and c_A are local concentrations of metal ions and additive at the interface and m is an empirical constant. If the additive has a large stabilizing effect (i.e., if small amounts of additive cause large changes in electrodeposition rates), m must be a large positive number. When an initially flat interface is deformed, arguments similar to those of the initial paragraph of Section 8 indicate that both metal ion and additive concentrations increase at points where the solid projects into the liquid. In view of Equation 6.75, transport of the metal ion must be destabilizing, while transport of the additive is stabilizing. Whether instability develops depends on the relative magnitudes of ion and additive effects (see Problem 6.9). As a minimum, the additive reduces the growth rate of unstable perturbations.

These examples provide some idea of the variety of phenomena that can influence the stability of reacting solid surfaces. No doubt numerous other reaction schemes showing interesting behavior can be devised.

10. INTERMEDIATE PHASE FORMATION

The phase behavior of a water-surfactant or an oil-water-surfactant system may be such that when two single-phase mixtures are brought in contact without mixing, and diffusion is allowed to occur, one or more new phases not initially present may form near the surface of contact. We shall call them "intermediate phases." Their formation constitutes another example of intriguing phenomena involving simultaneous diffusion and phase transformation.

Consider, for example, the behavior when the pure nonionic surfactant $C_{12}E_5$ is brought into contact with water at a temperature of 25°C, a few degrees below the cloud point. According to the phase equilibrium diagram (Figure 4.13), it is not possible for the pure liquid surfactant (L_2) and the aqueous solution (L_1) to form a simple interface at this temperature. Instead, an intermediate lamellar (L_α) phase develops between them. Such contacting or "penetration" experiments are widely used to identify the liquid crystalline phases present at various temperatures in water-surfactant systems. Another example already mentioned in Chapter 4 (Section 12) is formation of intermediate microemulsion or lamellar liquid crystalline phases at surfaces of contact between oils and surfactant-water mixtures which can, under suitable conditions, greatly increase the rate of removal of the oils from solids during cleaning processes by a solubilization-emulsification mechanism.

In this section and the next we develop criteria for intermediate phase formation and for the related phenomenon of spontaneous emulsification. First, we consider quasi-steady-state diffusion processes leading to intermediate phase formation some time after initial contact of the phases. In the next section, intermediate phase formation on initial contact is discussed.

Another situation relevant to detergency that was discussed in Chapter 4 is the behavior that occurs when a small amount of a mixture of an alkane and a long-chain alcohol or fatty acid is contacted with a surfactant solution that is dilute but above its CMC. Experiments (Lim and Miller, 1991a) show that when the alcohol content of the oil is below that of the excess oil phase in equilibrium with a microemulsion and excess water at the balanced condition or phase inversion temperature (PIT), no intermediate phase forms and the oil is solubilized into the surfactant solution, albeit at a very slow rate. However, when alcohol content exceeds the above value, a drop of the oil swells in the surfactant solution. Eventually the lamellar liquid crystal forms as an intermediate phase (Figure 6.20).

The conditions for formation of an intermediate phase as in Figure 6.20 can be found using a quasi-steady-state approach in which compositions both inside and outside an injected oil drop change slowly with time. This approach is justified whenever the time required for equilibration within the drop, which is on the order of R^2/D, with R the radius and D the diffusivity, is much smaller than the time until intermediate phase formation begins. The latter is several minutes in the experiments of Lim and Miller (1991a), while the former is only about 10 seconds for $R = 75$ μm and $D = 10^{-10}$ m²/sec.

Accordingly, one may neglect the accumulation term in the diffusion equation for each species. In the absence of convection and for a fluid of uniform density ρ, the equation for species i simplifies to

$$\frac{d}{dr}\left[r^2 \frac{d\omega_i}{dr}\right] = 0 ,$$ (6.76)

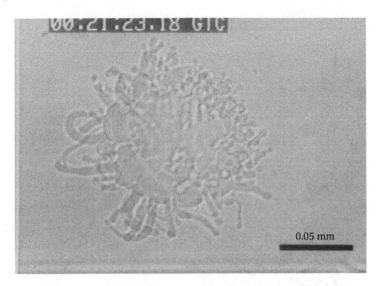

FIGURE 6.20 Video frame showing growth of the intermediate lamellar phase as myelinic figures approximately 21 minutes after contact of 0.05 wt. % $C_{12}E_5$ solution with a drop of 5.67:1 n-hexadecane/oleyl alcohol at 50°C. From Miller (1996) with permission.

where ω_i is the mass fraction of species i. The general solution for this equation is

$$\omega_i = b_i + (a_i/r). \qquad (6.77)$$

Because concentrations must remain finite at $r = 0$, the solution within the drop can be written

$$\omega_i = \omega_i(t). \qquad (6.78)$$

Outside the drop we use primes to denote the mass fractions and find

$$\omega_i' = \omega_{i\infty}' + [a_i(t)/r], \qquad (6.79)$$

where $\omega_{i\infty}'$ is the mass fraction of i in the bulk aqueous solution.

Equations 6.78 and 6.79 can be solved with appropriate boundary conditions at the interface $r = R(t)$. The species mass balance equations are given by

$$\frac{d}{dt}(4\pi R^3 \rho \omega_i / 3) = \rho' D_i' \frac{d\omega_i'}{dr}\Big|_R + 4\pi R^2 \rho' \omega_i'(R)\frac{dR}{dt} \quad , \qquad (6.80)$$

where ρ' and D_i' are the density and diffusion coefficient of species i in the aqueous solution. The other boundary condition is local equilibrium at the interface. Under these conditions the concentrations at the interface lie on the ends

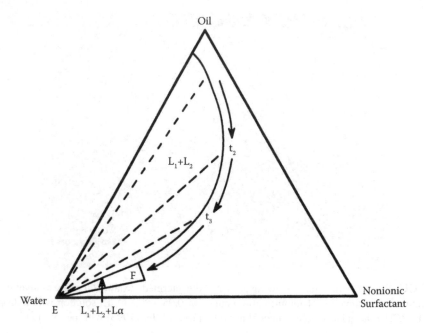

FIGURE 6.21 Schematic pseudoternary diagram of nonionic surfactant water–n-hexade-cane–oleyl alcohol system showing the change in oil drop composition with time. Reprinted with permission from Lim and Miller (1991a). Copyright 1991 American Chemical Society.

of a tie-line joining equilibrium compositions of the aqueous (L_1) and oil (L_2) phases (see Figure 6.21).

With this information, the qualitative behavior can be understood from the phase diagram of Figure 6.21, which represents the situation above the PIT. After a brief transient, the composition of the oil drop reaches the coexistence curve at a point near the water-oil side of the pseudoternary phase diagram, which corresponds to a water-in-oil microemulsion. As water and surfactant diffuse into the drop, its composition follows the coexistence curve, as indicated by the arrows in Figure 6.21, until it reaches the composition of the limiting tie-line F. During this time, drop diameter increases, as observed experimentally. Beyond point F, further transfer of surfactant into the drop is incompatible with two-phase coexistence, and an intermediate lamellar phase begins to form.

For the situation of Figure 6.20, where surfactant concentration in the aqueous phase is low and both hydrocarbon and long-chain alcohol have very low solubilities in water, the oil phase can be treated as a single pseudocomponent 1. Then Equation 6.80 simplifies to

$$\frac{d}{dt}\left(4\pi R^3 \omega_1 / 3\right) = 0 . \tag{6.81}$$

With initial conditions $R = R_o$ and $\omega_1 = \omega_{1o}$ at $t = 0$, this equation is readily integrated to obtain

$$\left(\frac{R}{R_o}\right)^3 = \left(\frac{\omega_{1o}}{\omega_1}\right) . \tag{6.82}$$

We expect surfactant concentration ω_2' in the L_1 phase to be very small at the interface—below the CMC of the pure surfactant, which is itself low for nonionic surfactants—when the system is above the PIT as in Figure 6.20. The phase diagram of Figure 6.21 has been drawn to correspond to this situation. With this assumption, the surfactant concentration becomes

$$\omega_2' = \omega_{2\infty}'[1 - (R / r)] . \tag{6.83}$$

As a further simplification we fit the L_2 portion of the coexistence curve by a polynomial:

$$\omega_2 = A_1 + A_2\omega_1 + A_3\omega_1^2 , \tag{6.84}$$

where A_i are constants. Then

$$\frac{d\omega_2}{dt} = \frac{d\omega_2}{d\omega_1} \cdot \frac{d\omega_1}{dt} = \left(A_2 + 2A_3\omega_1\right) \cdot \frac{d\omega_1}{dt} . \tag{6.85}$$

Substituting Equations 6.83 and 6.85 into Equation 6.80 applied to the surfactant, simplifying by setting $\omega_{2\infty}' = 0$, and invoking Equation 6.82 to eliminate R, we find

$$\frac{d\omega_1}{dt} = \frac{D_2'\omega_2'}{(R_o^2\omega_{1o}^{2/3} / 3) (A_3\omega_1^{1/3} - A_1\omega_1^{-5/3})} . \tag{6.86}$$

Integration of this equation from $t = 0$ to $t = t_F$, the time when the oil composition reaches point F of Figure 6.21 and the lamellar phase starts to form, leads to

$$t_F = K(R_o^2 / D_2'\omega_{2\infty}') , \tag{6.87}$$

where K is given by the following expression which depends only on the phase diagram:

$$K = \frac{A_3}{4}\left(\omega_{1F}^{4/3}\omega_{1o}^{2/3} - \omega_{1o}^2\right) + \frac{A_1}{2}\left[\left(\omega_{1o} / \omega_{1F}\right)^{2/3} - 1\right] . \tag{6.88}$$

Equation 6.87 predicts that the time t_F until liquid crystal formation begins is proportional to the square of the initial drop radius R_o and inversely proportional to the bulk surfactant concentration $\omega'_{2\infty}$. These predictions were in agreement with experiments for systems containing pure nonionic surfactants, n-hexadecane, oleyl alcohol, and water (Lim and Miller, 1991a). Moreover, for a hydrocarbon:alcohol ratio of 3:1 by weight and for solutions of $C_{12}E_7$ at 30°C, the phase diagram was determined and K calculated as 0.52. When the data were fitted to Equation 6.87, D_2' was found to be 1.3×10^{-10} m²/sec. The Stokes-Einstein equation was then used to estimate micelle radius r:

$$r = kT / \left(6\pi\mu D_2'\right)$$

(6.89)

The result was a micelle diameter of 3.5 nm, a reasonable value.

Equation 6.87 yields $R_o \propto \sqrt{t_F}$, a sign of diffusion-controlled mass transfer. Decanol can be contacted with a solution of sodium octanoate in a test tube for the system of Figure 4.14 such that an intermediate lamellar liquid crystalline phase results. Labeled L in the phase diagram of Figure 4.14, this phase has been shown to grow as $\sqrt{t}$ at short times by Kielman and van Steen (1979a,b), but without analysis. (The diffusion path method described in Section 11 was later used to successfully predict the lowest surfactant concentration at which the lamellar phase will form (Lim and Miller 1990)). It is important to note that in spite of the fact that the lamellar phase is anisotropic, no unusual effects were seen, probably because of the many defects associated with the presence of multiple small domains having different orientations.

In this situation, unlike that discussed previously in this section, it is not adequate to concentrate on conservation of mass of a single species, which greatly complicates analysis. This difficulty can be avoided by choosing the compositions of the two phases with different microstructures to be contacted in such a way that at equilibrium nothing is changed except that one component partially moves out of one phase to the other. Ma et al. (1989) present many such examples from the SDS-dodecanol-water system, the phase diagram of which looks very much like Figure 4.14, but without the reverse hexagonal phase F. The constraints imposed by this approach appear to produce non-Fickian diffusion. The authors suggest that when one microstructure breaks down to form another, say from inverse micelles to lamellar phase, different species must cooperate in some way, one factor likely being limits (not necessarily equilibrium ones) on allowable stoichiometry of each microstructure. The constraints imposed on which species can transfer between phases can eventually make this cooperation difficult and force the system to exhibit anomalous behavior.

It is usually assumed in a dynamic system that even though the system as a whole is not in equilibrium, local equilibrium holds at every point. The path of a dynamic process can then be shown on an equilibrium phase diagram as it evolves, such as the sequence of compositions shown by the arrows in Figure 6.21. Even supersaturation, discussed in the next section, can be treated this way.

However, when local equilibrium is not obeyed, it is possible to observe a microstructure for a given composition different from that shown on the equilibrium phase diagram. This lack of local equilibrium is not permanent, and the system "relaxes" back to local equilibrium at every point (de Groot and Mazur, 1984). Such behavior brings time and memory dependence to the dynamics in addition to the usual transport processes (Ma et al., 1989). Whereas these authors observed no unexpected microstructures, Egelhaaf and Schurtenberger (1999) reported that the micelles-to-vesicle transition which occurs on dilution of a lecithin/bile salt mixture exhibits an intermediate disk-shaped aggregate. This type of aggregate is not a stable shape at equilibrium in this system.

In spite of these difficulties, with one notable exception discussed in Section 6.12, simple diffusion is quite common, as discussed in this section and next. A similar analysis, as shown previously for a drop, can be made when the oil is present as a thin layer on a solid surface (Lim and Miller, 1991b), a more interesting situation for detergency. Moreover, the intermediate phase need not be the lamellar liquid crystal. Depending on the phase behavior and the initial compositions of the oil and aqueous phases, it could also be another liquid crystalline phase, a microemulsion, or the L_3 (sponge) phase discussed in Chapter 4. Finally, we note that sometimes more than one intermediate phase is seen. An example is given in the next section.

11. TRANSPORT-RELATED SPONTANEOUS EMULSIFICATION

Another phenomenon involving simultaneous transport and phase transformation is spontaneous emulsification. Figure 6.22 shows spontaneous emulsification in the aqueous phase that occurs when a mixture of n-propanol (65 wt. %) and 2,3-

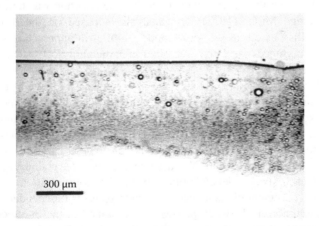

300 μm

FIGURE 6.22 Spontaneous emulsification in the aqueous phase observed 6 minutes after initial contact between water and a mixture of 65% n-propanol and 35% 2,3-dimethylpentane at 35°C. Reprinted from Miller (1988) with permission from Elsevier.

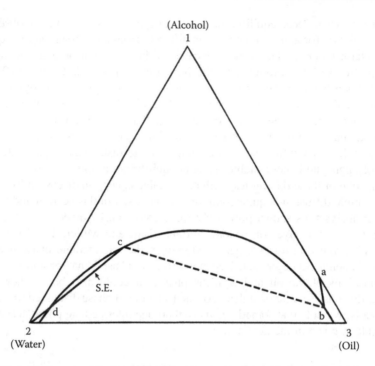

FIGURE 6.23 Sample diffusion path exhibiting spontaneous emulsification (S.E.). The initial compositions are point a (ω_{10}, ω_{20}) and pure 2 (ω'_{10}, ω'_{20}).

dimethyl pentane (35 wt. %) is carefully placed on water at 35°C with minimal mixing (Miller, 1988). It has long been recognized that spontaneous emulsification in such cases is associated with diffusion of the various species (Davies and Haydon, 1957), but theoretical analysis of the phenomenon was first carried out by Ruschak and Miller (1972). The analysis is based on the calculation of "diffusion paths," as discussed below, and is similar in approach to that used to predict the occurrence of isolated precipitation in metallurgical systems (Kirkaldy and Brown, 1963).

Consider a ternary system having the phase behavior shown in Figure 6.23. Many systems containing an oil, water, and a short-chain, slightly polar organic compound such as an alcohol or a fatty acid are of this type, including that in Figure 6.22. Now suppose that two semi-infinite phases having mass fractions (ω_{10}, ω_{20}) and (ω'_{10}, ω'_{20}) of species 1 and 2 are placed carefully in contact at position $x = 0$ and time $t = 0$. We assume that no convection develops, that transport of each species depends only on its own concentration gradient, that the diffusion coefficient of each species in each phase is independent of composition, that equilibrium between phases is maintained at the interface, and that the mass density of the system is uniform throughout. With these assumptions the concentration profiles can be found by solving the following four differential equations, two in each phase (Ruschak and Miller, 1972):

$$\frac{\partial \omega_i}{\partial t} = D_i \frac{\partial^2 \omega_i}{\partial x^2} \tag{6.90}$$

$$\frac{\partial \omega_i'}{\partial t} = D_i' \frac{\partial^2 \omega_i'}{\partial x^2}, \ (i = 1, 2).$$

ω_i and ω_i' are the mass fractions of species i in the two phases, and D_i and D_i' are the corresponding diffusion coefficients. The condition that the mass fractions in each phase must sum to unity can be used to determine ω_3 and ω_3' for the third component from the solution to Equation 6.90.

Since both the initial condition and the boundary conditions for $x \to \pm \infty$ are that the initial compositions must exist, a similarity solution is possible (Bird et al., 2002). It has the form

$$\omega_i = \omega_{io} + B_i \left(1 - erf \frac{x}{(4D_i t)^{1/2}} \right)$$

$$\omega_i' = \omega_{io}' + B_i' \left(1 + erf \frac{x}{(4D_i' t)^{1/2}} \right)$$
\qquad\qquad (6.91)

The constants B_i and B_i' must be determined from boundary conditions at the moving interface, the position of which we denote by $\varepsilon(t)$. Two of these conditions are the following component mass balances:

$$D_i \frac{\partial \omega_i}{\partial x}\bigg|_\varepsilon - D_i' \frac{\partial \omega_i'}{\partial x}\bigg|_\varepsilon = [\omega_i'(\varepsilon) - \omega_i(\varepsilon)] \frac{d\varepsilon}{dt}, \tag{6.92}$$

where $\omega_i(\varepsilon)$ and $\omega_i'(\varepsilon)$ are the mass fractions of species i in the two phases at the interface. These equations are consistent with the similarity solution of Equation 6.91 if the interfacial compositions are constant and if $\varepsilon(t)$ has the following form:

$$\varepsilon(t) = bt^{1/2}, \tag{6.93}$$

with b a constant.

The remaining boundary conditions are the phase equilibrium relationships at the interface given by the applicable phase diagram. It may be convenient in calculations to describe the coexistence curve and tie-lines by empirical equations such as those proposed by Hand (1930). Whatever the details of the method used, the problem is well posed once initial compositions, diffusion coefficients, and equilibrium data are specified. The solution yields the concentration profiles in both phases.

According to the similarity solution of Equation 6.91, the mass fractions ω_i are functions of the single variable $x/t^{1/2}$, which ranges from $-\infty$ to b in one phase and from b to $+ \infty$ in the other. Hence the set of compositions in the system is time independent, and this "diffusion path" can be plotted on the phase diagram as shown in Figure 6.23. The diffusion paths in the two phases (solid lines) are straight lines, as shown, if the diffusion coefficients of the various species are equal in each phase. Otherwise they have some curvature.

Note that in Figure 6.23, part of the diffusion path for the primed phase (viz., the segment cd) lies within the two-phase region of the phase diagram. Ruschak and Miller (1972) proposed that when such behavior is found, spontaneous emulsification should be expected in the phase or phases having supersaturated compositions. Their experiments for three toluene-water-solute systems having tie-lines of different slope confirmed the ability of this scheme to predict both when emulsion would form and in which phase or phases. As Figure 6.23 indicates, one factor that favors spontaneous emulsification is diffusion of the solute into the phase in which it is more soluble.

Later, video microscopy was used to view the interfacial region in more detail, as shown in Figure 6.22. With this technique it was possible to confirm that, in systems such as that in Figure 6.22, where there was no Marangoni flow, interfacial position was proportional to $t^{1/2}$, in accordance with Equation 6.93. Moreover, most droplets form a short distance into the aqueous phase from the interface, where supersaturation should be greatest, according to the diffusion path of Figure 6.23. That is, the experiments confirm that the droplets form as a result of local supersaturation, not by deformation of the interface leading to droplet break-off.

Even when there is convection, whether generated spontaneously by Marangoni flow or imposed by mechanical agitation, the condition for spontaneous emulsification remains unchanged (Ostrovskii et al., 1970). However, the amount of emulsification is greatly enhanced by convection, which continually brings fresh material from both bulk phases to the interface where contact occurs, resulting in supersaturation in accordance with the analysis given above.

In oil-water-surfactant systems, phase diagrams are typically much more complicated than that of Figure 6.23, as discussed in Chapter 4. Accordingly, a variety of phenomena are observed; for instance, spontaneous emulsification occurring simultaneously with the growth of "intermediate" phases at the interface which were not present initially (Raney et al., 1985). Diffusion path analysis can still be applied, however, provided that the intermediate phase forms on initial contact.

Figure 6.24a (Raney and Miller, 1987) shows the calculated diffusion path for a situation where an anionic surfactant-alcohol-brine mixture of composition D is contacted with a comparable volume of a hydrocarbon near the optimal salinity. The main features of the phase diagram, including the three-phase region and the existence of the initial mixture as a dispersion of the lamellar liquid crystal and brine, are consistent with known behavior. However, the precise locations of the phase boundaries were assumed because of the lack of experimental information. The diffusion path (dotted lines) indicates that both brine and a microemulsion should form as intermediate phases, as was in fact observed experimentally in such systems. Figure 6.24b is an expanded view of the oil

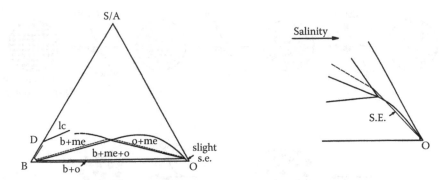

FIGURE 6.24 (a) Diffusion path at optimal salinity in a model oil-surfactant-brine system. (b) Enlargement of an oil corner showing local supersaturation leading to spontaneous emulsification. Reprinted with permission from Raney and Miller (1985). Copyright (1985) American Chemical Society.

corner showing why spontaneous emulsification of water in the oil phase is predicted. Such emulsification was also observed.

As mentioned in Chapter 4 (Section 12), extensive spontaneous emulsification of water in the oil phase is also observed when water-nonionic surfactant mixtures contact oil at temperatures above the PIT. Here the surfactant is preferentially soluble in oil, so that one has a solute diffusing into the phase in which it is more soluble, as in Figure 6.23. A plausible diffusion path has been proposed for this situation (Benton et al., 1986). One difference is that, as might be expected, the emulsions in the oil-water-surfactant systems are considerably more stable than in the oil-water-alcohol systems.

Spontaneous emulsification need not occur immediately upon contact, as in the above examples. When a drop of an oil containing 80 wt. % of a 90/10 mixture of n-hexadecane and octanol and 20 wt. % of the pure nonionic surfactant $C_{12}E_6$ is injected into water at 30°C, the drop initially takes up water, becoming a microemulsion. As octanol diffuses into the water, the surfactant/alcohol films in the microemulsion become more hydrophilic, and it inverts from water-in-oil to bicontinuous to oil-in-water. At some point during this process the microemulsion is no longer able to solubilize all the oil present and spontaneous formation of small oil droplets begins. The oil does not become a uniform layer at the interface because it is a nonwetting phase. Ultimately the microemulsion becomes so hydrophilic as it continues to lose octanol that it becomes miscible with water, leaving an oil-in-water emulsion with small droplets (Rang and Miller, 1998). A schematic diagram of the emulsification process is shown in Figure 6.25.

This behavior is useful when it is desired to disperse oils containing pesticides, drugs, or cosmetics to small droplets without application of high shear. Usually, gentle stirring is used to disperse the droplets so formed into water, and the process is called "self emulsification." Other systems were investigated as well. It was found that emulsification yielding small droplets could be achieved when inversion of a water-in-oil microemulsion was produced by other methods,

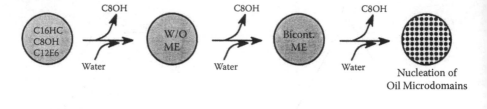

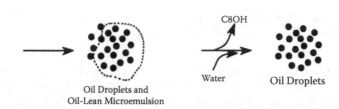

FIGURE 6.25 Schematic diagram showing mechanism of spontaneous emulsification for a drop of n-hexadecane containing suitable amounts of pure $C_{12}E_6$ and n-octanol.

such as a decrease in ionic strength when an anionic surfactant was used (Nishimi and Miller, 2000) or a chemical reaction that converted a lipophilic to a hydrophilic surfactant (Nishimi and Miller, 2001). While in these systems the droplets typically had diameters of a few micrometers, in one case a rather stable emulsion with droplets only about 60 nm in diameter was obtained for an oil with suitable amounts of n-hexadecane, $C_{12}E_6$, and oleyl alcohol (Rang and Miller, 1999). Independently Forgiarini et al., 2001 observed similar behavior in a commercial nonionic surfactant-water-oil system. The occurrence and properties of such "nanoemulsions" have been reviewed by Solans et al., (2005).

12. INTERFACIAL MASS TRANSFER RESISTANCE

The rate of evaporation of water decreases dramatically when a compact and coherent insoluble monolayer occupies the water-air interface (Gaines, 1966; La Mer, 1962). In this context, n-cetyl alcohol (n-hexadecanol) is a well-studied insoluble surfactant which forms such monolayers. A simple explanation is that a significant part of the interfacial area available for evaporation is blocked by the monolayer. Even at a liquid-liquid interface, a resistance to mass transfer is sometimes observed when a solute crosses the interface as it transfers from one liquid into another. This resistance appears to be independent of the convection in the liquid, suggesting that it lies solely at the interface (Chen and Lee, 2000).

A different kind of interfacial mass transfer resistance is seen when a surface active species crosses over from one liquid to another. The experiments of England and Berg (1971) and the model of Brenner and Leal (1978, 1982) were discussed earlier. A simple way of quantifying this resistance is to express the flux of species i crossing an interface as follows:

$$N_{ix} = k_i \left(K_i c_i^I - c_i^{II} \right) \Big|_s = -D_i^{II} \left. \frac{\partial c_i^{II}}{\partial x} \right|_s = -D_i^I \left. \frac{\partial c_i^I}{\partial x} \right|_s \qquad (6.94)$$

where accumulation at the interface is neglected. This assumption is reasonable at steady state or when the species is not surface active. Here, the superscripts I and II denote the two phases and K_i is the partition coefficient. The x direction is normal to the surface traversing from phase II to I. For the case where the interfacial conductance k_i becomes infinite, the second part of Equation 6.94 leads to the conventional condition

$$K_i c_i^I = c_i^{II} \qquad (6.95)$$

on S, used in Sections 10 and 11. One important premise of transport phenomena is that of local equilibrium; although the system is not at equilibrium, every point is locally at equilibrium, satisfying the equation of state. Local equilibrium does not hold in Equation 6.94 at the interface. As was mentioned earlier, the Fickian diffusion in Equation 6.94 also does not hold in complicated surfactant systems (Ma et al., 1989). We add here that one reason for this to occur is that the condition of local equilibrium is not satisfied in the bulk.

More complex situations such as solubilization of oils also exhibit nonequilibrium behavior near interfaces, but only at times. Carroll (1981) found that the rate of solubilization of a very insoluble oil drop by a micellar solution of nonionic surfactant (in moles per unit time per unit interfacial area) was independent of time and drop volume in a batch experiment. This behavior indicates that interfacial resistance, not mass transfer in the micellar solution, controls the rate of solubilization. His view was that first micelles completely demicellized to individual molecules in the immediate vicinity of the oil-water interface with the rate constant τ_2^{-1} discussed in Chapter 4. These surfactant molecules were then adsorbed at the interface and subsequently desorbed as aggregates containing solubilized oil. Oh and Shah (1993) also suggest that τ_2^{-1} dictates solubilization rates. Chan et al. (1976) and Shaeiwitz et al. (1981) reported that the rate of solubilization of solid fatty acids by micellar solutions was partly dependent on convective diffusive transport and partly on an interfacial resistance that they considered involved adsorption of acid-free micelles and emission of acid-containing micelles.

Although τ_2^{-1} could play a role in some systems, supporting data are sparse. Accordingly, we begin with a model that considers micellization-demicellization kinetics to be very fast, so that micelles are preserved as a whole. Solubilization takes place when a singly dispersed oil molecule collides with a micelle or when a micelle collides with the oil-water interface.

If the flux of oil at the interface due to solubilization is N_i, then Equation 6.94 yields

$$N_i = k_i(K_o c_m - c_i),$$
(6.96)

where c_i is the actual concentration of the oil at the oil-water interface and $K_o c_m$ is the equilibrium value. K_o is the partition coefficient and c_m is the concentration of micelles at the interface. Note that the oil is expected to be mainly that solubilized owing to its low molecular solubility. There is a dynamic equilibrium between the oil molecules in the two forms and significant exchange between solubilizates in different micelles. Both features are suggested by the nuclear magnetic resonance (NMR) self-diffusion coefficient measurements (Lindman and Stilbs, 1987).

The term within the brackets in Equation 6.96 is the effective driving force for this step for which k_i is the conductance due to yet unspecified mechanism. The second step is

$$N_i = k_L c_i,$$
(6.97)

where k_L is the effective convective-diffusive mass transfer coefficient of the oil and the concentration of oil in the bulk is assumed to be zero. In general, k_L depends on stirring speed, and in the absence of stirring becomes time dependent. Now,

$$N_i = K_o k c_m$$
(6.98)

is obtained by combining Equations 6.96 and 6.97, where the overall mass transfer coefficient is

$$\frac{1}{k} = \frac{1}{k_i} + \frac{1}{k_L}.$$
(6.99)

The right-hand side of Equation 6.99 represents two mass transfer resistances in series. When $k_i^{-1} \gg k_L^{-1}$, the mass transfer is controlled by the interfacial resistance, and when $k_i^{-1} \ll k_L^{-1}$, the mass transfer is controlled by the conventional convective diffusive transport and local equilibrium applies at the interface.

When oil and a surfactant solution are brought into contact at a plane interface, the thickness of the diffusion layer is proportional to the square root of the contact time in the absence of convection. In this case, interfacial resistance, if it exists, will always dominate at short times. For a single drop, the thickness of the diffusion layer reaches a steady-state value having the same order of magnitude as the drop radius, so that a sufficiently large interfacial resistance can control transport at all times. In any case, one has $K_o c_m \gg c_i$ when interfacial resistance controls and when little oil has been solubilized, and Equation 6.96 becomes

$$N_i \approx K_o k_i c_m.$$
(6.100)

In this situation, which can be identified by a time-independent solubilization rate in a batch experiment, Equation 6.100 can be used to calculate the coefficient k_i.

We now proceed to look at the mechanisms lumped into k_i. First, oil is transferred from the oil phase to the water phase by ordinary molecular dissolution. Its rate is given by $k_{Lo}(c_{os} - c_o)$, where k_{Lo} is an appropriate mass transfer coefficient for the thin interfacial region, and $c_{os} - c_o$ is the concentration difference of molecularly dissolved oil across this region. Here we have already taken c_o, the bulk concentration of dissolved oil, to be zero, with the result that the oil flux at the interface becomes $k_{Lo}c_{os}$ with c_{os} the solubility of individual oil molecules.

Yet another kind of oil flux exists at the interface. The micelles can interact with the oil-water interface in a way that is tantamount to collision, picking up oil molecules there. This flux will then be $k_c c_m$, where k_c is proportional to this "collision frequency" and the flux is proportional to the number of micelles. Hence, overall,

$$N_i = k_{Lo}c_{os} + k_c c_m. \tag{6.101}$$

Combining Equations 6.100 and 6.101 we get

$$k^* = K_o k_i = \frac{k_{Lo}c_{os}}{c_m} + k_c \tag{6.102}$$

for situations far from equilibrium. The term in k_{Lo} should also contain the "reaction" between dissolved oil molecules and micelles, a feature that is taken up later.

We now compare this model with experimental results. For an ionic surfactant, both micelles and the interface are highly charged, and the number of collisions between them is very low. Hence k_c in Equation 6.102 can be ignored. In solubilization experiments with SDS and toluene, Shah and Neogi (2002) found that except at low micelle concentrations c_m, measured values of $k^* = K_o k_i$ were inversely proportional to c_m, i.e., the solubilization rate was independent of surfactant concentration. They also showed that this relationship was true for other published experiments, with SDS solubilizing other oils such as decane and undecane with much lower but not negligible solubilities. Their results for k^* and literature values of τ_2^{-1} for SDS are shown in Figure 6.26. The dependence of the two on c_m is strikingly different, confirming that complete demicellization has little influence on the solubilization rate in this system. When the oil is highly insoluble, (e.g., squalane or triolein), c_{os} and the first term on the right-hand side of Equation 6.102 can also be ignored, and such oils cannot be solubilized by ionic surfactants.

On the other hand, k_c is significant for nonionic surfactants, and solubilization does take place for many alkanes and other oils rather insoluble in water, provided that appreciable solubilization of oil in the surfactant micelles occurs. As first

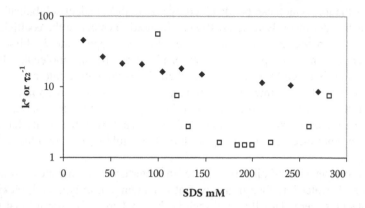

FIGURE 6.26 Measured values of k^* ($= K_o k_i$) plotted against surfactant concentration (black). Also shown are the demicellization rate constant τ_2^{-1} from Lessner et al. (1981) (white). Reprinted from Shah and Neogi (2002) with permission from Elsevier.

observed by Carroll (1981), k_c is independent of surfactant concentration for linear alkyl ethoxylates, but increases with increasing temperature and decreasing oil chain length for alkanes. The large difference in the magnitude of k_c between ionic and nonionic surfactants and the resulting influence on solubilization kinetics was pointed out by Kabalnov and Weers (1996), as well in experiments for systems with mixtures of ionic and nonionic surfactants (Ward, 1995).

Kabalnov and Weers (1996) proposed a model in which solubilization is the result of molecular dissolution, diffusion, and a "reaction" which involves irreversible uptake of oil molecules by micelles. The rate constant for this reaction is on the order τ_1^{-1}, τ_1 being the *first* time constant of micellar kinetics, which is often said to represent the characteristic time for exchange of a single surfactant molecule between the bulk solution and micelle (see Chapter 4), and with $\tau_1 \ll \tau_2$. This process would contribute to k_{Lo} in the above equations. Since the emphasis was on ionic surfactants, k_c was neglected. This model predicts a solubilization rate that is independent of time and proportional to $c_m^{1/2}$. However, it cannot be complete because it predicts that the solubilization rate increases indefinitely with increasing c_m, which conflicts with the constant rate found experimentally at high surfactant concentrations mentioned previously. Todorov et al. (2002) found that they had to supplement the reaction term with a constant interfacial resistance of unspecified origin to fit their data on solubilization of n-decane in SDS solutions. Then they obtained equations that predicted a constant solubilization rate at large c_m, as did Neogi (2003), who assumed a fast reaction to begin with. There is currently no complete explanation of the mechanism causing this interfacial resistance for ionic surfactants. Some extensions of the above models have been suggested (Dungan et al., 2003; Sailaja et al., 2003). Problem 6.13 considers features of some of the models.

The available evidence suggests that, except for rather soluble oils, the rate of solubilization is often controlled by interfacial resistance. As discussed previously, the solubilization rate for ionic surfactants becomes constant as surfactant concentration increases. Since most of the studies have used SDS, it is not clear how the asymptotic rate depends on surfactant structure. For many nonionic surfactants of the linear alcohol ethoxylate type, the second term of Equation 6.101 dominates, with the result that the solubilization rate is proportional to surfactant concentration. It is not clear whether this behavior is maintained as the length of the ethylene oxide chain increases, since steric repulsion could hinder micelle collisions with the oil-water interface and reduce k_c to very small values. In this case, the Kabalnov–Weers model of molecular dissolution and subsequent "reaction" with micelles could become dominant.

So far we have emphasized experiments involving single drops of oil or nearly flat oil-water interfaces in contact with micellar solutions. Peña and Miller (2002) showed that single-drop results for nonionic surfactants in situations where interfacial resistance controlled could be used to interpret certain transport processes in emulsions, in particular, solubilization of oil from emulsion drops and a phenomenon known as "compositional ripening." The latter refers to transport in emulsions having drops with different initial compositions in which one or more components of the oil leave some drops and enter others, traveling between them solubilized in surfactant micelles. In order to model such a process for the simplest case where only one oil in a binary mixture can be solubilized and so transported, it was necessary to generalize the second term of Equation 6.101 as follows for the more soluble oil (the first term in this equation was negligible):

$$N_i = k_c c_m (x_{Di} - f_i). \qquad (6.103)$$

Here, x_{Di} is the mole fraction of the more soluble oil in the drop and f_i is the solubilized oil content of the bulk micellar solution expressed as a fraction of the maximum possible solubilization. In the discussion above, $x_{Di} = 1$ (drops are a pure oil) and $f_i = 0$ (no oil solubilized in bulk solution), so that the second term of Equation 6.101 is recovered. Note that the flux N_i can be either positive or negative, depending on the sign of the quantity in parentheses. That is, the more soluble oil leaves drops having large x_{Di} but enters those with small x_{Di} when $f_i \neq 0$, as is the case during compositional ripening. Equation 6.103 was shown to hold in experiments with single drops of n-decane/squalane mixtures, both when decane was leaving and when it was entering the drop from micellar solutions of pure $C_{12}E_8$ partially saturated with decane (squalane being insoluble in the micelles). The results of these experiments were used with Equation 6.103 and a mass balance to interpret the compositional ripening experiments of Binks et al. (1998) for this system. Excellent agreement with experimental results was obtained when polydispersity of the initial emulsion was included in the calculations.

So far all discussion has been limited to micellar solutions. Constant rates of mass transfer have been reported in solutions containing large amounts of surfactants. Friberg et al. (1985) contacted a micellar solution of a nonionic surfactant with an oil in a test tube such that the overall composition corresponded to that of middle phase microemulsion. As discussed in Section 12, a few intermediate phases formed transiently. In particular, the micellar solution transformed into water droplets in lamellar liquid crystal at a constant rate. The modeling of these experiments is quite complex and laborious, and an effort was made to measure the concentration profiles by chopping up the test tube into smaller segments in the transverse direction (Friberg et al., 1988; Ma et al., 1988). This procedure is also quite laborious, but yields on analysis (Anandakrishnan et al., 1992) the "constant" rates related to the interfacial resistance for individual species and not just the overall values. The results appear to show that over individual 10-day periods, the rate coefficients are nearly constant, but they decrease significantly toward zero over a 60-day period.

Some work has been reported on the kinetics of dissolution of pure detergents. This rate for a commercial solid detergent has been found to be diffusion limited by Figge and Neogi (1996). Chen et al. (2000) also found in their studies on dissolution of pure nonionic surfactants that rates were limited by diffusion.

13. OTHER INTERFACIAL PHENOMENA INVOLVING DISPERSED PHASE FORMATION

Spontaneous emulsification is but one example of dispersed phase formation where diffusion effects are important, albeit one of interest to many workers in interfacial phenomena. Numerous other situations exist where a dispersed solid or fluid phase forms as a result of phase transformation or chemical reaction. Moving fronts can develop, for instance, between regions containing and free of the dispersed phase. Such fronts involving the growth or shrinkage of regions of fog were considered by Toor (1971), and their stability was analyzed by Miller and Jain (1973) using a method similar to that described in Section 8. It was found that a front where fog is consumed by evaporation of drops is stable, but a front where fog forms by invading a region of supersaturated vapor is unstable.

Another interesting phenomenon is formation of a dispersed phase in narrow, individual bands separated by regions containing no dispersed phase. The bands are often called "Liesegang rings." Figure 6.27 illustrates the formation of discrete bands of magnesium hydroxide ($Mg(OH)_2$) precipitate in test tubes where aqueous solutions of ammonium hydroxide (NH_4OH) (top) and magnesium sulfate ($MgSO_4$) (bottom) have been contacted and diffusion allowed to proceed. A small amount of a gelling agent has been added to prevent convection.

Theoretical descriptions of such behavior have been given in terms of a mechanism first put forward by Ostwald around the turn of the century, namely that a critical local supersaturation is required to initiate local precipitation (see Henisch, 1988; Klueh and Mullins, 1969; Prager, 1956; Wagner, 1950). But subsequent experimental studies have raised questions concerning the applicability of this mechanism (Kai et al., 1982). In particular, the solid has been observed to form as a colloidal dispersion before the appearance of bands. Also it has been found that bands sometimes develop, in contrast to predictions of the above theory, in systems of uniform initial composition (Flicker and Ross, 1974).

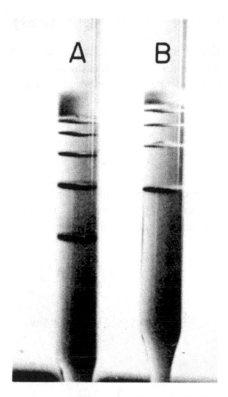

FIGURE 6.27 Liesegang ring formation. From Kai et al. (1982) with permission.

We focus for simplicity on this last situation and on the instability mechanism put forward by Lovett et al. (1978), which, according to the experiments, may also be pertinent in Liesegang ring formation where initial concentration gradients exist. First, we note that solution of the diffusion equation for a single spherical particle of a pure species A in an infinite expanse of a solvent having a bulk concentration c_∞ of A leads to the following expression for particle growth rate:

$$\frac{da}{dt} = \frac{Dv_{As}c_e}{a}\left(\frac{c_\infty}{c_e} - 1 - \frac{2\gamma v_{As}}{aRT}\right),\tag{6.104}$$

where a is the particle radius, D the diffusivity of A in the solvent, v_{As} the molar volume of the particle, γ the interfacial tension, and c_e the concentration of A in the solvent when it has been equilibrated with a large bulk phase of A. We note from Equation 6.104 that the concentration c_∞, for which there is no particle growth (i.e., for which da/dt vanishes), is larger for particles of smaller radius as a result of the term involving interfacial tension. In other words, small particles are more soluble than large particles, so that the latter should be expected to grow at the expense of the former, a process that reduces interfacial area and free energy. It is this basic behavior or "Ostwald ripening" that leads in the analysis below to the development of bands.

We consider a uniform initial dispersion where the values of c_∞^* and a^* for solute concentration and radius are such that da/dt, as given by Equation 6.104, is zero. Now suppose that c_∞ and a are perturbed in such a way that

$$c_\infty = c_\infty^* + c_{\infty p} e^{i\alpha x} e^{\beta t} \tag{6.105}$$

$$a = a^* + a_p e^{i\alpha x} e^{\beta t}, \tag{6.106}$$

where $c_{\infty p}$ and a_p are the perturbation amplitudes and α and β are the wavenumber and time factor as in previous stability analyses. The diffusion equation in the bulk solvent is, in the absence of convection, given by

$$\frac{\partial c_\infty}{\partial t} = D \frac{\partial^2 c_\infty}{\partial x^2} + r . \tag{6.107}$$

Here, r is the net rate of solute addition to the solvent by particle dissolution and is given by

$$r = -\frac{da}{dt} \cdot \frac{4\pi a^2}{v_{As}} \cdot n , \tag{6.108}$$

with n the number of particles per unit volume.

We can use Equation 6.104 to calculate da/dt following the perturbation:

$$\frac{da}{dt} = \beta a_p e^{i\alpha x + \beta t} = \frac{D v_{As} c_e}{a^*} \left[\frac{c_{\infty p}}{c_e} - \left(\frac{c_\infty}{c_e} - 1 \right) \frac{a_p}{a^*} \right.$$
$$\left. + \frac{4\gamma v_{As}}{RT} \frac{a_p}{a^{*2}} \right] e^{i\alpha x + \beta t} \tag{6.109}$$

From this equation we find

$$c_{\infty p} = a_p \left[\frac{\beta a^*}{D v_{As}} - \frac{2\gamma v_{As} c_e}{a^{*2} RT} \right]. \tag{6.110}$$

Finally, substituting Equations 6.105, 6.106, and 6.108 through 6.110 into Equation 6.107, we obtain a quadratic equation that yields the following solutions for the time factor β:

$$\beta = -\frac{b}{2} \pm \left[\frac{b^2}{4} + \frac{2\gamma\alpha^2 D^2 v_{As}^2 c_e}{a^{*3}RT} \right]^{1/2} \tag{6.111}$$

$$b = 4\pi Dna^* + D\alpha^2 - \frac{2\gamma v_{As}^2 c_e D}{a^{*3}RT} . \tag{6.112}$$

Inspection of this equation reveals that with $b > 0$ there is always one positive root for β (i.e., the system is always unstable). However, the positive root becomes very small in the limit of long wavelengths ($\alpha \rightarrow 0$), the expected result since solute must diffuse over relatively long distances in order for the perturbation to grow. Lovett et al. (1978) showed that when diffusion of the particles themselves is considered, perturbations of very short wavelengths are stable. That is, particle diffusion virtually eliminates particle concentration gradients and precludes discrete bands from forming. An intermediate wavelength λ_m exists where perturbation growth rate reaches a maximum value. For reasonable values of the physical properties, λ_m was calculated to be on the order of 1 cm and the time constant for growth on the order of 1 day. Both these values are generally consistent with experimental observations.

Although this analysis yields insight on how bands of precipitate may form, understanding the origin of Liesegang rings is not yet complete. For instance, several studies, including some quite old (Dhar and Chaterji, 1925) and some more recent (Antal et al., 1998; Kanniah et al., 1983) suggest that flocculation of a colloidal dispersion can play an important role in the development of visible precipitate bands.

REFERENCES

GENERAL REFERENCES ON PHENOMENA INVOLVING TRANSPORT NEAR INTERFACES

Berg, J.C., Interfacial phenomena in fluid phase separation processes, in *Recent Developments in Separation Science*, vol. 2, N.N. Li (ed.), CRC Press, Cleveland, 1972, p. 1.

Berg, J.C., Interfacial hydrodynamics: An overview, *Can. Metall. Q.*, 21, 121, 1982.

Birikh, R.V., Briskman, V.A., Verlarde, M.G., and Legros, J.-C., *Liquid Interfacial Systems, Oscillations and Instability*, Marcel Dekker, New York, 2003.

Edwards, D.A., Brenner, H., and Wasan, D.T., *Interfacial Transport Processes and Rheology*, Butterworths-Heinemann, Boston, 1991.

Levich, V.G. and Krylov, V.S., Surface-tension-driven phenomena, *Annu. Rev. Fluid Mech.*, 1, 293, 1969.

Lucassen-Reynders, E.H., Surface elasticity and viscosity in compression/dilation, in *Anionic Surfactants: Physical Chemistry of Surfactant Action,* E.H. Lucassen-Reynders (ed.), Marcel Dekker, New York, 1981, p. 173.

Miller, C.A., Stability of interfaces, in *Surface and Colloid Science,* vol. 10, E. Matijevic (ed.), Plenum Press, New York, 1978, p. 227.

Sawistowski, H., Interfacial phenomena, in *Recent Advances in Liquid-Liquid Extraction,* C. Hanson (ed.), Pergamon Press, Oxford, 1971, p. 293.

Sekerka, R.F., Morphological stability, *J. Crystal Growth,* 3/4, 71, 1968.

Slattery, J.C., *Interfacial Transport Phenomena,* Springer-Verlag, New York, 1990.

van den Tempel, M. and Lucassen-Reynders, E.H., Relaxation processes at fluid interfaces, *Adv. Colloid Interface Sci.,* 18, 281, 1983.

TEXT REFERENCES

Anandakrishnan, K., Neogi, P., and Friberg, S.E., Mass transfer analysis in a microemulsion system, *Colloids Surfaces,* 64, 197, 1992.

Antal, T., Draz, M., Magnin, J., Raez, Z., and Zrinyi, M., Derivation of the Matalon-Packter law for Liesegang patterns, *J. Chem. Phys.,* 109, 9479, 1998.

Armento, M.E. and Miller, C.A., Stability of combustion fronts in porous media, *Soc. Petrol. Eng. J.,* 17, 423, 1977.

Avsec, D., Thermoconductive eddies in air. Application to meteorology, *Publ. Sci. Tech. Minist. Air France,* 155, 1939.

Bainbridge, G.S. and Sawistowski, H., Surface tension effects in sieve plate distillation columns, *Chem. Eng. Sci.,* 19, 992, 1964.

Bascom, W.D., Cottington, R.L., and Singleterry, C.R., , Dynamic surface phenomena in the spontaneous spreading of oils on solids, *Adv. Chem. Ser.,* 43, 355, 1964.

Benard, H., Les tourbillons cellulaires dans une nappe liquide transportant de la chaleur par convection en régime permanent, *Rev. Gen. Sci. Pur. Appl.,* 11, 1261, 1309, 1900.

Benton, W.J., Raney, K.H., and Miller, C.A., Enhanced videomicroscopy of phase transitions and diffusional phenomena in oil-water-nonionic surfactant systems, *J. Colloid Interface Sci.,* 110, 363, 1986.

Berg, J.C., Interfacial phenomena in fluid phase separation processes," in *Recent Developments in Separation Science,* vol. 2, N.N. Li (ed.), CRC Press, Cleveland, 1972, p. 1.

Berg, J.C., Interfacial hydrodynamics: An overview, *Can. Metall. Q.,* 21, 121, 1982.

Berg, J.C. and Acrivos, A., The effect of surface active agents on convection cells induced by. surface tension, *Chem. Eng. Sci.,* 20, 737, 1965.

Berg, J.C., Acrivos, A., and Boudart, M., Evaporative convection, *Adv. Chem. Eng.,* 6, 61, 1966.

Binks, B.P., Clint, J.H., Fletcher, P.D.I., Rippon, S., Lubetkin, S.D., and Mulqueen, P.J., Kinetics of swelling of oil-in-water emulsions, *Langmuir,* 14, 5402, 1998.

Bird, R.B., Stewart, W.E., and Lightfoot, E.N., *Transport Phenomena,* 2nd ed., John Wiley, New York, 2002.

Birikh, R.V., Briskman, V.A., Verlarde, M.G. and Legros, J.-C., *Liquid Interfacial Systems, Oscillations and Instability,* Marcel Dekker, New York, 2003.

Bonfillon, A., Sicoli, F., and Langevin, D., Dynamic surface tension of ionic surfactant solutions, *J. Colloid Interface Sci.,* 168, 497, 1994.

Bonfillon, A. and Langevin, D., Electrostatic model for the viscoelasticity of ionic surfactant monolayers, *Langmuir*, 10, 2965, 1994.

Braun, R.L., Lewis, A.E., and Wadsworth, M.E., In-place leaching of primary sulfide ores, in *Solution Mining Symposium*, Aplan, F.F. (ed.), AIME, New York, 1974.

Brenner, H. and Leal, L.G., Interfacial resistance to interphase mass transfer in quiescent two-phase systems, *AIChE J.*, 24, 246, 1978.

Brenner, H. and Leal, L.G., Conservation and constitutive equations for adsorbed species undergoing surface diffusion and convection at a fluid-fluid interface, *J. Colloid Interface Sci.*, 88, 136, 1982.

Brian, P.L.T., Effect of Gibbs adsorption on Marangoni instability, *AIChE J.*, 17, 765, 1971.

Burkholder, H.C. and Berg, J.C., Effect of mass transfer on laminar jet breakup: part I. Liquid jets in gases, *AIChE J.*, 20, 863, 872, 1974.

Cannon, K.J. and Denbigh, K.G., Studies on gas-solid reactions—I: the oxidation rate of zinc sulphide, *Chem. Eng. Sci.*, 6, 145, 155, 1957a.

Cannon, K.J. and Denbigh, K.G., Studies on gas-solid reactions—II : Causes of thermal instability, *Chem. Eng. Sci.*, 6, 155, 1957b.

Carroll, B.J., The kinetics of solubilization of nonpolar oils by nonionic surfactant solutions, *J. Colloid Interface Sci.*, 79, 126, 1981.

Cazabat, A.M., Heslot, F., Troian, S.M., and Carles, P., Finger instability of thin spreading films driven by temperature gradients, *Nature*, 346, 6287, 1990.

Chan, A.F., Evans, D.F., and Cussler, E.L., Explaining solubilization kinetics, *AIChE J.*, 22, 1006, 1976.

Charles, P. and Cazabat, A.M., The thickness of surface-tension-gradient-driven spreading films, *J. Colloid Interface Sci.*, 157, 196, 1993.

Chen, L.-H. and Lee, Y.-L., Adsorption behavior of surfactants and mass transfer in single-drop extraction, *AIChE J.*, 46, 160, 2000.

Chen, B.-H., Miller, C.A., Walsh, J.M., Warren, P.B., Ruddock, J.N., and Garrett, P.R., Dissolution rates of pure nonionic surfactants, *Langmuir*, 16, 5276.

Colinet, P., Legros, J.C., and Verlarde, M.G., *Nonlinear Dynamics of Surface-Tension-Driven Instabilities*, Wiley, New York, 2001.

Cottington, R.L., Murphy, C.L., and Singleterry, C.R., Effect of non-polar additives on oil spreading on solids, with applications to nonspreading oils," in *Adv. Chem. Ser.*, 43, 341, 1964.

Coyle, R.W., Berg, J.C., and Nina, J.C., Liquid-liquid jet breakup under conditions of relative motion, mass transfer and solute adsorption, *Chem. Eng. Sci.*, 36, 19, 1981.

Daniel, R.C. and Berg, J.C., Dynamic surface tension of polydisperse surfactant solutions: a pseudo-single-component approach, *Langmuir*, 18, 5074, 2002.

Datwani, S.S. and Stebe, K.J., Surface tension of an anionic surfactant: equilibrium, dynamics, and analysis for Aerosol-OT, *Langmuir*, 17, 4287, 2001.

Davies, J.T. and Haydon, D.A., Spontaneous emulsification, *Proc. Int. Congr. Surf. Act. 2nd*, 1, 417, 1957.

Defay, R., and Petre, G. (1971) "Dynamic Surface Tension" in *Surface and Colloid Science.*, E. Matijevic (ed), *vol. 3*, Wiley. New York p. 27.

de Groot, S.R. and Mazur, P., *Non-Equilibrium Thermodynamics*, Dover, New York, 1984, p. 221; first published by North-Holland, Amsterdam, 1962.

Douglas, J.F., Johnson, H.E., and Granick, S., A simple kinetic-model of polymer adsorption and desorption, *Science*, 262, 2010, 1993.

Dhar, N.R. and Chatterji, N.G., Theories of Liesegang ring formation, *Kolloid-Z.*, 37, 2, 89, 1925.

Dukhin, S.S., Kretzschmar, G., and Miller, R., *Dynamics of Adsorption at Liquid Interfaces*, Elsevier, Amsterdam, 1995.

Dungan, S.R., Tai, B.A., and Gerhardt, N.I., Transport mechanisms in the micellar solubilization of alkanes in oil-in-water emulsions, *Colloids Surfaces A*, 216, 149, 2003.

Edwards, J., Aspects of addition agent behavior, *Trans. Inst. Met. Finishing*, 41, 169, 1964.

Egelhaaf, S.U. and Schurtenberger, P., Micelle-to-vesicle transition: A time-resolved structural study, *Physical Rev. Letts.*, 82, 2804, 1999.

England, D.C. and Berg, J.C., Transfer of surface-active agents across a liquid-liquid interface, *AIChE J.*, 17, 313, 1971.

Fainerman, V.B., Miller, R., Wustneck, R., and Makievski, A.V., Adsorption isotherm and surface tension equation for a surfactant with changing partial molar area. 1. Ideal surface layer, *J. Phys. Chem.*, 100, 7669, 1996.

Ferri, J.K. and Stebe, K.J., Soluble surfactants undergoing surface phase transitions: a Maxwell construction and the dynamic surface tension, *J. Colloid Interface Sci.*, 209, 1, 1999.

Ferri, J.K. and Stebe, K.J., Which surfactants reduce surface tension faster? A scaling argument for diffusion-controlled adsorption, *Adv. Colloid Interface. Sci.*, 85, 61, 2000.

Flicker, M. and Ross, J., Mechanism of chemical instability for periodic precipitation phenomena, *J. Chem. Phys.*, 60, 3458, 1974.

Figge, C.L. and Neogi, P., Dissolution of solid detergents, *Langmuir*, 12, 1107, 1996.

Forgiarini, A., Esquena, J., Gonzalez, C., and Solans, C., Formation of nano-emulsions by low-energy emulsification methods at constant temperature, *Langmuir*, 17, 2076, 2001.

Friberg, S.E., Ma, Z., and Neogi, P., Temporary liquid crystals in microemulsion systems, in *Surfactant-Based Mobility Control*, Smith, D.K. (ed.), American Chemical Society, Washington, D.C., 1988, p. 108.

Friberg, S.E., Mortensen, M., and Neogi, P., Hydrocarbon extraction into surfactant phase with nonuonic surfactants. I. Influence of phase equilibria for extraction kinetics, *Sep. Sci. Technol.*, 20, 285, 1985.

Gaines, G.L., , *Insoluble Monolayers at Gas-Liquid Interfaces*, Interscience, New York, 1966.

Gouda, J.H. and Joos, P., Application of longitudinal wave theory to describe interfacial. instability, *Chem. Eng. Sci.*, 30, 521, 1975.

Gunn, R.D. and Krantz, W.B., Reverse combustion instabilities in tar sands and coal, *Soc. Petrol. Eng. J.*, 20, 267, 1982.

Hand, D.B., Dineric distribution, *J. Phys. Chem.*, 34, 1961, 1930.

Hansen, R.S., Diffusion and the kinetics of adsorption of aliphatic acids and alcohols at the water-air interface, *J. Colloid Sci.*, 16, 549, 1961.

Hansen, C.M. and Pierce, P.E., Cellular convection in polymer coatings. Assessment, *Ind. Eng. Chem. Prod. Res. Dev.*, 19, 919, 1973.

Henisch, H.K., *Crystals in Gels and Liesagang Rings*, Cambridge University Press, Cambridge, 1988.

Hickman, K., Surface behaviour in the pot still, *Ind. Eng. Chem.*, 44, 1892, 1952.

Hinkbein, T.E. and Berg, J.C., Surface tension effects in heat transfer through thin liquid films, *Int. J. Heat Mass Transfer*, 21, 1241, 1970.

Jackson, R., Diffusion in ternary mixtures with and without phase boundaries, *Ind. Eng. Chem. Fundam.*, 16, 304, 1977.

Johnson, D.K., and Berg, J.C., "The Effect of Surfactants on Thin Film Behavior in Distillation Equipment," Preprint 62c for AIChE National Meeting, Houston, 1983.

Joos, P., *Dynamic Surface Tension*, VSP, Utrecht, The Netherlands, 1999.

Kabalnov, A. and Weers, J., Kinetics of mass transfer in micellar systems: surfactant adsorption, solubilization kinetics, and ripening, *Langmuir*, 12, 3442, 1996.

Kanniah, N., Gnanam, F.D., and Ramasamy, P., Revert and direct Liesegang phenomenon of silver iodide: Factors influencing the transition point, *J. Colloid Interface Sci.*, 94, 412, 1983.

Kataoka, D.E. and Troian, S.M., A theoretical study of instabilities at the advancing front of thermally driven coating films, *J. Colloid Interface Sci.*, 192, 350, 1997.

Kai, S., Muller, S.C., and Ross, J., Measurements of temporal and spatial sequences of events in periodic precipitation processes, *J. Chem. Phys.*, 76, 1392, 1982.

Kenning, D.B.R., Two-phase flow with nonuniform surface tension, *Appl. Mech. Rev.*, 21, 1101, 1968.

Kirkaldy, J.S. and Brown, L.F., Diffusion behavior in ternary multiphase systems, *Can. Metall. Q.*, 2, 89, 1963.

Kielman, H.S. and van Steen, P.J.F., Influence of mesomorphic phase formation on detergency, in *Surface Active Agents*, Proceedings of Conference Soc. Chem. Industry Nottingham, England, 1979a.

Kielman, H.S. and van Steen, P.J.F., Rheological investigation of mesomorphic layers formed at the liquid/liquid interface, *J. Physique* 40, C3-447, 1979b.

Klueh, R.L. and Mullins, W.W., Periodic precipitation (Liesegang phenomenon) in solid silver: II. Modification of Wagner's mathematical analysis, *Acta Met.*, 17, 69, 1969.

Knapp, R. and Aris, R., On the equations for the movement and deformation of a reaction front, *Arch. Rat. Mech. Anal.*, 44, 165, 1972.

La Mer, V.K. (ed.), *Retardation of Evaporation by Monolayers: Transport Processes*, Academic Press, New York, 1962.

Leaist, D.G., Binary diffusion of micellar electrolytes, *J. Colloid Interface Sci.*, 111, 230, 240, 1986a.

Leaist, D.G., Diffusion of ionic micelles in salt solutions: sodium dodecyl sulfate + sodium chloride + water, *J. Colloid Interface Sci.*, 111, 240, 1986b.

Lessner, E., Teubebner, M., and Kahlweit, M., Relaxation experiments in aqueous solutions of ionic micelles. 2. Experiments on the system water-sodium dodecyl sulfate-sodium perchlorate and their theoretical interpretation, *J. Phys. Chem.*, 85, 3167, 1981.

Levich, V.G. and Krylov, V.S., Surface-tension-driven phenomena, *Annu. Rev. Fluid Mech.*, 1, 293, 1969.

Lim, J.-C. and Miller, C.A., Dynamic behavior and detergency in systems containing nonionic surfactants and mixtures of polar and nonpolar oils, *Langmuir*, 7, 2021, 1991a.

Lim, J.-C., and Miller, C.A., Dynamic behavior in systems containing nonionic surfactants and polar oils and its relationship to detergency, in *Surfactants in Solution*, vol. 11, K.L. Mittal and D.O. Shah (eds.), Plenum Press, New York, 1991b, p. 491.

Lim, J.-C. and Miller, C.A., Predicting conditions for intermediate phase formation in surfactant systems, *Progr. Colloid Polym. Sci.*, 83, 29, 1990.

Lin, S.-Y., McKeigue, K., and Maldarelli, C., Diffusion-limited interpretation of the induction period in the relaxation in surface tension due to the adsorption of straight chain, small polar group surfactants: theory and experiment, *Langmuir*, 7, 1055, 1991.

Linde, H., Schwartz, P., and Wilke, H., Dissipative structures and nonlinear kinetics of the Marangoni instability, in *Dynamics and Instability of Fluid Interfaces*, Sorensen, T.S. (ed.), Springer-Verlag, Berlin, 1979, p. 75.

Lindman, B. and Stilbs, P., Molecular diffusion in micromulsions, in *Microemulsions: Structure and Dynamics,* Friberg, S.E. and Bothorel, P. (eds.), CRC Press, Boca Raton, Fla., 1987, p. 119.

Lovett, R., Ortoleva, P., and Ross, J., Kinetic instabilities in first order phase transitions, *J. Chem. Phys.*, 69, 947, 1978.

Lu, J.R., Hromadova, M., Thomas, R.K., and Penfold, J., Neutron reflection from triethylene glycol monododecyl ether adsorbed at the air-liquid interface: the variation of the hydrocarbon chain distribution with surface concentration, *Langmuir*, 9, 2417, 1993.

Lucassen, J. and Barnes, G.T., Propagation of surface tension changes over a surface with limited area, *J. Chem. Soc. Faraday Trans. I*, 68, 2129, 1972.

Lucassen, J., and van den Tempel, M., Longitudinal waves on visco-elastic surfaces, *J. Colloid Interface Sci.*, 41, 491, 1972.

Lucassen-Reynders, E.H., Surface elasticity and viscosity, in *Anionic Surfactants*, Lucassen-Reynders, E.H. (ed.), Marcel Dekker, New York, 1981, p. 173.

Ma, Z., Friberg, S.E., and Neogi, P., Observation of temporary liquid crystals in water-in-oil microemulsion systems, *Colloids Surfaces*, 33, 249, 1988.

Ma, Z., Friberg, S.E., and Neogi, P., Single-component mass transfer in a cosurfactant-water-surfactant system, *AIChE J.*, 35, 1678, 1989.

MacLeod, C.A. and Radke, C.J., Surfactant exchange kinetics at the air/water interface from the dynamic tension of growing liquid drops, *J. Colloid Interface Sci.*, 166, 73, 1994.

Maru, H.C., Mohan, T.V., and Wasan, D.T., Dilational viscoelastic properties of fluid interfaces—I, *Chem. Eng. Sci.*, 34, 1283, 1979.

Maru, H.C. and Wasan, D.T., Dilational viscoelastic properties of fluid interfaces—II: experimental study, *Chem. Eng. Sci.*, 34, 1295, 1979.

Miller, C.A., Stability of moving surfaces in fluid systems with heat and mass transport. II. Combined effects of transport and density difference between phases, *AIChE J.*, 19, 909, 1973.

Miller, C.A., Stability of moving surfaces in fluid systems with heat and mass transport. III. Stability of displacement fronts in porous media, *AIChE J.*, 21, 474, 1975.

Miller, C.A., Solubilization and intermediate phase formation in oil-water-surfactant systems, *Tenside Surf. Deter.*, 33, 191, 1996.

Miller, C.A., ,Stability of interfaces, in *Surface and Colloid Science*, vol. 10, Matijevic, E. (ed.), Plenum Press, New York, 1978, p. 227.

Miller, C.A. and Jain, K., Stability of moving surfaces in fluid systems with heat and mass transport — I : Stability in the absence of surface tension effects, *Chem. Eng. Sci.*, 28, 157, 1973.

Miller, C.A., Spontaneous emulsification produced by diffusion—a review, *Colloids Surfaces*, 29, 89, 1988.

Miller, R. and Kretzschmar, G., Adsorption kinetics of surfactants at fluid interfaces, *Adv. Colloid Interface Sci.*, 37, 97, 1991.

Miller, R., Aksenenko, E.V., Liggieri, L., Ravera, F., Ferrari, M., and Fainerman, V.B., Effect of the reorientation of oxyethylated alcohol molecules within the surface layer on equilibrium and dynamic surface pressure, *Langmuir*, 15, 1328, 1999.

Mulqueen, M., Datwani, S.S., Stebe, K.J., and Blankenstein, D., Dynamic surface tensions of aqueous surfactant mixtures: experimental investigation, *Langmuir*, 17, 7494, 2001.

Narayanan, R. and Schwabe, D. (eds.), *Interfacial Fluid Dynamics and Transport Processes*, Springer-Verlag, Berlin, 2003.

Narsimhan, G. and Uraizee, F., Kinetics of adsorption of globular proteins at an air-water interface, *Biotechnol. Prog.*, 8, 187, 1992.

Nelson, N.K., Jr. and Berg, J.C., The effect of chemical reaction on the breakup of liquid jets, *Chem. Eng. Sci.*, 37, 1067, 1982.

Neogi, P., Tears-of-wine and related phenomena, *J. Colloid Interface Sci.*, 105, 94, 1985.

Neogi, P., Diffusion in a micellar solution, *Langmuir*, 10, 1410, 1994.

Neogi, P., A model for interfacial resistance observed during solubilization with micellar solution, *J. Colloid Interface Sci.*, 261, 542, 2003.

Nield, D.A., Surface tension and buoyancy effects in cellular convection, *J. Fluid Mech.*, 19, 341, 1964.

Nishimi, T. and Miller, C.A., Spontaneous emulsification of oil in aerosol-OT/water/hydrocarbon systems, *Langmuir*, 16, 9233, 2000.

Nishimi, T., and Miller, C.A., Spontaneous emulsification produced by chemical reactions, *J. Colloid Interface Sci.*, 237, 259, 2001.

Noskov, B.A., Kinetics of adsorption from micellar solutions, *Adv. Colloid Interface Sci.*, 95, 237, 2002.

Oh, S.G. and Shah, D.O., The effect of micellar lifetime on the rate of solubilization and detergency in sodium dodecyl sulfate solutions, *J. Am. Oil Chem. Soc.*, 70, 673, 1993.

Ostrovskii, M.V., Barenbau, R.K., and Abramson, A.A., Conditions for the generation of emulsion during mass transfer in a liquid-liquid system, *Colliod J. USSR*, 32, 565, 1970.

Palmer, H.J., The hydrodynamic stability of rapidly evaporating liquids at reduced pressure, *J. Fluid Mech.*, 75, 487, 1976.

Palmer, H.J. and Berg, J.C., Convective instability in liquid pools heated from below, *J. Fluid Mech.*, 47, 779, 1971.

Pearson, J.R.A., On convection cells induced by surface tension, *J. Fluid Mech.*, 4, 489, 1958.

Peña, A.A. and Miller, C.A., Transient behavior of polydisperse emulsions undergoing mass transfer, *Ind. Eng. Chem. Res.*, 41, 6284, 2002.

Perez de Ortiz, E.S. and Sawistowski, H., Interfacial stability of. binary liquid-liquid systems. I. Stability analysis, *Chem. Eng. Sci.*, 28, 2051, 1973.

Prager, S., Periodic precipitation, *J. Chem. Phys.*, 25, 279, 1956.

Rabinovich, L.M., Problems in modelling and intensification of mass transfer with interfacial instability and self-organization, in *Mathematical Modelling of Chemical Processes*, Rabinovich, L.M. (ed.), CRC Press, Boca Raton, Fla., 1991, chap. 4.

Rang, M.-J. and Miller, C.A., Spontaneous emulsification of oil drops containing surfactants and medium-chain alcohols, *Prog. Colloid Polym. Sci.*, 109, 101, 1998.

Rang, M.-J. and Miller, C.A., Spontaneous emulsification of oils containing hydrocarbon, nonionic surfactant, and oleyl alcohol, *J. Colloid Interface Sci.*, 209, 179, 1999.

Raney, K.H., Benton, W.J., and Miller, C.A., Diffusion phenomena and spontaneous emulsification in oil-water surfactant systems, in *Macro- and Microemulsions*, Shah, D.O. (ed.), American Chemical Society Symposium Series 272, 1985, p. 193 ff.

Raney, K.H. and Miller, C.A., Diffusion path analysis of dynamic behavior of oil-water-surfactant systems, *AIChE J.*, 33, 1791, 1987.

Rubin, E. and Radke, C.J., Dynamic interfacial tension minima in finite systems, *Chem. Eng. Sci.*, 35, 1129, 1980.

Ruckenstein, E. and Berbente, C., The occurrence of interfacial turbulence in the case of diffusion accompanied by chemical reaction, *Chem. Eng. Sci.*, 19, 329, 1964.

Ruschak, K.J. and Miller, C.A., Spontaneous emulsification in ternary systems with mass transfer, *Ind. Eng. Chem. Fundam.*, 11, 534, 1972.

Sailaja, D., Suhasini, K.L., Kumar, S., and Gandhi, K.S., Theory of rate of solubilization into surfactant solutions, *Langmuir*, 19, 4014, 2003.

Sanfeld, A., Steinchen, A. Hennenberg, M., Bisch, P.M., van Lansweerde Gallez, D., and Vedove, W.D., Mechanical, chemical, and electrical constraints and hydrodynamic interfacial stability, in *Dynamics and Instability of Fluid Interfaces*, Sorensen, T.S. (ed.), Springer-Verlag, Berlin, 1979, p. 168.

Sawistowski, H., Interfacial phenomena, in *Recent Advances in Liquid-Liquid Extraction*, Hanson, C. (ed.), Pergamon Press, Oxford, 1971, p.293.

Scriven, L.E. and Sternling, C.V., Sternling, "On cellular convection driven by surface-tension gradients: effects of mean surface tension and surface viscosity, *J. Fluid Mech.*, 19, 321, 1964.

Seshan, P.K., Electrodeposition cells: a theoretical investigation into their performance and deposit growth stability, Ph.D. dissertation, Carnegie-Mellon University, Pittsburgh, Pa., 1975.

Shaeiwitz, J.A., Chan, A.-C., Cussler, E.L., and Evans, D.F., The mechanism of solubilization in detergent solutions, *J. Colloid Interface Sci.*, 84, 47, 1981.

Shah, R. and Neogi, P., Interfacial resistance in solubilization kinetics, *J. Colloid Interface Sci.*, 253, 443, 2002.

Solans, C., Izquierdo, P., Nolla, J., Azemar, N., Garcia-Celma, M.J., Nano-emulsions, *Current Opinion Colloid & Interface Sci.*, 10, 102, 2005.

Sorensen, T.S., Instabilities induced by mass transfer, low surface tension in and gravity at isothermal and deformable fluid interfaces, in *Dynamics and Instability of Fluid Interfaces*, Sorensen, T.S. (ed.), Springer-Verlag, Berlin, 1979, p. 1.

Sorensen, T.S. and Hennenberg, M., Instability of a spherical drop with surface chemical reactions and transfer of surfactants, in *Dynamics and Instability of Fluid Interfaces*, Sorensen, T.S. (ed.), Springer-Verlag, Berlin, 1979, p. 276.

Smith, K.A., On convective instability induced by surface-tension gradients, *J. Fluid Mech.*, 24, 401, 1966.

Sternling, C.V. and Scriven, L.E., Interfacial turbulence: hydrodynamic instability and the Marangoni effect, *AIChE J.*, 5, 514, 1959.

Subramanian, R.S. and Balasubramaniam, R., *The Motion of Bubbles and Drops in Reduced Gravity*, Cambridge University Press, 2001.

Sutherland, K.L., The kinetics of adsorption at liquid surfaces, *Aust. J. Sci. Res.*, A5, 683, 1952.

Todorov, P.D., Kralchevsky, P.A., Denkov, N.D., Broze, G., and Mehreteab, A., Kinetics of solubilization of n-decane and benzene by micellar solutions of sodium dodecyl sulfate, *J. Colloid Interface Sci.*, 245, 371, 2002.

Toor. H.L., Fog formation in boundary value problems, *AIChE J.*, 17, 5, 1971.

van Hunsel, J. and Joos, P., Steady-state dynamic interfacial tensions of 1-alkanols during mass transfer across the hexane/water interface, *Langmuir*, 3, 1069, 1987.

van Voorst Vader, F., Erkens, T.F., and van den Tempel, M., Measurement of dilatational surface properties, *Trans. Faraday Soc.*, 60, 1170, 1964.

Vedove, W.D., and Sanfeld, A., Hydrodynamic and chemical stability of fluid-fluid reacting interfaces: I. General theory for aperiodic regimes, *J. Colloid Interface Sci.*, 84, 318, 328, 1981.

Vidal, A. and Acrivos, A., Effect of nonlinear temperature profiles on the onset of convection driven by surface-tension gradients, *Ind. Eng. Chem. Fundam.*, 7, 53, 1968.

Vlahovska, P.M., Horozov, T., Dushkin, C.D., Kralchevsky, P.A., Mehreteab, A., and Broze, G., Adsorption from micellar surfactant solutions: nonlinear theory and experiment, *J. Colloid Interface Sci.*, 183, 223, 1997.

Vollhardt, D., Morphology and phase behavior of monolayers. *Adv. Colloid Interface Sci.*, 64, 143, 1996.

von Gottberg, F.K., Hatton, T.A., and Smith, K.A., Surface instabilities due to interfacial chemical reaction, *Ind. Eng. Chem. Res.*, 34, 3368, 1975.

Wagner, C., Mathematical analysis of the formation of periodic precipitations, *J. Colloid Sci.*, 5, 85, 1950.

Wang, K.H.T., Ludviksson, V., and Lightfoot, E.N., Hydrodynamic stability of Marangoni films. II. Preliminary analysis of the effect of interphase mass transfer, *AIChE J.*, 17, 1402, 1971.

Ward, A.F. and Tordai, L., Time-dependence of boundary tensions of solutions I. The role of diffusion in time-effects, *J. Chem. Phys.*, 14, 453, 1946.

Ward, A.J., Kinetics of solubilization in surfactant-based systems, in *Solubilization in Surfactant Aggregates*, 1995, p. 237.

Weinheimer, R.M., Evans, D.F., and Cussler, E.L., Diffusion in surfactant solutions, *J. Colloid Interface Sci.*, 80, 357, 1981.

Wollkind, D.J. and Segel, L.A., Nonlinear stability analysis of the freezing of a dilute binary alloy, *Philos. Trans. R. Soc. Lond.* A268, 351, 1970.

Yortsos, Y.C., Effect of lateral heat losses on the stability of thermal displacement fronts in porous media, *AIChE J.*, 28, 480, 1982.

Zuiderweg, F.J. and Harmens, A., The influence of surface phenomena on the performance of distillation columns, *Chem. Eng. Sci.*, 9, 89, 1958.

PROBLEMS

6.1   Adapt the derivation of Section 3 to apply to the case of a solute with some surface activity which desorbs from a thin layer of liquid into a gas phase. Derive the condition for marginal stability for the case of a flat interface ($N_{cr} = 0$). Neglect surface diffusion effects.

6.2   Consider a thin layer of liquid where the temperature is uniform at each lateral position but where a uniform lateral temperature gradient (dT/dx) is imposed. Assuming that film thickness is everywhere equal to h and that, at steady state, a pressure gradient develops so that there is no net lateral flow of liquid, calculate the pressure gradient in the film and the velocity distribution. Comment on whether the uniform film thickness and the existence of a pressure gradient in the film are consistent features of this problem.

6.3   Starting with Equation 6.1, derive the differential interfacial mass balance of Equation 6.2.

6.4   In the falling meniscus method for measuring dynamic interfacial tensions, a vertical tube of nonuniform diameter is fabricated, having at its top a small circular opening of radius r with a sharp edge as shown here. The tube is placed in a large pool of liquid with the small hole a height h above the surface of the pool. A fresh surface is created at the hole at time $t = 0$. Assuming that surface tension decreases with time as a result of diffusion of a surface active material to the interface, find the value of the surface tension for which the liquid level in the tube will fall precipitously. Ignore deviation of the meniscus shape from a spherical conformation, although this factor should be considered in accurate work (see Defay and Petrie, 1971).

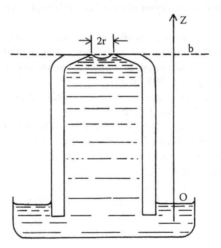

PROBLEM 6.4 Principle of the falling meniscus method. From Dhar and Chaterjee (1925) with permission.

6.5   Suppose that in the arrangement of Figure 6.13 the barriers are moved apart in such a way that tangential velocity along the surface between them is given by

$$v_x = ax,$$

where a is a constant. If a steady state is reached where interfacial tension is time independent, derive an expression for surfactant concentration c_o in the solution just beneath the surface. Assume that c_o and the surface concentration are related by a linear isotherm:

$$\Gamma = Ac_o.$$

Also assume that concentration gradients are negligible except in the immediate vicinity of the interface. This technique for studying dynamic interfacial tension was proposed by van Voorst Vader et al. (1964).

Note that since the adsorption isotherm and the corresponding surface equation of state (see Section 5 of Chapter 2) can be used to calculate c_o for an experimental value of interfacial tension γ, the expression you have derived for c_o could be used to obtain the diffusivity D.

6.6   Prove the statement made in the text that for a soluble surfactant the quantity $\Gamma_{os}(-\,d\gamma/d\Gamma_s)$ in Equation 5.71 must be replaced by the first expression for E in Equation 6.51. Then show that in the low viscosity limit E is given by the second expression (i.e., that containing the parameter B).

6.7   Show that the equilibrium melting temperature T_M of a pure material varies with interfacial curvature in accordance with the following equation (cf. the last term of Equation 6.66 and the accompanying discussion):

$$T_M = T_o + \frac{2H\gamma T_o}{\lambda\rho},$$

where T_o is the equilibrium melting temperature of a flat interface and the sign convention is such that curvature is negative when the center of curvature is in the solid phase.

6.8   Analyze the stability of a reacting solid surface when the reaction scheme is that described by Equations 6.73 and 6.74. Assume that the liquid is stirred so that the concentrations of A and B have their bulk values c_{Ao} and c_{Bo} outside a stagnant film of thickness d and that the concentration profiles in the film are linear before perturbation. Neglect interfacial tension effects.

(a)   Show that the concentration perturbation c_{Ap} for A in the stagnant film has the form $[g_A(z)e^{i\alpha x + \beta t}]$ with

$$g_A = A_1(\cosh aq_A z + \coth aq_A d \sinh \alpha q_A z),$$

with $q_A^2 = 1 + (\beta/D_A\alpha^2)$. Write the corresponding expression for c_{Bp}.

(b) Apply suitable boundary conditions and derive the following expression for the time factor β in the limit $q_A, q_B \to 1$:

$$\beta = \frac{\dfrac{k}{c_s}c_{As}^a c_{Bs}^{b_l}\left[\dfrac{b_1}{c_{Bs}}\dfrac{(c_{Bs}-c_{Bo})}{d}-\dfrac{a}{c_{As}}\dfrac{(c_{Ao}-c_{As})}{d}\right]}{1+kc_{As}^a c_{Bs}^{b_l}\left[\dfrac{a^2\tanh\alpha d}{c_{As}D_A\alpha}-\dfrac{b_1(b_2-b_1)\tanh\alpha d}{c_{Bs}D_B\alpha}\right]},$$

where c_{As} and c_{Bs} are the concentrations of A and B at the initial plane interface.

6.9 Analyze the stability of an initially plane interface where a metallic phase is growing as a result of electrodeposition and where the electrodeposition rate is given by Equation 6.75. Assume that the effect of the applied electrical potential is included in the constant k and that the organic additive is slightly soluble in the metal, its distribution coefficient between metal and the electrodeposition bath being b. For simplicity, neglect the stabilizing effect of interfacial tension. Show that the time factor β is given by

$$\beta = \frac{\dfrac{G_M}{c_{MI}}\left(1-\dfrac{V}{\omega_M D_M}\right)-\dfrac{mG_A}{c_{AI}}\left(1-\dfrac{V(1+b)}{\omega_A D_A + bV}\right)}{\dfrac{c_{Ms}c_{AI}^m}{kc_{MI}}+\dfrac{c_{Ms}}{c_{MI}\omega_M D_M}-\dfrac{mb}{\omega_A D_A + bV}}.$$

6.10 Section 9 dealt with prediction of diffusion paths and spontaneous emulsification for a particular unsteady state situation. We consider here determination of the diffusion path at steady state (Jackson, 1977).

Consider one-dimensional, steady-state, ternary diffusion along a capillary tube as shown in part (a) of the accompanying figure. Two immiscible liquids occupy the two halves of the tube with the position of the interface between them taken as the origin of the coordinate system ($z = 0$). Compositions A and D at the ends of the tube are known. Compositions B and C at the interface are not known initially. But if local equilibrium is assumed at the interface, B and C must be at the ends of a tie-line on the (known) ternary phase diagram as shown in part (b) of the figure. The question is which tie-line? Once the tie-line is determined, the concentration profiles are known and the question of whether spontaneous emulsification occurs can be settled.

Let $\omega_i(z)$ be the mass fraction of component i at position z for $i = 1, 2, 3$. If the density within each liquid phase is uniform, the mass flux j_i with respect to the mass average velocity has the form

$$j_i = -\rho D_i \frac{d\omega_i}{dz},$$

where D_i is the diffusivity of component i. Note that we have neglected the dependence of j_i on the concentration gradients $(d\omega_i/dz)$ of the other two components.

(a) Show that $j_i(z)$ in the left half of the capillary (between A and B) is given by

$$\omega_i(z) = \frac{n_i}{n} + \left(\omega_{iA} - \frac{n_i}{n}\right) \exp\left[\frac{n(z+L)}{D_i}\right], \quad i = 1, 2, 3,$$

where n_i is the mass flux of i with respect to fixed coordinates,

$$n = n_1 + n_2 + n_3$$

and D_i is independent of composition.

(b) If $D_1 = D_2 = D_3 = D$, show that $\omega_1(z)$ is a linear function of $\omega_2(z)$ in the left half of the capillary tube. Hence all compositions in this region must be on a straight line joining A and B.

(c) Derive an equation similar to that of (a) for the right half of the capillary tube, assuming that the density is ρ' and that the diffusivities are D_i'. As in (b), we shall assume that $D'_1 = D'_2 = D'_3 = D'$.

(d) What boundary conditions apply at the interface? Show that if L, ρ, ρ', D, D', ω_{iA}, ω_{iD}, and the complete phase diagram are known, these conditions are sufficient to solve for all unknown constants in your equations and hence to determine the concentration profiles and fluxes.

(e) By manipulating the equations of (a) and (e), one can show that if $\rho = \rho'$ and $D = D'$:

$$\frac{\omega_{1B} - \omega_{1D}}{\omega_{2B} - \omega_{2D}} = \frac{\omega_{1C} - \omega_{1A}}{\omega_{2C} - \omega_{2A}}.$$

In view of this result, can you suggest a simple way to locate points B and C on the phase diagram given the location of points A and D?

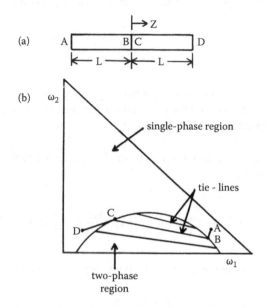

PROBLEM 6.10 Steady-state diffusion path in ternary system.

6.11   Consider the problem discussed in Equation 6.50. By itself it cannot be solved because of the nonlinearity, but if the approximation that $\Gamma_{A\infty} - \Gamma_A \cong \Gamma_{A\infty}$, i.e., the system is dilute, is made, then together with the conservation equation in the bulk:

$$\frac{\partial c_A}{\partial t} = D_A \frac{\partial^2 c_A}{\partial x^2},$$ (6.11.i)

where bulk concentration c_A is zero at $x = \infty$, an approximate solution can be obtained. However, it is more useful to look at two special cases. (a) Desorption controlling: Set adsorption terms to zero and ignore the contribution from diffusion. Integrate to show that Γ_A falls exponentially. (b) Diffusion controlled (local equilibrium): Set the adsorption-desorption terms equal to zero in Equation 6.50 to get Henry's law

$$\Gamma_A = \left(\frac{k_a \Gamma_{A\infty}}{k_d} \right) c_A \big|_{x=0} = K c_A \big|_{x=0}.$$

Take the Laplace transform of Equation 6.11.i, subject to the initial condition that c_A is zero, and solve to get $\bar{c}_A = A.e^{-x\sqrt{s/D}}$, where the above boundary condition has been used, the overbar indicates Laplace transformed quantity, and s is the variable of transformation. Return to Equation 6.50: ignoring the adsorption-desorption part, use Henry's law and take the Laplace transform to get A. Use an initial condition that the initial surface concentration is Γ_{Ao}. Use Henry's law again to calculate Γ_A in the Laplace domain (i.e., $\bar{\Gamma}_A$). This expression can be inverted using tables or an expansion can be made for large values of s (i.e., small values of time t) and then inverted. The result shows that Γ_A falls linearly with $t^{1/2}$.

6.12 Consider a drop of oil of initial radius R_o surrounded by a micellar solution into which the drop dissolves with time. Make a material balance using Equations 6.98 and 6.99. Use the fact that Sherwood number $k_L 2R/D = 2$ in a quiescent fluid to integrate to get the time for disappearance of the drop to be

$$\frac{1}{K_o}\left[\frac{R_o}{k_i} + \frac{R_o^2}{2D}\right] = c_m t_F \ .$$

Obtain two approximations when the interfacial resistance is and is not controlling.

6.13 (a) Consider a flat oil ($x < 0$)-water interface. Oil dissolves molecularly and reacts with a micelle. The conservation equation for the oil in water was given by Kabalnov and Weers (1996) as

$$0 = D_o\frac{d^2 c_o}{dx^2} - k_+ c_o c_m \ , \tag{6.13.i}$$

where D_o is the diffusivity of oil in water and c_o its concentration. As before, c_m is the concentration of micelles. Show that the solution to Equation 6.13.i, subject to the boundary conditions that $c_o = c_{os}$ (saturation) at $x = 0$, and $c_o \to 0$ as x goes to infinity, leads to a flux

$$N_i = D_o\left.\frac{dc_o}{dx}\right|_{x=0} = c_{os}\sqrt{D_o k_+ c_m} \ . \tag{6.13.ii}$$

Combining with Equation 6.100:

$$N_i = c_{os}\sqrt{D_o k_+ c_m} \ . \tag{6.13.iii}$$

That is, the fluxes vary with the square root of c_m, a result they verified from experiments. Here, k_+ is the reaction rate constant where $k_+ c_m = \tau_1^{-1}$ and experiments show that $\tau_1^{-1} \propto c_m$. Here, τ_1 is the first time constant in demicellization, which is smaller and represents the rate at which a single amphiphile is ejected from a micelle. In contrast, τ_2, the second time constant, is used in Figure 6.26. It is also clear from Equation 6.13.iii that in the limit of large c_m the reaction rate is infinite compared to the diffusion, and no reasonable expression results. Such a limiting case is quite relevant, and in such a situation a model such as Equation 6.101 may be the only recourse.

(b) Substitute Equations 6.99 and 6.102 into Equation 6.98. Set $k_c = 0$ (ionic surfactant). Nondimensionalize N_i with $k_{Lo}K_o c_{os}$ where $c_m{}^* = k_{Lo}c_{os}/k_L$ is a reference concentration. Make a log-log plot of dimensionless N_i against $c_m/c_m{}^*$. What is the range of slopes, and what would be a representative value?

(c) An ionic micelle cannot approach an oil-water interface which is similarly charged. If the closest distance of approach is r_m, the radius of the micelle, show in a steady-state model that

$$N_i = D_o \frac{c_{os}}{r_m} \ , \tag{6.13.iv}$$

where D_o is the diffusivity of molecularly dissolved oil.

7 Dynamic Interfaces

1. INTRODUCTION

An important feature of all mass transfer operations and of a significant number of reaction systems in chemical engineering is the critical role played by interfacial phenomena. Liquid-liquid and gas-liquid systems are characterized by convective-diffusive transfer at interfaces that keep distorting (e.g., in distillation, gas absorption, and liquid-liquid extraction). One fluid phase in such systems is often dispersed in another. The dynamic behavior of drops and bubbles, for example their shapes under various flow conditions and their breakage and coalescence, has been studied for many years, one goal being to predict mass transfer rates in dispersed systems (Azbel, 1981; Clift et al., 1978; Möbius anad Miller, 1998).

Interfacial phenomena are also important in other applications involving fluid flow, e.g., microfluidic devices (Stone and Kim, 2001; Tabeling, 2005). In some cases only the shapes of fluid-fluid interfaces need be studied, for instance, in the coating problems. Here, the material to be coated is in the form of a flat plate or wire that is withdrawn continuously from a pool of liquid. The liquid adheres to the surface as a thin film of constant thickness and is dried to form the coating. Considering that the coat thicknesses are very small and that the uniformity of the coat is often essential to the product, there exists a great need for precision (Derjaguin and Levi, 1964; Kistler and Schweizer, 1997). Knowledge of the thickness of the liquid layer (i.e., the shape of the liquid interface) (see Figure 7.3) as it deposits on the solid surface, is needed. This is well known as the dip coating problem. Similar problems where the final products or processes depend on the shapes of moving fluid-fluid interfaces are widely known: in extrusion of polymer melts, in fiber spinning, in formation of droplets from jets, in spreading of ink drops, oil slicks, etc.

In spite of many attempts, no general strategy has emerged for solving the equations of fluid mechanics to obtain the shapes of fluid interfaces (Higgins et al., 1977). A solid-fluid interface is rigid and can be made to have regular shapes (e.g., planar, spherical, cylindrical). However, liquid-fluid interfaces are rarely simple in shape. Moreover, their locations and shapes are determined by force balances (i.e., the solutions to the momentum equations) and are not known beforehand. This last difficulty has proven to be so enormous that even numerical solutions to a complete problem can be obtained only with difficulty.

Indeed, the difficulty is even greater. Not only is the interfacial position influenced by the flow, but the presence of the interface alters the flow. In the hydrostatics discussed in Chapter 1, it was seen that the pressure is discontinuous across the interface separating phases A and B by

$$p_A - p_B = -2H\gamma, \tag{7.1}$$

385

where H is the mean curvature of the interface and p is the pressure evaluated at the interface. It may be anticipated that since pressure differences give rise to a flow, the shape of an interface will affect fluid flow.

Interfacial phenomena are also important in other applications involving fluid flow, e.g., microfluidic devices (Stone and Kim, 2001; Tabeling, 2005).

2. SURFACES

As indicated above, dynamic interfaces frequently have complex shapes. In this section we provide additional information on the geometric properties of surfaces needed to apply some of the basic principles discussed previously to interfaces of arbitrary shape. For simplicity, only certain pertinent results are given; the reader may consult Slattery (1990) for a detailed account.

Knowledge of how to describe a surface is essential. Consider a surface given by

$$z = f(x, y) \tag{7.2}$$

in rectangular coordinates. Equation 7.2 can be rewritten as

$$g(x, y, z) = z - f. \tag{7.3}$$

On differentiating a curve in a plane we obtain the tangent line. The differentiation of a variable in three dimensions is done vectorially through the gradient operator

$$\nabla = e_x \frac{\partial}{\partial x} + e_y \frac{\partial}{\partial y} + e_z \frac{\partial}{\partial z} . \tag{7.4}$$

Here e_x, e_y, and e_z are the unit vectors in the x, y, and $z-$ directions, respectively. On a surface, a tangent plane (and not just a single tangent) can be drawn. The quantity ∇g is perpendicular to the plane at that point. The unit vector

$$n = \frac{\nabla g}{|\nabla g|} \tag{7.5}$$

is called the unit normal to the surface (and in the present case the outward unit normal). Note that $|\nabla g| = (\nabla g \cdot \nabla g)^{1/2}$ and $n \cdot n = 1$. If two mutually perpendicular unit vectors t_1 and t_2 are chosen on the tangent plane, then the unit vector set (n, t_1, t_2) describes the orientation of the surface and the vector or the tensor quantities describe the interfacial properties.

As a special case, the surface

$$z = h(x) \tag{7.6}$$

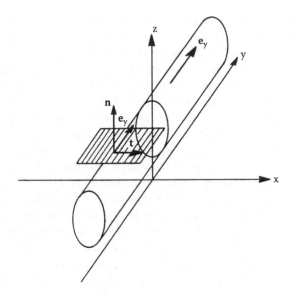

FIGURE 7.1 The equation of the surface is given by Equation 7.6. There are no variations in the y direction. The two tangents t and e_y are shown on the tangent plane (shaded). The unit normal is n.

can be considered. It is sketched in Figure 7.1. There is no variation in the y direction. The unit normal is found from Equation 7.5 to be

$$n = \left[-\frac{dh}{dx} \, e_x + e_z \right] \left[1 + \left(\frac{dh}{dx} \right)^2 \right]^{-1/2}. \tag{7.7}$$

One tangent vector, e_y, is evident. The other is obtained from the condition that it is perpendicular to n and also perpendicular to e_y, that is,

$$t = \left[e_x + \left(\frac{dh}{dx} \right) e_z \right] \left[1 + \left(\frac{dh}{dx} \right)^2 \right]^{-1/2}. \tag{7.8}$$

It should be noted that $n \cdot n = e_y \cdot e_y = t \cdot t = 1$, as well as $n \cdot t = t \cdot e_y = e_y \cdot n = 0$. Further, any linear combination of e_y and t is also a tangent, that is, perpendicular to n. However, the choice of e_y and t as tangents here is very convenient, since there is no variation in the y direction and e_y will not enter into the calculations.

On any surface it is possible to draw two sets of curves which are everywhere perpendicular to one another. It can be seen in Figure 7.1 that lines parallel to the y axis on the surface (direction e_y) and the curves formed by the intersection of the x-z planes with the surface (direction t) form one such set. These are the lines of curvature. The problem arises that neither these two tangents nor the associated lines of curvature can be fixed uniquely. If we take the curvatures c_1 and c_2 at a point along a pair of lines of curvature, then $c = 1/2(c_1 + c_2)$ is the same for all pairs of lines of curvature. However, for a particular pair we find c_1 to be the maximum and c_2 to be the minimum. These are the principal curvatures and also lead to the principal directions at that point (Weatherburn, 1961). In Figure 7.1 it can be seen that the curvature along t is a maximum and along e_y is zero, a minimum. This provides the uniqueness we have been looking for and the pair is called the principal pair. The inverses, that is, $r_1 = c_1^{-1}$ and $r_2 = c_2^{-1}$, are the two principal radii of curvature, and $H = 1/2(c_1 + c_2)$ is the mean curvature. The quantities r_1, r_2, and H were introduced in Chapter 1, and the Gaussian curvature $K = c_1 c_2$ in Chapter 5. The importance of H and K lies in the fact that they are invariants, that is, they do not depend on the coordinate system and the invariance holds even when the two curvatures are not the principal pair.

It is not only necessary to know about the variation in the shape of a surface, but also to know the variations in the properties on the surface. If the scalar $\phi(x, y, z)$ is a property of a system in the bulk (say, the temperature), then its gradient is $\nabla\phi$. If ϕ is evaluated on the surface $\phi_s = \phi[x, y, f(x, y)]$, then its variation is given through the surface gradient operator as $\nabla_s \phi_s$.

The operator ∇_s is the component of the operator ∇ on the tangent plane. Like the unit tensor $I = e_x e_x + e_y e_y + e_z e_z$ in three dimensions, one may write n unit tensor I_s for a surface, where $I_s + t_1 t_1 + t_2 t_2 = I - nn$. The component ∇_s is obtained from ∇ with the projection $\nabla_s = [I_s \cdot \nabla]_s$. Similarly one obtains the component of the three-dimensional velocity v on the surface by projection. The result is $v_t = [I_s \cdot v]_s$. The brackets signify that it needs to be evaluated at the surface. Now, v_t is two dimensional as it lies on the tangent plane, and it has only two components. One may also look for a surface component of a second-order tensor such as the stress tensor τ in the form of $\tau_s = [I_s \cdot \tau]_s$. Unlike τ, which has nine components, τ_s has only four and all forces there lie on the tangent plane.

The second-order tensors are characterized by three invariants, that is, it is possible to combine the nine components in three ways to get quantities that are independent of the coordinate systems and express some fundamental properties. For the surface component there are only two such invariants. The first of these can be written as tr (τ_s), where tr is short for trace and the operation that sums the diagonal elements of the tensor. The second is $1/2\{[\text{tr }(\tau_s)]^2 - \tau_s : \tau_s\}$, where a double dot product has been introduced. Since the trace itself is an invariant, some authors drop this term from the second invariant. In addition, the second invariant of this symmetric surface tensor is the same as the third invariant in three dimensions, which is the determinant of τ_s (see the remark after Equation 7.E1.8). There is a very important second-order surface tensor in the form of

$$b = - \nabla_s n. \tag{7.9}$$

The first invariant leads to

$$2H = \text{tr }(b) = -\nabla_s \cdot n \tag{7.10}$$

and the second to

$$K = 1/2[\text{tr }(b)^2 - b{:}b].\tag{7.11}$$

It is also possible to relate b to the two principal curvatures (see Slattery, 1990).

EXAMPLE 7.1

Obtain the expressions for the two curvatures for a surface of revolution as given in Equations 1.52 and 1.53 in Chapter 1.

In cylindrical coordinates, the surface vector for a surface of revolution given by $z = h(r)$ is given by (Happel and Brenner, 1983)

$$R = re_r + he_z\tag{7.E1.1}$$

$$dR = dr\ e_r + rd\theta\ e_\theta + dh\ e_z,$$

where variations in the z direction have been constrained. The two tangent vectors are (Aris, 1962)

$$a_r = \frac{\partial R}{\partial r} = e_r + \frac{dh}{dr}\ e_z$$
$$a_\theta = \frac{\partial R}{\partial \theta} = r\ e_\theta\tag{7.E1.2}$$

The two vectors are not normalized. The related quantities are $a_r \cdot a_r = a_{rr} = (a^{rr})^{-1}$ $= 1 + (h')^2$ and $a^r = a^{rr} a_r$ and $a_\theta \cdot a_\theta = a_{\theta\theta} = (a^{\theta\theta})^{-1} = r^2$ and $a^\theta = a^{\theta\theta}a_\theta$, which define the dual set, where h' is dh/dr. The unit normal (which is perpendicular to both a_θ and a_r) is obtained from Equation 7.5 as

$$n = -\frac{h'}{(1+h'^2)^{1/2}}e_r + \frac{1}{(1+h'^2)^{1/2}}e_z.\tag{7.E1.3}$$

The surface gradient is best written in terms of r and θ, and then the differentiation $\nabla_s n$ is carried out where the relations

$$\frac{\partial e_r}{\partial \theta} = e_\theta\ and\ \frac{\partial e_\theta}{\partial \theta} = -e_r\tag{7.E1.4}$$

from Happel and Brenner (1983) are used. The results are recast in terms of surface vectors

$$b = -\frac{\partial n}{\partial r} a^r - \frac{\partial n}{\partial \theta} a^\theta .$$ (7.E1.5)

Substituting Equation 7.E1.3 into Equation 7.E1.5 and following the same steps as before, one has

$$b = \frac{h''}{(1+h'^2)^{3/2}} a^r a_r + \frac{h'}{r(1+h'^2)^{1/2}} a^\theta a_\theta .$$ (7.E1.6)

Note that $a^r \cdot a_r = 1$, etc. It leads to

$$2H = tr(b) = \frac{h''}{(1+h'^2)^{3/2}} + \frac{h'}{r(1+h'^2)^{1/2}}$$ (7.E1.7)

$$K = \tfrac{1}{2}[tr(b)^2 - b:b] = \det(b) = \frac{h'h''}{r(1+h'^2)^2} ,$$ (7.E1.8)

where det stands for determinant, which in three dimensions gives the third invariant, but is redundant in two.

An additional problem is given in Problem 7.1, and some relations for surface operations on simple surfaces are given in Table 7.1.

3.   BASIC EQUATIONS OF FLUID MECHANICS

In terms of the formalism presented in the previous section, the boundary conditions derived in Chapter 5 (Section 2) are rewritten for the particular situation of interest here, which is the case of two immiscible fluid media, labeled A and B, separated by an interface. The shape of the interface is given vectorially as $a(r, t)$ and the rate at which it moves is given by $\dot{a} = da/dt$. For simplicity, the fluids are chosen to be incompressible and Newtonian, thus the equations of motion and continuity become

$$\rho_i \left[\frac{\partial v_i}{\partial t} + v_i \cdot \nabla v_i \right] = \mu_i \nabla^2 v_i - \nabla p_i + \rho_i \hat{F}_i$$ (7.12)

and

$$\nabla \cdot v_i = 0,$$ (7.13)

where $i = A$ in phase A and B in phase B. $\hat{F}_i$ is the body force and ρ_i and μ_i are the densities and viscosities. v_i and p_i are the velocities and pressures, respectively.

TABLE 7.1
Expressions involving the surface gradient operator in simple coordinate systems

Planar surface: $z = A$

$$\nabla_s g = \frac{\partial g}{\partial x} \boldsymbol{e}_x + \frac{\partial g}{\partial y} \boldsymbol{e}_y$$

$$\nabla_s \cdot \boldsymbol{p}_s = \frac{\partial p_{sx}}{\partial x} + \frac{\partial p_{sy}}{\partial y}$$

$$\nabla_s^2 g = \frac{\partial^2 g}{\partial x^2} + \frac{\partial^2 g}{\partial y^2}$$

Cylindrical surface: $r = A$

$$\nabla_s g = \frac{1}{A} \frac{\partial g}{\partial \theta} \boldsymbol{e}_\theta + \frac{\partial g}{\partial z} \boldsymbol{e}_z$$

$$\nabla_s \cdot \boldsymbol{p}_s = \frac{1}{A} \frac{\partial p_{s\theta}}{\partial \theta} + \frac{\partial p_{sz}}{\partial z}$$

$$\nabla_s^2 g = \frac{1}{A^2} \frac{\partial^2 g}{\partial \theta^2} + \frac{\partial^2 g}{\partial z^2}$$

Spherical surface: $r = R$

$$\nabla_s g = \frac{1}{R} \frac{\partial g}{\partial \theta} \boldsymbol{e}_\theta + \frac{1}{R \sin \theta} \frac{\partial g}{\partial \varphi} \boldsymbol{e}_\varphi$$

$$\nabla_s \cdot \boldsymbol{p}_s = \frac{1}{R \sin \theta} \frac{\partial}{\partial \theta} (p_{s\theta} \sin \theta) + \frac{1}{R \sin \theta} \frac{\partial}{\partial \varphi} p_{s\varphi}$$

$$\nabla_s^2 g = \frac{1}{R^2 \sin \theta} \frac{\partial}{\partial \theta} \left(\sin \theta \frac{\partial g}{\partial \theta} \right) + \frac{1}{R^2 \sin^2 \theta} \frac{\partial^2 g}{\partial \varphi^2}$$

Equations 7.12 and 7.13 are subject to appropriate boundary conditions away from the interface, while at the interface we have, first of all,

$$n \cdot \rho_A (v_A - \dot{a}) = n \cdot \rho_B (v_B - \dot{a}) . \qquad (7.14)$$

This equation is obtained on neglecting terms in Γ in the interfacial mass balance equation (Equation 5.32). The two terms are equal to the flux (mass transfer) through the interface, in the absence of which Equation 7.14 reduces to

$$\begin{aligned} n \cdot v_A &= n \cdot \dot{a} = \dot{a}_n \\ n \cdot v_B &= n \cdot \dot{a} = \dot{a}_n \end{aligned} \qquad (7.15)$$

The tangential velocities are continuous at the interface, the generalization of Equation 5.40,

$$t \cdot v_A = t \cdot v_B = t \cdot \dot{a} = \dot{a}_t . \qquad (7.16)$$

The interfacial momentum balance equation (Equation 5.35) will now be written for the case of negligible surface excess mass and momentum and two Newtonian fluids. The stress tensor in a Newtonian fluid is written following Bird et al. (2002) as $pI - \mu \, [\nabla v + (\nabla v)^T]$, where I is the identity tensor and the superscript T represents the transpose of a tensor. When the inertial forces are considered as well, the total force on an interface exerted by the ith phase is $n \cdot (\rho_i(v_i - \dot{a})v_i + p_i I - \mu_i[\nabla v_i + (\nabla v_i)^T])$. The dot product with n denotes that the forces act on a surface characterized by n. It is in making a force balance on the interface that the effects of interfacial tension make themselves felt. The balance is known to be (see Chapter 5)

$$n \cdot (\rho_A(v_A - \dot{a})v_A + p_A I - \mu_A(\nabla v_A + (\nabla v_A)^T)$$

$$= n \cdot (\rho_B(v_B - \dot{a})v_B + p_B I - \mu_B(\nabla v_B + (\nabla v_B)^T.)$$

$$-\nabla_s \gamma - n 2 H \gamma . \qquad (7.17)$$

The last two terms on the right-hand side in Equation 7.17 show that the forces are discontinuous across the interface due to the interfacial tension, which resists deformation of the interface. One may resolve Equation 7.17 into a tangential component by taking the dot product with t. The result, on using Equations 7.14 and 7.16 and the relations $n \cdot I = n$, $n \cdot t = 0$, is

$$-\mu_A n \cdot [\nabla v_A + (\nabla v_A)^T] \cdot t = -\mu_B n \cdot [\nabla v_B + (\nabla v_B)^T] \cdot t$$

$$-\mathbf{\nabla}_s \gamma \cdot t, \text{ or}$$

$$\tau_A = \tau_B - \mathbf{\nabla}_s \gamma \cdot t, \tag{7.18}$$

where τ denotes the viscous shear part of the stress at the interface, and t is the unit tangent on the tangent plane. In the absence of mass transfer, the balance in the normal direction yields

$$p_A - \mu_A \mathbf{n} \cdot [\nabla v_A + (\nabla v_A)^T] \cdot \mathbf{n} = p_B - \mu_B \mathbf{n} \cdot [\nabla v_B + (\nabla v_B)^T] \cdot \mathbf{n}$$

$$-2H\gamma, \text{ or}$$

$$N_A = N_B - 2H\gamma, \tag{7.19}$$

where N denotes the normal stress at the interface, which includes the pressure.

As mentioned previously, the fluid mechanical problems must be solved in order to determine the shapes of the interfaces. However, solving a problem demands that the boundary conditions for the fluid flow problem be satisfied first, with the equations of the boundaries still unknown. Combined with this coupling of the interfacial shape and the fluid mechanical problem is yet another difficulty, which is that the shapes of the interfaces are not "regular" like those of spheres, cylinders, or planes. A cursory look at introductory or advanced texts in fluid mechanics (Bird et al., 2002; Happel and Brenner, 1983; Slattery, 1981) shows that symmetry in the geometry is critical to solving a problem in a straightforward (although at times lengthy) manner. For example, flow in the annulus between infinitely long coaxial cylinders may be considered simple because of the symmetry in the problem. However, flow around a sphere settling inside a cylinder, even when it is falling along the axis of the cylinder, is a difficult problem to solve.

Consequently analytical methods are mostly confined to creeping flows. Roughly, there are two types of problems that can be solved. The first of these deals with interfaces that show small deviations from simple geometric forms, as for instance the case of a "slightly deformed sphere" settling in an infinite fluid. The second type constitutes cases where interfacial position changes, but only very slowly. Then its variation can be neglected to the first approximation and the lubrication theory approximation or the slender body approximation applied. It should be noted that both the above methods yield approximate solutions.

The cases of a sphere and slightly deformed sphere in a uniform flow field are considered first in Sections 4 and 5. The mathematical method used conventionally in these problems is the regular asymptotic expansion. The reader is introduced to this method. In Section 6, the dip coating problem under the lubrication theory approximation is examined. (The closely related slender body approximation is outlined in Problem 7.5.) A more sophisticated method of matched asymptotic expansions is used to solve this problem and its main features are explained there. More realism is brought to the study of interfacial dynamics by incorporating the nonideal nature of surfaces. This is done through a physic-ochemical approach in Section 7 and through a surface rheological approach in Section 8. Application of these approaches to drainage of thin films is outlined in Section 9. The interconnected topics of dynamic contact angles are analyzed in Section 10, the slip condition in Section 11 through a problem that needs a complete analysis in matched asymptotic expansions, and ultrathin films are examined in Section 12, where a brief discussion on molecular phenomena in spreading is included. Material on coating flows and waves on surfaces beyond that considered in Chapter 5, including the wave motion of fluids in turbulent flow, has been omitted.

4.   FLOW PAST A DROPLET

The fluid mechanical problem of solving for the flow around a droplet is a standard one and has formed the starting point of many of the methods and theories on how interfacial effects influence fluid flow. A droplet of material A is immersed in an infinite medium B. The latter flows upward past the droplet and exerts a drag force on the droplet. This drag force is balanced by gravity, so that the droplet stays in place and steady state is reached. Far from the droplet, the velocity in phase B assumes a constant (upward) value U.

The problem is solved at very low Reynolds numbers, when the left-hand side in the momentum equation (Equation 7.12) may be neglected. As shown in Figure 7.2, a z axis passing through the center of the drop is chosen in the same direction as the velocity U. Consequently, far away from the drop $v_B \rightarrow Ue_z$. Although this condition suggests the use of a cylindrical coordinate system (z, r, θ), a spherical coordinate system (r, θ, ϕ) with the origin at the center of the drop is used instead in order to make it easier to satisfy the boundary conditions on the drop surface. The drop is assumed to have a spherical shape of radius a. From the symmetry of the figure, $v_\phi = 0$ (no swirl) and v_θ, v_r are functions of r and θ only. Together with the previous conditions, these simplifications allow one to write the r and θ components of the momentum equation (Equation 7.12) and the continuity equation (Equation 7.13). On eliminating the pressure by cross differentiation of the components of the momentum equation and subtracting, much as in Chapter 5 (Section 2.1), and on defining a stream function ψ to assure that continuity is satisfied, one obtains the biharmonic stream function equation (Equation 7.20). The velocity components are given in terms of the stream function in Equations 7.21 and 7.22:

$$E^4\psi_i = 0 \qquad (7.20)$$

$$v_{ri} = -\frac{1}{r^2 \sin\theta}\frac{\partial\psi_i}{\partial\theta} \qquad (7.21)$$

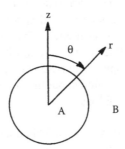

FIGURE 7.2 The spherical coordinate system is employed with the origin at the center of the drop of phase A. Far away from the drop the velocity field is $U e_z$ in phase B.

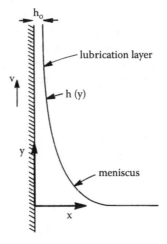

FIGURE 7.3 The arrangement of the dip coating experiment is shown. The plate is being withdrawn at a constant velocity V. The film profile h reaches a constant value h_o in the region far from the meniscus.

$$v_{\theta i} = \frac{1}{r \sin\theta} \frac{\partial \psi_i}{\partial r} , \qquad (7.22)$$

where

$$E^2 = \frac{\partial^2}{\partial r^2} + \frac{\sin\theta}{r^2} \frac{\partial}{\partial \theta} \left(\frac{1}{\sin\theta} \frac{\partial}{\partial \theta} \right)$$

and $E^4 = E^2(E^2)$. Stream functions fully specify the fluid flow at low Reynolds numbers when the velocity component in one direction is zero and all the fluid mechanical quantities are independent of the coordinate in that direction. Because of the low Reynolds numbers, this kind of flow is called creeping flow.

Bird et al. (2002) discuss the biharmonic operator E^4 and show how Equation 7.20 can be solved for flow around a solid sphere and how quantities like the drag force on the sphere can be evaluated. The solution for the case of a fluid sphere is given in Levich (1962) using a somewhat different method of attack. Following the lines of development pursued here, Equations 7.20 through 7.22 apply in both the inner and outer fluids, but the boundary conditions are different. We first list the boundary conditions

(i) $\psi_B \to -1/2\, Ur^2 \sin^2\theta$ as $r \to \infty$. This condition is that $v_B \to Ue_z$ as $|r| \to \infty$, mentioned earlier.

(ii) $v_{B\theta} = v_{A\theta}$ on $r = a$, from Equation 7.16.

(iii) $v_{rB} = v_{rA} = 0$ on $r = a$, from Equations 7.14 and 7.15.

(iv) $\tau_{r\theta A} = \tau_{r\theta B}$ on $r = a$, from Equation 7.18, where it has been assumed that γ is constant on the surface. An additional condition that all fluid mechanical quantities in A are finite at $r = 0$ is imposed as well. The boundary conditions are rewritten in terms of the stream functions using Equations 7.21 and 7.22. The expressions for τ in terms of the velocities are also invoked. When the results are substituted into the general solutions of Equation 7.20 for fluids A and B, one obtains

$$\psi_A = \frac{Ua^2 \sin^2\theta}{4}\frac{\mu_b}{\mu_A+\mu_B}\left[\left(\frac{r}{a}\right)^2 -\left(\frac{r}{a}\right)^4\right] \tag{7.23}$$

$$\psi_B = -\frac{Ua^2\sin^2\theta}{4}\left[2\left(\frac{r}{a}\right)^2 - \frac{3\mu_A+2\mu_B}{\mu_A+\mu_B}\left(\frac{r}{a}\right)+\frac{\mu_B}{\mu_A+\mu_B}\left(\frac{a}{r}\right)\right]. \tag{7.24}$$

The drag force F is the quantity of interest and is equal to the net gravitational force

$$F = 2\pi Ua\mu_B\left[\frac{3\mu_A+2\mu_B}{\mu_A+\mu_B}\right]e_z = \left(\rho_A-\rho_B\right)ge_z. \tag{7.25}$$

This equation can be solved for the velocity U.

The analysis usually stops here without the normal stress balance given by Equation 7.26. Here $N_i = p_i - 2\mu_i(\partial v_{ri}/\partial r)$ and the balance yields the relation

$$p_A^o - p_B^o = \frac{2\gamma}{a} , \tag{7.26}$$

where p_A^o and p_B^o are the two constants of integration that appear in the pressure terms. The other terms in the normal stress balance cancel out on using Equation 7.25. The value of $H = -a^{-1}$ is used in Equation 7.26. Equation 7.26 can be used to evaluate one of the pressures if the other is known or held to be the datum pressure. It is to be noted that if $U \to 0$, v_A and $v_B \to 0$, and p_A and p_B reduce to the constants that appear in Equation 7.26. One may say that p_A^o and p_B^o are the static pressures.

5.   ASYMPTOTIC ANALYSIS

It is fortunate that in the previous problem the normal stress balance is satisfied by the spherical drop profile $r = a$. However, no satisfactory proof that this is the only solution that has been given. Indeed, one may anticipate some deviation in shape when inertial effects are considered. If one assumes that this deviation is small so that the profile is given by $r = a(1 + \xi)$ with $|\xi| \ll 1$, then from the symmetry, ξ is a function of 0 alone. If the drop is to have the same volume, then

$$\frac{4}{3}\pi a^3 - \int_0^\pi \{\int_0^{2\pi} [\int_0^{a(1+\xi)} r^2 dr] d\varphi \} \sin\theta \, d\theta = 0 .$$

This expression reduces to

$$1 - \frac{1}{2}\int_{-1}^{1}(1+\xi)^3 dq = 0 ,$$

where $q = \cos\theta$, and further to

$$\int_{-1}^{1} \xi \, dq = 0 . \tag{7.27}$$

Similarly, if the drop is to have its center of mass at the origin, then

$$\int_0^\pi \{\int_0^{2\pi} [\int_0^{a(1+\xi)} r\cos\theta r^2 \, dr] d\varphi\} \sin\theta \, d\theta = 0 ,$$

which reduces to

$$\int_{-1}^{1} \xi \, q dq = 0 \, . \tag{7.28}$$

Finally, for $|\xi| \ll 1$, one has (Levich, 1962; Taylor and Acrivos, 1964)

$$2H = -\frac{1}{a}\left[2 - 2\xi - \frac{d}{d\xi}\left\{\left(1 - q^2\right)\frac{d\xi}{dq}\right\}\right]. \tag{7.29}$$

Taylor and Acrivos (1964) found an approximate expression for ξ applicable for small values of the Reynolds number N_{Re} ($= Ua\rho_B/\mu_B$) and capillary number N_{Ca} ($= U\mu_B/\gamma$). They first obtained the creeping flow solution $\psi^{(0)}$ that satisfied all boundary conditions, those at the drop-fluid interface being satisfied at $r = a$. The normal stress balance (Equation 7.19), which was not used in this initial procedure, was then applied, with N_i evaluated at $r = a$ to obtain a first approximation $\xi^{(0)}$ satisfying the conditions given by Equations 7.27 and 7.28. Equation 7.29 was used for the curvature. Since the normal stress condition is satisfied exactly by the creeping flow solution given by Equations 7.23 and 7.24 (see Equation 7.26), it was found that $\xi^{(0)} = 0$. If this term had been nonzero, then it would have been proportional to N_{Ca}.

Hence, in order to determine the deviation from spherical shape, it was necessary to include higher order terms proportional to U^2 in the stream function. However, such a term emphasizes inertial forces which are not accounted for in the biharmonic equation (Equation 7.21). Proudman and Pearson (1957) have shown how to correct the biharmonic equation for small effects of inertia and obtained the solution for a perfect sphere. The stream function takes the form $\psi_i \sim \psi_i^{(0)} + N_{Re} \, \psi_i^{(1)}$.

Taylor and Acrivos (1964) made the solution satisfy the boundary conditions on the surface $r = a$, but satisfied the normal stress boundary condition on the surface of a slightly deformed sphere $a(1 + \xi^{(1)})$ by using a Taylor series expansion to express any function $f(r)$ at the interface as $f(a) + (\partial f/\partial r)_a \xi^{(1)}$. Since the normal stress boundary condition contains the curvature and the curvature is given by Equation 7.29, we get a differential equation in $\xi^{(1)}$ as a function of q. The solution is

$$\xi^{(1)} = -N_{Re}N_{Ca}\lambda P_2(q) \tag{7.30}$$

$$\lambda = \frac{1}{4(\kappa+1)^3}\left\{\left(\frac{81}{80}\kappa^3 + \frac{57}{20}\kappa^2 + \frac{103}{40}\kappa + \frac{3}{4}\right) - \frac{\sigma-1}{12}(\kappa+1)\right\}, \tag{7.31}$$

where P_2 is the Legendre polynomial of order two, and in addition, $\kappa = (\mu_A/\mu_B)$ and $\sigma = (\rho_A/\rho_B)$.

Better approximation is sought in the form of stream function

$$\psi_i \sim \psi_i^{(0)} + N_{Re}\,\psi_i^{(1)} + N_{Re}N_{Ca}\psi_i^{(2)} . \tag{7.32}$$

This equation is now made to satisfy the Navier-Stokes equations, which the first two terms already satisfy, leading to the solution for $\psi^{(2)}$. It is subjected to the boundary conditions at $a(1 + \xi^{(1)})$ except for the normal stress balance, which is satisfied at $a(1 + \xi^{(2)})$. Taylor series expansions are used as before. It is quite fortuitous that these balances can be made. The normal stress balance leads to

$$\xi^{(2)} = -N_{Re}N_{Ca}\lambda P_2 + N_{Re}N_{Ca}^2 \frac{3\lambda(11\kappa+10)}{70(\kappa+1)}P_3 . \tag{7.33}$$

The complexity of the equations involved made them stop here.

From Equation 7.33 one finds that a drop is spherical for creeping flow but becomes spheroidal (pill shaped) at somewhat higher Reynolds numbers (P_2 term) and a spherical cap at still higher velocities (P_3 term). If only $\xi^{(1)}$ is considered, then as $\kappa \to \infty$, $\xi^{(1)} = -0.25\,N_{Re}N_{Ca}P_2$, while for $\kappa = 0$ and $\sigma = 0$ (gas bubble) $\xi^{(1)} = -0.21\,N_{Re}N_{Ca}P_2$. That is, the shape is relatively independent of the physical properties. It should also be noted that in related problems, the series solution obtained is rarely as complex as in this case. The term $\xi^{(0)}$, which is obtained from $\psi^{(0)}$, usually suffices. In the case of drops, $\xi^{(0)}$ turns out to be zero and higher order approximations are needed. See Problem 7.2 for further details.

In this low Reynolds number, or small U, problem the successive correction terms are proportional to higher powers of U. Hence one may be tempted to assume that as the last term goes to zero, the solution is a series solution. However, the construction of the solution is contingent on the small magnitude of U (or N_{Ca}, N_{Re}, and $N_{Re}N_{Ca}$). This gives rise to only an asymptotic expansion of the solution. The expansion has the following features:

(1) The solution is approximate and its construction is based on a small parameter that exists in the problem.
(2) The practical aspect of a series solution is its speed of convergence, because, after all, in order to get numbers from the solution, the series has to be truncated. It is only the asymptotic series that addresses itself to estimating the magnitude of the remainder directly. The magnitude of the remainder in an asymptotic expansion is the magnitude of the first term dropped. This is not true in an exact infinite series in general. From Equation 7.32, the remainder here is on the order of U^4 (or $N_{Re}^2N_{Ca}^2$ or $N_{Re}N_{Ca}^3$). However, there is little to suggest that

by calculating more terms and decreasing the remainder in this fashion
the series obtained will be convergent. In fact, divergent series are
sometimes obtained.

(3) In conclusion, progressively adding terms to an asymptotic series is
     of little value and even dangerous if the series is not convergent.
     Besides being prudent, it is at least very desirable from the view of
     mathematical simplicity to stop at one or two terms. Indeed, a very
     satisfactory approximation is obtained when the assumed smallness
     of U is justified.

Based on the method just described, one can suggest a general solution
scheme. When an interface deviates slightly from a regular shape where one of
the coordinates ζ_k has a constant value A, the solution proceeds as follows:

(a) Solve for $\psi = \psi^{(0)}$ as if the interface were indeed at $\zeta_k = A$, using all
     boundary conditions except the normal stress balance.
(b) Write the curvature as $2H \sim 2(H^{(0)} + H^{(1)})$, where $H^{(0)}$ corresponds to
     the shape given by $\zeta_k = A$ and $H^{(1)}$ is the correction when the interface
     deviates slightly from the regular shape. Equation 7.19 is now nondi-
     mensionalized by dividing by $\mu_B U/A$ to give

$$\overline{N}_A^{(0)} - \overline{N}_B^{(0)} = -\frac{2}{N_{Ca}}[\overline{H}^{(0)} + \overline{H}^{(1)}], \tag{7.34}$$

where the overbars denote dimensionless quantities. Rearranging, one has

$$-\overline{H}^{(1)} = \frac{N_{Ca}}{2}\left[\overline{N}_A^{(0)} - \overline{N}_B^{(0)} + \frac{2}{N_{Ca}}\overline{H}^{(0)}\right]. \tag{7.35}$$

The left-hand side is usually in a differential form. The right-hand side is N_{Ca}
times a function. The constant of integration in $[\overline{N}_A^{(0)} - \overline{N}_B^{(0)}]$ cancels
with $2\overline{H}^{(0)}/N_{Ca}$. If this cancellation does not take place, the asymptotic scheme
will not apply, since $\overline{H}^{(1)}$, the correction, becomes comparable to $\overline{H}^{(0)}$. Table 7.2
provides some expressions for $\overline{H}^{(0)}$ and $\overline{H}^{(1)}$.

This method of successive approximation is called Piccard's iteration, and
need not be confined to iterations using the normal stress balance. The kinematic
boundary condition, which is the one that deals with the velocities normal to the
interfaces (Equation 7.14), has been used as well (Bhattacharji and Savic, 1965).
In fact, in numerical solutions, such iterations have to be carried out until numerical convergence is reached. It has been observed (Silliman and Scriven, 1980)
that iterations using normal stress boundary conditions converge faster if the
capillary number is low and those using kinematic boundary conditions converge

TABLE 7.2
Expressions for curvature of surfaces slightly deformed from regular shapes

Surface	$\overline{H}^{(0)}$	$\overline{H}^{(1)}$
Sphere of radius $A[1 + \xi(q, \phi)]$	-1	$\left[\xi + \frac{1}{2}\frac{\partial}{\partial q}\left((1-q^2)\frac{\partial\xi}{\partial q}\right)\right.$ $\left. + \frac{1}{2(1-q^2)}\frac{\partial^2\xi}{\partial\varphi^2}\right]$
Cylinder of radius $A[1 + \xi(\theta, z)]$	$-1/2$	$\frac{1}{2}\left[\xi + \frac{\partial^2\xi}{\partial\theta^2} + A^2\frac{\partial^2\xi}{\partial z^2}\right]$
Almost planar $z = A[1 + \xi(x, y)]$	0	$A^2\left[\frac{\partial^2\xi}{\partial x^2} + \frac{\partial^2\xi}{\partial y^2}\right]$

faster when the capillary number is high. Note that in numerical solutions it is possible to start from the basic momentum balance equation (Equation 7.12), and not necessarily from the creeping flow equation (Equation 7.18). That is, the solution is not restricted to small capillary numbers.

In this section, minimal attention has been paid to the methods for solving the stream function equations since the basic purpose has been to illustrate the method for obtaining the correction ξ. It should be noted that the solutions in the form of stream functions for many regular geometric shapes are available and can be used to calculate the first correction $\xi^{(0)}$. In cylindrical and spherical coordinates, the general solutions have been provided by Haberman and Sayre (1958) and their method can be applied to many of the separable coordinate systems (Happel and Brenner, 1983).

6. DIP COATING

In this problem a flat plate is withdrawn with a steady velocity V from a pool of liquid. When V is sufficiently large, entrainment will occur, as shown in Figure

7.3. Sufficiently far away from the meniscus the film reaches a constant thickness. The problem differs from previous ones by the important fact that it is not easy to visualize the profile as a small deviation from a simple geometric shape.

The lubrication theory approximation is used to formulate this problem. In this scheme it is assumed that the film is almost flat. Consequently, if $x = h(y)$ is the profile shape, then $|dh/dy| \ll 1$. Thus the boundary conditions applicable on $h(y)$ are used by treating h to be almost a constant. This procedure also allows the use of the assumption that, to a good approximation, the flow is fully developed, with $v_x \sim 0$ and v_y a function of x alone. Then the equations of motion become

$$0 = -\frac{\partial p}{\partial y} + \mu \frac{\partial^2 v_y}{\partial x^2} - \rho g \qquad (7.36)$$

$$0 = -\frac{\partial p}{\partial x} . \qquad (7.37)$$

Equation 7.37 says that p is a function of y alone. Consequently Equation 7.36 can be rewritten as

$$\frac{\partial^2 v_y}{\partial x^2} = \frac{1}{\mu}\left(\frac{\partial p}{\partial y} + \rho g\right) , \qquad (7.38)$$

where the right-hand side is not a function of x.

Equation 7.38 is to be solved subject to the boundary conditions

$$v_y|_{x=0} = V \text{ (no slip)} \qquad (7.39)$$

$$\frac{\partial v_y}{\partial x}\bigg|_{x=h} = 0 \text{ (zero shear at the air-liquid interface).} \qquad (7.40)$$

The solution is

$$v_y = V + \frac{1}{2\mu}\left(\frac{\partial p}{\partial y} + \rho g\right)(x^2 - 2xh) . \qquad (7.41)$$

The pressure p is evaluated from the normal stress balance:

$$p|_{x=h} = -\gamma \frac{d^2 h}{dy^2} , \qquad (7.42)$$

which, in view of Equation 7.37, is the pressure for any x at a given y. Note that the curvature has been obtained from Equation 1.52 with $|dh/dy| \ll 1$ Further,

$$\frac{\partial p}{\partial y} = -\gamma \frac{d^3 h}{dy^3} . \qquad (7.43)$$

Substituting Equation 7.43 in Equation 7.41, one has

$$v_y = V + \frac{1}{2\mu} \left(\rho g - \gamma \frac{d^3 h}{dy^3} \right) (x^2 - 2xh) . \qquad (7.44)$$

One writes now an equivalent form of the continuity equation. If Q is the volumetric flow rate in the film, then under steady state it is a constant for all values of y. Hence

$$Q = \int_0^h v_y \, dx = Vh - \frac{h^3}{3\mu} \left(\rho g - \gamma \frac{d^3 h}{dy^3} \right) . \qquad (7.45)$$

This third-order ordinary differential equation needs three boundary conditions for solution and a fourth one to evaluate the unknown constant Q. Far away from the meniscus the thickness reaches a steady value (i.e., the film has a flat profile). There the curvature is zero and the second term within the parentheses in Equation 7.45 is hence zero. If the steady value h_o is very small, the term $Vh_o \gg (\rho g h_o^3/3\mu)$, and consequently

$$Q = Vh_o. \qquad (7.46)$$

As noted previously, the slope and the curvature go to zero at large y, the boundary conditions become

$$h \rightarrow h_o \text{ for } y \rightarrow \infty \qquad (7.47)$$

$$\frac{dh}{dy} \rightarrow 0 \text{ for } y \rightarrow \infty \qquad (7.48)$$

$$\frac{d^2 h}{dy^2} \rightarrow 0 \text{ for } y \rightarrow \infty. \qquad (7.49)$$

Equations 7.45 (with gravitational effects neglected) and 7.47 through 7.49 may be nondimensionalized to

$$\frac{d^3 L}{d\lambda^3} = \frac{1-L}{L^3}$$ (7.50)

$$L \to 1 \text{ as } \lambda \to \infty$$ (7.51)

$$\frac{dL}{d\lambda} \to 0 \text{ as } \lambda \to \infty$$ (7.52)

$$\frac{d^2 L}{d\lambda^2} \to 0 \text{ as } \lambda \to \infty,$$ (7.53)

where $L = (Vh/Q)$ and $\lambda = (3\mu V^4/\gamma Q^3)^{1/3}y$. Equation 7.50 can be solved numerically to yield L as a function of λ. More important is the quantity

$$\left(\frac{d^2 L}{d\lambda^2} \right)_{L \to \infty} = \alpha ,$$ (7.54)

where α is evaluated from this solution to be 0.63.

If the meniscus region is examined, it is seen that $h \to \infty$ as $y \to 0$. Consequently the average value of the velocity there is approximately Q/h, which is very small, as Q is a constant and $h \to \infty$. Thus dynamic effects are very small and the profile is very close to that of a static meniscus. The method of determining static meniscus profiles is given in Chapter 1. Starting from the solution of L^* as a function of λ (Landau and Lifshitz, 1959), one obtains

$$\left(\frac{d^2 L^*}{d\lambda^2} \right)_{L^* \to 0} = \left(\frac{2\rho g}{\gamma} \right)^{1/2} \frac{Q}{V^{5/3}} \left(\frac{\gamma}{3\mu} \right)^{2/3} .$$ (7.55)

The symbols have the same meaning as before except that the asterisk on L^* is used to denote the fact that L^* represents the static meniscus. It is assumed now that the right-hand side in Equation 7.55 is equal to given by Equation 7.54, leading to

$$N_B = 0.86 \, N_{Ca}^{4/3} ,$$ (7.56)

where the capillary number $N_{Ca} = \mu V/\gamma$ and the Bond number $N_B = \rho g h_0^2/\gamma$. Equation 7.56 is known to be valid for $N_{Ca} < 0.01$. A range of experimental results and theory have been reviewed by Quéré (1999).

The principle behind equating the right-hand side of Equation 7.54 to the right-hand side of Equation 7.55 can be explained as follows. The lubrication theory approximation is valid away from the meniscus and cannot satisfy the boundary conditions in the meniscus region. Consequently the solution contains a number of unknowns equal to the number of boundary conditions in the meniscus region. Similarly the solution in the meniscus region cannot satisfy the boundary conditions in the lubrication layer. If the solution in the lubrication layer is extrapolated into the meniscus region ($L \rightarrow \infty$), and if the solution in the meniscus region is extrapolated into the lubrication layer ($L^* \rightarrow 0$), they should agree with each other. The conditions under which they "agree" with each other leads to evaluation of the unknowns. The process is called matching. The matched solution is the sum of the two, less the parts common to both solutions. It should be noted that the two solutions are the first terms of the two asymptotic series and the solution is referred to as the matched asymptotic solution.

This solution, which was obtained by Landau and Levich (1942), matches only the second derivatives, $d^2L/d\lambda^2$ to $d^2L^*/d\lambda^2$. The solutions $L(\lambda)$ and $L^*(\lambda)$ are the first terms in the asymptotic solutions in these two regions, but it is very difficult to proceed any further in this problem. The reason is that the only length scale in this problem, h_o, is a variable. More progress can be made where there is at least one fixed length scale (Ruschak and Scriven, 1977), such as in the flow through a slot (see Problem 7.3). Nevertheless, the ability to get additional terms is only a fleeting victory since the differential equations governing those terms appear unsolvable. More details of the method of matched asymptotic expansions are given in Section 10. Numerical methods have greatly enlarged the domain of solutions to free surfaces problems.

The lubrication theory approximation and the closely related slender body approximation (see Problem 7.5) have been applied successfully in many cases. The critical assumption is that the variation of the velocity profile in the direction of flow is ignored in the first approximation. The flow under this approximation becomes one-dimensional in most cases, or at least retains a simple form.

Coating flow processes are plagued by instabilities. Although the effects of gravity should be important in dip coating, as in the flow of a liquid down an incline (Chapter 5, Section 9), which will induce surface waves in the direction of flow, it turns out that a common instability shows waves running perpendicular to the flow. The phenomenon is called ribbing. A review by Ruschak (1985) can be consulted. Shown in Figure 7.4 are regions of instability seen experimentally by Coyle et al. (1990) in reverse roll coating, for which instabilities have also been predicted theoretically. These put an upper limit on the range of operation in coating systems, and hence production rates. Newer coating systems that reduce these difficulties (including the reverse roll coating) and numerical methods used to predict coat thicknesses as functions of physical properties and operating conditions have been discussed by Kistler and Schweizer (1997) and Kistler and Scriven (1983). Practically all solutions deal with Newtonian fluids. Some discussion of real coating fluids can be found in Benkreira (1993), Gutoff et al. (1995), and Kistler and Schweizer (1997). In view of the preference for solventless coatings, some of these methods are now dated. It is estimated at this time that some solvent-type coatings will remain, some water-based coatings

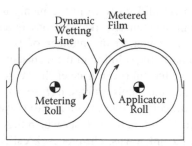

Half-submerged metered film flow

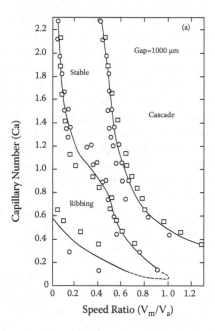

FIGURE 7.4 The first figure shows a "half submerged metered flow" system. The coat is from the applicator roll. The gap width is an important system parameter. The next figure shows the region of stable operation for a fixed gap width (the highest value investigated). The capillary number is that of the applicator. The symbols denote two different viscosities: circles, 0.4 Pa·sec; squares, 0.2 Pa·sec. Decreasing the gap width tends to collapse all three curves into one set of close vertical lines (i.e., the region of stable operation at high speeds is decreased). Very low capillary numbers offer stable flows. From Coyle et al. (1990). Reproduced with permission. Copyright 1990 American Institute of Chemical Engineers.

will continue, and the rest will turn to powder coating, that is, dry coating that is baked to yield the final coat (Marrion, 1994). Finally, we note that there is a great deal of information available on the chemistry of coating materials (Calbo, 1992; Paul, 1996). These are often suspensions with many additives that change dramatically as they set. Their formulation illustrates how Marangoni instability, colloidal instability, etc., can be eliminated in real systems.

7. SPHERICAL DROP REVISITED

Previously the drag on a spherical droplet falling through an infinite fluid was obtained at Equation 7.25. On balancing this force with gravity, the terminal velocity $U^* = U$ is obtained as

$$U^* = \frac{2a^2 g \rho_B}{3\mu_B} (\sigma - 1) \frac{(\kappa + 1)}{(3\kappa + 2)},$$
(7.57)

where $\sigma = \rho_A/\rho_B$ and $\kappa = \mu_A/\mu_B$. When $\kappa \to \infty$, $U^* \to U_\infty^* = [2\rho_B g a^2 (\sigma - 1)/9\mu_B]$, which is the Stokes sedimentation velocity for a solid sphere. This limit is expected since the inner fluid becomes undeformable when the viscosity is very large. Similarly if $\kappa \to 0$, $U^* \to U_0 = [\rho_B g a^2 (\sigma - 1)/3\mu_B]$. The limit $\kappa \to 0$ is closely approximated by gas bubbles rising in a liquid. The important feature is that $U_o^* > U_\infty^*$. However, from experiments, the terminal velocity of a small gas bubble in water is often seen to be given better by the Stokes velocity U_∞^* than by U_o^* as predicted by theory. It has been postulated that this behavior is due to the fact that under nonequilibrium conditions the equilibrium surface tension γ is supplemented by a dynamic contribution $\hat{\gamma}$. The dynamic surface tension γ vanishes when equilibrium is attained. Since the velocities on the interface vary from point to point, so will $\hat{\gamma}$, leading to a surface stress given by $\nabla_s \hat{\gamma}$. The origin of this stress was discussed earlier.

There are two different ways of accounting for $\hat{\gamma}$ or $\nabla_s \hat{\gamma}$. The first consists of postulating a physical mechanism, which in this case is the effect of the presence of minute quantities of surface active impurities. The second is to define an effective surface viscosity, a matter considered further in Section 8.

With the first of these models, $\hat{\gamma}$ is the change in γ due to adsorption of the impurities. The transport of these materials in bulk is governed at steady state by the conservation equation

$$v\nabla c = \nabla \cdot (D \nabla c),$$
(7.58)

where c is the concentration in the bulk phase and D is the diffusion coefficient. The mass transfer to the droplet surface is given by $J^* = -D(\partial c/\partial r)_a$. The term J^* can also be expressed in terms of the adsorption-desorption process as follows:

$$J^* = Q(\Gamma, c(a)) - P(\Gamma),$$
(7.59)

where Q is the rate of adsorption, which is dependent on the surface concentration Γ and the concentration $c(a)$ at the interface, and on P, the rate of desorption.

Finally, one has the equation of conservation of the adsorbed species, which at steady state simplifies from Equation 6.2 to

$$\nabla_s \cdot (\dot{a}_t \Gamma) = \nabla_s \cdot (D_s \nabla_s \Gamma) + J^* ,\tag{7.60}$$

where J^* is the "rate of production" of the species on the interface due to transport from the bulk to the interface. D_s is the surface diffusion coefficient and $\dot{a}_t$ is the tangential velocity. At a clean interface it can be obtained from Equations 7.23 and 7.24 as $\dot{a}_t = v_\theta^* e_\theta$, where this expression is modified to

$$v_\theta^* = -\frac{U \sin \theta}{2(1 + \kappa)}\tag{7.61}$$

when surfactant is present, although v_θ^* remains proportional to sin θ.

The general procedure for solution with a contaminated interface is as follows. Equation 7.21 is solved to obtain the general form of the stream function in liquid B (see Problem 7.6). The resulting velocity distribution is then used to solve Equation 7.58 for the concentration distribution. Unknown constants appearing in these general solutions are found using suitable boundary conditions. Among these are conditions (i) through (iv) following Equation 7.22, with the tangential stress condition modified to include the surface tension gradient. Also the interfacial mass balance condition for the surfactant is needed along with the condition that surfactant concentration is some known value c_0 far from the drop. This method of solution based on a physicochemical mechanism is from Frumkin and Levich (1947) and has been discussed in detail by Levich (1962). One simplified version of the solution is given below.

In this special case it is considered that the adsorption-desorption is the rate limiting step, consequently $c(a) = c_0$, the bulk concentration. It is further assumed that Γ deviates only a little from the equilibrium Γ_o given by

$$Q(\Gamma_o, c_o) - P(\Gamma_o) = 0\tag{7.62}$$

to $\Gamma = \Gamma_o + \Gamma'$. Equation 7.59 can be linearized to give

$$J^* = -\alpha \Gamma',\tag{7.63}$$

where

$$\alpha = \left(\frac{\partial P}{\partial \Gamma} - \frac{\partial Q}{\partial \Gamma} \right)_{\Gamma_o, c_o} .\tag{7.64}$$

Substituting Equations 7.61 and 7.63 into Equation 7.60 and approximating on the left-hand side of Equation 7.60 with Γ_o, one has, on using the relations from Table 7.1

$$-\frac{\Gamma_o}{a\,\sin\theta}\frac{d}{d\theta}[bU\,\sin\theta\,\cos\theta] =$$
$$\frac{D_s}{a^2\,\sin\theta}\frac{d}{d\theta}\left(\sin\theta\,\frac{d\Gamma'}{d\theta}\right) - \alpha\Gamma'$$
(7.65)

where b is an unknown constant and surface diffusivity D_s has been assumed constant. If D_s is negligible,

$$\Gamma' = \frac{2\Gamma_o Ub}{\alpha a}\cos\theta .$$
(7.66)

In this case $\nabla_s\hat{\gamma}$ becomes

$$\nabla_s\,\hat{\gamma} = \left(\frac{\partial\gamma}{\partial\Gamma}\right)_{\Gamma_o}\frac{1}{a}\frac{d}{d\theta}\left[\frac{2\Gamma_o Ub\cos\theta}{\alpha a}\right]e_\theta ,$$
(7.67)

where the term $\partial\gamma/\partial\Gamma$ at Γ_o can be determined explicitly if the adsorption isotherm is known. Equation 7.67 is now substituted into Equation 7.18 to obtain the boundary condition on the shear stresses. When this equation and the other fluid mechanical conditions are used to solve for U and b, the result is

$$U = 3\,U_\infty\frac{1+\kappa+e}{2+3\kappa+3e} ,$$
(7.68)

where U_∞ is the Stokes velocity and

$$e = -\frac{2\Gamma_o}{3\mu_B\alpha a}\left(\frac{\partial\gamma}{\partial\Gamma}\right)_{\Gamma_o} .$$
(7.69)

A simple interpretation of Equation 7.68 is obtained on noting that $U \to U_\infty$ as $e \to \infty$. This limit is attained when the contaminant is surface active and surface elasticity $[-\Gamma_o(\partial\gamma/\partial\Gamma)]$ is very large. More specific information on the action of surfactants is revealed on noting that the velocity at the interface v_θ^* is

$$v_\theta^* = -\frac{3}{2}\frac{U_\infty\,\sin\theta}{2+3\kappa+3e} ,$$
(7.70)

which obviously goes to zero as $e \to \infty$. The physical picture that emerges is that adsorbed surfactant is swept upward due to tangential flow along the interface. Desorption occurs near the top of the drop. Thus the surfactant concentration is high at the top and low at the bottom of the drop. This sets up a force proportional to

$$\nabla_s \, \gamma = \left(\frac{\partial \gamma}{\partial \Gamma} \right) \nabla_s \Gamma \; .$$

It is directed from the top, the region of higher surfactant concentration and lower surface tension, toward the bottom. Obviously the direction of this force, being against the direction of v_θ^*, tends to decrease this velocity. When v_θ^* is zero (i.e., $e \to \infty$), the surface is immobile and $U \to U_\infty$. This situation is analogous to the case of an inextensible surface considered in the damping of capillary waves in Chapter 5 (Section 3).

A general conclusion is also reached that the presence of surfactants makes the interface more rigid and provides resistance to lateral interfacial deformation, the measure of the latter being the interfacial velocities. It should be noted that the normal stress balance is satisfied with this solution, the same result as was found in Section 4 in the absence of surfactants.

8.   SURFACE RHEOLOGY

Although the model of the preceding section reveals the basic mechanism by which surfactants retard drop motion, it appears that it has little predictive abilities. To predict U, for instance, it is necessary to know a host of physical properties for components that are essentially contaminants. Nevertheless, there is a great need to know the "stiffness" of fluid-liquid interfaces in real systems, which are almost always contaminated. Unless enough information is available to formulate suitable boundary conditions, the correct fluid mechanical problem cannot be solved, nor can useful quantities like the drag force, pressure drop, etc., be extracted. Frequent lack of availability of the requisite data has led investigators to a phenomenological form of describing the "stiffness" of an interface.

If an interface is stiff, then it has a resistance to deformation. In three dimensions the compressibility is linked to $\nabla \cdot v$, which is zero for incompressible fluids. Further, the rate of strain $1/2[\nabla v + (\nabla v)^T]$ also describes the deformation. The resulting stress for a Newtonian fluid is $[p + (2/3\mu - \kappa)\nabla \cdot v]I - \mu \, [\nabla v + (\nabla v)^T]$, where κ is the bulk viscosity and conventions from Bird et al. (2002) are being followed. For incompressible fluids, the stress reduces to $pI - \mu \, [\nabla v + (\nabla v)^T]$. The surface stress, by analogy, would be $[-\gamma + (2/3\hat{\epsilon} - \hat{\kappa} \, \nabla_s \cdot \dot{a} \,) \, \mathbf{I}_s - \hat{\epsilon} \, (\nabla_s \dot{a} + (\nabla_s \dot{a} \,)^T)]$. The negative sign in front of γ denotes tension, as opposed to p, which is usually compressive. Here, $\mathbf{I}_s$ is the surface identity tensor defined previously, while $\hat{\epsilon}$ and $\hat{\kappa}$ are two appropriate constants.

Conventionally the surface viscous stress T_s is written in a slightly different form (Scriven, 1960; Slattery, 1990):

$$T_s = -(\eta - \varepsilon)(\nabla_s \cdot \dot{a}) I_s - \varepsilon \left[I_s \cdot \nabla_s \dot{a} + (\nabla_s \dot{a})^T \cdot I_s \right], \qquad (7.71)$$

where η is the surface dilational viscosity and ε is the surface shear viscosity. The term in γ is dropped in Equation 7.71 because its effects on force balances have already been included in Equations 7.18 and 7.19. Use of a stress tensor T_s provides a more general treatment than the use of a dynamic surface tension in the form of a scalar γ (i.e., the surface stress has the form $-\gamma I_s + T_s$ instead of $-(\gamma + \hat{\gamma}) I_s$. However, their actions as surface stresses in the forms of $\nabla_s \cdot T_s$ and $-\nabla_s \hat{\gamma}$ are similar. The first also contains terms in the direction perpendicular to the surface. In particular we have

$$\tau_A - \tau_B = [-\nabla_s \gamma - \eta \nabla_s (\nabla_s \bullet \dot{a}_t) - \varepsilon \nabla_s^2 \dot{a}_t] \bullet t . \qquad (7.72)$$

The last two terms on the right-hand side provide the effects of the surface viscosities. The effects of the component $\dot{a}_n n$ have been ignored for simplicity. If there is no local expansion or contraction of the interface, one has

$$\nabla_s \bullet \dot{a}_t = 0 , \qquad (7.73)$$

which means that the effects of the surface dilational viscosity η are not felt.

With $\varepsilon = 0$, Boussinesq (1913) obtained for a fluid sphere

$$U = U_\infty \frac{1 + \kappa + e}{2 + 3\kappa + 3e} , \qquad (7.74)$$

where $e = (2\eta / a\mu_B)$. As $e \to \infty$, $U \to U_\infty$ and it can be also shown that $v_\theta^* \to 0$ everywhere on the surface.

With Equations 7.71 and 7.72, the system of equations relevant to a given experimental setup can be solved to give the surface rheological coefficients. The deep channel viscometer of Mannheimer and Schechter (1970) is popular and also provides a very nice instance where the complete solution to the equations of motion is available. A sketch of the apparatus is shown in Figure 7.5. A cylindrical annulus is filled with two liquids A and B. The annulus is held stationary and the floor rotated with a constant angular velocity Ω. Also, r_i and r_o are the inner and the outer radii of the annulus in the cylindrical coordinate system and z_o designates the position of the interface. It is seen here that $v_{\theta A}$ and $v_{\theta B}$ are the only nonzero velocity components in the two phases, that they are functions of r and z only, and that the pressures satisfy the hydrostatic equilibrium equations. The solution to the equations of motion subject to the conventional boundary conditions are

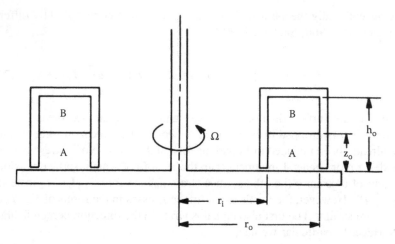

FIGURE 7.5 The deep channel viscometer. The longitudinal section of a radially symmetric figure is shown.

$$V_{\theta B} = \sum_{i=0}^{\infty} (A_i \ \sinh E_i D \ + \ B_i \ \cosh E_i D)(\sinh E_i Z - \tanh E_i H \ \cosh E_i Z).$$

$$\psi_1(RE_i)/(\sinh E_i \ D \tanh E_i H \cosh E_i D) \qquad\qquad (7.75)$$

$$V_{\theta A} = \sum_{i=0}^{\infty} (A_i \ \sinh E_i Z \ + \ B_i \ \cosh E_i Z) \ \psi_1(RE_i) , \qquad (7.76)$$

where

$$V_\theta = \frac{v_\theta}{\Omega r_o} , \ Z = \frac{z}{(r_o - r_i)} , \ R = \frac{r}{(r_o - r_i)} , \ H = \frac{h_o}{(r_o - r_i)}$$

$$D = \frac{z_o}{(r_o - r_i)} , \ A = \frac{r_o}{(r_o - r_i)} , \ B = \frac{r_i}{(r_o - r_i)} \ \text{ and}$$

$$B_i = \frac{2[B^2\psi_o(BE_i) - A^2\psi_o(AE_i)]}{AE_i \ \ [A^2\psi_0^2(AE_i) - B^2\psi_0^2(BE_i)]} .$$

$$\psi_o(RE_i) = J_0(RE_i)\, Y_1\,(AE_i) - J_1\,(AE_i)\, Y_0\,(RE_i),\ \text{and}$$

$$\psi_1(RE_i) = J_1\,(RE_i)\, Y_1\,(AE_i) - J_1\,(AE_i)\, Y_1\,(AE_i).$$

J_υ and Y_υ are Bessel functions of the first and second kind and order υ, and the E_i's are the roots of $\psi_1(RE_i) = 0$. Equations 7.75 and 7.76 have been derived using all boundary conditions except the tangential stress balance at $z = z_0$.

In view of the fact that $V_{\theta A} = V_{\theta B} = V_\theta^*$ at $z = z_0$ where $V_\theta^* = (\,v_\theta^* /\Omega r_0)$ and that v_θ^* is not a function of θ, the term $\nabla_s \cdot \dot{a}_t$, where $\dot{a}_t = v_\theta^*\, e_\theta$, can be shown to be zero and Equation 7.73 becomes

$$\varepsilon \left[\frac{d^2 V_\theta^*}{dR^2} + \frac{1}{R}\frac{dV_\theta^*}{dR} - \frac{V_\theta^*}{R^2} \right] = -\mu_B \frac{\partial V_{\theta B}}{\partial Z}\Big|_{Z=D} + \mu_A \frac{\partial V_{\theta A}}{\partial Z}\Big|_{Z=D}. \qquad (7.77)$$

When $\mu_B/\mu_A = 0$ (i.e., when the upper fluid is a gas), the interfacial velocity is given by

$$V_\theta^* = \frac{4}{\pi} \sum_{n=1}^{\infty} \frac{\sin (2n-1)\pi Y}{[(2n-1)\pi S \sinh (2n-1)\pi D + \cosh (2n-1)\pi D]}, \qquad (7.78)$$

where $Y = (r - r_i)/(r_o - r_i)$ and $S = [\varepsilon/\mu_A(r_o - r_i)]$. For deep canals, only the first term in the series need be retained.

Combining the resulting equations for surfactant covered and surfactant free surfaces, one obtains the equation

$$\varepsilon \simeq \frac{(r_o - r_i)\mu_A}{E_1} \left(\frac{t_c}{t_c^*} - 1 \right) \coth E_1 D. \qquad (7.79)$$

Here, t_c is the time needed for a point on the interface at the centerline $r = 1/2(r_i + r_o)$ to travel an arc of a given length and t_c^* is the time needed for a corresponding surface with $\varepsilon = 0$. These times are measured by following small particles placed on the interface. Obviously a calibration with a pure liquid needs to be performed. For all practical purposes E_1 in Equation 7.79 can be approximated as π, and for deep channels $\coth \pi D \approx 1$. The surface viscosity ε is expressed as surface poise (sp) = 1 (mN·s/m) and ranges from very low values to 1 sp or more. An interface can be considered rigid at $\varepsilon \sim 5$ sp. More discussion can be found in Edwards et al. (1991) and Slattery (1990). These cover the other equipment, including the use of surface waves discussed in Chapter 5, magnitudes of surface viscosities of soluble and insoluble surfactants and emphasize the fact that these viscosities can be strong functions of the rates of surface deformation, and sometimes the equipment and experimental methods used.

It becomes apparent now that the effects of surface viscosities can be included in the boundary conditions for all problems involving dynamic interfaces, including those considered in Chapters 5 and 6.

9. DRAINAGE OF THIN LIQUID FILMS

Chapter 5 (Section 6) described the conditions when a thin liquid film could become unstable and rupture and thereby cause coalescence of bubbles or drops. Instability was possible when the film became thin enough (less than 100 nm) for the disjoining pressure effects to be significant. However, considerable time may be required for the film to drain to this thickness, so that the rate of drainage has an important influence on the coalescence rate. The literature on experimental and theoretical aspects of thin film drainage is extensive (Exerowa and Kruglykov, 1998; Ivanov and Dmitrov, 1988).

More than a century ago Reynolds (1886) solved for the rate of decrease in the thickness h of the liquid layer separating two solid, parallel disks being pushed toward each other. His analysis is readily adapted to the case of a small circular film of radius R with sufficient surfactant present to eliminate lateral flow at the film surfaces:

$$\mathrm{d}/\mathrm{d}t(1/h^2) = (4/3\mu R^2)(\Delta p_c - \Pi_{vw} - \Pi_{el}), \qquad (7.80)$$

where Δp_c is the pressure difference between the film and the bulk continuous phase. Since Π_{vw} is negative for foams and emulsions where the Hamaker constant A_H is positive, van der Waals forces promote thinning, the expected result since thinning transfers molecules to the bulk liquid, where their potential energy due to interaction with surrounding molecules is lower than in the film. On the other hand, Π_{el} is positive and electrical repulsion opposes drainage. Indeed, if Π_{el} becomes equal to $\Delta p_c - \Pi_{vw}$, an equilibrium film thickness is reached and drainage stops. In terms of Figure 5.7, the equilibrium thickness is that for which $\Pi = \Delta p_c$. If the surface potential ψ_0 is known, values of equilibrium film thickness can be used to estimate A_H.

Experiments on drainage of individual films can also be used to estimate A_H (Sheludko, 1967). In aqueous films with high electrolyte contents and in films of organic liquids, Π_{el} is negligible and $\Pi_{vw}(h)$ can be obtained from Equation 7.80 and data on the time dependence of h. The results confirm that Π_{vw} varies inversely with h^3 (cf. Equation 5.111) and provide values of A_H.

If the adsorbed surfactant is unable to completely prevent lateral motion at the film surfaces, it is still possible to obtain an analytical solution for the rate of drainage of a film of uniform thickness. The most recent of these analyses (Singh et al., 1996), which employs an improved boundary condition on the surface velocity at the film perimeter based on results of numerical simulations of drainage of nonuniform films, yields the following expression for the ratio V^*

$= V/V_{Re}$ of the actual rate of film drainage to that predicted by the Reynolds theory for the case of an insoluble surfactant (see Problem 7.10):

$$V^* = \{[4N_\mu N_\alpha (2I_1(\theta) - \theta I_o(\theta))/(N_\Gamma N_\alpha + 1)^2 (I_1(\theta) - \theta I_o(\theta))]$$

$$(7.81)$$

$$+ 1 - (N_\Gamma N_\alpha + 1)^{-1}\}^{-1}$$

where I_o and I_1 are modified Bessel functions of the first kind, $N_\alpha = h/R$, and $\theta = [(N_\Gamma N_\alpha + 1)/N_\mu N_\alpha]^{1/2}$. N_Γ and N_μ are dimensionless groups describing the effects of surface elasticity and surface viscosity, respectively:

$$N_\Gamma = [-\Gamma(d\gamma/d\Gamma)R/3\mu D_s] \tag{7.82}$$

$$N_\mu = [(\eta + \varepsilon)/3\mu R], \tag{7.83}$$

where μ is the viscosity of the liquid in the film and D_s is the surface diffusivity of the surfactant at the film surfaces. Calculations reveal the remarkable result that the dependence of V/V_{Re} on surface elasticity and viscosity can be expressed in terms of a single dimensionless group N_R, which is a linear combination of N_μ and N_Γ:

$$N_R = N_\mu + 0.27 \, N_\Gamma. \tag{7.84}$$

Figure 7.6 shows that the rate of film drainage is rapid for small values of N_R and approaches that given by the Reynolds theory for large N_R where lateral flow at the film surfaces is minimal. From the horizontal line on Figure 7.6 one sees that V/V_{Re} increases as the film thins at constant N_R. The actual rate of thinning V decreases as h (and N_α) decreases, but V_{Re} decreases even faster.

A draining film of uniform thickness $h(t)$ as assumed in the above analyses is, of course, an idealization. According to the Young-Laplace equation, the pressure p in the film depends on the curvature of the film surfaces, and, as discussed in Chapter 5, the disjoining pressure π depends on h. Since gradients in p and Π are required for outward flow, a radial gradient in h must exist. In fact, experiments show that a thick region or "dimple" typically develops near the center of the film. The dimple is separated from the meniscus by a thin "barrier ring" (see Figure 7.7). This behavior can be understood as follows. When two drops or bubbles approach one another along their line of centers, their surfaces start to flatten near the point of minimum separation, forming a film. The radial gradient in film thickness decreases as the flattening continues, and a point is eventually reached where the rate of liquid flow from the center to the edge of the film falls below the rate of flow from the edge into the meniscus. As a result, the film thins at its edge, forming a barrier ring with an enclosed dimple.

If the surfactant film is coherent and has a high surface shear viscosity ε, the film will continue to drain while maintaining its axisymmetric shape, although

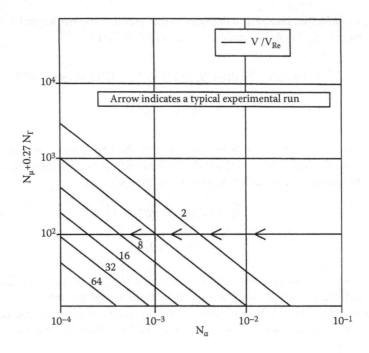

FIGURE 7.6 Contours of the mobility factor V/V_{Re} as a function of the material properties N_R and the dimensionless film thickness N_α. Reprinted from Singh et al. (1996). Copyright 1996 with permission from Elsevier.

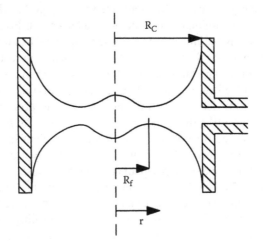

FIGURE 7.7 Dimple and barrier ring as in Sheludko-Exerowa cell.

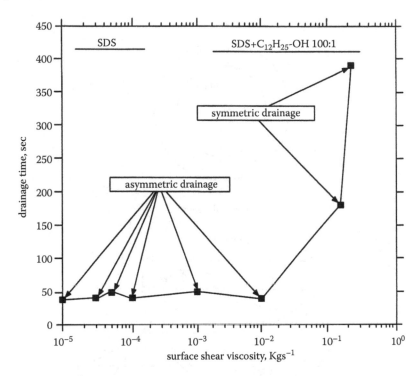

FIGURE 7.8 Drainage time as a function of surface shear viscosity for SDS and SDS-$C_{12}H_{25}OH$ (100:1). Film radius approximately 100 μm. Reprinted with permission from Joye et al. (1994). Copyright 1994 American Chemical Society.

drainage may become very slow because liquid leaving the dimple must pass through the thin barrier ring region, which has a large resistance to flow. Instability and subsequent rupture may occur in the barrier ring region if the criteria described in Chapter 5 are met.

In contrast, if the surfactant film has a low surface shear viscosity, another type of hydrodynamic instability may develop in which the film shape becomes asymmetric and the liquid in the dimple rapidly escapes into the meniscus, leaving a film that is relatively flat. It is under these circumstances that the uniform film models described above provide a first approximation of the subsequent drainage. Because the dimple disappears quickly, overall drainage time is much faster when asymmetric drainage occurs, as shown by Figure 7.8 for films made with sodium dodecyl sulfate (SDS)/dodecanol mixtures.

Let us consider the mechanism by which asymmetric drainage arises. Suppose that a small, wavy perturbation occurs so that the barrier ring becomes thicker at some points and thinner at others (see Figure 7.9). At the former points, the overall resistance to outward flow decreases. The resulting increase in flow rate produces an outward surface velocity that drives surfactant into the meniscus, so that local surface concentration Γ in the film decreases and surface tension γ increases. The opposite occurs (i.e., Γ increases and γ decreases) at points where

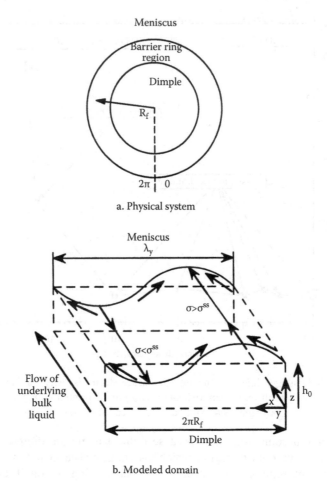

FIGURE 7.9 Schematic view of perturbations in thickness h, surface velocity, and surface tension σ in an unstable film. Reprinted with permission from Joye et al. (1994). Copyright 1994 American Chemical Society.

the barrier ring is thin. Thus a lateral surface tension gradient arises that causes flow from thin to thick portions of the barrier ring, a destabilizing effect. Because the resulting flow pattern in the surface involves shear, the instability is opposed by surface shear viscosity. It is also opposed by the lateral curvature gradient, which produces a pressure gradient from thick to thin portions of the barrier ring. This last effect dominates for perturbations of short wavelengths λ, causing them to be stable. However, perturbations having large λ are unstable and asymmetric drainage is observed if surface shear viscosity is low (e.g., for pure SDS films) (Figure 7.8). The least stable wavelength is the longest one possible (i.e., the perimeter of the barrier ring).

A linear stability analysis has been developed that predicts instability leading to the asymmetric drainage condition when a dimensionless parameter $\Theta < 1$ (Joye et al., 1994):

$$\Theta = (\varepsilon D_s R_c^2 / 256\varphi^2 \gamma R^4)\,(1 + N_\mu N_\alpha + N_\Gamma N_\alpha)\,, \qquad (7.85)$$

where φ is the dimensionless pressure gradient at the barrier ring before perturbation, which is expected to have an order of magnitude near unity. Also, R_c is the radius of the meniscus outside the film, typically the radius of the drops approaching one another or the radius of the capillary tube in which the film is formed in laboratory experiments. The criterion $\Theta < 1$ is satisfied by all the points in Figure 7.8 where asymmetric drainage was observed experimentally. Clearly, low values of surface shear viscosity ε and large values of film radius R promote instability.

A similar instability is seen near the base and along the sides of the films formed when a vertically oriented rectangular frame is withdrawn from a surfactant solution and allowed to drain. Mysels et al. (1959) described this behavior and referred to it as "marginal regeneration." The origin of the instability is the same as for asymmetric drainage discussed above (Joye et al., 1994), although gravity influences the behavior, which occurs once growth of unstable perturbations produces substantial differences in thickness within the film.

By far the most common method for studying the drainage of individual foam films is that developed by Sheludko and Exerowa and illustrated schematically in Figure 7.7 (see Exerowa and Kruglyakov, 1998; Sheludko, 1967). A small quantity of surfactant solution is placed in a short capillary tube having a radius R_c that is typically about 1 mm. Liquid is withdrawn through a small tube attached to the side of the capillary until the two surfaces approach one another near the axis of the capillary and flatten to form a film. As more liquid is withdrawn, film radius increases and a dimple develops, as described above. When the film has reached the desired radius, withdrawal is halted and drainage of the film continues, driven by capillary forces and, when it becomes sufficiently thin, disjoining pressure effects. Usually film radii studied are about an order of magnitude less than R_c (i.e., on the order of 100 μm).

In order to collect information on film thickness and profile the cell is placed on a microscope stage and monochromatic light is directed toward the film. Light waves reflected from the top and the bottom surfaces of the film interfere. The resulting interference patterns provide information on variations in film thickness with position. For films having thicknesses considerably below the wavelength of the light used, there are no interference patterns, but the intensity of the reflected light can be used to determine thickness. A comprehensive discussion of experimental procedures using this technique and its modifications can be found in Exerowa and Kruglyakov (1998).

10. DYNAMIC CONTACT LINES

The study of nonequilibrium contact lines is the logical extension of the study of dynamic interfaces. Very little is known about nonwetting liquids, and unless otherwise stated, the following discussion pertains to wetting liquids. Dynamic contact lines can be grouped in two categories. The first is that of spontaneous spreading, where there are no external forces and the contact line moves by the action of the surface tension forces. The rates of spreading are very small—in the range of 10^{-2} mm/sec. Most of the shapes of spreading drops under these conditions appear to be spherical caps (Lin et al., 1996; Schonhorn et al., 1966) and hence show contact angles very removed from zero. The method used is microscopy with an error of $\pm$ 14 μm in linear measurements. Where sufficient accuracies have been used for measurements (ellipsometry or scanning electron microscopy) (Bascom et al., 1964; Lin et al., 1996), the results show that the profiles of the menisci change sharply very close to the contact lines, indeed at such small film thicknesses that optical microscopy will miss this feature completely. A thin film, often called the precursor, extends outward and does indeed show that it makes a slope of zero with the solid surface, in accordance with the fact that the liquid is a wetting liquid. Consequently, in such cases investigators refer to their measurements using optical microscopy as the "apparent dynamic contact angles," suggesting that these are the contact angles formed by extrapolating the macroscopic profiles to the horizontal surface.

The second type of spreading is forced spreading, where forces from outside are imposed on the system and give rise to the movement of the contact line. The special case of a slug of one fluid displacing an immiscible fluid in a tube is shown in Figure 7.10. The displacing fluid wets the solid surface and when the slug velocity is zero, the apparent contact angle is zero (Figure 7.10a). As the slug velocity is increased (Figure 7.10b and c), the apparent contact angle increases, ultimately reaching 180° (Hoffman, 1975), and the menisci appear to have the shape of spherical caps. In this case it is difficult to imagine how the apparent contact angle could be so large and the profile could still maintain 0° at the contact line, which the previous case of spontaneous spreading suggests. Indeed, it is more plausible to suggest that in this case the hydrodynamic forces are so high that Young's law breaks down and the dynamic contact angle represents a new, but fundamental force balance.

Thus dynamic contact angles of wetting liquids tend to their equilibrium value of $\lambda = 0$ as contact line velocities decrease to zero. They also become much larger than λ as contact line velocities increase. Up to this point it appears that the contact line always exists. That is not true. At sufficiently large velocities (Figure 7.10d), entrainment of one of the phases occurs and the contact line disappears (Hansen and Toong, 1971; Rose and Heins, 1962). Blake and Ruschak (1979) have shown for a plate moving into the liquid (or out of the liquid) that there is a velocity at which dynamic contact angles increase to reach 180° (or decrease to reach 0°). Entrainment velocities are low when the displacing fluid in Figure 7.10d has a low viscosity, such as air. Beyond this velocity the contact line shows

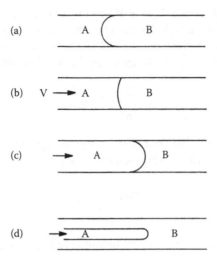

FIGURE 7.10 The nature of displacement of the fluid B by the fluid A. Case (a) is the static case. In case (c), the velocity of displacement is higher than in case (b). At very high displacement velocities, a lubrication layer of phase B is left behind, as shown in (d).

a jagged appearance, as well as some signs of entrainment that are not sustained. Steady entrainment occurs only at much larger velocities. Both in view of the misleading measurements of the dynamic contact angles in spontaneous spreading and in view of entrainment, the concept of a revised Young's law for dynamic systems appears suspect. Nevertheless, Jiang et al. (1979) show that for many geometries studied, the apparent dynamic contact angles obey,

$$\frac{\cos \lambda - \cos \theta}{\cos \lambda + 1} = \tanh \left(4.96 \, N_{Ca}^{0.702}\right), \qquad (7.86)$$

where θ is the dynamic contact angle and N_{Ca} is the capillary number.

de Gennes (1985) divided the contact line region into three parts, of which the wedge-shaped film contacting the apparent dynamic contact angle is considered below and shown in Figure 7.11. The liquid wedge is a wetting liquid and

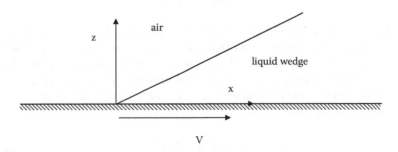

FIGURE 7.11 Schematic view of a dynamic contact angle.

advances on the solid surface. To keep the meniscus stationary, the plate is plunged into the liquid at a velocity V. Using lubrication theory approximation and the boundary conditions that the tangential velocity at the solid surface is V, tangential stress at the liquid-air interface is zero, and the net flow rate is zero, one has

$$v_x = \frac{3V}{2h^2}(z^2 - 2hz) + V \ , \tag{7.87}$$

where h is the local film thickness. The viscous dissipation per unit volume (Bird et al., 2002) in this region can be found to be equal to

$$\frac{9\mu V^2}{h^4}\ (z-h)^2.$$

This is integrated over the volume by first integrating over z from 0 to h, then h is approximated as θx and integration is carried out over x from ℓ, a microscopic length scale and a lower bound, which is used to exclude the details of the actual region containing the contact line or even suggest a molecular scale, to L, the macroscopic length scale. The result is

$$\frac{3\mu V^2}{\theta} \ln | \frac{1}{\varepsilon} |,$$

where $\varepsilon = \ell/L$, a small quantity. It is then equated to the rate of surface work done. This rate is obtained as the dot product of the unbalanced force $\gamma_{SV} - \gamma_{SL} - \gamma_{LV}\cos\theta$ and the velocity, which is the rate of spreading. Expanding for small θ, the surface force becomes $S + \frac{1}{2}\gamma\theta^2$, where the LV subscripts have now been dropped. S is the spreading pressure, which is a force that operates in the small region containing the actual contact line and not in the region being considered here. In the latter region, $V \times \frac{1}{2}\gamma\theta^2$ gives us the rate of surface work. Equating it to viscous dissipation one has

$$V = \frac{\gamma}{6\mu}\left[\ln | \frac{1}{\varepsilon} |\right]^{-1} \theta^3 \ , \tag{7.88}$$

which is known as the Hoffman (1975)–Voinov (1976)–Tanner (1979) rule (see Kistler, 1993). The inverse logarithm term is very insensitive to ε. Hoffman obtained the above dependence of V on θ^3 from his experimental data, and Voinov and Tanner from approximate solutions. The pressure gradient in Equation 7.87 is given by $3\mu V/h^2$. Substituting for pressure with the Laplace pressure leads to the equation of the film profile of

$$h^2 \frac{d^3h}{dx^3} = \frac{3\mu V}{\gamma} \ . \tag{7.89}$$

Recently, an exact solution for Equation 7.89 has been obtained that upholds the approximate solutions discussed above, but shows some others to be incorrect (Duffy and Wilson, 1997). The apparent dynamic contact angle is simply defined as the constant slope sufficiently far away. The predicted contact angle at the contact line, $h = 0$, is seen to be infinite. The curvature at the contact line is also infinite, and not zero as implied in the system studied in Equation 7.87. de Gennes overcame these problems by ignoring films with thicknesses less than some value (which in the previous problem occurs at $x = \ell$). He also showed that the viscous dissipation in the precursor film not considered in the wedge, and one that actually makes a zero contact angle at the contact line, exactly balances the contribution from S, which has not been considered in the wedge.

It is very important to note that the dynamics of the precursor can be treated separately from the bulk. This uncoupling is possible only because the precursor moves faster than the bulk, as seen in the experiments (Bascom et al., 1964). More detailed fluid mechanics, where the two regions are treated separately, also show this behavior: the location of the contact line for the bulk varies with $t^{1/10}$ (Neogi and Miller, 1982a,b), but that for the precursor with $t^{1/2}$ (Lopez et al., 1976).

For nonwetting liquids, most investigators believe that the menisci shapes for the drop and in the capillary are still spherical caps, and a generalized correlation governing the dynamic contact angle exists which can be used to describe both wetting and nonwetting liquids. The nonwetting case has its start from a nonzero λ and is shifted accordingly, as shown in Equation 7.86. However, there is a postulate that the kinetics of wetting liquids form an envelope to the kinetics of nonwetting liquids (Neogi, 2001). That is, if Equation 7.86 is drawn for N_{Ca} as a function of θ for different values of λ, then we a get a family of curves. The curve for the wetting case should form an envelope to this family and not be the limit obtained as λ goes to zero, which is the case in Equation 7.86.

Example 7.2

Show that Equation 7.86 also satisfies the dependence of the velocity on dynamic contact angle as given by the Hoffman–Voinov–Tanner rule.

For small capillary numbers, the right-hand side becomes $4.96\, N_{Ca}^{0.702}$. For small θ and $\lambda = 0$, the left-hand side is $\theta^2/4$. Hence one has

$$V = 0.0142 \frac{\gamma}{\mu} \theta^{2.85} \; ;$$

that is, the power is 2.85, close to 3.0.

11. SLIP

The problems involving dynamic contact lines contain additional difficulties that arise from the fact that these lines describe the intersection of two surfaces. The physical problem can require two different types of boundary conditions to be satisfied, one condition on each of the surfaces. However, both need to be satisfied at the contact line, leading to an overspecification and physically unrealistic results. Consider the case of a drop spreading on a solid surface. At the air-liquid interface, the zero shear stress boundary condition applies, whereas at the solid-liquid interface the no-slip boundary condition is applicable. In Figure 7.12, approximate velocity profiles in a drop near the contact line region are shown. Since the contact line O is moving outward, the average velocity over the drop thickness must be nonzero. As the thickness decreases, the velocity gradient must increase as the contact line is approached to preserve the nonzero value of the average velocity. In fact, the solution to this problem shows that the velocity gradient becomes infinite at the contact line (Huh and Scriven, 1971). This means that the shear stress, pressure gradient, viscous dissipation, etc., at the contact line are infinite! It turns out that the total force on the solid, obtained by integrating the stresses on the surface, is also infinite. It is the unbounded behavior of the overall quantities that is disturbing, as it disallows predictions of useful quantities even approximately. For the simple case considered in Figure 7.12, it suffices to say that this singular behavior of the shear stress at the contact line arises due to the fact that a material point on the dynamic contact line is required both to move on the solid surface and to obey the no-slip boundary condition.

In a more general context, Dussan and Davis (1974) showed that the two velocities on the two interfaces in the limit when the contact line is approached are different, leading to a discontinuity in the velocity field and an infinite velocity gradient at the contact line. The problem treated in the previous section and illustrated in Figure 7.11 also shows this behavior. Note that the velocity gradient dv_x/dz in Equation 7.87 at $z = 0$ (on the solid surface) and $h = 0$ (at the contact line) is infinite. One great simplification used there is the assumption that the profile is wedge-shaped, $h = \theta.x$. That considerable difficulties lie in determining the actual profile is discussed after Equation 7.89.

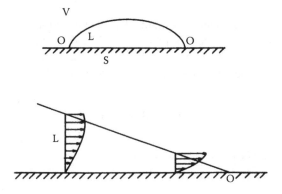

FIGURE 7.12 (Top) A liquid drop (L) is lying on a solid surface (S). The third phase is a vapor (V). The contact line is shown as 0. (Bottom) The velocity profiles are shown under dynamic conditions (i.e., the point O is moving to the right). The velocity gradient gets sharper as the point O is approached.

It is suggested that to eliminate this problem a slip boundary condition be used in the vicinity of the contact line. This slip velocity is a tangential velocity v_t on the liquid-solid interface and eliminates the need to have an infinite velocity gradient at the contact line in order to reach a finite average velocity. An empirical form for slip may be chosen as

$$\beta \, \frac{\partial v_t}{\partial x_n} = v_t \,. \tag{7.90}$$

Here, β is a constant. The left-hand side of Equation 7.90 is proportional to the shear stress on the solid surface; the equation thus represents a physical argument that if the forces on the surface are sufficiently large they can move the liquid molecules next to the surface. Since as $\beta \rightarrow 0$, the familiar no-slip boundary condition is obtained, β has to be very small in reality for the slip to be undetectable in conventional systems.

The fact that β has the dimensions of length has an important implication. Wherever the representative length scale of the fluid mechanical problem greatly exceeds β, the effects of slip are not felt. Thus for a spreading drop of volume V, the length scale in the bulk of the drop is $V^{1/3} >> \beta$, and consequently the effects of the slip in the major part of the drop are not felt. This disparity between two length scales which appear in a single problem is seen often in fluid mechanics. In the contact line region, the thickness of the drop h approaches β. Hence, in this region the slip dominates the fluid mechanics. From the previous discussion it becomes clear that the method of matched asymptotic solutions is needed to solve this problem. One asymptotic solution applies to the major part of the drop and is characterized by the length scale $V^{1/3}$. The other asymptotic solution is characterized by β and applies to the contact line region. Here the ratio $\beta/V^{1/3}$ is the required small quantity on which the asymptotic solutions are built.

In search of a basis for slip, one observes that in the immediate neighborhood of the contact line region the continuum approach breaks down. Consequently the molecular activity there, like adsorption, relaxation, reorientation, etc., could be important. However, if these molecular effects are to be used to predict the slip velocity, they should not yield a length scale comparable to molecular dimensions because such dimensions are not admissible under the continuum treatment. Neogi and Miller (1982a) and Ruckenstein and Dunn (1977) considered the possibility that molecules at the solid-liquid interface can actually move in the tangential direction. Their results can be eventually expressed as a slip velocity that is dependent on surface diffusivity,

$$v_t = -\frac{D_s}{n_L kT} \frac{\partial \Phi}{\partial x_t} \,, \tag{7.91}$$

on the solid surface. Here, kT is the product of the Boltzmann constant and the absolute temperature, n_L is the number of molecules in the liquid per unit volume, and x_t is the coordinate in the direction tangential to the interface. D_s/kT is the mobility of the liquid on the solid surface and $\Phi = p + \phi$, where p is the pressure and ϕ is the molecular potential of the type discussed in Chapter 2 (Section 3) and Chapter 5 (Section 6), except that it is evaluated on the solid surface. D_s is viewed as the surface diffusivity. Equation 7.91 says that the velocity at the interface is equal to the product of the mobility and a thermodynamic force expressed as the gradient of a potential in the tangential direction.

Neogi and Miller (1982a) also solved the equations of motion for a drop sufficiently small that the effects of gravity could be neglected. At large times of spreading, the drop is flat and thin and the lubrication theory approximation can be used. For this problem, their method of matched asymptotic expansions yielded drop shapes and the rates of spreading. The latter is reported by them as a function of time, but can also be expressed as a function of θ, the dynamic contact angle (see Problem 7.12d). The results are same as the Hoffman–Voinov–Tanner rule (Equation 7.88), including the inverse logarithm with $\varepsilon = (3\mu D_s/n_L kT)^{1/2}/V^{1/2}$, as shown in Equation 7.91. Thus one has results from mechanisms that invoke surface diffusion equivalent to those that involve molecular dimensions. Even though the slip length can be small, such that $\varepsilon \sim 10^{-6}$, its effect on hydrodynamics is more pronounced as

$$\left[\ln \left| \frac{1}{\varepsilon} \right| \right]^{-1} \sim 0.06.$$

The above method eliminates the infinite stresses and pressure gradients at the contact lines. It also provides an expression for the rate of spreading. In spite of these favorable properties, the slip condition of Equation 7.91 has limitations that arise from the fact that most solid surfaces have surface irregularities larger than the scale of molecular effects. Even machine polished metal surfaces have irregularities of some 0.1 to 10 μm, compared to the range of the molecular effects that are confined below 0.1 μm, as discussed in Chapter 3. Thus a slip over the length scale dictated by roughness is necessary to handle the more realistic problems.

Neogi and Miller (1982b) modeled surface roughness by assuming the rough surface to be the surface of a porous medium. The slip velocity is given by Darcy's law for flow through a porous medium on the surface of the solid. The slip length in this case is found to be $(3k)^{1/2}$, where k is the permeability of the porous medium. It was also found to the first approximation that $3k \sim \sigma^2$, where σ^2 is the variance of the surface irregularities. The spreading velocity was found to be proportional to

$$\left[\ln \left| \frac{1}{\varepsilon} \right| \right]^{-1},$$

where $\varepsilon = (3k)^{1/2}/V^{1/3}$. The lubrication theory approximation and matched asymptotic expansions were again used. The available experimental data are limited to short times of spreading where lubrication theory does not apply. However, under simple adjustments for this difference they fitted their result to the data on the rate of spreading to obtain an estimate of surface irregularities of 1 to approximately 5 μm, whereas approximately 3 μm was reported in the original papers (Schonhorn et al., 1966; van Oene et al., 1969).

An important mechanism for movement of a contact line is provided by Wayner (1991). It has been suggested for a long time that liquid evaporates from the bulk region and condenses ahead of the contact line, thus making it advance. The mechanism was called "surface distillation." Wayner quantified the process by considering the fact that pressure affects the saturation vapor pressure. The Kelvin equation in Chapter 1 is a good example where the capillary pressure makes this difference, to which one can also add the disjoining pressure. It is then possible to show that the liquid will vaporize from the thicker part of the film and condense on the thinner part of the film.

Lin et al. (1996) compiled the information in Table 7.3. Small drops of about 5 μl were allowed to spread on a clean glass slide (an example of a rough surface) and on freshly cleaved mica. If the spreading was too fast, then no measurements could be taken. When the drop did not spread, as much as 5 hours time was allotted to check if it would spread any farther. The equipment did not allow any measurements of the contact angle below 3°. The liquids were all wetting, as the substrates were high energy.

It is evident from the table that viscosities remaining equal, liquids spread faster on a rough surface than on a smooth surface, and the more volatile liquids spread faster than less volatile liquids. For polymer melts on mica, the only mechanism available is that of surface diffusion. It is observed there that the lower molecular weight polystyrene spreads faster. The surface diffusivity of polydimethylsiloxane appears so high that the surface roughness makes almost no difference. Lin et al. (1996) quenched the high molecular weight polystyrene and saw using scanning electron microscopy that the drops were actually spreading, in that they showed precursor films, even though they did not appear to move over a long time.

The main reason for postulating slip is to be able to solve fluid mechanical problems involving contact lines and many such problems have been solved using both asymptotic analysis and numerical methods. Hocking (1979) observed that the fluid mechanical effect of the contact line singularity is spread over a wide area, but when this singularity is eliminated by using slip, the hydrodynamic influence of the contact line shrinks to a very small domain. The existence of the small slip length makes problems very difficult to solve numerically, and numerical methods will not be considered here. However, the method of matched asymptotic expansions has proven to be very useful in obtaining solutions to the equations of motion in problems involving dynamic contact angles. It is worth examining the standard, and extremely illuminating texts on this method (Cole,

TABLE 7.3
Rates of spreading of drops of different liquids

Liquid	Boiling point (°C)	Viscosity (× 10³ Pa·sec)	Wetting behavior on	
			Mica	Glass
Deionized water	100	0.89	Wetting	Wetting
			Spreads very fast with dynamic contact angle reaching 3° in a few seconds	
Butyl acetate	125	0.69	Same as above	
Squalane	176	21.7	Nonwetting, 5 hour wait	Wetting spreads slowly, data by Nieh et al. (1996)
n-Hexadecane	287	3.0	Practically nonwetting, 5 hour wait	Wetting, spreads too fast
Dibutyl phthalate	350	16.5	Nonwetting, 5 hour wait	Wetting spreads slowly, data by Nieh et al. (1996)
Polystyrene (monodispersed, molecular weight 300, spreading at 160°C)	Nonvolatile		Fast wetting, like water	Not checked
Polydimethylsiloxane (molecular weight 17,000)	Nonvolatile		Wetting, spreads slowly, rates by Lin et al. (1996) on mica only slightly slower than on glass	
Polystyrene (monodispersed, molecular weight 4000, spreading at 160°C)	Nonvolatile		Nonwetting, 5 hour wait	Wetting spreads slowly, data by Lau and Burns (1973)

1968; van Dyke, 1975), which is of great importance in fluid mechanics. A simple illustration is given below.

A thin liquid film lies on the solid surface which forms the floor of a narrow horizontal slit. Through the slit, air is blown at a steady rate. The air is seen to exert a constant shear stress on the liquid surface, thus the film thickness varies linearly with the distance from the leading edge, which is also the contact line (Derjaguin et al., 1944; Levich, 1962). Very close to the contact line the profile changes to retain the equilibrium contact angle at the contact line. The equations of motion and continuity under the lubrication theory approximation reduce to (Neogi, 1982) (see Problem 7.13)

$$\frac{\partial H}{\partial \tau} = -\frac{\partial}{\partial X}\left[H\left\{(H^2 + \varepsilon^2)\frac{\partial^3 H}{\partial X^3} + N_{Ca}H\right\}\right], \qquad (7.92)$$

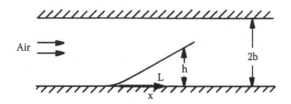

FIGURE 7.13 Arrangement of the blown-off liquid (L) film. The slit width is $2b$ and h is the local film thickness. The thickness h is shown on an exaggerated scale.

where $H = h/b$, $X = x/b$, $\varepsilon^2 = (3\mu D_s/n_L kT)/b^2$, $\tau = \gamma t/2\mu b$, and the capillary number $N_{Ca} = 9\mu_o V/2\gamma$. The arrangement showing the thickness h and the coordinate x in the horizontal direction can be seen in Figure 7.13. The term ε is due to the slip velocity and is a small quantity. V is the mean air velocity and μ_o is the viscosity of the air. The initial and boundary conditions are written as

(1) $\partial H/\partial X \rightarrow \theta(\tau)$ as $X \rightarrow \infty$. This condition arises out of experimental conditions and will not be imposed.

(2) $\theta(0) = \theta_o$, the initial slope.

(3) $\partial H/\partial X = \lambda$ for $X = X_o$ (i.e., the equilibrium contact angle is retained at the contact line). The approximation that $\tan \lambda \approx \lambda$ for small contact angles, assumed here, has been made.

(4) $H = 0$ for $X = X_o$, at the contact line.

(5) $X_o(0) = 0$. The initial position of the contact line forms the origin.

In solving this problem with the matched asymptotic expansions it is seen that there are two driving forces, that is, the air drag and the curvature. The air drag acts everywhere and leads to a slope of θ. However, near the contact line region the curvature effects overwhelm this force to change the slope from θ to λ. Thus a bulk or an outer region can be identified where the air drag dominates and only conditions (1) and (2) apply. To obtain an asymptotic expansion, one rescales the variable X at distances far from the contact line X_o to $p = [(X - X_o)/u(\varepsilon)]$. Here, u is large; in fact, it becomes infinite as $\varepsilon \rightarrow 0$. Thus, to examine the profile away from the contact line it is sufficient to analyze the region where p is comparable to 1. One chooses H as $H \sim uH_o + u_1H_1 + \ldots$, such that as $\varepsilon \rightarrow 0$, $u_1(\varepsilon)/u \rightarrow 0$ (i.e., $u_1H_1 \ll uH_o$). Substituting this series into Equation 7.92 and using the rescaled variable p for small ε, one has as $\varepsilon \rightarrow 0$:

$$\frac{\partial H_o}{\partial \tau} = -N_{Ca}\frac{\partial}{\partial p}[H_o^2] . \tag{7.93}$$

The coefficient of H_o in the series is found to be u if a nontrivial equation such as Equation 7.93 is to be obtained. Equation 7.93 can be solved by the separation of variables method to obtain as one solution

$$H_o = \theta p \tag{7.94}$$

$$\theta = \frac{1}{\dfrac{1}{\theta_o} + 2N_{Ca}\tau} . \tag{7.95}$$

uH_o constitutes the first approximation to the outer solution. Note that a similar substitution has been made in the boundary conditions and the limit $\varepsilon \to 0$ has been taken to get the boundary conditions for H_o.

Just as the outer solution stresses the role of the air drag, the inner solution (i.e., the solution in the vicinity of the contact line) has to be dominated by the slip and the curvature effects, and boundary conditions (3) through (5) apply. The series chosen is

$$H \sim \varepsilon \hat{H}_o + \varepsilon \left(\frac{dX_o}{d\tau}\right)\hat{H}_1 + \dots, \tag{7.96}$$

where $dX_o/d\tau$, the rate of movement of the contact line, is assumed to go to zero as $\varepsilon \to 0$. $\hat{H}_o$ and $\hat{H}_1$ are functions of $s = (X - X_o)/\varepsilon$, a rescaled variable that allows one to examine the region close to the contact line. Substituting the rescaled variables into Equation 7.96 and taking the limit $\varepsilon \to 0$, one has

$$\frac{\partial}{\partial s}[\hat{H}_o \, (\hat{H}_o^2 + 1) \frac{\partial^3}{\partial s^3} \hat{H}_o] = 0 . \tag{7.97}$$

Similarly the boundary conditions reduce to $(\partial \hat{H}_o / \partial s) = \lambda$ and $\hat{H}_o = 0$ at $s = 0$. The solution to Equation 7.97 becomes

$$\hat{H}_o = \lambda s \tag{7.98}$$

when subject to boundedness. Now, substituting Equation 7.96 into rescaled Equation 7.92, using Equation 7.97 to eliminate the leading term, and dividing the entire expression by $dX_o/d\tau$, one has

$$-\frac{\partial}{\partial s} \hat{H}_o = -\frac{\partial}{\partial s}[\hat{H}_o \, (\hat{H}_o^2 + 1)\frac{\partial^3}{\partial s^3}\hat{H}_1] \tag{7.99}$$

on taking the limit $\varepsilon \to 0$. Similarly, suitable steps lead to conditions that $(\partial \hat{H}_1 / \partial s) = 0$ at $s = 0$ and $\hat{H}_1(0) = 0$. The solution subject to Equation 7.98 and the condition of boundedness is

$$\hat{H}_1 = \frac{1}{2\lambda^3}(\lambda^2 s^2 - 1)\tan^{-1}(\lambda s) + \frac{s}{2\lambda^2} - \frac{s}{2\lambda}\ln(1 + \theta^2 s^2) + Bs^2 \dots, \tag{7.100}$$

where B is a constant of integration.

Now, if the inner solution $\varepsilon \hat{H}_o + \varepsilon (dX_o / d\tau) \hat{H}_1 + ...$ and the outer solution $uH_o + ...$ have been correctly constructed they should match. That is, when extrapolated into an intermediate region they should agree. A suitable coordinate for the intermediate region is chosen here as z, where $p = \eta z/u$ and $s = \eta z/\varepsilon$. η is a small quantity such that as $\varepsilon \to 0$, $\eta/u \to 0$ and $\eta/\varepsilon \to \infty$. Consequently $p \to 0$ and $s \to \infty$ when z is on the order of 1, and the two solutions are forced to go into the domains of one another. The difference between the two solutions is examined:

$$uH_o - \left[\varepsilon \hat{H}_o - \varepsilon \left(\frac{dX_o}{d\tau}\right) \hat{H}_1\right] = \frac{\eta z}{\dfrac{1}{\theta_o} + 2N_{Ca}\tau}$$

$$- \left[\lambda \eta z + \frac{dX_o}{d\tau}\left\{-\frac{\eta z}{\lambda^2} \ln|\frac{1}{\varepsilon}| + \left(\frac{\pi}{4\lambda} + B\right)\frac{\eta^2 z^2}{\varepsilon} + \textit{higher order terms...}\right\}\right]$$

The largest of the terms omitted are proportional to $\eta \ln \eta$. Obviously the leading terms cancel if

$$\frac{1}{\dfrac{1}{\theta_0} + 2N_{Ca}\tau} - \lambda + \frac{1}{\lambda^2}\frac{dX_o}{d\tau}\ln|\frac{1}{\varepsilon}| = 0$$

$$B = -\frac{\pi}{4\lambda}$$

The first equality provides the rate of spreading as

$$\frac{dX_o}{d\tau} \sim \lambda^2 [\ln|\frac{1}{\varepsilon}|]^{-1} (\lambda - \theta) , \qquad (7.101)$$

which may be integrated subject to the initial condition. The remainder is ignored as it is now small, on the order of

$$[\ln|\frac{1}{\varepsilon}|]^{-1}\eta\ln\eta ,$$

and provides the magnitude of the small errors in the estimated profiles and the rates of spreading. Note that both $[\ln|\frac{1}{\varepsilon}|]^{-1}$ and $\eta\ln\eta \to 0$ as $\varepsilon \to 0$.

In the matching process it is observed that the outer solution is contained in the inner solution, hence the latter provides the "overall" solution. It should be noted that the justification for the choice of asymptotic expansions, scaling variables, etc., lies in the fact that the two asymptotic expansions match. Note that

Equation 7.101 is very similar to Equation 7.88. In fact, the use of slip leads to the same mathematical expression as observed when a cutoff is employed. This is not surprising because the liquid ahead of the cutoff does not exert a force or backpressure on the cutoff point and hence must be slipping perfectly according to how the problem has been set up.

This process, where inner, outer, and sometimes intermediate regions are required, is also important in numerical solutions, which are generally used to address forced spreading. A coarse grid is sufficient for the main flow, but a fine grid is needed near the contact line. Use of slip in forced spreading poses a greater problem. If a slip boundary condition is used (see Silliman and Scriven, 1980; Tilton, 1988), then it requires that the slip velocity assume values close to the basic flow velocity at the contact line. If slip velocities can acquire such high values, then it poses an unsatisfactory question as to why it is not detected in other cases. de Gennes and coworkers (Brochard and de Gennes, 1979, 1992; Brochard-Wyart et al., 1992) have claimed for years that polymer melts effectively slip when they untangle as they flow near a solid surface. Some of their calculations indicate that very high slip velocities may indeed occur. Some have suggested that under forced spreading conditions the true dynamic contact angle is 180°, when it becomes unnecessary to use a slip (Mahadevan and Pomeau, 1999; Pismen and Nir, 1982). Others (Neogi and Adib, 1985) suggest that the equilibrium contact angle cannot be retained at the contact line under forced flow without leading to an unstable contact line region. Contact line instability under dynamic conditions, in general, appears to attract considerable attention from investigators, although a complete view is lacking at present.

Finally, we note that the speed of wetting at times can be unacceptably low because of large viscous drag exerted by the solid substrate. In various places (Problem 2.19, Figure 2.13) we have pointed out that if the solid substrate is rough, air can be trapped inside this roughness, for both patterned and random roughness. When the air is trapped it is lot easier to move liquids in the form of drops (Miwa et al., 2000) or through channels (Ou et al., 2004). This is of great interest in situations where small amounts of liquids need to be moved on microchips.

12. THIN AND ULTRATHIN FILMS

So far we have been downplaying the role of the precursor films in the spreading process. The precursor films eventually thin into adsorbed layers that are fully mobile under surface diffusion and hence do not require the use of slip. The entire problem could be solved and then a cutoff made at a point suitably selected as the "contact line." Even if such a solution can be constructed, and it seems not to be possible at this time, a great many practical problems remain. Most surfaces are rough, with roughness greater than the thickness of thin/ultrathin films. It is difficult to deal with thin films here. Nonwetting liquids can be made to spread, but cannot support thin films and the adsorption is minimal. Further, forced spreading rates can be much higher than the rates at which the thin film spreads spontaneously ahead of the bulk. As a result, the slip/cutoff proves to be very useful. Even in the case of spontaneous spreading of a wetting liquid on a smooth surface where the precursor film exists, it does no harm to use slip/cutoff to decouple the two regions if the intention is to find the wetting rate of the bulk liquid. The reason why such a decoupling can be carried out is that the thin film moves ahead faster than the bulk, as discussed earlier.

Nevertheless, situations exist where the spreading of thin films is important on its own. Better resolution than optical microscopy, with about ±14 μm error in linear measurement shows that a thin precursor film exists ahead of this region, a film sufficiently thin (below 100 nm) that disjoining pressure plays a role (Bascom et al., 1964). Lopez et al. (1976) used Hamaker-type disjoining pressure as a driving force and showed that such contact lines spread as $t^{1/2}$ instead of the slower $t^{1/10}$ in the bulk liquid.

Even further ahead lies the region of ultrathin films. In general, they appear to be thick adsorbed films, the rate of movement of which is governed by an effective surface diffusion (Heslot et al., 1989; O'Connor et al., 1996). If a very small source is layered on a solid surface initially and it spreads under surface diffusion, then the governing equation (from Equation 6.2) is

$$\frac{\partial \Gamma}{\partial t} = \frac{D_s}{r} \frac{\partial}{\partial r} \left(r \frac{\partial \Gamma}{\partial r} \right) , \qquad (7.102)$$

which has the solution (Neogi, 1996)

$$\Gamma = \frac{M}{8\pi D_s t} \exp\left(-\frac{r^2}{4 D_s t} \right) , \qquad (7.103)$$

where M is the total moles deposited. If the smallest concentration that can be measured occurs at r^*, then at large times $r^* \propto \sqrt{t}$, which is the same time dependence as seen in thin films spreading under Hamaker-type disjoining pressure.

A very peculiar kind of spreading has also been observed, in that the liquid profile is stratified, going up in steps of one molecular layer thickness (Fraysee et al., 1993). Quantitative models exist (de Gennes and Cazabat, 1990).

Finally, de Gennes (1985) predicted that for a thin drop of wetting liquid where the surface ahead of the drop remains dry, the drop will eventually stop spreading and reach an equilibrium configuration, which he called "pancake shape." Experimental verification has been reported by Cazabat et al. (1994). The calculated value of the spreading coefficient from de Gennes theory is about 0.06 mN/m, very small but positive, as it should be for a wetting liquid (see Chapter 2). Others have also discovered such an equilibrium configuration and called the system "nonwetting," which is also not unreasonable (Min et al., 1995). Based on Table 7.3, one can also say that such drops may not have reached equilibrium.

It appears that for answers to the questions of what the structure of the actual contact line is under dynamic conditions and what causes it to spread, we will have to move from phenomenology to molecular physics. Early work goes back to Thompson and Robbins (1989). Recent works by van Remoortere et al. (1999)

and Voue et al. (2000) emphasize that problems in wetting kinetics using molecular dynamics simulations are solvable and need to be investigated.

Equations of motion of individual molecules are governed by Newton's laws. For the ith particle

$$m_i \frac{d^2 x_i}{dt^2} = F_i \, , \qquad (7.104)$$

where m_i is the mass, x_i is the position, t the time, and F_i is the force. The force is made up of the sum of all intermolecular attractions, and in this case not only from all other liquid molecules but also those of the solid. As the molecules of the solid substrate do not move, their equations of motion need not be solved, but the solid can be crystalline or amorphous. Hence some attention needs to be paid to how they are organized in space.

A typical order of magnitude for the "number of molecules" of interest comes from the number in a mole, or Avogadro's number, that is, on the order of 10^{24}! That many differential equations cannot be solved. van Remoortere et al. (1999) consider no more than 10^4 molecules (liquid and solid combined) and solve the equations of motion (for liquids) for 5×10^{-10} seconds in real time. Because very few molecules are used, the drop is two dimensional, almost like a cardboard cutout, because it is very thin in the perpendicular direction. Yet the results are very novel. The very tiny drop spreads, showing diffusion-like behavior. Waves are seen on the surface which are reminiscent of thermocapillary waves, suggesting that we may be observing a fundamental mechanism of drop movement at this scale.

REFERENCES

GENERAL REFERENCES

Edwards, D.A., Brenner, H., and Wasan, D.T., *Interfacial Transport Process and Rheology*, Butterworths-Heinemann, Boston, 1991.

Exerowa, D. and Kruglyakov, P.M., *Foams and Foam Films*, Elsevier, Amsterdam, 1998.

Ivanov, I.B. (ed.), *Thin Liquid Films*, Marcel Dekker, New York, 1988.

Leal, L.G., *Laminar Flow and Convective Transport Processes: Scaling Principles and Asymptotic Analysis*, Butterworths-Heinemann, Boston, 1992.

Levich, V.G., *Physicochemical Hydrodynamics*, Prentice-Hall, Englewood Cliffs, N.J., 1962.

Russel, W.B., Saville, D.A., and Schowalter, W.R., *Colloidal Dispersions*, Cambridge University Press, Cambridge, 1989.

Slattery, J.C., *Interfacial Transport Phenomena*, Springer-Verlag, New York, 1990.

van de Ven, T.G.M., *Colloidal Hydrodynamics*, Academic Press, New York, 1989.

TEXT REFERENCES

Acrivos, A. and Lo, T.S., Deformation and breakup of a single slender drop in an extensional flow, *J. Fluid Mech.*, 86, 641, 1978.

Aris, R., *Vectors, Tensors, and the Basic Equations of Fluid Mechanics,* Dover, New York, 1989; first published by Prentice-Hall, Englewood Cliffs, N.J., 1962, p. 193.

Azbel, D., *Two-Phase Flows in Chemical Engineering,* Cambridge University Press, Cambridge, 1981.

Bascom, W.D., Cottington, R.L., and Singleterry, C.R., Dynamic surface phenomena in the spontaneous spreading of oils on solids, in *Contact Angle, Wettability, and Adhesion,* Fowkes, F.M. (ed.), American Chemical Society, Washington, D.C., 1964, p. 355.

Benkreira, H. (ed.), *Thin Film Coating,* Royal Society of Chemistry, Cambridge, 1993.

Bhattacharji, S. and Savic, P., Real and apparent non-Newtonian behavior in viscous pipe flow of suspension driven by a fluid pressure, in *Heat Transfer and Fluid Mechanics Institute Proceedings,* Charwat, A.F., et al. (eds.), Stanford University Press, Stanford, Calif., 1965, p. 248.

Bird, R.B., Stewart, W.E., and Lightfoot, E.N., *Transport Phenomena,* 2nd ed., John Wiley, New York, 2002.

Blake, T.D. and Ruschak, K.J., A maximum speed of wetting, *Nature,* 282, 489, 1979.

Boussinesq, J., Existence of a superficial viscosity in the thin transition layer separating one liquid from the other contiguous fluid, *Comp. Rend.,* 282, 983, 1035, 1124, 1913.

Boussinesq, J., Existence of surface viscosity in the thin, transition layer separating a liquid from a contiguous fluid, Application of surface viscosity formulas to the surface of a spherical drop falling with uniform velocity into a fluid of less specific gravity, Velocity of the fall of a spherical drop into a viscous fluid of less specific gravity, *Ann. Chim.,* 29, 349, 357, 364, 1914.

Brochard, F. and de Gennes, P.G., Shear-dependent slippage at a polymer/solid interface, *Langmuir,* 8, 3033, 1992.

Brochard, F. and de Gennes, P.G., Conformations of melt polymers in very small pores, *J. Phys.,* 40, L-399, 1979.

Brochard-Wyart, F., de Gennes, P.G., and Pincus, P., Suppression of sliding at the interface between incompatible polymer melts, *C. R. Acad. Sci. Paris,* 314 II, 873, 1992.

Cazabat, A.M., Fraysse, N., Heslot, F., Levinson, P., Marsh, J., Tiberg, F., and Valignat, M.P., Pancakes, *Adv. Colloid Interface Sci.,* 48, 1, 1994.

Calbo, L.J. (ed.), *Handbook of Coatings Additives,* vols. 1 and 2, Marcel Dekker, New York, 1992.

Clift, R., Grace, J.R., and Weber, M.E., *Bubbles, Drops, and Particles,* Academic Press, New York, 1978.

Cole, J.D., *Perturbation Methods in Applied Mechanics,* Blaisdell, Waltham, Mass., 1968.

Coyle, D.J., Macosko, C.W., and Scriven, L.E., The fluid dynamics of reverse roll coating, *AIChE J.,* 36, 161, 1990.

de Gennes, P.G., Wetting: statics and dynamics, *Rev. Mod. Phys.,* 57, 827, 1985.

de Gennes, P.G. and Cazabat, A.M., Spreading of a stratified incompressible droplet, *C. R. Acad. Sci. Paris,* 310 II, 1601, 1990.

Derjaguin, B.V. and Levi, S.M., *Film Coating Theory,* Focal Press, London, 1964.

Derjaguin, B., Strakhovsky, G., and Malysheva, D., Measurement of the viscosity of wall-adjacent boundary layers of liquids by the blow-off method, *Acta Phys. Chim.,* 19, 541, 1944.

Duffy, B.R. and Wilson, S.K., A third-order differential equation arising in thin-film flows and relevant to Tanner's law, *Appl. Math. Lett.,* 10, 63, 1997.

Dussan, E.B. and Davis, S.H., On the motion of a fluid-fluid interface along a solid surface, *J. Fluid Mech.,* 65, 71, 1974.

Edwards, D.A., Brenner, H., and Wasan, D.T., *Interfacial Phenomena and Rheology,* Butterworths-Heinemann, Boston, 1991.

Exerowa, D. and Kruglyakov, P.M., *Foams and Foam Films,* Elsevier, Amsterdam, 1998.

Fraysee, N., Valignat, M.P., Cazabat, A.M., Heslot, F., and Levinson, P., The spreading of layered microdroplets, *J. Colloid Interface Sci.,* 158, 27, 1993.

Frumkin, A.N. and Levich, V.G., Effect of surface-active substances on movements at the boundaries of liquid phases, *Zhur. Fiz. Khim.,* 21, 1183, 1947.

Gutoff, E.B., Cohen, E.D., and Kheboian, G.I., , *Coating and Drying Defects: Troubleshooting Operating Problems,* Wiley Interscience, New York, 1995.

Haberman, W.L. and Sayre, R.M., , *Motion of Rigid and Fluid Spheres in Stationary and Moving Liquids Inside Cylindrical Tubes,* U.S. Dept. of Navy, David Taylor Model Basin, Report no. 1143, 1958.

Hansen, R.S. and Toong, T.V., Interface behavior as one fluid completely displaces another from a small-diameter tube, *J. Colloid Interface Sci.,* 36, 410, 1971.

Happel, J. and Brenner, H., *Low Reynolds Number Hydrodynamics,* Martinus Nijhoff, The Hague, 1983; first published by Prentice-Hall, Englewood Cliffs, N.J., 1965.

Heslot, F., Cazabat, A.M., and Levinson, P., Dynamics of wetting of tiny drops: ellipsometric study of the late stages of spreading, *Phys. Rev. Letts.,* 62, 1286, 1989.

Higgins, B.G., Silliman, W.J., Brown, K.A., and Scriven, L.E., Theory of meniscus shape in film flows. A synthesis, *Ind. Eng. Chem. Fundam.,* 16, 393, 1977.

Hocking, L.M., A moving fluid interface. Part 2. The removal of the force singularity by a slip flow, *J. Fluid Mech.,* 77, 209, 1979.

Hoffman, R.L., A study of the advancing interface. I. Interface shape in liquid-gas systems, *J. Colloid Interface Sci.,* 50, 228, 1975.

Huh, C. and Scriven, L.E., Hydrodynamic model of steady movement of a solid/liquid/fluid contact line, *J. Colloid Interface Sci.,* 35, 85, 1971.

Ivanov, I.B. and Dmitrov, D.S., Thin film drainage, in *Thin Liquid Films,* Ivanov, I.B. (ed.), Marcel Dekker, New York, 1988, p. 379.

Jacazio, G., Probstein, R.F., Sonin, A.A., and Yung, Y., Electrokinetic salt rejection in hyperfiltration through porous material: theory and experiment, *J. Phys. Chem.,* 76, 4015, 1972.

Jiang, T.-S., Oh, S.-O., and Slattery, J.C., Correlation for dynamic contact angle, *J. Colloid Interface Sci.,* 69, 74, 1979.

Joye, J.-L., Hirasaki, G.J., and Miller, C.A., Asymmetric drainage in foam films, *Langmuir,* 10, 3174, 1994.

Kistler, S.F., Hydrodynamics of wetting, in *Wettability,* Berg, J.C. (ed.), Marcel Dekker, New York, 1993, p. 311.

Kistler, S.F. and Schweizer, P.M. (eds.), *Liquid Film Coating: Scientific Principles and Their Technological Implications,* Chapman and Hall, London, 1997.

Kistler, S.F. and Scriven, L.E., Coating flows, in *Computational Analysis of Polymer Processing,* Pearson, J.R.A. and Richardson, S.M. (eds.), Applied Science, London, 1983, p. 243.

Koh, W.-H. and Anderson, J.L., Electroosmosis and electrolyte conductance in charged microcapillaries, *AIChE J.,* 21, 1176, 1975.

Landau, L.D. and Levich, V.G., Dragging of a liquid by a moving plate, *Acta Phys. Chim. URSS,* 17, 42, 1942.

Landau, L.D. and Lifshitz, E.M., *Fluid Mechanics,* Addison Wesley, Reading, Mass., 1959, p. 235.

Lau, W.W.Y. and Burns, C.M., Kinetics of spreading. Polystyrene melts on plane glass surfaces, *J. Colloid Interface Sci.,* 45, 295, 1973.

Levich, V.G., *Physicochemical Hydrodynamics*, Prentice-Hall, Englewood Cliffs, N.J., 1962.

Lin, C.-M., Ybarra, R.M., and Neogi, P., Three- and two-dimensional effects in wetting kinetics, *Adv. Colloid Interface Sci.*, 67, 185, 1996.

Lopez, J., Miller, C.A., and Ruckenstein, E., Spreading kinetics of liquid drops on solids, *J. Colloid Interface Sci.*, 56, 460, 1976.

Marrion, A.R. (ed.), *The Chemistry and Physics of Coatings*, Royal Chemical Society, Cambridge, 1994.

Mannheimer, R.J. and Schecter, R.S., An improved apparatus and analysis for surface rheological measurements, *J. Colloid Interface Sci.*, 32, 195, 1970.

Mahadevan, L. and Pomeau, Y., Rolling droplets, *Phys. Fluids*, 11, 2449, 1999.

Min, B.G., Choi, J.W., Brown, H.R., Yoon, D.Y., O'Connor, T.M., and Jhon, M.S., Spreading characteristic of thin liquid films of perfluoropolyalkylethers on solid surfaces. Effects of chain-end functionality and humidiy, *Tribology Letts.*, 1, 225, 1995.

Miwa, M., Nakajima, A., Fujishima, A., Hashimoto, K., and Watanabe, T., Effects of the surface roughness on sliding angles of water droplets on superhydrophobic surfaces, *Langmuir*, 16, 5754, 2000.

Möbius, D. and Miller, R. (eds.), *Drops and Bubbles in Interfacial Research*, Elsevier, Amsterdam, 1998.

Mysels, K.J., Shinoda, K., and Frankel, S., , *Soap Films*, Pergamon Press, New York, 1959.

Neogi, P., The film "blow-off" experiments, *J. Colloid Interface Sci.*, 89, 358, 1982; Volume 89, no. 2 (1982), in the article, "The film 'blow-off' experiments," by P. Neogi, pages 358–365, *J. Colloid Interface Sci.*, 90, 554, 1982.

Neogi, P., Tears-of-wine and related phenomena, *J. Colloid Interface Sci.*, 105, 94, 1985.

Neogi, P., Dynamics of an adsorbed patch and a model for spreading of films of ultralow thicknesses, *J. Chem. Phys.*, 105, 8909, 1996.

Neogi, P., The relation between dynamics of wetting and nonwetting liquids, *J. Chem. Phys.*, 115, 3342, 2001.

Neogi, P. and Adib, F., Dynamic contact lines in rotating liquids, *Langmuir*, 1, 747, 1985.

Neogi, P. and Miller, C.A., Spreading kinetics of a drop on a smooth solid surface, *J. Colloid Interface Sci.*, 86, 525, 1982a.

Neogi, P. and Miller, C.A., Spreading kinetics of a drop on a rough solid surface, *J. Colloid Interface Sci.*, 92, 338, 1982b.

Nieh, S.-Y., Ybarra, R. and Neogi, P., Wetting kinetics of polymer solutions. Experimental observations, *Macromolecules*, 29, 320, 1996.

O'Connor, T.M., Back, Y.R., Jhon, M.S., Min, B.G., Yoon, D.Y., and Karis, T.E., Surface diffusion of thin perfluoropolyalkylether films, *J. Appl. Phys.*, 79, 5788, 1996.

Ou, J., Perot, B., and Rothstein, J.P., Laminar drag reduction in microchannels using ultrahydrophobic surfaces, *Phys. Fluids*, 16, 4635, 2004.

Paul, S. (ed.), *Surface Coatings: Science and Technology*, Chichester, New York, 1996.

Payne, L.E. and Pell, W.H., The Stokes flow problem for a class of axially symmetric bodies, *J. Fluid Mech.*, 7, 529, 1960.

Pismen, L. and Nir, A., Motion of a contact line, *Phys. Fluids*, 25, 3, 1982.

Proudman, I. and Pearson, J.R.A., Expansions at small Reynolds numbers for the flow past a sphere and a circular cylinder, *J. Fluid Mech.*, 2, 237, 1957.

Quéré, D., Fluid coating on a fiber, *Annu. Rev. Fluid Mech.*, 31, 347, 1999.

Reynolds, O., On the theory of lubrication and its application to Mr. Beauchamp Tower's experiments, including an experimental determination of the viscosity of olive oil, *Philos. Trans. R. Soc. Lond. A*, 177, 157, 1886.

Rose, W. and Heins, R.W., Moving interfaces and contact angle rate-dependency, *J. Colloid Sci.*, 17, 39, 1962.

Ruckenstein, E. and Dunn, C.S., Slip velocity during wetting of solids, *J. Colloid Interface Sci.*, 59, 135, 1977.

Ruschak, K.J., Coating flows, *Annu. Rev. Fluid Mech.*, 17, 65, 1985.

Ruschak, K.J. and Scriven, L.E., Developing flow on a vertical wall, *J. Fluid Mech.*, 81, 305, 1977.

Sheludko, A., Thin liquid films, *Adv. Colloid Interface Sci.*, 1, 391, 1967.

Schonhorn, H., Frisch, H.L., and Kwei, T.K., Kinetics of wetting of surfaces by polymer melts, *J. Appl. Phys.*, 37, 4967, 1966.

Scriven, L.E., Dynamics of a fluid interface. Equation of motion for Newtonian surface fluids, *Chem. Eng. Sci.*, 12, 98, 1960.

Silliman, W.J. and Scriven, L.E., Separating flow near a static contact line: slip at a wall and shape of a free surface, *J. Comput. Phys.*, 34, 287, 1980.

Singh, G., Hirasaki, G.J., and Miller, C.A., Effect of material properties on the drainage of symmetric, plane parallel, mobile foam films, *J. Colloid Interface Sci.*, 184, 92, 1996.

Slattery, J.C., *Interfacial Transport Phenomena*, Springer-Verlag, New York, 1990.

Slattery, J.C., *Momentum, Energy and Mass Transfer in Continua*, 2nd ed., Robert E. Krieger, Huntington, N.Y., 1981.

Stone, H.A. and Kim, S., Microfluidics: basic issues, applications and challenges, *AIChEJ*, 47, 1250, 2001.

Tabeling, P., *Introduction to Microfluidics*, transl. by S. Chen, Oxford University Press, 2005.

Tanner, L.H., The spreading of silicone oil drops on horizontal surfaces, *J. Phys. D Appl. Phys.*, 12, 1473, 1979.

Taylor, T.D. and Acrivos, A., On the deformation and drag of a falling viscous drop at low Reynolds number, *J. Fluid Mech.*, 18, 466, 1964.

Thompson, P.A. and Robbins, M.O., Simulations of contact-line motion: slip and the dynamic contact angle, *Phys. Rev. Letts.*, 63, 766, 1989.

Tilton, J.N., The steady movement of an interface between two viscous liquids in a capillary tube, *Chem. Eng. Sci.*, 43, 1371, 1988.

van Dyke, M., *Perturbation Methods in Fluid Mechanics*, Parabolic Press, Stanford, Calif., 1975.

van Oene, H., Chang, Y.F., and Newman, S., Rheology of wetting by polymer melts, *J. Adhesion*, 1, 54, 1969.

van Remoortere, P., Mertz, J.E., Scriven, L.E., and Davis, H.T., Wetting behavior of a Lennard-Jones system, *J. Chem. Phys.*, 110, 2621, 1999.

Voinov, O.V., Hydrodynamics of wetting, *Fluid Dynamics*, 11, 714, 1976.

Voue, M., Rovillard, S., de Connick, J., Valignat, M.P., and Cazabat, A.M., Spreading of liquid mixtures at the microscopic scale: a molecular dynamics study of the surface-induced segregation process, *Langmuir*, 16, 1428, 2000.

Vrentas, J.S. and Duda, J.L., Effect of axial diffusion of vorticity on flow development in circular conduits. Part II. Analytical solution for low Reynolds numbers, *AIChE J.*, 13, 97, 1967.

Wayner, P.C., Jr., The effect of interfacial mass transport on flow in thin liquid films, *Colloids Surfaces*, 52, 71, 1991.

Weatherburn, C.A., *Differential Geometry of Three Dimensions*, Cambridge University Press, Cambridge, 1961.

PROBLEMS

7.1 Obtain the expressions for the two principal curvatures in $z = h(x)$. Here, $R = xe_x + ye_y + he_z$ and $dB = dx\, e_x + dy\, e_y + dh\, e_z$, where all unit vectors are constants. Calculate first $\mathbf{n}$ and then ∇_s.

7.2 (a) Instead of the series $\psi_i \sim \psi_i^{(0)} + N_{Re}\psi_i^{(1)} + N_{Re}N_{Ca}\psi_i^{(2)}\cdots$, in the problem in Section 6, consider using a series

$$\psi_i \sim \psi_i^{(0)} + N_{Re}\psi_i^{(1)} + u\psi_i^{(2)} + \cdots , \tag{7.2.i}$$

where u is small but unknown. For Equation 7.2.i to be an asymptotic series, it is also required that as $N_{Re}, u \to 0$, $u/N_{Re} \to 0$. The momentum equation is

$$N_{Re}D^4\psi_i = E^4\psi_i, \tag{7.2.ii}$$

where

$$D^4\psi_i = \frac{1}{r^2\sin^2\theta}\frac{\partial(\psi_i, E^2\psi_i)}{\partial(r,\theta)} - \frac{2E^2\psi_i}{r^2\sin^2\theta}\left(\frac{\partial\psi_i}{\partial r}\cos\theta - \frac{1}{r}\frac{\partial\psi_i}{\partial r}\sin\theta\right)$$

and the Jacobian is given by

$$\frac{\partial(\psi_i, E^2\psi_i)}{\partial(r,\theta)} = \frac{\partial\psi_i}{\partial r}\frac{\partial E^2\psi_i}{\partial\theta} - \frac{\partial\psi_i}{\partial\theta}\frac{\partial E^2\psi_i}{\partial r}.$$

When the series of Equation 7.2.i is substituted into Equation 7.2.ii, the result is

$$N_{Re}D^4\psi_i^{(0)} + N_{Re}^2 D^4\psi_i^{(1)} + N_{Re}u D^4\psi_i^{(2)} + \cdots$$
$$= E^4\psi_i^{(0)} + N_{Re}E^4\psi_i^{(1)} + uE^4\psi_i^{(2)} + \cdots \tag{7.2.iii}$$

$$E^4\psi_i^{(0)} = 0. \tag{7.2.iv}$$

On taking the limit $N_{Re} \to 0$, $u \to 0$ is obtained. Substituting Equation 7.2.iv into Equation 7.2.iii and dividing by N_{Re}, one has

$$D^4\psi_i^{(0)} + N_{Re}D^4\psi_i^{(1)} + uD^4\psi_i^{(2)} + \cdots$$

$$= E^4 \psi_i^{(1)} + \left(\frac{u}{N_{Re}} \right) E^4 \psi_i^{(2)} + \ldots \ . \qquad (7.2.\text{v})$$

On taking the limit $N_{Re}, u \to 0$ and using the property that $u/N_{Re} \to 0$ in that limit, Equation 7.2.v becomes

$$D^4 \psi_i^{(0)} = E^4 \psi_i^{(1)} \ . \qquad (7.2.\text{vi})$$

Continue this process to obtain

$$E^4 \psi_i^{(2)} = 0 \ , \qquad (7.2.\text{vii})$$

under the assumption that $N_{Re}^2/u \to 0$ as $N_{Re}, u \to 0$.

(b) Examine the normal stress balance

$$(\bar{N}_A^{(0)} + N_{Re} \ \bar{N}_A^{(1)} + u\bar{N}_A^{(2)}) - (\bar{N}_B^{(0)} + N_{Re} \ \bar{N}_B^{(1)} + u\bar{N}_B^{(2)})$$

$$= -\frac{2}{N_{Ca}}[\bar{H}^{(0)} + v_1 \bar{H}^{(1)} + v_2 \bar{H}^{(2)}] \text{ on } r = 1 + v_1 \xi^{(1)} + v_2 \xi^{(2)} \ , \quad (7.2.\text{viii})$$

where $\bar{N}_A^{(j)}$ are the normal stresses obtained from $\psi_i^{(j)}$, v_1 and v_2 are small quantities with $v_2 \ll v_1$ and $\bar{H}^{(0)}$ is the curvature of the shape given by $r = 1$, $v_1 \bar{H}^{(1)}$ of $v_1 \xi^{(1)}$ and $v_2 \bar{H}^{(2)}$ of $v_2 \xi^{(2)}$. (It can be verified from Table 7.2 that the order of will be the same as H.) Taylor's series expansion can be used to simplify Equation 7.2.viii: for example,

$$\bar{N}_A^{(0)} \Big|_{r=1+v_1\xi^{(1)}+v_2\xi^{(2)}} \sim \bar{N}_A^{(0)} \Big|_{r=1} + \frac{\partial \bar{N}_A^{(0)}}{\partial r} \Big|_{r=1} (1 - 1 - v_1\xi^{(1)} - v_2\xi^{(2)})$$

$$+ \frac{1}{2} \frac{\partial^2 \bar{N}_A^{(0)}}{\partial r^2} \Big|_{r=1} (1 - 1 - v_1\xi^{(1)} - v_2\xi^{(2)})^2 \cdots \ .$$

Substituting such expansions into Equation 7.2.viii and using the method shown in part (a), where in addition to the limits $N_{Re}, u \to 0$, it is also necessary to take the limit $N_{Ca} \to 0$, show that

$$\bar{N}_A^{(0)} - \bar{N}_B^{(0)} = -\frac{2\bar{H}^{(0)}}{N_{Ca}} \ , \qquad (7.2.\text{ix})$$

which is the result obtained in Equation 7.34. In obtaining the second term, show that $\psi_i^{(1)}$ affects the shape of the drop only when $\upsilon_1 = N_{Re}N_{Ca}$, and thus

$$\bar{N}_A^{(1)} - \bar{N}_B^{(1)} = -2\bar{H}^{(1)} . \tag{7.2.x}$$

To obtain the next term, show that it is necessary to have $u = N_{Re}N_{Ca}$ and $\upsilon_2 = N_{Re}N_{Ca}^2$, and thus

$$\left(\bar{N}_A^{(2)} - \xi^{(1)} \frac{\partial \bar{N}_A^{(0)}}{\partial r} - \bar{N}_B^{(2)} + \xi^{(2)} \frac{\partial \bar{N}_B^{(0)}}{\partial r} \right)\Big|_{r=1} = -2\bar{H}^{(2)} . \tag{7.2.xi}$$

At this level the curvature $\bar{H}^{(2)}$ is affected both by $\psi_i^{(2)}$ and by the distortion $\xi^{(1)}$. Note that it is not absolutely essential for u, υ_1, and υ_2 to have the above values. However, any other values would either lead to inconsistencies or would necessitate calculations of higher order terms in $\psi_i^{(j)}$ or $\xi^{(j)}$. In the latter case, some of the terms are repeated and become redundant. Lastly, it is necessary to stress that in the asymptotic expansions for ψ_i, ξ, and H, the order of a term like $u\psi_i^{(2)}$ is given by the small quantity u, and the order of $\psi_i^{(2)}$ itself is on the order of one. That is, the functions $\psi^{(j)}$, $\xi^{(j)}$, $\bar{H}^{(j)}$, $\bar{N}_A^{(j)}$, $\bar{N}_B^{(j)}$, etc., have been chosen to be on the order of one and hence are unaffected by the limits N_{Re}, u, $N_{Ca} \to 0$. One exception here is $\bar{H}^{(0)}$, which from Equation 7.2.ix should be infinite. But when it is worked out, the problem actually disappears.

(c) Provide the order of the next terms in the two series

$$\psi_i \sim \psi_i^{(0)} + N_{Re}\psi_i^{(1)} + N_{Re}N_{Ca}\psi_i^{(2)} + \cdots \tag{7.2.xii}$$

$$\xi \sim N_{Re}N_{Ca}\xi^{(1)} + N_{Re}N_{Ca}^2\xi^{(2)} + \cdots . \tag{7.2.xii}$$

7.3   A jet leaves a tube of radius d under creeping flow.

(a) What would be the profile of the jet as the first approximation?

(b) Assuming that the solution to this problem is known (Vrentas and Duda, 1967), how would a normal stress balance be used to obtain the first correction to the assumed profile? What is the necessary condition under which this correction is valid?

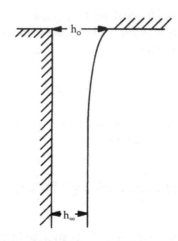

PROBLEM 7.4 The shape of the falling film.

7.4   Consider the flow of a thin film down a wall, as shown in the accompanying figure. Write the relevant equations of motion under the lubrication theory approximation, including the effect of surface tension. Obtain the thickness h far from the entrance. If h is the thickness at any distance from the entrance which can be approximated as

$$h \sim h_\infty + \varepsilon h' + \cdots ,$$

where $\varepsilon = (h_o - h_\infty)/h_\infty$ is small, obtain a differential equation for the correction term h'. Here, h_o is the film thickness at the entrance. What are the boundary conditions for h'? Integrate the differential equation subject to the boundary conditions to obtain h' as completely as possible.

7.5   An air bubble is subjected to extensional flow characterized by $v_z = Gz$ and constant p, as shown in the accompanying figure. The bubble is long and slender, the surface of which is described as $r = R(z)$.

(a)   From the continuity equation and with the velocity $\left(v_r - \dfrac{dR}{dz} v_z\right)$ on the bubble surface zero, obtain v_r.

(b)   From the normal stress balance at the surface, show that

$$z\frac{dR}{dz} - R\left(\frac{p}{2\mu G} - 1\right) = \frac{\gamma}{2\mu G} .$$

Integrate the equation to obtain $R(z)$ (Acrivos and Lo, 1978).

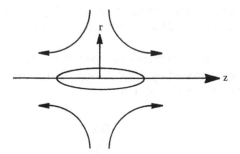

PROBLEM 7.5 The shape of the air bubble.

7.6   In the problem discussed in Section 7 on the role of surface active agents on the terminal velocity of a drop, if the surface active agent is insoluble in both phases and resides only at the interface between the drop and the infinite medium, how will Equation 7.60 be modified?

The term e can be determined through the following detailed procedure. Starting with the solution

$$\psi_B = \sin^2 \theta \left(\frac{Ar^4}{10} - \frac{1}{2} Br + Cr^2 + \frac{D}{r} \right)$$

$$\psi_A = \sin^2 \theta \left(\frac{Er^4}{10} - \frac{1}{2} Fr + Gr^2 + \frac{H}{r} \right),$$

satisfy the condition of boundedness at $r = 0$ for ψ_A and conditions (i) through (iii) given after Equation 7.22. Satisfy next Equation 7.18, where

$$\nabla_s \gamma \cdot t = a^{-1} \frac{\partial \gamma}{\partial \Gamma} \frac{\partial \Gamma}{\partial \theta} .$$

Use now a theorem by Payne and Pell (1960) to calculate the drag force as

$$\lim_{r \to \infty} \frac{8 \pi \mu_B \, r (\psi_B - \psi_B^\infty)}{\omega^2} ,$$

where ω is the radial distance perpendicular to the direction of U (i.e., $r \sin\theta$ in this case) and ψ_B^∞ is the value of the stream function at $r \to \infty$ (boundary condition (i)). Then calculate U and e. Is the normal stress balance satisfied? (The pressures are $p_B = [p_{B_0} + \mu_B B \cos\theta/r^2 - \rho_B rg \cos\theta]$ and $p_A = [p_{A0} - 2E\mu_A\, r \cos\theta - \rho_A rg \cos\theta]$.)

7.7 In the same class of problems as above, the surface concentration $\Gamma_0 + \Gamma'$ may be controlled by the diffusion from the infinite fluid to the interface. Under those conditions

$$J^* = \frac{D\Delta c}{\delta},$$

where δ is the thickness of the boundary layer and $\Delta c = c_0 - c(a)$.

Since the rates of adsorption-desorption are not controlling, the surface concentration is in equilibrium with $c(a)$. Thus

$$\Delta c = \left(\frac{\partial c}{\partial \Gamma}\right) \Gamma',$$

where $\partial c/\partial \Gamma$ is the inverse of the slope of the adsorption isotherm. Obtain the expression for e in this case. Evaluate e when the adsorption is given by the Langmuir isotherm. Is there a maximum or a minimum in e with respect to c_0? Assume an ideal solution.

7.8 Can the effect of surface active agents, as given in Section 7, describe the observations made in the deep channel viscometer experiments? Discuss.

7.9 As discussed in Chapter 3, measurement of the velocity of a charged colloidal particle in an applied electric field can be used to estimate the surface potential and hence the magnitude of the electrical interaction between nearby particles in a colloidal dispersion.

(a) Suppose that an electric field E_x is applied in a direction parallel to a flat charged solid surface. The tangential component of the equation of motion leads, in the absence of a tangential pressure gradient, to the following equation applicable in the double-layer region

$$0 = \rho_e E_x + \mu \frac{d^2 v_x}{dz^2}, \tag{7.9.i}$$

where z is the coordinate in the direction perpendicular to the surface. Poisson's equation in this case is

$$\nabla^2 \psi = \frac{\partial^2 \psi}{\partial z^2} = -\frac{\rho_e}{\varepsilon} . \qquad (7.9.ii)$$

We note that $\partial^2 \psi / \partial x^2$ has been neglected in this equation since the potential varies rapidly in the z direction in the double-layer region. Substitute Equation 7.9.i into Equation 7.9.ii and integrate twice to show that the velocity v_x at the outer limit of the double layer is given by

$$v_x = -\frac{\varepsilon E_x \zeta}{\mu} , \qquad (7.9.iii)$$

where ζ is the potential at the z coordinate where $v_x = 0$, the "plane of shear."

(b) Use the result of (a) to estimate the steady velocity of a spherical nonconducting particle of radius a when $\kappa a \gg 1$ (i.e., in the limit of a very thin double layer). Use a moving coordinate system such that the fluid flows past the fixed sphere with a velocity U far away from the sphere, as shown in Figure 7.2. The general solution for the fluid flow around a solid sphere for this case can be obtained from Problem 7.6 with the boundary condition at $r \sim a$ given by Equation 7.9.iii. That is, the local tangential velocity just outside the (thin) double layer is given by Equation 7.9.iii, with $E_x = \partial \psi / \partial x$, the local tangential field. Note that the potential distribution outside the double layer is found by solving $\nabla^2 \psi = 0$ subject to the condition that the electric field is uniform far from the particle.

Calculate the drag force on the sphere using the method given in Problem 7.6. From physical arguments, assign an appropriate value to this drag force and obtain U in terms of known parameters. Neglect the role of gravity. The result is the electrophoretic velocity in the von Smoluchowski limit referred to in Chapter 3.

(c) Rearrange the physics of this problem as solved in the parts (a) and (b) above to show that this is a matched asymptotic solution with $(\kappa a)^{-1}$ as the small parameter.

7.10 Consider a porous membrane separating two reservoirs, both containing the same 1:1 electrolyte at concentration c_o. An electric potential difference $\Delta V = (V$ at $z = L) - (V$ at $z = 0)$ is applied across the membrane, causing the liquid to flow through the membrane of thickness L. The walls of pores of radius a are charged, leading to a total potential of $u = \phi(z) + \psi(r, z)$, where z is the axial and r is the radial direction in a pore. The pore is assumed to be slender. Hence in most instances the variations in z direction can be ignored in the presence of variations in the r direction. This leads to a Poisson equation of the form

$$\frac{1}{r}\frac{\partial}{\partial r}\left(r\frac{\partial\psi}{\partial r}\right) = -\frac{1}{\varepsilon}\rho_e \ . \tag{7.10.i}$$

ϕ is due to the imposed field $-\Delta V/L$. The flow is assumed to be mainly in the z direction, leading to

$$0 = \rho_e E_z + \frac{\mu}{r}\frac{\partial}{\partial r}\left(r\frac{\partial}{\partial r}v_z\right) \ , \tag{7.10.ii}$$

where the field $E_z = -\Delta V/L$. Show that Equations 7.10.i and 7.10.ii lead to the solution in the form

$$v_z = -\frac{\Delta V\varepsilon}{\mu L}(\psi - \psi_s) \ , \tag{7.10.iii}$$

where ψ_s is the value of the potential at the wall. For a small charge effect and no overlap of the double layers at the center of the pore, Equation 7.10.i becomes the Debye-Huckel equation:

$$\frac{1}{r}\frac{d}{dr}\left(r\frac{d\psi}{dr}\right) = \kappa^2\psi \ . \tag{7.10.iv}$$

Show that the solution to Equation 7.10.iv leads to

$$v_z = \frac{\Delta V\varepsilon\psi_s}{\mu L}\left[1 - \frac{I_o(\kappa r)}{I_o(\kappa a)}\right] . \tag{7.10.v}$$

Comment on the direction of flow. For $\kappa a = 5$, plot the term inside the square brackets in Equation 7.10.v as a function of $\xi = r/a$. This phenomenon is called electroosmosis. There are many who have contributed to the present development. We mention the work by Koh and Anderson (1975), where both theory and comparison with experiments are available.

Show that the average velocity over the pore is

$$<v_z> = \frac{\Delta V \varepsilon \psi_s}{\mu L} \left[1 - \frac{2}{\kappa a} \frac{I_1(\kappa a)}{I_0(\kappa a)} \right].$$ (7.10.vi)

Now, if in addition to the potential difference a pressure difference Δp = (p at $z = L$) – (p at $z = 0$) is also applied, then a term corresponding to Hagen-Poiseuille flow of $-\Delta p a^2/8\mu L$ is added to the right-hand side in Equation 7.10.vi. For zero flow rate, what should be the pressure drop? Show that with decreasing values of radius a, the pressure drop increases more rapidly than the concomitant potential drop. Often it is Δp that is imposed and ΔV, called the streaming potential, is measured.

7.11   For the case described in Problem 7.10, investigate the situation where there is no applied electric field, but the concentration of the electrolyte in one reservoir (I) is greater than in the other (II), $c_I > c_{II}$. The pressures are similarly unequal, $p_I > p_{II}$. In the pores, both flow and mass transfer take place. The double layers in the pores overlap, as a result $c_{\pm} = c(z).\exp(\mp\bar\psi)$, where $\bar\psi = ev\psi/kT$, as usual. The fluxes through the pore are

$$N_{\pm} = uc_{\pm} - D \left[\frac{\partial c_{\pm}}{\partial z} \mp c_{\pm} \frac{\partial \bar\varphi}{\partial z} \right].$$ (7.11.i)

Note that one approximation (slender pore) has been made in writing the gradient of the electrostatic potential. All notations follow Problem 7.10. Here u is assumed to be a parabolic velocity profile with an unknown coefficient: $u = 2V(1 - \xi^2)$. $V = Q/A$ and $A = \pi a^2$, where Q is the flow rate. Show that Equation 7.11.i simplifies to

$$N_{\pm} = uc_{\pm} - D \exp(\mp\bar\psi) \left[\frac{dc}{dz} \mp c \frac{d\bar\varphi}{dz} \right].$$ (7.11.ii)

Since there is no applied electric potential difference, there is no current and hence $<N_+> = <N_-> = N_s$, where the angular brackets denote averages over the cross section and N_s is the flux of the salt, a constant at steady state.

    Average the two equations in Equation 7.11.ii over the cross-section by first multiplying with 2ξ and integrating over ξ from 0 to 1. Take into account that only u and $\bar\psi_s$ are dependent on r (or ξ). Show using the condition of zero current that

$$N_s = +2cQK_1 - 2ADK_3(dc/dz) + 2ADcK_4(d\bar\varphi/dz)$$ (7.11.iii)

$$0 = -2cQK_2 + 2ADK_4(dc/dz) - 2ADcK_3(d\overline{\varphi}/dz) , \qquad (7.11.\text{iv})$$

where

$$K_1 = 4\int_0^1 \cosh \overline{\psi}(1 - \xi^2)\xi d\xi; \quad K_3 = 2\int_0^1 \cosh \overline{\psi}\xi \, d\xi$$

$$K_2 = 4\int_0^1 \sinh \overline{\psi}(1 - \xi^2)\xi d\xi; \quad K_4 = 2\int_0^1 \sinh \overline{\psi}\xi \, d\xi .$$

Equations 7.11.iii and 7.11.iv are two coupled first-order ordinary differential equations in dc/dz and $d\overline{\psi}/dz$. These can be separated and integrated to yield

$$\frac{c_{\text{II}}}{c_I} = 1 - R = \frac{e^{Pe/k_1}}{(1 + \kappa_2) + \kappa_2 e^{Pe/\kappa_1}} , \qquad (7.11.\text{v})$$

where R is called the rejection coefficient and the streaming potential is

$$\overline{\varphi}_{\text{II}} - \overline{\varphi}_1 = \kappa_4 \ln\left[\frac{(1 - \kappa_2)e^{-Pe/\kappa_1} + \kappa_2}{e^{-Pe/\kappa_1}}\right] , \qquad (7.11.\text{vi})$$

where

$$\kappa_1 = (K_3^2 - K_4^2)(K_1K_3 - K_2K_4); \quad \kappa_2 = K_3/(K_1K_3 - K_2K_4)$$

$$\kappa_3 = K_4/K_2; \quad \kappa_4 = K_4/K_3.$$

In addition, the Peclet number $Pe = QL/AD$, and $c_{\text{II}} = N_a/2Q$. More simplifications are possible. Where double layers overlap, the solution to the Poisson-Boltzmann equation

$$\frac{1}{r}\frac{d}{dr}\left(r\frac{d\psi}{dr}\right) = \kappa^2 \sinh \psi$$

for a cylindrical pore at high values of surface potentials becomes $\overline{\psi} \approx \overline{\psi}_s$. This allows simplifications for $K_i s$ and $\kappa_i s$. Eventually, show that the maximum rejection is given by

$$R_{max} = 1 - \text{sech } \bar{\psi}_s . \qquad (7.11.\text{vii})$$

The phenomenon above describes reverse osmosis. Here, a liquid with a higher concentration of electrolyte is driven through a membrane (pore sizes are on the order of 3 nm), and the exiting solvent contains much less electrolyte. The reverse osmosis membranes usually have an R_{max} of 0.995. Calculate the corresponding $\bar{\psi}_s$. The present treatment is from Jacazio et al. (1972), who also compared theory to experiments.

7.12 Consider the drainage of a thin, circular, horizontal liquid film having a half-thickness h that is independent of radial position r, but does vary with time t.

(a) Assuming that inertial, gravitational, and disjoining pressure effects are negligible, show by integrating the equation of motion over the film and invoking overall continuity that

$$V = -\frac{dh}{dt} = \frac{h}{r}\frac{\partial}{\partial r}(rU) - \frac{h^3}{3\mu}\frac{1}{r}\frac{\partial}{\partial r}\left(r\frac{\partial P}{\partial r}\right), \qquad (7.12.\text{i})$$

where U is the surface velocity and P is the difference between the local film pressure and the pressure in the meniscus outside the film. Other notations are the same as in Section 7.9. Integrate this equation with suitable boundary conditions at $r = 0$ to obtain

$$\frac{\partial P}{\partial r} = \frac{3\mu}{h^3}\left[hU(r) - \frac{r}{2}V\right]. \qquad (7.12.\text{ii})$$

(b) If the surfactant is insoluble, show that a surfactant mass balance at the film surface yields

$$\frac{\partial \Gamma}{\partial t} + \frac{1}{r}\frac{\partial}{\partial r}(rU\Gamma) = D_s \frac{1}{r}\frac{\partial}{\partial r}\left(r\frac{\partial \Gamma}{\partial r}\right). \qquad (7.12.\text{iii})$$

Simplify this equation by using the quasistatic approximation $\left[\frac{\partial \Gamma}{\partial t} \approx 0\right]$ and taking $\Gamma = \Gamma_o + \Gamma_1$, where $\Gamma_1 \ll \Gamma_0$. Integrate this equation with the appropriate boundary condition at $r = 0$ to obtain

$$\frac{\partial \Gamma}{\partial r} = U(r)\frac{\Gamma_o}{D_s} \ . \tag{7.12.iv}$$

(c) The tangential component of the interfacial momentum balance takes the form

$$\frac{\partial \gamma}{\partial \Gamma}\frac{\partial \Gamma}{\partial r} + (\eta + \varepsilon)\frac{\partial}{\partial r}\left[\frac{1}{r}\frac{\partial}{\partial r}(rU)\right] = h\left(\frac{\partial P}{\partial r}\right) . \tag{7.12.v}$$

The first and second terms of this equation represent the effects of gradients of surface tension and surface viscosity, respectively. Using the velocity profile across the thickness of the film found in deriving the result of (a), show that the last term in the equation represents the shear stress at $z = h$.

(d) Use the results of (a) and (b) to replace $\partial P/\partial r$ and $\partial \Gamma/\partial r$ in Equation 7.12.v. Show that it can be rearranged to give

$$r^2\frac{d^2U}{dr^2} + r\frac{dU}{dr} - r^2U\left(\frac{3\mu}{h} - \frac{\Gamma_o(d\gamma/d\Gamma)}{D_s(\eta+\varepsilon)}\right) = -\frac{3\mu V}{2h^2}r^3 . \tag{7.12.vi}$$

The left side of the equation has the form of a modified Bessel function of zero order. After defining suitable dimensionless variables, this equation can be solved to obtain Equation 7.81 (Singh et al., 1996). The boundary conditions employed in the solution are that $U = 0$ for $r = 0$ and $dU/dr = 0$ at the film periphery $r = R$. The latter condition is based on the results of numerical simulation of drainage with nonuniform thickness. It was found that dU/dr was zero at the barrier ring (i.e., the point of minimum thickness between the central part of the film and the surrounding meniscus).

7.13 A capillary tube of radius a is placed vertically in a large pool of liquid (see Figure 1.10). Assuming perfect wetting, and that at any time t during the approach to equilibrium the average velocity is given by the expression for steady state flow in a circular tube, derive a differential equation to predict how liquid height h depends on t from the time the tube is initially inserted until equilibrium is reached. Specify the boundary condition or conditions and show that the result is

$$1 - \frac{h}{h_e} = \exp\left[-\left(\frac{h+\theta}{h_e}\right)\right] ,$$

where $\theta = \dfrac{\rho g a^2 t}{\mu}$, a = tube radius, and h_e = equilibrium height.

7.14   A drop is spreading slowly on a solid surface. At large times of spreading, the drop appears to be thin and flat and the lubrication theory
approximation can be used. Obtain the appropriate equations of motion
(do not neglect gravity).

(a) Integrate the continuity equation to show that the integral form is

$$\frac{\partial h}{\partial t} = -\frac{1}{r}\frac{\partial}{\partial r}\left(r\int_0^h v_r dz \right) , \qquad (7.14.\text{i})$$

where r is the radial direction, z is the direction normal to the solid
surface in a cylindrical coordinate system, and $z = h(r, t)$ describes
the shape of the drop.

(b) From the above equations, obtain a partial differential equation
describing the profile h. Write the initial and the boundary conditions to solve for the case where surface tension effects can be
neglected and the drop spreads under its own weight. Assume that
the solution is of the form $r_o^{-2} Z\,(r/r_o)$, where $r_o(t)$ describes the
position of the contact line. Show that the pressure gradient $\partial p/\partial r$
at $r = r_o$ is infinite. Why?

(c) When the drops are small, the effects of gravity can be neglected
and the drops are driven by surface tension. In this case, show only
that

$$\frac{dr_o}{dt} \propto t^{-(9/10)} \qquad (7.14.\text{ii})$$

for a figure of revolution

$$2H = \frac{\dfrac{d^2h}{dr^2}}{\left[1+\left(\dfrac{dh}{dr}\right)^2\right]^{3/2}} + \frac{\dfrac{1}{r}\dfrac{dh}{dr}}{\left[1+\left(\dfrac{dh}{dr}\right)^2\right]^{1/2}} . \qquad (7.14.\text{iii})$$

(d) If the basic shape of the drop is that of a spherical cap, obtain the
expression for the volume V as a function of r_o and contact angle
θ. Obtain the approximate form when θ is small and use this in

Equation 7.14.ii to derive the Hoffman-Voinov-Tanner rule that $dr_o/dt \propto \theta^3$ (Equation 7.88).

7.15   Derive Equation 7.86. Consider Figure 7.13, where it can be assumed that $h \ll 2b$. Solve the equations of motion for the air flowing through the slit. Since the viscosity of air is much less than the viscosity of the liquid, ignore the thickness h and assume no slip at both the walls. Obtain the tangential stress at the liquid-air interface. With these boundary conditions, follow the procedure of Problem 7.12 to obtain the equations of motion under the lubrication theory approximation. Neglect gravity, derive the appropriate integral continuity equation as in Problem 7.12 and nondimensionalize to obtain Equation 7.86.

7.16   Consider the situation discussed in Figure 7.13. Suppose that there is no air current, but that the liquid contains a volatile species that evaporates and gives rise to a variation in surface tension. As a first approximation, assume $\partial\gamma/\partial x$ to be a constant and formulate the problem under the lubrication theory approximation as in Problem 7.14. Nondimensionalize the resulting equation for the profile using $[(1/\gamma)(\partial\gamma/\partial x)]^{-1}$ as the macroscopic length scale. Solve the problem using the method of matched asymptotic expansions as in Section 10, both when $\partial\gamma/\partial x$ is positive and when it is negative (Neogi, 1985).

7.17   The concept of order is very important in asymptotic analysis since one is always involved in comparing the magnitudes of various terms. Consider functions $f(\varepsilon)$ and $g(\varepsilon)$, when as $\varepsilon \to 0$, $f \to 0$ and $g \to 0$.

It is necessary to find which is "greater." If in the limit $\varepsilon \to 0$, $f/g \to \infty$, then ord $(f) >$ ord (g) or $g \sim o(f)$. If in the limit $\varepsilon \to 0$, $f/g \to$ a finite value, then ord $(f) =$ ord (g) or $f \sim o(g)$ and $g \sim o(f)$. If in the limit $\varepsilon \to 0$, $f/g \to 0$, then ord $(f) <$ ord (g) or $f \sim o(g)$.

Show that $\exp[-1/\varepsilon] \sim o(\varepsilon^\upsilon)$ and $\varepsilon^\upsilon \sim O\left(\left[\ln\left|\frac{1}{\varepsilon}\right|\right]^{-1}\right)$, for $\upsilon > 0$, however large or small the values of υ. Consequently one has $\exp[-1/\varepsilon]$ "transcendentally small" and $\left[\ln\left|\frac{1}{\varepsilon}\right|\right]^{-1}$ "transcendentally large" compared to ε.

8 Size, Shape, Structure, Diffusivity, and Mass Transfer

1. INTRODUCTION

In this chapter we examine some issues in mass transfer. The reader has already been introduced to some of the key aspects. In Chapter 3 (Section 7), flocculation kinetics of colloidal particles is considered. It shows the importance of diffusivity in the rate process, and in Equation 3.72, the Stokes-Einstein equation, the effect of particle size on diffusivity is observed, leading to the need to study sizes, shapes, and charges on colloidal particles, which is taken up in Chapter 3 (Section 4). Similarly some of the key studies in mass transfer in surfactant systems — dynamic surface tension, surface elasticity, contacting and solubilization kinetics — are considered in Chapter 6 (Sections 6, 7, 10, and 12 with some related issues considered in Sections 11 and 13). These emphasize the roles played by different phases, which are characterized by molecular aggregation of different kinds. In anticipation of this, the microstructures are discussed in detail in Chapter 4 (Sections 2, 4, and 7). Section 2 also includes some discussion on micellization-demicellization kinetics.

In this chapter we focus our attention on key optical methods and nuclear magnetic resonance (NMR), which have been indispensable for quantitative descriptions of size and structure, and diffusivity, where size and structure play an important role. Whereas in the previous chapters we have tended to focus on the overall dynamics, we concentrate here at the smallest scale needed to understand what the fundamental building blocks are in those systems. With the exception of NMR, the other methods are restricted to transparent systems. This can sometimes be a drawback, as in the study of water-in-crude oil emulsions, which are black in color. These are very important systems industrially and require de-emulsification. NMR techniques for measurement of drop size distributions in such emulsions, while beyond the scope of this chapter, have been reviewed by Peña and Hirasaki (2003).

2. PROBING WITH LIGHT

Light provides a valuable tool for probing the structure of matter. The simplest device is the compound microscope. The objective lens magnifies the object and places its virtual image on the focal plane of the ocular (lens) (Slayter, 1970). Seen through the objective, the final image is infinitely large to the eye, but upside down. From left to right in Figure 8.1 they appear in order: object, objective,

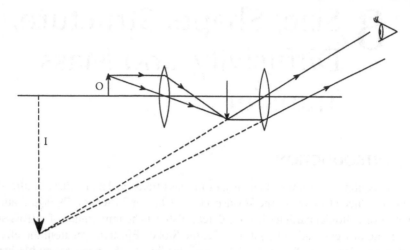

FIGURE 8.1 The formation of the enlarged image is shown using methods from geometrical optics: one ray from the tip of the object O runs parallel to the horizontal up to the lens and then passes through its focal point. The other ray passes undeviated through the center of the lens. The two intersect (when extrapolated) at the tip of the image I.

focal plane of the ocular, ocular, and the eye. The image is infinitely far to the left. There have of course been many improvements. Geometric optics, used to explain magnification, assumes that light can be focused to a point. Because of the wave nature of light, the best that can be achieved is a volume of dimension λ, the wavelength, as the closest approximation to a point. Consequently the microscope begins to falter when the object reaches this dimension, and many colloids and micelles are even smaller than λ. It becomes necessary to look at the wave nature of light and what it does in the presence of small objects.

If light is traveling in the z direction with a speed v, then an electric field is found in the x direction and a magnetic field in the y direction. That is, all three are mutually perpendicular and are shown schematically in Figure 8.2. The electric field is generally the one that is analyzed, since it is easy to locate the magnetic field if the electric field is known. The electric field also fluctuates in time and is hence written as

$$E = E_o e^{-i\omega t}. \tag{8.1}$$

Here, $i = \sqrt{-1}$ and $e^{iq} = \cos(q) + i\sin(q)$, which shows the oscillating character. At any time, the wave train along the z direction (the direction of propagation) will also show oscillating amplitude. Thus a better representation is

$$E = (E_{ox}e^{ik_x x}e_x + E_{oy}e^{ik_y y}e_y)e^{-i\omega t}, \tag{8.2}$$

(a)

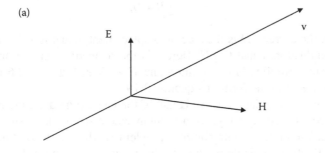

(b)

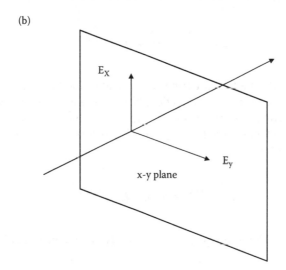

FIGURE 8.2 Schematic of the basic components of light.

where k_x and k_y are wavenumbers in the x and y directions and e_x and e_y are unit vectors in the x and y directions. Generally the wavenumbers are the same in all directions and Equation 8.2 becomes (see Problem 8.1)

$$E = (E' + iE'')e^{i(kr-\omega t)} . \tag{8.3}$$

A phase angle θ is sometimes added in the exponent. In addition, $k = 2\pi/\lambda$, where λ is the wavelength and $v = \lambda\omega/2\pi$. The magnetic field H can be defined in a similar manner. The electric and magnetic field have to satisfy Maxwell equations

$$\nabla \times E = -\partial B/\partial t \tag{8.4}$$

$$\nabla \times H = J + \partial D/\partial t \tag{8.5}$$

$$\nabla \cdot D = \rho_e \tag{8.6}$$

$$\nabla \cdot \boldsymbol{B} = 0, \tag{8.7}$$

where $\boldsymbol{J}$ is the current, $\boldsymbol{D}$ is the electric displacement, equal to $\varepsilon \boldsymbol{E}$, and $\boldsymbol{B}$ is the magnetic induction, equal to $\mu \boldsymbol{H}$. Here ε is the dielectric constant and μ is the magnetic susceptibility. For constant ε and $\boldsymbol{E} = -\nabla \psi$, Equation 8.6 becomes a Poisson equation in the form of Equation 3.19.

The simplest problem to solve is that of the electric field around an electron (represented as a point charge) in a vacuum where radiation of known electric field is incident on it. A more complex problem is when, instead of an electron, we have a fluctuating dipole $\boldsymbol{p} = \boldsymbol{p}_o e^{-i\omega t}$. This is our model for a colloidal particle, microemulsion droplet, micelle, etc. The scattered field is

$$\boldsymbol{E} = -\frac{k^2}{r} \boldsymbol{e}_r \times (\boldsymbol{e}_r \times \boldsymbol{p}_o) e^{i(kr-\omega t)} , \tag{8.8}$$

where r is measured from the scattering center, that is, the location of the dipole, and $\boldsymbol{e}_r$ is the unit vector in the $r-$ direction. Now, it is assumed that the scatterer does not have a permanent dipole, and the dipole is induced through the electric field of the incident radiation $\boldsymbol{E}_o$, or $\boldsymbol{p}_o = \alpha \boldsymbol{E}_o$, where α is the polarizability. Equation 8.8 becomes

$$\boldsymbol{E} = -\frac{\alpha k^2}{r} \boldsymbol{e}_r \times (\boldsymbol{e}_r \times \boldsymbol{E}_o) e^{i(kr-\omega t)} . \tag{8.9}$$

The solution to the Maxwell equations also yields the magnetic field as well:

$$\boldsymbol{H} = -\frac{\alpha k^2}{r} \boldsymbol{e}_r \times \boldsymbol{E}_o e^{i(kr-\omega t)} . \tag{8.10}$$

The radiant energy flux is given by Poynting vector:

$$\boldsymbol{S} = \frac{v}{8\pi} Re(\boldsymbol{E}) \times Re(\boldsymbol{H}) , \tag{8.11}$$

where Re stands for the real part of the vector. What follows is very detailed development involving complex quantities and vector algebra and is omitted. The material is covered in many books which emphasize classical electrodynamics, such as that by Stone (1963), which can be consulted for details. First, the time averaged value of S is taken to yield $<S>$, both of which are in the r direction. Then taking the dot product with $\boldsymbol{e}_r$, we have the irradiance i^*. If we draw a sphere of radius r around the scattering center, then the flux of energy (watts per unit area) at a point is the irradiance there. Note that the flux travels outward in the

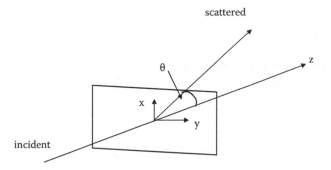

FIGURE 8.3 A view of the frame of reference in scattering.

r direction and perpendicular to the area, which is the area of the surface on the sphere. After considerable work one has

$$i^* = \frac{\alpha^2 k^4 v}{r^2 4\pi} < \hat{E}_{ox}^2 > (1 + \cos^2 \theta) .$$ (8.12)

This is the key result of Rayleigh scattering, that is, scattering by a point particle. Here $\hat{E}_0 = Re(E_0)$; as before, subscript 0 denotes that it is the property of the incident radiation, and the angular brackets indicate the time averages. It has been assumed that the phase angles in the x and y directions are equal and $<\hat{E}_{0x}^2> = <\hat{E}_{0y}^2>$. Lastly, θ is the angle between the incident beam (z direction) and the scattered beam (r direction). The basic geometry is shown in Figure 8.3. The same treatment for the incoming intensity of light leads to

$$I_o = \frac{v}{2\pi} < \hat{E}_{ox}^2 > .$$ (8.13)

Consider now the term α^2. It has two contributions. The first is a steady one due to the solvent. The second is a fluctuating one that arises because colloidal particles drift in and out of the small volume under consideration in the path of the incident light. This volume has to be very small for the fluctuations to be significant. Let $\alpha = <\alpha> + \delta\alpha$, where $<\alpha>$ is the base and $\delta\alpha$ is the fluctuation such that its time averaged value, denoted by the angular brackets, is zero. Then it is possible to show that $<\alpha^2> = <\alpha>^2 + <(\delta\alpha)^2>$. Consequently, if from the scattering from a colloidal suspension we can subtract the contribution from the solvent, then the result is expressed by replacing α^2 in Equation 8.12 with $<(\delta\alpha)^2>$. This is possible only when the suspension is so dilute that the scattering from one particle is not intercepted by another.

The relationship between α and the concentration of scattering centers is derived next (Tanford, 1961). The Lorentz-Lorenz formula (another aspect of electromagnetic theory) is

$$4\pi N\alpha = \frac{3(\tilde{n}^2 - 1)}{\tilde{n}^2 + 2} , \qquad (8.14)$$

where N is the number of scattering centers per unit volume and $\tilde{n} = n/n_o$, with n is the refractive index of the solution and n_o is that of the solvent. It follows that for $\tilde{n}$ close to one,

$$4\pi N\delta\alpha = 2\tilde{n}\left(\frac{d\tilde{n}}{dc}\right)\delta c , \qquad (8.15)$$

where c is the concentration in mass per unit volume. Replacing α^2 in Equation 8.12 with $<(\delta\alpha)^2>$ and invoking Equation 8.15, then multiplying by N and dividing with Equation 8.13, one has

$$\frac{i_\theta}{I_o} = \frac{2\pi^2(1 + \cos^2 \theta)\tilde{n}^2(d\tilde{n}/dc)^2}{\lambda^4 r^2 N} <(\delta c)^2> . \qquad (8.16)$$

The concentration fluctuations are related to fluctuations in free energy F, where for small fluctuations the Taylor series expansion is

$$\delta F = \left(\frac{\partial F}{\partial c}\right)_{T,V} \delta c + \frac{1}{2!}\left(\frac{\partial^2 F}{\partial c^2}\right)_{T,V} (\delta c)^2 +\dots . \qquad (8.17)$$

At equilibrium $(\partial F/\partial c)_{T,V} = 0$, thus

$$\delta F = \frac{1}{2}\left(\frac{\partial^2 F}{\partial c^2}\right)_{T,V} (\delta c)^2 +\cdots . \qquad (8.18)$$

The probability of such a fluctuation is given by the Boltzmann factor $\exp[-\delta F/kT]$, where k is the Boltzmann constant. Truncating the series in Equation 8.18 after the first term, the average is expressed as

$$< (\delta c)^2 >= \frac{\int_0^\infty (\delta c)^2 \exp\ [-\delta F / kT]d(\delta c)}{\int_0^\infty \exp\ [-\delta F / kT]d(\delta c)} = kT / (\partial^2 F / \partial c^2) . \qquad (8.19)$$

All that is left at this stage is the evaluation of $\partial^2 F/\partial c^2$, which is available from solution thermodynamics in the form of

$$\left(\frac{\partial^2 F}{\partial c^2}\right)_{T,V} = \frac{kT}{cN} N_A \left(\frac{1}{M} + 2Bc + 3Cc^2 + \cdots\right), \tag{8.20}$$

where N_A is Avogadro's number, B is the second virial coefficient dependent on two-particle interaction potential, and C is the third virial coefficient, which is rarely needed. Both B and C are defined here in terms of mass concentrations and M is the molecular weight.

Substituting Equations 8.19 and 8.20 into Equation 8.16 and rearranging, one obtains the Rayleigh ratio

$$R_\theta = \frac{r^2 i_\theta}{I_o(1 + \cos^2\theta)} = \frac{Kc}{1/M + 2Bc + 3Cc^2}, \tag{8.21}$$

where the optical constant K is given by

$$K = 2\pi^2 \tilde{n}^2 (\partial \tilde{n}/\partial c)^2 / N_A \lambda^4. \tag{8.22}$$

Equation 8.21 is still not in a form suitable for comparing with experimental data. For that we define turbidity as

$$\tau = -\ln(I/I_o), \tag{8.23}$$

which for low values of turbidity, (i.e., small concentrations of scattering centers) and I close to I_o reduces to

$$\tau \cong 1 - \frac{I}{I_o}. \tag{8.24}$$

I/I_o is the fraction of total light scattered into the forward hemisphere, and is also expressed as

$$\int_0^\pi \frac{i_\theta}{I_o} 2\pi r^2 \sin^2\theta d\theta = \frac{16\pi}{3} R_\theta \cong 1 - \frac{I}{I_o}. \tag{8.25}$$

Combining Equations 8.21, 8.24, and 8.25, one has in terms of quantities that are all experimentally accessible,

$$\frac{Hc}{\tau} = \frac{Kc}{R_\theta} = \frac{1}{M} + 2Bc + 3Cc^2 + \cdots, \tag{8.26}$$

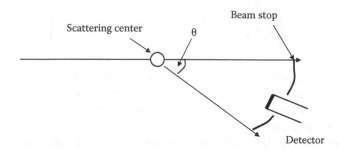

FIGURE 8.4 Schematic of the sampling of scattered light.

where $H = 16\pi K/3$. As indicated in Figure 8.4, the intensity is often measured as a function of position. In fact, this feature is very important in complex systems, as will be discussed later. The turbidity results for micellar solutions of sodium lauryl sulfate (NaLS, another name for SDS) in water, with and without added electrolyte sodium chloride (NaCl), are shown in Figure 8.5. Some effort has been made to subtract the scattering from singly dispersed amphiphiles, as seen in the quantity indicated on the y axis. The lines are extrapolated to stop at the critical micelle concentration (CMC). Hence the last point can be used as the intercept in Equation 8.26. This intercept gives us the molecular weight and hence the aggregation number. The aggregation number appears to increase with the electrolyte concentration, but not with increasing NaLS concentration (because of which the lines in Figure 8.5 remain straight). The slopes are proportional to the second virial coefficient, which are all seen to be positive. Positive second virial coefficients are associated with repulsion between two particles, which we know here to be the electrostatic repulsion. Consequently it is not surprising that the slope and second virial coefficient decrease when the electrolyte concentration is increased. As a measure of this effect (from Figure 8.5), we have with no added electrolyte values for CMC, aggregation number, and second virial coefficients of 234 mg/100 ml, 80 and 245.4 ($\pm$0.5) $\times$ 10^{-4} mole/[g $\cdot$ (100 ml/mg)], respectively. At a NaCl concentration of 0.1 M, they change to 42.2 mg/100 ml, 112 and 8.4 ($\pm$0.5) $\times$ 10^{-4} mole/gm $\cdot$ (100 ml/mg), respectively. Light scattering provides a mass averaged molecular weight, and it is one of the very few methods that does so. Most methods for determining molecular weights give us the mole average value. The two differ in polydispersed systems.

Rayleigh analysis suffers from the fact that an assumption has been made that the particles are points and that the refractive index of the scatterer and the liquid are quite close. In effect

$$\frac{a}{\lambda} << 1 \qquad\qquad (8.27)$$

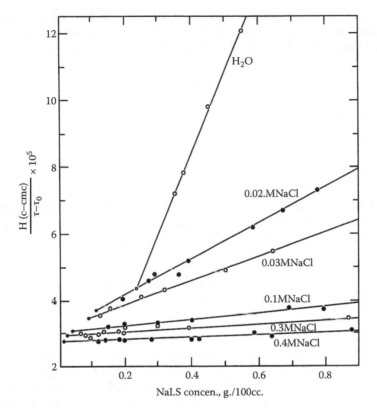

FIGURE 8.5 Light scattering from a dilute solution of sodium lauryl sulfate in a water and salt solution. Data have been plotted using Equation 8.26. Reprinted from Phillips and Mysels (1955). Copyright 1955 American Chemical Society.

and

$$|\tilde{n} - 1| \ll 1 , \tag{8.28}$$

where a is the radius of the scatterer. An improved treatment called Rayleigh-Gans-Debye theory relaxes the limitations somewhat.

The key step in moving from a point scatterer to, say, a scatterer of radius a is that while the detector position is unchanged (r, θ), the different points of the scatterer all send scattered radiation to the detector. However, the paths from different parts of the scatterer all vary in length, a feature that introduces phase lag. When contributions from all points are summed (integrated) one has

$$E = -\frac{k_f^2}{r} e_r \times (e_r \times E_o) \, e^{i(k_f r - \omega_i t)} \overline{\delta\alpha}(q, t) , \tag{8.29}$$

where k_f is the wavenumber of the scattered radiation ("forward"), ω_i is the frequency of the incident radiation, k_i is the incident wave vector (z direction), and k_f is the scattered wave vector (r direction). The scattering vector is

$$q = k_i - k_f, \tag{8.30}$$

which has magnitude

$$q = \frac{4\pi\tilde{n}}{\lambda}\sin\frac{\theta}{2} . \tag{8.31}$$

Note that q is based on a change of wavelength on collision, but the change is considered to be small, such that its average value, the denominator of Equation 8.31, is unchanged. The Fourier transform is defined as

$$\overline{\delta\alpha}(q,\ t) = \int_v d^3r'\ e^{iq\cdot r'}\delta\alpha(r',\ t) . \tag{8.32}$$

If we assume that the fluctuation in the polarizability is located at a point, (i.e., we have a point scatterer), then mathematically $\delta\alpha(r',t) = \delta\alpha(t)\cdot\delta(r')$, where the second on the right-hand side is the Dirac delta function. Using the properties of the delta function, Equation 8.32 can be integrated to give $\delta\alpha\ (q,t) = \delta\alpha(t)$. Substituting $\tilde{n}$ into Equation 8.29, one has Equation 8.9. Equation 8.29 was obtained from Berne and Pecora (1976) after changes in notations have been made to correspond to ours.

As mentioned before, it has been assumed that the frequency of incident radiation ω_i, or wavelength, does not change significantly on scattering. Since the frequency determines the energy, this means we have assumed that there is no significant loss of energy due to collision. This is called quasielastic scattering. Following the same procedure as before, one obtains to the first approximation (Hiemenz and Rajagopalan, 1997)

$$\frac{Kc}{R_\theta} = \left[1 + \frac{16\pi^2}{3\lambda^2}\ R_G^2\ \sin^2\frac{\theta}{2} + \right]\left[\frac{1}{M} + 2Bc + \right], \tag{8.33}$$

where R_G is the radius of gyration. Again, $\tilde{n}$ is assumed to be close to one, however, the range of validity now is enlarged to

$$\frac{a}{\lambda}\left|\tilde{n} - 1\right| \ll 1 . \tag{8.34}$$

Equations 8.27 and 8.28 are not satisfied if $a/\lambda = 0.3$ and $\tilde{n} - 1 = 0.3$, but these values lead to a value of 0.09 for the left-hand side in Equation 8.34, which is satisfied. Thus the Rayleigh–Gans–Debye result can be extended to systems not covered by Rayleigh scattering alone. Further, for $\tilde{n}$ very close to one, much larger values of a, the length scale characterizing the dispersed phase, can be tolerated.

The experimental verification of Equation 8.33 and determination of the radius of gyration is done through the use of Zimm plots. Here, Kc/R_θ is plotted against $\sin^2\theta/2$ at constant c, or concentration c at constant θ, on the same plot by using a single variable $\sin^2\theta/2 + Fc$, where F is an arbitrary constant used to bring the two terms to within the same order of magnitude. Hiemenz and Raja-gopalan (1997) can be consulted for details, which lead to the evaluation of the molecular weight M, second virial coefficient B, and, in addition, the radius of gyration R_G. To calculate these from the data, the limits $c = 0$ and $\theta = 0$ have to be obtained on the graph. The $c = 0$ line can be measured, but the $\theta = 0$ line has to be extrapolated. Now, $\theta = 0$ is the location of the beam stop. As the turbidity here is small, most of the incident radiation passes unchanged and is located at this point. Hence the task of screening a strong background radiation to obtain the scattering at small values of θ, that is, "small angle scattering," is a daunting one that needs good instrumentation, laser light (incident radiation is at the same phase angle), and polarized light (more on that in the next section).

If we know the molecular weight M and the density of the particle, then for a spherical particle it should be possible to predict the radius of gyration R_G. If the predicted value does not compare with the one from the Zimm plot, then the particle is not spherical (Anacker, 1970). Their shapes are determined using results from large angle scattering, as discussed extensively by Davis (1996). Although the results of scattering experiments are very rewarding, they remain among the most difficult to perform.

3. MORE LIGHT

In static light scattering, considered above, if one is interested only in R_G, then Equation 8.33 can be used to calculate dissymmetry:

$$d(\theta) = \frac{R_\theta(\theta)}{R_\theta(180° - \theta)} = 1 + \left(\frac{4\pi}{\lambda\tilde{n}}\right)^2 \frac{R_G^2}{3} \cos(\theta). \qquad (8.35)$$

Dissymmetry has been plotted in Figure 8.6 against θ for sodium dodecyl sulfate (SDS) at 20 g/L and 0.6 M NaCl at three temperatures. The slopes lead to R_G, the values of which are shown there. Also shown are values of R_H, the hydrody-namic radius found by dynamic light scattering, which is discussed later. The fact that the two are very different was used to infer that SDS micelles are cylindrical.

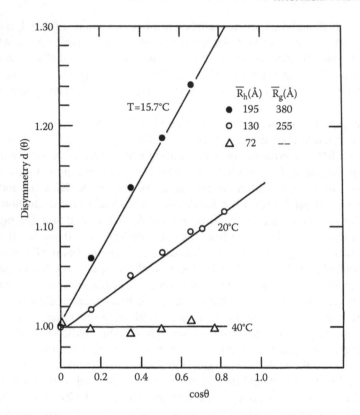

FIGURE 8.6 Plots of dissymmetry versus cos θ for 2 g/dl SDS solution in 0.6 M NaCl at 40°C, 20°C, and 15.7°C. Reprinted from Young et al. (1978) with permission. Copyright 1978 American Chemical Society.

Davis (1996) provides an extensive discussion on scattering theory based on statistical mechanical considerations, where the electromagnetic origins have been abbreviated. Equation 8.33 is actually valid for small scattering vectors defined in Equations 8.30 and 8.31. At larger scattering vectors the differences between spheres, cylinders, and ellipsoids become apparent, as discussed by Davis (1996).

All of these results apply to dilute solutions where the scattering from one particle is not picked up by a second. Working with dilute solutions is, of course, not a problem unless the shape and size change with concentration. The assumption that ñ is close to one needs to be relaxed for inorganic colloids such as gold sols. This was first attempted by Mie (1908) by solving fully the boundary value problem posed by the Maxwell equations. The present-day results and practices are described by Hiemenz and Rajagopalan (1997).

One can use these principles to study the kinetics of flocculation. Assume that the rate of flocculation in a dilute sample is very small, such that over the time scale of interest only single particles (n_1) and doublets (n_2) are observed. Their number densities change according to

$$\frac{dn_1}{dt} = -kn_1^2 \tag{8.36}$$

$$\frac{dn_2}{dt} = -\frac{1}{2}\frac{dn_1}{dt}. \tag{8.37}$$

Integrating, one has

$$n_1 + n_2 = \frac{n_{1o}(2 + n_{1o}kt)}{2(1 + n_{1o}kt)}, \tag{8.38}$$

where initial conditions of $n_1 = n_{1o}$ and $n_2 = 0$ have been used.

If the particles are sufficiently small, such that both single particles and doublets are Rayleigh scatterers, then $n_1 + n_2$ is proportional to turbidity τ and n_{1o} is proportional to the initial turbidity τ_o. Equation 8.38 becomes

$$\frac{1}{2\tau/\tau_o - 1} = 1 + n_{1o}kt. \tag{8.39}$$

Thus $(2\tau/\tau_o - 1)^{-1}$ versus t will give a straight line. At infinite time, τ becomes $\tau_o/2$.

It is assumed here that although the doublets are larger, gravity is still unimportant. If this is not the case, then the rate of settling will exceed the rate of flocculation, $n_2 = 0$ and

$$\tau_o/\tau = 1 + n_{1o}kt. \tag{8.40}$$

$1/\tau$ versus t is shown as straight lines in Figure 8.7 for two emulsions of 1% v/v of toluene in water. One (1) contains a dissolved polymer in the continuous medium and shows slow flocculation and the other (2) contains the dissolved polymer in the dispersed phase and shows fast flocculation.

Energy is lost in transmission not only through scattering, but also by absorption, when it exists. The refractive index is expressed as $n = m(1 - i\kappa)$. When m and κ are both real, κ can be attributed to the sum of scattering and absorption. For small changes in intensity,

$$(I_o - I)/I_o = \varepsilon\ell, \tag{8.41}$$

where ℓ is the length of the path traversed by the ray of light through the solution/suspension. When the loss is due only to scattering, $\varepsilon\ell$ is the turbidity

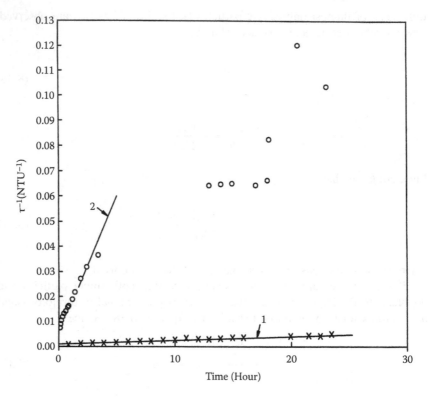

FIGURE 8.7 Inverse turbidity (turbidity in NTU, National Turbidity Unit) versus time from Equation 8.40. Two cases of 1% v/v toluene-in-water emulsion are shown: (1) 0.1 wt. % dextran in water, and (2) only solute is 0.1 wt. % polystyrene in toluene. Reprinted from Rivard et al. (1992). Copyright 1992 with permission from Elsevier.

τ. When it is due solely to absorption, $\varepsilon = \dfrac{4\pi}{\lambda}\dfrac{d\kappa}{dc}c$, which is Beer's law, $d\kappa/dc$ is a constant, and these relations provide a means for measuring concentration c.

4. DIFFRACTION

Terms such as scattering, diffraction, and interference are used by different authors somewhat differently (Lipson and Lipson, 1981). If, instead of an object (even as small as a "point") blocking light, we have light going through a pinhole, the waves produce a diffraction pattern on the screen, as shown in Figures 8.8 and 8.9. Generally, instead of dealing with the field E, it is more conventional to deal with the magnitude of the vector, called the displacement, normalized such that its product with the complex conjugate ("square") provides the intensity. The total radiation at an angle θ in Figure 8.8 is

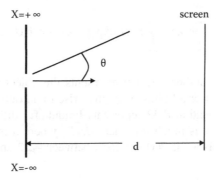

FIGURE 8.8 Basic scheme for diffraction.

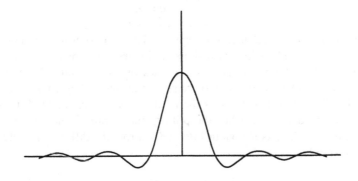

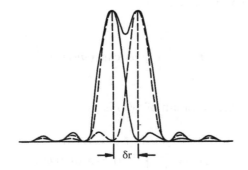

FIGURE 8.9 Diffraction patterns for single (displacement) and twin (intensity) discs of radius *a*.

$$\psi(k) = \int_{-\infty}^{\infty} f(x) \ \exp \ (-i \ kx \ \sin \ \theta) dx \ , \qquad (8.42)$$

where k is the wavenumber and $f(x)$ represents the object which lies on the x plane. A summation over all phase lags gives rise to a Fourier transform, just as in Equation 8.29. Equation 8.42 represents Fraunhofer diffraction, valid when the scale of the object is much less than $\sqrt{\lambda d}$, where d is the distance to the screen. If the object is a slit, $f(x) = 1$ for x between $-a/2$ and $+a/2$ and 0 for all other x, and one has

$$\psi(k) = a \frac{\sin \ (au \ / \ 2)}{au \ / \ 2} \ , \qquad (8.43)$$

where $u = k \sin\theta$. ψ is sketched in Figure 8.9. The first point where the intensity $|\psi(u)|^2$ is zero is $\sin \ \theta_1 = \lambda/a$. Also shown in Figure 8.9 is the case of diffraction from two slits. The plot of intensities for the case of two circular holes of diameter D is very similar. Obviously the two are not resolvable if they come too close to one another. The closest distance is the limit of resolution. Rayleigh postulated that this limit be defined as when the primary maximum of one lies on the first minimum of the other, as shown in Figure 8.9. From the diffraction pattern, this is calculated to be

$$\sin\theta_{min} = 1.22\lambda/D. \qquad (8.44)$$

Equation 8.44 will be revisited shortly.

If one diffraction pattern (Equation 8.42 and Figure 8.8) is allowed a second diffraction, then the resulting image is

$$F(x) = \int_{-\infty}^{\infty} \psi(u) \ \exp \ (-iux) du \ . \qquad (8.45)$$

Now the inverse Fourier transform of $\psi(u)$ is

$$f(x) = \frac{1}{2\pi} \int_{-\infty}^{\infty} \psi(u) \ \exp \ (iux) du \ . \qquad (8.46)$$

Comparing Equations 8.45 and 8.46, we have

$$F(x) = 2\pi f(-x). \qquad (8.47)$$

That is, the Fourier transform is self-inverting, and the scale of the resurrected image is different and turned upside down. Abbe theorized (Lipsom and Lipsom, 1981) that the first lens in an optical microscope gives rise to a diffraction pattern, and the second lens gives rise to another, hence the final image looks the "same" as the object. When there are no lenses, as in x-ray diffraction, only $\psi(u)$ can be determined. Note that u has dimensions of inverse length. Consequently ψ represents an inverse image. Neighboring points in the object are a great distance apart in ψ. This is good news for magnification, but from a practical point of view, we cannot work with very large distances. The images are cutoff by what is called the aperture, which cuts off images in the inverse space and hence images of points closest to one another are lost and resolution is reduced. If α is the angle that the opening of the aperture makes in Figure 8.8, then $\alpha = \theta_{min}$ in Equation 8.44. Equation 8.44 becomes

$$D = \frac{1.22\lambda}{\sin \alpha} \; . \tag{8.48a}$$

A better description of resolution leads to an added factor of two, and to (Lipsom and Lipsom, 1981)

$$D = \frac{0.61\lambda}{\sin \alpha} \; . \tag{8.48b}$$

If the refractive index of the space between the object and the diffraction pattern is μ and not one, then

$$D = \frac{0.61\lambda}{\mu \sin \alpha} \; , \tag{8.48c}$$

which is Rayleigh's formula for the limiting resolution. Because of the term μ in the denominator on the right-hand side, an "oil immersed microscope" has a better resolution. Resolution also improves with decreasing wavelengths. Hence, in optical microscopes, $\lambda \sim 0.5~\mu m$, calculated resolution $\sim 0.3~\mu m$, and actual resolution $\sim 0.3~\mu m$. In x-rays, $\lambda \sim 0.1$ nm, calculated resolution ~ 0.06 nm, and actual resolution is much larger.

Diffraction patterns of periodic structures as seen in lamellar and hexagonal liquid crystals (Figures 4.10 and 4.11) are well known from small-angle x-ray studies (Fontell, 1974). Small angles (less than $10°$) imply small dimensions from the center of the screen in Figure 8.9 (the two can be related to one another through the geometry of the instrument) and hence larger dimensions in the sample, dimensions where the liquid crystalline order is seen. The order manifests itself on the screen as peaks in intensity as we travel outward (increasing angle). They are densest at small angles and cease after a maximum angle. If this maximum occurs at one, then the peaks in lamellar liquid crystals are distributed as

$$\cdots \frac{1}{3} : \frac{1}{2} : \frac{1}{1}$$

and for hexagonal liquid crystals as

$$\cdots\cdots \frac{1}{\sqrt{12}} : \frac{1}{\sqrt{7}} : \frac{1}{\sqrt{4}} : \frac{1}{\sqrt{3}} : 1 \, .$$

That is, the occurrence of peaks quickens as the beam stop is approached. For obvious reasons "1" leads to the smallest repeat distance, or d values. Similar results are available for other forms of liquid crystals, such as cubic, and hence this method provides a means for identifying the type of liquid crystals present. It is important to state here that no structure is observed at large angles, (i.e., at small distances in the sample), hence the term "liquid crystals." That is, the hydrocarbon chain region inside the bilayers or long cylinders is disordered like a liquid, not crystalline.

In Figure 8.10a, the d values in the dilution path in Figure 8.10b are shown for the lamellar phase of bilayer thickness tetradecyltrimethyl ammonium bromide (TTMAB)-pentanol-water. Extrapolated values at zero water give the bilayer thickness. Note that pentanol:surfactant ratios are held constant. Figure 8.10c shows on extrapolation that pentanol swells the liquid crystals, i.e., bilayer thickness decreases with a corresponding increase in the area per surfactant molecule. This swelling action in TTMAB is very similar to that in tetradecyltriethyl ammonium bromide (TTEAB).

Qualitative identification of liquid crystals is straightforward. Light can be polarized by passing it through a crystal. Here $\langle \hat{E}_{ox}{}^2 \rangle = \langle \hat{E}_{oy}{}^2 \rangle$, as stated for instance in the derivation of Equation 8.12. Without this feature, the scattered light assumes a more complex dependency on angles than just the polar angle in Equation 8.12. This form of radiation is "circularly" polarized, as the tip of the field vector in Figure 8.1 describes a circle around the z axis. A polarizer plate has an axis such that it would admit light only with a phase angle zero.

Now, refractive index is defined as the ratio between the speed of light in the medium and that in a vacuum. What happens to the polarized light as it enters an anisotropic medium? Loosely speaking, one component travels at a speed proportional to the refractive index $n_{||}$ and another component travels at a speed proportional to $n_{\perp}$. When the light exits the anisotropic medium, the speeds go back to being equal but a phase shift is introduced (Slayter, 1970):

$$\Delta\theta = 2\pi(n_{||} - n_{\perp})\ell/\lambda, \qquad (8.49)$$

where ℓ is the path length. Consequently if we place a polarizer in front of an isotropic sample and a polarizer with an axis at $90°$ to the first one at the rear, then light will enter the first polarizer but cannot exit through the crossed polarizer

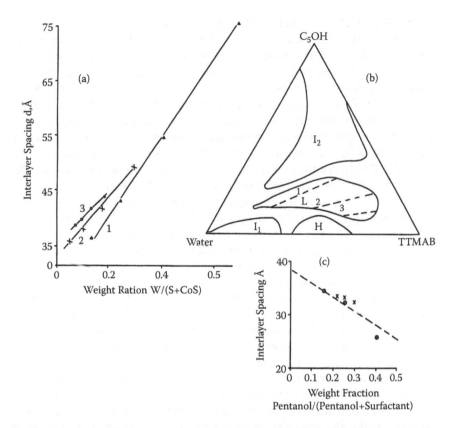

FIGURE 8.10 (a) Interlayer spacings (repeat distances) in lamellar liquid crystals from x-ray diffraction at constant surfactant:pentanol ratios and varying water content. (b) The compositions are shown on the phase equilibrium diagram. (c) The intercepts from (a) have been replotted as circles and the data for TTEAB as x's. Reproduced from Friberg et al. (1985). Copyright 1985 with permission from Elsevier.

at the rear. If we place a liquid crystal sample between the polarizers then the light will show some phase shift. Generally, liquid crystals have lots of defects, etc., because of which the phase shifts are distributed, and there will be some light that is phase shifted by 90°, that is, $\Delta\theta = 90°$, which will pass through. The pair of polarizers with a 90° difference in the axis is referred to as "crossed polarizers," or as the polarizer-analyzer pair, and liquid crystals between crossed polarizers show up with a ghostly glow in the dark. Light can also be linearly polarized, in which case $\langle \hat{E}_{oy}^2 \rangle = 0$ and the field traces a straight line on the x axis.

For improved magnification, one exploits the dual nature of matter, that particles also behave as waves. The wavelengths are obtained from the momentum as $mV = 2\pi h/\lambda$, where m is the mass of the particle, V is its velocity, h is Planck's constant, and λ is the resulting wavelength. For electrons, the wavelength works out to be about 0.005 nm, calculated resolution 0.003 nm, and the actual resolution more like 2 nm. Since electric and magnetic fields can be used like lenses, unlike x-ray results, which are available only in the inverse space, electron microscopy [scanning (SEM) and transmission (TEM)] gives us results in real space. However, nothing shows up unless the material is "electron rich;" to achieve this the specimens are often stained with oxides of heavy metals, commonly osmium tetroxide. A concern that must be addressed is that the stain may distort the microstructure. To increase the resolution, the applied electric field is increased, which increases the velocity V and decreases the wavelength λ. To study liquids, the specimen is frozen, broken, and metals are vapor deposited on the fracture surface. SEM is used to observe and measure inclusions such as vesicles using a method called freeze fracture. TEM is currently the only way to measure the size distribution of inorganic nanoparticles (Taleb et al., 1997).

A method widely used for investigation of biological and surfactant systems involves freezing the sample so rapidly that the microstructure is not distorted by the formation of ice crystals. In the freeze fracture technique, a thin layer of sample is placed between parallel plates of a highly conducting metal and then rapidly frozen, for instance, by immersing in liquid propane. Then a microtome is used to fracture the frozen material in the region near plates (where cooling is fastest) and a thin metal or metal/carbon layer is deposited on the surface so exposed. The replica of a surface formed in this way can be examined by SEM or TEM. This technique has been used, for example, to observe the microstructure of liquid crystalline phases, dispersions of vesicles, bicontinuous microemulsions, and L_3 phases in surfactant systems.

An alternate approach is Cryo-TEM, in which a thin layer of a liquid phase is carefully deposited on a TEM grid covered by a perforated carbon film, which is then plunged into liquid ethane maintained at approximately $-183°C$. In places where the liquid film spanning the perforations is on the order of 0.1 μm or thinner, it solidifies without forming ice crystals. The vitrified sample is transferred to the electron microscope at cryogenic temperatures and viewed directly (Danino et al., 2001). Vesicles, microemulsions, and micelles of various shapes and sizes are among the surfactant aggregates that have been imaged in this way.

Neutrons are massive compared to electrons and scatter because they have a very small quadrupole moment. Scattering experiments with light, x-rays, and neutrons involve different wavelengths of the incident radiation (the wavelength for particles such as neutrons is $2\pi h/mV$, as indicated previously). They can be used individually or in combination to study a variety of colloidal systems, polymer solutions, and surfactant-containing phases. Neutron scattering has been used to investigate microemulsions (Chen, 1986). A special feature of neutron scattering stems from the large difference in scattering behavior between the isotopes ^1H and ^2H. By using various combinations of water (H_2O), deuterium oxide (D_2O or heavy water), and ordinary and deuterated surfactant and oils (if present), one can design a set of experiments to provide information on where each type of molecule is located in micelles and microemulsion drops. Neutron scattering and related methods are being used to measure fluctuations in lamellae (Salamat et al., 2000), so important to the stability of liquid crystals, as discussed in Chapter 4.

Neutron reflection from fluid interfaces has revealed detailed arrangements of molecules in adsorbed layers (Lu et al., 2000). Basically neutrons can be totally reflected from interfaces in a manner similar to the total internal reflection of light at interfaces where a suitable refractive index distribution exists. Close to the critical angle for total reflection, the intensity of the weak scattered beam provides information on the refractive index distribution normal to the interface. For neutrons, a refractive index can be defined in terms of the density of the various nuclei present and their scattering cross sections. As in neutron scattering, the use of ordinary and deuterated compounds is a powerful tool in identifying the location and arrangement of molecules.

5. DIFFUSION

A starting point for looking at diffusive fluxes is the flux equation for a binary system:

$$J_1^* = c_1(v_1 - v^*), \tag{8.50}$$

where J_1^* is the molar flux with respect to mole average velocity, v_1 is the velocity of species 1, v^* is the mole average velocity, and c_1 is the molar concentration (Bird et al., 2002). Equation 8.50 can be rewritten as

$$N_1 = c_1 v_1 = c_1 v^* + J_1^*, \tag{8.51}$$

where N_1 is the molar flux with respect to fixed coordinates. The first term on the right-hand side is the convective transport and the second term is the diffusive transport. The diffusive flux is written as

$$J_1^* = - D_{12} c \nabla x_1, \tag{8.52}$$

where D_{12} is the diffusivity of 1 in 2, c is the total molar density, and x_1 is the mole fraction of species 1. When the solute (1) is dilute, c is almost a constant and Equation 8.52 becomes

$$J_1^* = - D_{12} \nabla c_1. \tag{8.53}$$

There is another way of looking at this problem, leading to the Stokes–Einstein equation for diffusivity, already derived earlier in Equation 3.72. This is rederived below because new terms, notations, and concepts have been encountered since. For a Brownian particle, a driving force on the particle is the gradient of chemical potential, which is opposed by the hydrodynamic drag, that is,

$$F_D = - \nabla \hat{\mu}_1, \tag{8.54}$$

where $F_D = 6\pi\mu a v_1$, following Stokes, where μ is the viscosity of the liquid and a is the radius of the particle assumed to be spherical. Similarly, for a single particle in dilute systems, $\hat{\mu}_1 = \hat{\mu}_1^\circ + kT\ell n\hat{c}_1$, where carats indicate that particles have been used instead of moles. Substituting these into Equation 8.54 and rearranging, we get

$$\hat{N}_1 = \hat{c}_1 v_1 = -\frac{kT}{6\pi\mu a}\nabla\hat{c}_1 , \qquad (8.55)$$

or in moles

$$N_1 = -D_{12}\nabla c_1. \qquad (8.56)$$

Equation 8.56 is the same as Equations 8.51 and 8.53 combined, where the missing convective velocity in Equation 8.55 can be easily incorporated when required. The important conclusion is

$$D_{12} = \frac{kT}{6\pi\mu R_H} , \qquad (8.57)$$

and Equation 8.57 is the same as Equation 3.72. Better expressions for the chemical potential can be used to include the effect of solution nonideality on diffusivity, which is discussed below. In addition, the force in Equation 8.54 in an anisotropic system will vary with the direction leading to anisotropic diffusivities, such as $D_\parallel$ (parallel) and $D_\perp$ (perpendicular). Equation 8.57 is called the Stokes–Einstein equation and shows that the size of the particle is related to the diffusivity, and vice versa. We replaced a with R_H, which is called the hydrodynamic radius, a more general term encompassing nonspherical particles. The Stokes–Einstein equation is suitable for solutes that are considerably larger than the solvent molecules, such that the solvent can be viewed as a continuum and Stokes drag can be used. One important example is that of globular proteins, which are only approximately spherical. It is possible to fit their shapes with flat surfaces (triangles, quadrilaterals, etc.) and then calculate both translational diffusivity (the one discussed above) and rotational diffusivity (Brune and Kim, 1993). Conversely, knowledge of the diffusivity leads to some information on the size of the particle. Equation 8.57 is used to calculate the hydrodynamic radius R_H (see Figure 8.6). As noted earlier, we can generalize chemical potential, the gradient of which—we emphasize again—is the driving force. For nonideal systems, the flux is expressed as

$$J_1^* = -\frac{D_{12}c_1}{RT}\nabla\mu_1 , \qquad (8.58)$$

leading to

$$J_1^* = - D_{12}\nabla c_1, \tag{8.59}$$

where

$$D_{12} = D_{12}[1 + \frac{\partial \ell n \gamma_1}{\partial \ell n c_1}], \tag{8.60}$$

and γ_1 is the activity coefficient. It is the term on the left-hand side, D_{12}, that is measured experimentally, and some knowledge of solution thermodynamics is required to extract D_{12}. It is D_{12} that is relevant to many of the important theoretical results in diffusivity, such as Equation 8.57, and the relationship that $D_{12} = D_{21}$ in binary systems. For ideal systems and dilute systems γ, does not depend on c_1 and $D_{12} = D_{12}$.

In real systems, there is more than one diffusing species involved. For instance, in colloids and proteins there are surface charges, counterions, added electrolyte, acids, bases, buffers, etc. If the concentrations are small, then they can be looked at independently, that is, each is treated as a binary with the solvent as component 2. Other difficulties arise when there is convection in the system, since the velocity used in fluid mechanics is the mass average velocity, not the mole average velocity used in Equations 8.50 and 8.51. Again, this is not much of a constraint in dilute solutions.

Methods of measuring diffusivities of solutes in liquids range from simple to very complex. Stokes diaphragm cells are discussed by Cussler (1976) and the Taylor–Aris dispersion method is described by Sherwood et al. (1975), among many others. These are the simplest methods. It should be kept in mind that for most small molecules the diffusivities lie in a very narrow range of 1 to 2×10^{-5} cm^2/sec. Micelles, proteins, and other polymers, etc., have lower diffusivities because of their large sizes. Weinheimer et al. (1981) measured the diffusivity of the total surfactant above the CMC using Taylor-Aris dispersion. The equations for conservation of species in this case are interesting:

$$\frac{\partial c_m}{\partial t} + v \cdot \nabla c_m = D_m \nabla^2 c_m + R_m \tag{8.61}$$

$$\frac{\partial c_1}{\partial t} + v \cdot \nabla c_1 = D_1 \nabla^2 c_1 + R_1, \tag{8.62}$$

where c_1 and c_m are singly dispersed amphiphiles and micelles, D_i, $i = 1, m$ their diffusivities, and R_i their production rates due to micellization-demicellization reactions. Using stoichiometry, $NR_m = -R_1$ where N is the aggregation number. Because micellization-demicellization reactions are very rapid, the impulse is to assume local equilibrium and set the rates of reactions to zero, which appears to be quite reasonable. Either equation is solved now and the concentration of other species is obtained using the equilibrium relation between the two, which is what the condition of zero reaction rates leads to. It leads to a view that micelles retain their identities (as in colloids) except when the total surfactant concentration falls below the CMC, when they demicellize. It is indeed correct that c_1 and c_m are almost at local equilibrium, but as small deviations from equilibrium are multiplied by large rate constants, it follows that the values of overall rates become comparable to the convective-diffusive fluxes and R_m and R_1 cannot be set to zero (Neogi, 1994). Weinheimer et al. (1981) eliminated these rates by multiplying Equation 8.61 by N and adding the two. They then wrote the resultant equation in terms of the overall surfactant concentration $c = c_1 + Nc_m$. Their analysis is actually valid for the more complex case of an ionic surfactant and allows one to define an effective diffusivity in terms of D_1, D_m, and the concentrations of c_1 (= CMC), c_m, and the added electrolyte. Because of the concentration dependence of the effective diffusion coefficient, their experimental procedure is a nonstandard one (sometimes called "differential") that allows them to measure diffusivity as a function of concentration. Not only do they successfully compare their results using Taylor-Aris dispersion with theory, later investigators have also verified their theoretical result using other experimental techniques (Leaist, 1986a,b). The diffusivities measured for SDS in 0.1 M NaCl are rather low, 1.2 to 1.5 × 10^{-6} cm^2/sec, about the lowest that Taylor-Aris dispersion is good for. Optical methods (Cussler, 1976) and the use of an ultracentrifuge (Tanford, 1961) can take measurements below this range. The more common methods are dynamic light scattering (DLS) and nuclear magnetic resonance (NMR) methods, discussed in the next sections.

Before looking into these, consider first what the fluxes are like in a multicomponent system. The generalized Stefan-Maxwell equation (Mason and Viehland, 1978) is

$$c_i \nabla \mu_i = \sum_j \frac{RT}{D_{ij}} (x_i N_j - x_j N_i).$$ (8.63)

Here, diffusivities D_{ij} are binary values encountered earlier and can be functions of concentrations. Summing Equation 8.63 over all i, the right-hand side becomes zero and the left-hand side also becomes zero on using the Gibbs-Duhem equation. Consider now a four-component system: cyclohexane (1), benzene (2), cyclohexane with a radioactive carbon tracer (1*), and benzene with a radioactive tracer (2*). (Radioactive ^{14}C and ^{3}H are common.) The last two are very dilute ("infinitely dilute") but are easy to detect quantitatively. We write out the equations for 1, 1*, and 2*, assuming that (1) $x_{1*} \approx 0$ and $x_{2*} \approx 0$, but that their gradients are not neglected, (2) $D_{1*2} = D_{12}$, etc., and (3) $D_{12} = D_{21}$, etc. This leads to

$$N_{1*} = -\left(\frac{x_1}{D_{11*}} + \frac{x_2}{D_{12}}\right)^{-1} \nabla c_{1*}$$ (8.64)

and

$$N_{2*} = -\left(\frac{x_1}{D_{12}} + \frac{x_2}{D_{22*}}\right)^{-1} \nabla c_{2*} .$$ (8.65)

Note the effective diffusivities (as measured in a diaphragm cell for instance) are a combinations of various diffusivities and concentrations, which for 1^* becomes D_{12} as $x_1 \rightarrow 0$ and for 2^* also becomes D_{12} as $x_2 \rightarrow 0$. This was verified experimentally by Mills (1965). If we assume that there is no convection, that is, $N_1 + N_2 + N_{1*} + N_{2*} = 0$, the last equation becomes

$$\nabla c_1 = x_1\left(\frac{1}{D_{11*}} - \frac{1}{D_{12}}\right) N_{1*} - \frac{N_1}{D_{12}} .$$ (8.66)

If N_{1*} can be ignored in the presence of N_1, which is a reasonable assumption, then the effective diffusivity for 1 is simply D_{12}.

The self-diffusivities for the radioactive tracers are set equal to those corresponding to nonradioactive species, for example, $D_{11*} = D_{11}$. Although the self-diffusion coefficients purport to be those in the absence of other species, we emphasize that in measurements as well as in numerical simulations, some labeling has to be done (MacElroy, 1996), which renders the system at least binary. It is very important to be able to distinguish among different diffusivities to be able to identify the basic building blocks for studies of dynamics in complex systems such as protein and micellar solutions and microemulsions. They usually have more than two components and are probed using more than one tracer, thus increasing the components even more. The tracer diffusivities are measured, leading to a great need to interpret the results.

6. DYNAMIC LIGHT SCATTERING

We return to Equation 8.9 for Rayleigh scattering in scalar form:

$$E_f = -\frac{k_f^2}{r} E_o e^{i(kr-\omega t)} \cdot \frac{\partial \alpha}{\partial c} \cdot \delta c(r, t)$$ (8.67)

where the subscript f is used to describe the forward scattering. Further, instead of the polarizability α in Equation 8.9, we use concentration fluctuation using the developments that follow. In static light scattering, δc was subsequently replaced with its weighted magnitude. However, the time fluctuation itself should have some information to provide us. Random variables of this kind, say $A(t)$, are used to calculate an autocorrelation function,

$$u(t) = <A(0)A(t)> = \lim_{T \to \infty} \frac{1}{T} \int_0^T A(t') A(t + t')dt' \ . \tag{8.68}$$

Note that an average is being defined and, in general, angular brackets will be used to denote averages. Generally autocorrelation functions decay with time exponentially, $u(t) \propto \exp(t/\tau)$, and it is τ that provides information on the dynamics. Note that $u(0)$ provides the mean squared magnitude of A.

Instead of proceeding directly to obtain the autocorrelation using Equation 8.67, in one common application (heterodyne) the scattered field is mixed with incident field/light and the resulting field is $E_s = E_o + E_f$. Now the "square" of the field (quotation marks are being used to signify that the field, being vectorial and complex, is not so easily squared to get a magnitude, but a simplified description here should suffice) is proportional to irradiance, $i(t)$. This, being the energy flux, is measurable. For the study of dynamics, its correlation is sought ("photon correlation spectroscopy"). Thus for $|E_o| >> |E_f|$,

$$<i(0)i(t)> \approx i_o^2 + 2i_o i_f. \tag{8.69}$$

Here, i_o is the irradiance corresponding to the incident radiation, $i_f(0)$ is the irradiance of the scattered light, and $i_f(t)$ is proportional to $<\delta c(0)\delta c(t)>$. What follows in most derivations is a lengthy discourse in statistics and the many justifications made on statistical grounds. The fluctuations are assumed to be summable over contributions from individual molecules, which requires that the system be dilute. This correlation function F_s for an individual molecule is related to another function G_s, called the van Hove (1954) correlation function (who also gave the proof):

$$F_s(q,t) = \int d^3R \ e^{iq \cdot R} G_s(R,t), \tag{8.70}$$

where

$$G_s(R,t) = <(\delta R \ [r_j(t) - r_j(0)]>. \tag{8.71}$$

δ is the Dirac delta function, and the subscript j denotes a tagged particle. G_s is simply the probability distribution function that a particle will see a displacement R. Such a displacement is governed by the diffusion equation

$$\frac{\partial G_s(R, \ t)}{\partial t} = D\nabla^2 G_s \ . \tag{8.72}$$

We take a Fourier transform of Equation 8.72 to get

$$\frac{\partial}{\partial t} F_s(q,t) = - q^2 D F_s,\qquad(8.73)$$

where Equation 8.70 has been used. Thus we get the important result that the autocorrelation is proportional to $\exp[-q^2 D/t] = \exp[-t/\tau]$, which yields the time constant of the system. A term in t^2 is added inside the exponent to account for polydispersity (Olivier and Sorensen, 1990). In complex systems D is a function of q (see caption to Figure 8.12).

In summary, the intensity of the scattered beam mixed with the incident beam is measured and autocorrelated (heterodyne). This autocorrelation function has an exponential decay. The time constant can be used to calculate the mutual diffusion coefficient. Berne and Pecora (1976) provide more detailed derivations and further treatments of all special cases.

In Figure 8.11a, diffusion data by Gibbs et al. (1991) from ultracentrifuge results ("synthetic boundery method"), DLS and NMR spin echo methods (see Section 5), are compared for ovalbumin, a globular protein. The conditions of the experiments are such that the charge effects are minimal or nonexistent, (i.e., they are conducted at the isoelectric point or the pH where the net charge on the protein is zero) and added electrolytes (such as buffer) are very low. That is, we have an almost binary system. The NMR data look the smoothest; the DLS data are more scattered, but are quite close to this set. The ultracentrifuge data show very large deviations. The investigators measured the osmotic pressure of the solution as a function of ovalbumin concentration to calculate the term within the square brackets in Equation 8.60. That is, they measured D_{12} and proceeded to calculate D_{12}. In Figure 8.11b, D_{12} is seen to agree much better with the other data. Figure 8.11 fulfills the important task of comparing types of data, including those from the NMR technique discussed in Section 7. It also points to the very important subject that had consumed researchers over a decade, that is, concentration dependence.

Anderson and Reed (1976) determined that if hydrodynamics alone were considered

$$D = D_o[1 - 1.83\phi],\qquad(8.74)$$

where D_o is Stokes-Einstein diffusivity (Equation 8.57) and ϕ is the volume fraction of the dispersion, assumed here to be monodispersed spheres. The authors provide a more complete result where charge effects are included. Fair et al. (1978) provide extensive DLS diffusivity data for bovine serum albumin (BSA), another globular protein. At isoelectric pH, the net charge on the protein is zero, at isoionic point there are no charges except charges from the solvent (in water H^+ and OH^-), and at high ionic strengths the double-layer repulsion is almost eliminated (Tanford, 1961). Various combinations of these were used. The coefficients were reported for isoelectric, isoionic, about neutral pH, and high or zero added ionic strengths for BSA. All cases showed good agreement with the theory of Anderson and Reed (1976). Isoelectric and isoionic conditions had coefficients that were large and negative. Positive coefficients were observed at neutral pH.

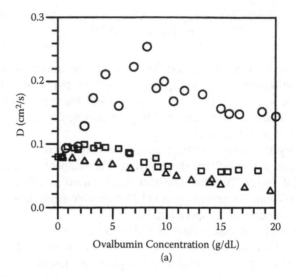

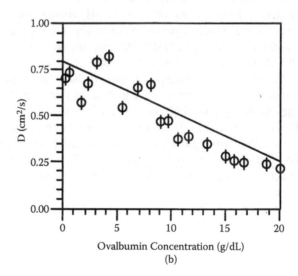

FIGURE 8.11 (a) Circles represent diffusivities from the synthetic boundary method, squares from DLS, and triangles are NMR self-diffusion coefficients. The system is ovalbumin with very little electrolyte and close to the isoelectric point. (b) Mutual diffusivities corrected for the solution nonideality compared to NMR results (line). Reprinted from Gibbs et al. (1991) with permission. Copyright 1978 American Chemical Society.

At the level of the Stokes-Einstein equation, infinite dilution is assumed. For calculation of the hydrodynamic drag, analysis of a single sphere is sufficient and particle-particle forces such as the charge effect need not be considered. When small concentration effects are included, up to the level indicated in Equation 8.74, the drag on a particle needs to be considered in the presence of a second particle and binary particle-particle interaction forces need to be included. In DLS measurements, should one look similarly at scattering from one particle as rescattered by a second? That is, should one include some form of binary interactions? Questions like these were posed by Pusey and Tough (1985).

Equation 8.74 is in error when ternary interactions are important. The error is on the order of φ^2. That is, if φ is 0.3, the error in D/D_o is on the order of 0.09, about the highest limit of acceptability and also the limit reached in the experiments of Fair et al. (1978). Pusey and coworkers have taken up the issue to a point where the multiple scattering can be subtracted even at very high concentrations of spheres. The (q-dependent) short-time diffusivity is seen to differ from the long-time values, but in a scalable manner. The biggest surprise is that diffusivity is inversely proportional to viscosity, as suggested by the Stokes-Einstein equation, over a very wide range of concentrations, shown in Figure 8.12 (Pusey et al., 1997; Serge and Pusey, 1997).

Returning to the Stokes-Einstein equation, Stokes drag on a sphere is inappropriate not only at high concentrations when the presence of other spheres distort Stokes flow. It is also inappropriate when the sphere is close to a wall, and for the same reason. This is important in particle collection, discussed in Chapter 3 (Section 7), where changes in hydrodynamics give rise to both anisotropy, (i.e., $D_\perp \neq D_\parallel$) and position dependence of both on h, the perpendicular

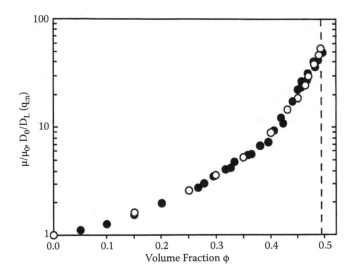

FIGURE 8.12 A DLS extension of the Stokes–Einstein equation that diffusivity is inversely proportional to viscosity at large concentrations of polymethyl methacrylate spheres in *cis*-decalin. Here μ is the zero shear viscosity and μ_o is the viscosity of the solvent. D_o is the Stokes–Einstein diffusivity and $D_L(q_m)$ is the long-term diffusivity evaluated at a scattering vector q_m, where the structure factor is a maximum. It is one way of evaluating diffusivity under conditions where the diffusing particle "sees" a cage best. A glass transition is seen a little before $\phi = 0.5$, as shown with the dashed line. Reprinted from Pusey et al. (1997). Copyright 1997 with permission from Elsevier.

distance from the plane wall. One additional result from the solution of Equation 8.72 is that

$$D = \frac{1}{6}\frac{d}{dt} \langle \mathbf{R} \cdot \mathbf{R} \rangle . \tag{8.75}$$

Anisotropic diffusivity can be accommodated if the scalar product is not used. However, this extension is usually invalid. Frej and Prieve (1993) measured R as a function of time for a single particle near a solid surface using an optical technique and calculated $D_\perp$ as a function of h. The theory behind Equation 8.75 is such that it can be extended to $D_\parallel$, in which direction the system is homogeneous, but not $D_\perp$. Nevertheless, such an extension was used to calculate $D_\perp$ from the experimental data, and these values agreed well with those calculated using hydrodynamics, $F_\perp$ in Equation 8.54. Their measurements show the smallest measured value to be as low as 0.015 of that in the bulk (Stokes–Einstein) at about 100 nm separation distance.

When colloids aggregate to form clusters, they are rarely closely packed. The mass of a cluster of particles of radius a is given by $M \propto R^3$ for a close packed cluster and $\propto R$ when the "cluster" is a single strand, the furthest removed from close packed, where R is an appropriate dimension describing the cluster. A generalization in that case is that $M \propto R^{d_f}$, where d_f is the fractal dimension that lies between one and three. A particle diffuses to a loosely packed cluster and eventually into it and then flocculates with one of the particles in the cluster. This flocculation step is often called reaction. If the reaction rate is very large, the approaching particle never gets a chance to make it to the core of the cluster. This is the diffusion-limited colloid aggregation (DLCA) and gives rise to a very spread out aggregate. When the reaction rate is slow, a more close packed aggregate results. Of course, this is the reaction-limited colloid aggregate (RLCA). A general result is

$$M \propto (R_G / a)^{d_f} , \tag{8.76}$$

where $d_f > 2$ for RLCA (about 2.1) and $d_f < 2$ (about 1.8) for DLCA. Equation 8.33 shows that

$$R_\theta \propto \left[1 + \frac{1}{3}(R_G q)^2 \right]^{-1} ,$$

and hence for $R_G q \ll 1$, $I(q) \propto (R_G q)^{-2}$ (after ignoring a constant), where I is the intensity, which is proportional to R_θ. However, if we average the intensity, given that we have a population distribution known from theory, a result that $I(q) \propto q^{-d_f}$ is obtained. This has been demonstrated experimentally using static light scattering (Lin et al., 1989; Wilcoxon et al., 1989). They were able to distinguish between the two indices for DLCA and RLCA, and to show using SEM that the DLCA particles were loosely packed while the RLCA particles were tightly packed. They also performed dynamic light scattering to measure diffusivities $\bar{D}_{eff}$ as a function of time and scattering vector q. Here, in the limit of $q \to 0$, a diffusivity $\bar{D}$ is obtained. The Stokes-Einstein equation (Equation 8.57) can now be used to calculate an appropriate $\bar{R}$, which is known to be equivalent to R_G. Experimental results indicate that $\bar{R} \propto t^{1/d_f}$, hence from Equation 8.76, $M \propto t$.

For polydispersed systems, the correlation function in DLS (see the discussion following Equation 8.73) is

$$C(t) = B + A \exp\left(-\mu_1 t + \frac{1}{2}\mu_2 t^2\right) , \qquad (8.77)$$

where μ_2 is related to polydispersity. As a caution, we note that t here is the time of relaxation in DLS and not the time of flocculation in the experiments. In fact, it is sufficiently short for the relaxation to appear instantaneous. Its normalized value μ_2 / μ_1^2, obtained at a single scattering vector q, was found to be a constant independent of time, suggesting not only a similarity in aggregate shapes (fractals), but also a similarity in the population distribution ("self-preserving" distribution). Brodie and Cohen (1992) measured the population distribution. They included in their analysis another fractal dimension d_h to define the effect of cluster size on diffusivity. For DLCA, they found $d_h \approx d_f \approx 2$. For RLCA, the answers were less clear. In both cases, the average aggregate size increased linearly with time after a short initial phase. There are many other investigators who have contributed to this area and we have discussed the results of only a few. Note that fast aggregation is the same as DLCA and requires a small stability ratio $W \sim 1$. Slow aggregation is same as RLCA, and Brodie and Cohen (1992) use W in hundreds, where at least 10^4 is required for the system to be considered stable (see Chapter 3, Section 6).

Dynamic light scattering is also used to measure the diffusivity of micelles, as mentioned in connection with Figure 8.5. However, according to the discussion on diffusion of surfactants above the CMC, and Equations 8.61 and 8.62, the van Hove function is complicated by the fact that micellization-demicellization reactions cannot be ignored in the conservation of species equations. This makes the problem very complex, even for simple reactions (Berne and Pecora, 1976). Interpretation of the reported values of diffusivities remains open to question. For microemulsions, one expects that the solubilizate core makes the breakdown of the microemulsion droplets very difficult and the measured values of diffusivities should be acceptable. Of these measurements, those by Bedwell and Gulari (1984) and Gulari et al. (1980) are discussed below. DLS is about the only way drop dimensions can be determined, and in the systems studied, drop diameters from 6 to 300 nm were measured. These investigators considered aerosol OT (or AOT) in water and heptane. AOT has two hydrocarbon tails and forms water-in-oil microemulsions. Both diameters and polydispersities (measured using Equation 8.77) were seen to increase as a phase transition was approached on increasing the temperature. For this system, the molecular weight measured by static light scattering agreed very well with R_H from DLS. Use of an electrolyte (NaCl) in (solubilized) water led to lower attraction between droplets. The second system considered was sodium dodecyl benzene sulfonate (SDBS) in hexadecane, water, and pentanol, which forms both water-in-oil and oil-in-water microemulsions in different composition regions. In the oil-in-water system, the polydispersity was explained as arising from a mixture of micelles and microemulsions. For one case of SDBS, and in only that case, it could not be established that the droplets were spherical. Concentration dependence of diffusivity was most often explained through a model that assumed that droplets behave as hard spheres.

As discussed in Chapter 4 (Section 8), DLS has also been used to determine phase transitions and criticality in microemulsion systems. It has also been used to measure interfacial tensions in microemulsion systems (Langevin, 1987). Measurements show that values are ultralow as required by theory (Chapter 4, Section 10).

7.   NMR SELF-DIFFUSION COEFFICIENT

Nuclear magnetic resonance methods have become very useful in measuring diffusion coefficients, but, as in DLS, it becomes more difficult to determine what is being measured as the system gets more complex. The measured diffusivity is most often called the "NMR self-diffusion coefficient," but in many instances it is the traditional diffusivity, discussed in Section 4.

The method rests on the fact that many atomic nuclei have permanent magnetic dipoles (commonly ^1H, hence "proton NMR"). The vector magnetization is

$$M = \sum_j \mu_j \, ,$$

where μ_j is the dipole moment of the jth nucleus. Note that the direction of μ_j can be different from that of M. The dynamics is described by (Farrar and Becker, 1971)

$$\frac{d}{dt} M = \gamma M \times H \, , \tag{8.78}$$

where γ is the magnetogyric ratio, a constant for a given nucleus. If a steady field $H = H_o$ ("d.c. field") is applied, then at steady state only the solution that $M = M_o$, parallel to H_o, is acceptable, and, of course, it is also proportional to H_o. However, the individual μ_j (embedded in a molecule) spins about H_o tracing a cone of fixed angle and spinning with an angular velocity

$$\omega_o = \gamma H_o, \tag{8.79}$$

where $\omega_o/2\pi$ is the Larmor frequency.

The method of describing the dynamics leads to shifting the frame of reference to one rotating with an angular velocity ω,

$$\frac{d}{dt} M = \left(\frac{\partial}{\partial t} M\right)_{rot} + \omega \times M \,,$$

which when substituted into Equation 8.78 reads

$$\left(\frac{\partial}{\partial t} M\right)_{rot} = \gamma M \times H_{eff} \,, \tag{8.80}$$

where the effective field is $H_{eff} = H_o + H_1 + \omega/\gamma$. The applied field has been split into a steady H_o in the z direction and an oscillating field ("radiofrequency" or "RF") H_1 in the x-y plane. If $\omega = \omega_o$, the first and the last terms cancel on using Equation 8.79. Thus one is left with H_1 only in the x' direction where the initial magnetization is in the z' direction. Here, the primes denote the new coordinates in the rotating coordinate system. The field H_1 tips the magnetization toward the y direction through an angle

$$\theta = \gamma H_1 t_p, \tag{8.81}$$

where t_p is the time of application of H_1. This is shown schematically in Figure 8.13. When the magnetization tips by $90°$ (y'-direction), the field H_1 is shut off and magnetization relaxes back to the z'-direction with a time constant T_1. This would have kept the magnetization in the y'-z' plane. However, collisions among molecules, exchange of energy, etc., lead to some magnetization in the x'-direction as well. Magnetization in both the x' and y'-directions decreases with a time constant T_2. Projection of the magnetization on the x'-y' plane shows that it fans out for some time, which is called dephasing. This is shown schematically in Figure 8.13. The mechanism behind it, called precession, is being ignored here for simplicity. Farrar and Becker (1971), among others, can be consulted for details.

In the program described above, it is difficult to separate the effects of T_2 from the effects of an inhomogeneous field. The latter is related to diffusion because diffusion deals with gradients. Without developing the ideas behind the steps, we introduce the form of a Fourier transform pulsed-gradient spin echo (PGSE) in Figure 8.14. The RF field is stopped after $90°$ excitation, allowed to dephase, and the magnetization is then inverted by $180°$, where it refocuses. This is the echo. Stilbs (1987) discusses some important features of the method. Whereas H_1 should oscillate with Larmor frequency, a very important fact of, say, ^1H resonance, is that Larmor frequencies vary with the position of ^1H in the molecule ("chemical shift"), which is fortuitous or else it would not be possible to distinguish among chemical species. Since the RF is off-resonance, the decay in signal is not exponential as shown in Figure 8.13, but a product of exponential and an oscillating (cosine) function. Decomposition among signals

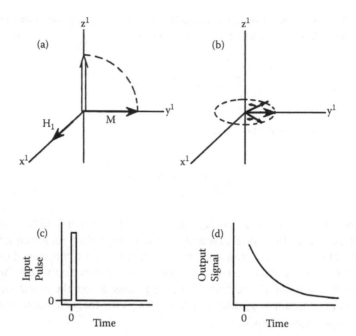

FIGURE 8.13 (a) Schematic of initial excitation. Rotating frames are shown. The short range of the RF burst is depicted on the time axis in (c). (b) Dephasing on x'-y' plane is shown. The signal (magnetization in the y' direction) is shown as a function of time in (d).

is accomplished through a Fourier transform. For reasons dealing with effective instrumentation, the gradient G in the d.c. field H_o is switched on only for a short time. The time scheme of the operation is shown in Figure 8.15.

The basic equation governing the magnetization in this scheme is (Karger et al., 1988)

$$\frac{\partial}{\partial t}M = \gamma M \times H - \frac{M_x e_x + M_y e_y}{T_2} - \frac{M_z - M_o}{T_1}e_z + D\nabla^2 M , \quad (8.82)$$

where H has a timed history described in Figure 8.15. The d.c. field remains steady while both RF bursts and gradients come and go. The solution is expressed as a ratio between the peak echo observed at $t = 2\tau$ and the signal following the initial excitation at $t = 0$,

$$A(2\tau) = \exp\left[-\frac{2\tau}{T_2}\right] \cdot \exp\left[-\gamma^2 G^2 D\delta^2\left(\Delta - \frac{\delta}{3}\right)\right], \quad (8.83)$$

where G is the gradient, and τ, δ, and Δ are the various times as shown in Figure 8.15. The time scales of experiments usually run into a few hundreds

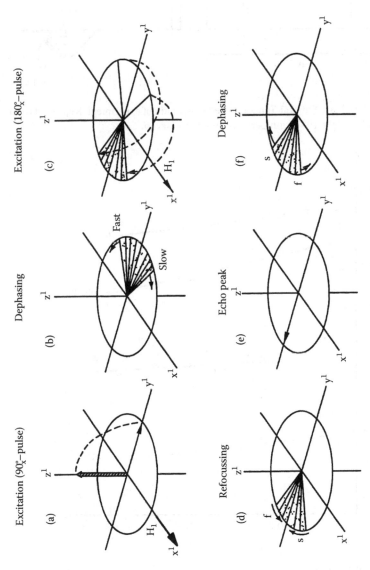

FIGURE 8.14 The basic 90°_x–180°_x spin-echo experiment. Reprinted from Stilbs (1987). Copyright 1987 with permission from Elsevier.

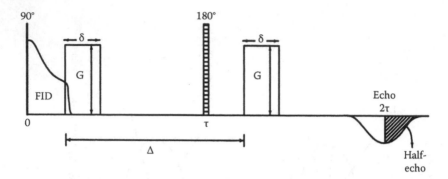

FIGURE 8.15 The basic Stejskal-Tanner pulsed field gradient experiment. Reprinted from Stilbs (1987). Copyright 1987 with permission from Elsevier.

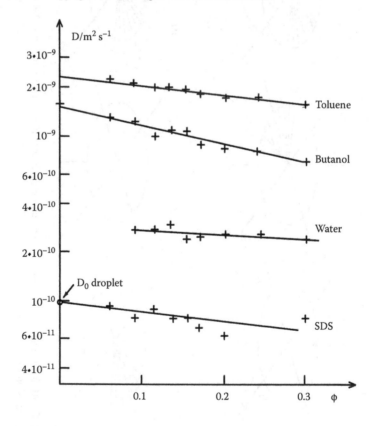

FIGURE 8.16 NMR diffusion coefficients in a water-in-oil microemulsion. The DLS diffusivity D_0 at infinite dilution is shown. The volume fraction ϕ is computed using a water droplet model for the microemulsion drops. Reprinted from Guéring and Lindman (1985) with permission. Copyright 1978 American Chemical Society.

of milliseconds, the shortest for all methods discussed here. Nevertheless, the root mean square (RMS) deviation, $<R \cdot R>^{1/2}$ from Equation 8.75 is in micrometers.

Consider a solution of two species (1 and 2) that respond to the magnetic field and a solvent that does not. The Stefan–Maxwell equation (Equation 8.63), with the constraint that there is no convection, leads to

$$\nabla c_1 = -N_1 \left(\frac{x_1 + x_s}{D_{1s}} + \frac{x_2}{D_{12}} \right) - N_2 \left(\frac{x_1}{D_{1s}} - \frac{x_1}{D_{12}} \right), \qquad (8.84)$$

which is truly a multicomponent diffusion with coupled fluxes. For a three-component mixture it is possible to isolate an effective diffusivity that has a very complex form (Cussler, 1976). However, it is seen that when the added components are present in trace amounts, (i.e., x_1 and x_2 can be ignored and x_s is about 1.0). Equation 8.84 predicts that the flux of species 1 is the product of the mutal diffusion coefficient D_{1s} and ∇c_1. These are often not the conditions in experiments. Stilbs (1987) reports proton resonance on 13 components, where the solution is "one drop" of each component in 0.3 ml deuterioacetone. Deuterium (^{2}H) has a frequency so removed from protons that it is transparent here. The important feature is that none of the components can be called dilute. The main aim in many of these studies has been to generate well-behaved and separable responses. For that, isotopes ^{2}H, ^{13}C, etc., can be added, although sometimes they are not needed since naturally occurring isotope concentrations are enough. Often these conditions result in diffusivities that are not easily interpreted, but shed light on the nature of some very complex systems as shown below.

Consider the protein diffusivities shown in Figure 8.11. As discussed in Section 4, this is a case where there are practically two species, the protein (p) and water (w). Proton resonance is used. The two mobilities are so disparate that their signals can be separated. For the binary system only D_{pw} is measured. As a result, it compares quite well with the other diffusivities, which are known to be binary diffusivities.

In the case of micellar solutions, it is not possible to separate the diffusivity of the surfactant as a whole from singly dispersed amphiphiles and micelles. Lindman and Brun (1973), using radiolabeled tracer studies, and Lindman et al. (1984), using magnetic resonance with ^{13}C labeling and heavy water as a solvent, showed that

$$D^* = D_1 p_1 + D_m p_m, \qquad (8.85)$$

where D_1 is the diffusivity of single amphiphiles and D_m that of micelles, and p_1 is the mass fraction of surfactant in singly dispersed form and p_m is that present as micelles. Neogi (1994) has shown under what assumptions it is possible to derive Equation 8.85. His starting equations were Equations 8.61 and 8.62. Care was taken to include electrostatic effects.

Diffusivity data for individual species in a water-in-oil microemulsion are shown in Figure 8.16. To explain the great difference between the diffusivities of water and toluene, it is assumed that the RMS deviation $<R \cdot R>^{1/2}$ from Equation 8.75 in the dispersed phase (water) is a lot smaller than in the continuous phase (toluene), leading to a lower value of diffusivity for water. The nature of the diffusivity of SDS suggests that all SDS is on the surface of the microemulsion droplets, and it reflects the diffusivity of the droplets. This is borne out by the fact that there is agreement with the DLS result.

Nuclear magnetic resonance diffusivities along a salinity scan (see Chapter 4, Section 8) are shown in Figure 8.17. At low salinities we have an oil-in-water microemulsion, which becomes the middle phase microemulsion, and eventually a water-in-oil microemulsion at large salinities. Note that the surfactant diffusivity of the bicontinuous phase is higher than that of the dispersed phase in both oil-in-water and water-in-oil microemulsions because SDS enters a continuous geometry in the middle phase. Butanol, the cosurfactant, is seen here to prefer the continuous phase.

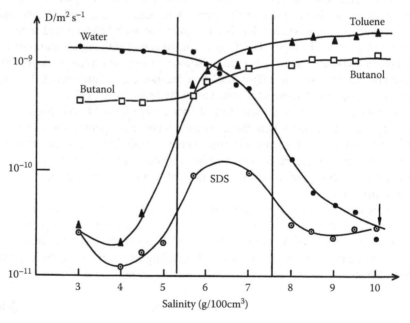

FIGURE 8.17 NMR diffusion coefficients across the phase inversion using a salinity scan. Reprinted from Guéring and Lindman (1985) with permission. Copyright 1978 American Chemical Society.

REFERENCES

GENERAL REFERENCES

Berne, B.J. and Pecora, R., *Dynamic Light Scattering*, John Wiley & Sons, New York, 1976.

Farrar, T.C. and Becker, E.D., *Pulsed and Fourier Transform NMR*, Academic Press, Orlando, 1971.

Hiemenz, P.C. and Rajagopalan, R., *Principles of Colloid and Surface Chemistry*, 3rd ed., Marcel Dekker, New York, 1997.

Kerker, M., *The Scattering of Light and Other Electromagnetic Radiation*, Academic Press, New York, 1969.

Lindman, B. and Stilbs, P., Molecular diffusion in microemulsions, in *Microemulsions: Structure and Dynamics*, Friberg, S.E. and Bothorel, P. (eds.), CRC Press, Boca Raton, Fla., 1987, p. 119.

Lipson, S.G. and Lipson, H., *Optical Physics*, 2nd ed., Cambridge University Press, Cambridge, 1981.

Stone, J.M., *Radiation and Optics*, McGraw-Hill, New York, 1963.

Tanford, C., *Physical Chemistry of Macromolecules*, John Wiley & Sons, New York, 1961.

van de Hulst, H.C., *Light Scattering by Small Particles*, Wiley, New York, 1957.

Zana, R. (ed.), *Surfactant Solutions: New Methods of Investigation*, Marcel Dekker, New York, 1987.

TEXT REFERENCES

Anacker, E.W., Micelle formation of cationic surfactants in aqueous media, in *Cationic Surfactants*, Jungermann, E. (ed.), Marcel Dekker, New York, 1970, p. 203.

Anderson, J.L. and Reed, C.C., Diffusion of spherical macromolecules at finite concentration, *J. Chem. Phys.*, 64, 3240, 1976.

Bedwell, B. and Gulari, E., Electrolyte-moderated interactions in water/oil microemulsions, *J. Colloid Interface Sci.*, 102, 88, 1984.

Berne, B.J. and Pecora, R., *Dynamic Light Scattering*, John Wiley & Sons, New York, 1976.

Bird, R.B., Stewart, W.E., and Lightfoot, E.N., *Transport Phenomena*, 2nd ed., John Wiley & Sons, New York, 2002, p. 536.

Brodie, M.L. and Cohen, R.J., Measurements of cluster-size distributions arising in salt-induced aggregation of polystyrene microspheres, *J. Colloid Interface Sci.*, 153, 493, 1992.

Brune, D. and Kim, S., Predicting protein diffusion coefficients, *Proc. Natl. Acad. Sci. USA*, 90, 3835, 1993.

Chen, S.H., Small angle neutron scattering studies of the structure and interaction in micellar and microemulsion systems, *Annu. Rev. Phys. Chem.*, 37, 351, 1986.

Cussler, E.L., *Multicomponent Diffusion*, Elsevier Science, New York, 1976.

Danino, D., Bernheim-Groswasser, A., and Talmon, Y., Digital cryogenic transmission electron microscopy: an advanced tool for direct imaging of complex fluids, *Colloids Surfaces A*, 183–185, 113, 2001.

Davis, H.T., *Statistical Mechanics of Phases, Interfaces, and Thin Films*, Wiley-VCH, New York, 1996, p. 659.

Fair, B.D., Chao, D.Y., and Jamieson, A.M., Mutual translational diffusion coefficients in bovine serum albumen solutions measured by quasielastic laser light scattering, *J. Colloid Interface Sci.*, 66, 323, 1978.

Farrar, T.C. and Becker, E.D., *Pulsed and Fourier Transform NMR*, Academic Press, Orlando, 1971.

Frej, N.A. and Prieve, D.C., Hindered diffusion of a single sphere very near a wall in a nonuniform force field, *J. Chem. Phys.*, 98, 7552, 1993.

Fontell, K., X-ray diffraction by liquid crystals. Amphiphilic systems, in *Liquid Crystals and Plastic Crystals*, vol. 2, Gray, G.W. and Winsor, P.A. (eds.), Ellis Horwood, Chichester, 1974, p. 80.

Friberg, S.E., Venable, R.L., Kim, M., and Neogi, P., Phase equilibria in water, pentanol, tetradecyltrialkylammonium bromide systems, *Colloids Surfaces*, 15, 285, 1985.

Gibbs, S.J., Chu, A.S., Lightfoot, E.N., and Root, T.W., Ovalbumin diffusion at low ionic strength, *J. Phys. Chem.*, 95, 467, 1991.

Guéring, P. and Lindman, B., Droplet and bicontinuous structures in microemulsions from multicomponent self-diffusion measurements, *Langmuir*, 1, 464, 1985.

Gulari, E., Bedwell, B., and Alkhafaji, S., Quasi-elastic light-scattering investigation of microemulsions, *J. Colloid Interface Sci.*, 77, 202, 1980.

Hiemenz, P.C. and Rajagopalan, R., *Principles of Colloid and Surface Chemistry,* 3rd ed., Marcel Dekker, New York, 1997, p. 193.

Karger, J., Pfeifer, H., and Heink, W., Principles and application of self-diffusion measurements by nuclear magnetic resonance, *Adv. Magn. Reson.*, 12, 1, 1988.

Langevin, D., Low interfacial tensions in microemulsion systems, in *Microemulsions: Structure and Dynamics*, Friberg, S.E. and Bothorel, P. (eds.), CRC Press, Boca Raton, Fla., 1987.

Leaist, D.G., Binary diffusion of micellar electrolytes, *J. Colloid Interface Sci.*, 111, 230, 1986a.

Leaist, D.G., Diffusion of ionic micelles in salt solutions: Sodium dodecyl sulfate + sodium chloride + water, *J. Colloid Interface Sci.*, 111, 240, 1986b.

Lin, M.Y., Lindsay, H.M., Weltz, D.A., Ball, R.C., Klein, R., and Meakin, P., Universality in colloid aggregation, *Nature (Lond.)*, 339, 360, 1989.

Lindman, B. and Brun, B., Translational motion in aqueous sodium *n*-octanoate solutions, *J. Colloid Interface Sci.*, 42, 388, 1973.

Lindman, B., Payal, M.-C., Kamenka, N., Rymden, R., and Stilbs, P., Micelle formation of anionic and cationic surfactants from Fourier transform proton and lithium-7 nuclear magnetic resonance and tracer self-diffusion studies, *J. Phys. Chem.*, 88, 5048, 1984.

Lipson, S.G. and Lipson, H., *Optical Physics*, 2nd ed., Cambridge University Press, Cambridge, 1981.

Lu, J.R., Thomas, R.K., and Penfold, J., Surfactant layers at the air/water interface: structure and composition, *Adv. Colloid Interface Sci.*, 84, 143, 2000.

MacElroy, J.M.D., Diffusion in homogeneous media, in *Diffusion in Polymers*, Neogi, P. (ed.), Marcel Dekker, New York, 1996.

Mason, E.A. and Viehland, L.A., Statistical-mechanical theory of membrane transport for multicomponent systems: Passive transport through open membranes, *J. Chem. Phys.*, 68, 3562, 1978.

Mie, G., Contributions to the optics of turbid media, especially colloidal metal solutions, *Ann. Physik*, 25, 377, 1908.

Mills, R., The intradiffusion and derived frictional coefficients for benzene and cyclohexane in their mixtures at 25°, *J. Phys. Chem.*, 69, 3116, 1965.

Neogi, P., Diffusion in a micellar solution, *Langmuir*, 10, 1410, 1994.

Olivier, B.J. and Sorensen, C.M., Evolution of the cluster size distribution during slow colloid aggregation, *J. Colloid Interface Sci.*, 134, 139, 1990.

Peña, A.A. and Hirasaki, G.J., Enhanced characterization of oilfield emulsions via NMR diffusion and transverse relaxation experiments, *Adv. Colloid Interface Sci.*, 105, 103, 2003.

Phillips, J.N. and Mysels, K.J., Light scattering by aqueous solutions of sodium lauryl sulfate, *J. Phys. Chem.*, 59, 325, 1955.

Pusey, P.N., Segre, P.N., Dynamics of concentrated colloidal suspensions, Behrend, O.P., Meeker, S.P., and Poon, W.C.K., *Physica A*, 235, 1, 1997.

Pusey, P.N. and Tough, R.J.A., Particle interactions, in *Dynamic Light Scattering*, Pecora, R. (ed.), Plenum Press, New York, 1985, p. 85.

Rivard, D.C., Stricker, L.A., and Neogi, P., Turbidimetry of two aqueous phase emulsions and related systems, *J. Colloid Interface Sci.*, 149, 521, 1992.

Salamat, G., de Vries, R., Kaler, E.W., Satija, S., and Sung, L., Undulations in salt-free charged lamellar phases detected by small angle neutron scattering and neutron reflectivity, *Langmuir*, 16, 102, 2000.

Segre, P.N. and Pusey, P.N., Dynamics and scaling in hard-sphere colloidal suspensions, *Physica A*, 235, 9, 1997.

Slayter, E.M., *Optical Methods in Biology*, Wiley-Interscience, New York, 1970.

Sherwood, T.K., Pigford, R.L., and Wilke, C.R., *Mass Transfer*, McGraw-Hill, New York, 1975.

Stilbs, P., Fourier transform pulsed-gradient spin-echo studies of molecular diffusion, *Prog. NMR Spectrosc.*, 19, 1, 1987.

Stone, J.M., *Radiation and Optics*, McGraw-Hill, New York, 1963, p. 30 ff.

Taleb, A., Petit, C., and Pelini, M.P., Synthesis of highly monodisperse silver nanoparticles from AOT reverse micelles: A way to 2D and 3D self-organization, *Chem. Mater.*, 9, 950, 1997.

Tanford, C., *Physical Chemistry of Macromolecules*, John Wiley & Sons, New York, 1961, p. 275.

van Hove, L., Correlations in space and time and born approximation scattering in systems of interacting particles, *Phys. Rev.*, 95, 249, 1954.

Weinheimer, R.M., Evans, D.F., and Cussler, E.L., Diffusion in surfactant solutions, *J. Colloid Interface Sci.*, 80, 357, 1981.

Wilcoxon, J.P., Martin, J.E., and Schaefer, D.W., Aggregation in colloidal gold, *Phys. Rev. A*, 39, 2675, 1989.

Young, C.Y., Missel, P.J., Mazer, N.A., Benedek, G.B., and Carey, M.C., Deduction of micellar shape from angular dissymmetry measurements of light scattered from aqueous sodium dodecyl sulfate solutions at high sodium chloride concentrations, *J. Phys. Chem.*, 82, 1375, 1978.

PROBLEMS

8.1   Equation 8.2 can be written as

$$E = (E_{ox}e_{\mathbf{x}} + E_{oy}e_{\mathbf{y}})e^{(i\ kr-\omega t)}.$$

Substitute complex numbers for E_{ox} and E_{oy} and show that Equation 8.3 results. Obtain the real part of E and calculate its time average over $t = 0$ to $2\pi/\omega$.

8.2   The cross product between two vectors lead to a third vector perpendicular to the first two. Show that this is sufficient to make E and H in Equations 8.9 and 8.10 perpendicular to one another.

8.3   A "thermal" electron has an energy of kT. At 300 K, calculate its velocity, wavelength, and resolution. Since the resolution is not as good as that for x-rays, recalculate these values if the energy is $10^6 \, kT$. Ignore relativistic effects. A few constants are

$k = 1.38 \times 10^{-16}$ ergs/K
$h = 6.624 \times 10^{-27}$ erg·sec
$m = 9.1066 \times 10^{-28}$ g.

8.4   Sum Equation 8.63 over i and show that it leads to the Gibbs-Duhem equation. Equation 8.63 is often written with a term $c_i F_i$, where F_i is the force on the ith species. If the Gibbs-Duhem equation is assumed, show that the above summation leads to a force balance.

8.5   Consider a sphere of radius a and uniform density ρ. Its moment of inertia about the z axis is defined as $\pi y^2 dz \cdot \rho$ (differential mass)·y^2 (square of the moment arm) integrated from $z = -R$ to $+R$. In addition $z^2 = R^2 - y^2$. Calculate the radius of gyration where the moment of inertia is MR_G^2 and obtain the fractal dimension from $M \propto R_G^{d_f}$. Repeat for a cylinder of radius a and length R, where the axis is along the middle of the rod. Note that in section 6, the rod is a circle just like in a sphere. In both cases, the variable is R (or R_G) only.

8.6   Compare on Figure 8.15 quantitatively the SDS diffusivities with that given by Anderson and Reed (1976) for hard spheres (Equation 8.76). Why would a hard sphere model be appropriate in this case?

8.7   Consider a binary system in Equation 8.63. If the volumetric flow rate is zero, $\bar{v}_1 N_1 + \bar{v}_s N_s = 0$, where the subscript s denotes solvent and $\bar{v}_i$ is the partial molar volume of the ith species. Obtain the expression for N_1 as a function of the gradient of c_1 only. Note that $c_i \bar{v}_i$ is the volume fraction.

Index

A

Adhesion
  adhesive failure, 66
  wettability, effect of, 61, 98
  work of, 65–68
Adsorption, 85–90
  adsorption limited transport, 330
  BET isotherm, *see* BET isotherm
  contact angles, effect on, 80–82
  fluid flow, effect on, 262–268, 335–339,
    407–408
  Gibbs adsorption equation, 11, 103
  Henry's law, 86, 88
  Heterogeneous surfaces, 93
  isotherms, 85–90, 102–103, 198–201, 267,
    337–339
  Langmuir isotherm, 87, 88,103
  spreading pressure, 90, 198–201
  surface area from, 90
  polymers, 139–140, 148–149
Aerosols, *see* colloidal dispersions
Aggregation number, 170–171,175–178
Asymptotic solutions
  convergence, 399–400
  matched, 393, 405, 426–432, 452
  order, 429, 452
  regular, 393, 398–401, 439–441

B

BET isotherm, 87–89, 106–107
Benard cells, 310–311, 318–319
Bending energy, 49–51, 176, 203, 210,
    215–217, 243–244
Bicontinuous structure, 216–217
Biharmonic equation, 394, 401, 439
Bond number, 316, 404

C

Capillary rise, 31-33, 56–57, 450

Capillary waves, 247
  damping, 262–268, 336–339
  diffusion, effect of, 335–339
  experimental, 265–267, 336–339
  maximum damping, 301–302
  oscillations, 260–262, 264–265, 291–292,
    304–305, 336–339
  stability criteria, 248–249, 257–258, 271,
    291–298
  wave or dispersion equations, 258–262, 264,
    291
Chemical reaction at interfaces, 345–349
  electrochemical reactions, 349
  heat of reaction, effect of, 346
  stability criterion, 346–348
Cloud point, 187–189
Coagulation, *see* flocculation
Coalescence, *see* flocculation
Coatings, 61, 385, 401–407
Cohesion, work of, 65–68
Colloidal dispersions, 1–2, 109–110, 128, 155
  attractive forces, 110–116
  charge density, 128
  colloidal stability, 132–149
  counter ion valence, effect of, 134–137
  depletion flocculation 142
  depletion stabilization 142
  diffusion coefficient, 150
  dispersed, dispersions, 247
  electrical forces, 118–128
    electrolyte concentration, effect of, 133,
      134–135
  emulsions, 109, 218–223
  entropic stabilization, 141, 147–148
  flocculation, *see* flocculation
  fractals, 482–483
  kinetics, flocculation, *see* flocculation
  liquid crystals in emulsions, 218, 221
  mass transfer effects on stability of, 307,
    320–328
  osmotic pressure, effect of, 141
  particle size, effect of, 134
  phase behavior, effect on emulsions, 218,
    220–221

Milton Keynes UK
Ingram Content Group UK Ltd.
UKHW020007071024
449327UK00031B/2688